Engineering Mathematics with MATLAB®

This textbook takes a streamlined, practical approach, designed to make engineering mathematics accessible and manageable for undergraduate students and instructors alike. Students will gain a fundamental understanding within the scope of a two-semester course.

This textbook introduces students to the fundamental principles of engineering mathematics through concise explanations, systematically guiding them from the basics of first-order, second-order, and higher-order ordinary differential equations (ODEs), Laplace transforms, and series solutions of ODEs. It then transitions to more advanced topics, including linear algebra, linear system of ODEs, vector differential calculus and vector integral calculus, Fourier analysis, partial differential equations (PDEs), and concludes with complex numbers, complex functions, and complex integration. The book presents fundamental principles systematically with concise explanations. It features categorized key concepts, detailed solutions, and alternative methods to connect material to prior knowledge. Exercises are thoughtfully organized, balancing problem-solving practice with real-world applications in fields like mechanical engineering, electrical engineering, chemical engineering, and so on. Notably, this book incorporates MATLAB® to enhance understanding. MATLAB®-based examples simplify complex calculations, offering visualizations that connect theory and practice. Chapters also include optional advanced topics, providing deeper insights for motivated learners.

Designed with practicality in mind, this book offers a balanced approach to mastering engineering mathematics, with a manageable workload aligned to academic schedules. It is an invaluable resource for instructors seeking effective teaching tools and for students aiming to build strong mathematical foundations that they can apply to their own engineering discipline.

Chul Ki Song is an Emeritus Professor at the School of Mechanical Engineering, Gyeongsang National University, Korea. He earned his Ph.D. degree from Seoul National University, Korea. He previously worked for KIA Motors.

Jong-Ryeol Kim is a Professor in the Department of Electrical Engineering, Sejong University, Korea. He earned his Ph.D. degree from KAIST, Korea. He previously worked for Samsung Electronics.

Engineering Mathematics with MATLAB®

Chul Ki Song and Jong-Ryeol Kim

CRC Press
Taylor & Francis Group
Boca Raton London New York

CRC Press is an imprint of the
Taylor & Francis Group, an **informa** business

Designed cover image: Engineering 3d Computer Aided Design
Concept With Metallic Gears On A Virtual Reality Blueprint Surface
With Glowing Programming Code And Infographic Overlay High-Res
Stock Photo - Getty Images

MATLAB® and Simulink® are trademarks of The MathWorks, Inc.
and are used with permission. The MathWorks does not warrant
the accuracy of the text or exercises in this book. This book's use or
discussion of MATLAB® or Simulink® software or related products
does not constitute endorsement or sponsorship by The MathWorks of
a particular pedagogical approach or particular use of the MATLAB®
and Simulink® software.

First edition published 2026
by CRC Press
2385 NW Executive Center Drive, Suite 320, Boca Raton FL 33431

and by CRC Press
4 Park Square, Milton Park, Abingdon, Oxon, OX14 4RN

CRC Press is an imprint of Taylor & Francis Group, LLC

© 2026 Chul Ki Song and Jong-Ryeol Kim

ISBN: 9781032979847 (hbk)
ISBN: 9781041002550 (pbk)
ISBN: 9781003608912 (ebk)

DOI: 10.1201/9781003608912

Typeset in Sabon
by KnowledgeWorks Global Ltd.

Contents

Preface

Engineering mathematics is a foundational subject that all engineering students must study. Over the years, many excellent textbooks have been published to facilitate understanding. However, some of these books include more content than can be realistically covered in two semesters, and their verbose explanations often obscure the focus. Additionally, exercises in many textbooks are excessive in number and inconsistent in difficulty, making it challenging to efficiently master the concepts.

This textbook addresses these challenges and aims to provide an engineering mathematics resource tailored to the practical needs of university education. Its key features are as follows:

1. Clear and Concise Explanations:
   The core principles and concepts are explained succinctly and systematically, with key ideas categorized under **Remark** for better organization. This approach helps students grasp the material more easily and enables instructors to teach more effectively.
2. Detailed Solutions:
   Each theoretical concept is accompanied by solved examples (**Solution**) that are thoroughly explained. To reinforce understanding, alternate solution methods (**Another Solution**) are provided, connecting the material to high school calculus or previously covered topics in the textbook.
3. Verification and Checkpoints:
   Complex calculations and the accuracy of answers are clarified in designated sections labeled as **Check**.
4. Integration with MATLAB®:
   Numerical analysis methods are included, and MATLAB® is utilized to simplify the problem-solving process. MATLAB®-based examples (**M_ Example**) are provided to deepen students' understanding and practical skills.
5. Categorized Exercises:
   Problems are organized by example type, with additional applied problems related to mechanical, electrical, electronic, and chemical engineering included in each chapter.

6. Concept-Focused Problems:
   The exercises emphasize clarity and understanding rather than overly complicated calculations, allowing students to develop a firm grasp of concepts.
7. Optional Advanced Topics:
   Challenging chapters are marked as (*optional) and include detailed explanations for interested students.
8. Content Optimization for Two Semesters:
   The number of exercises has been adjusted to align with the realities of university coursework, making the material manageable within a two-semester timeframe.

It is my hope that this textbook will become a valuable resource for engineering students. I extend my heartfelt gratitude to my son, Jaeyong, who reviewed and solved all the problems in this book, and to the team at CRC Press/Taylor & Francis Group, whose dedication made this publication possible.

Chul Ki Song
Main author

MATLAB® is a registered trademark of The MathWorks, Inc. For product information, please contact:
The MathWorks, Inc.
3 Apple Hill Drive
Natick, MA 01760-2098 USA
Tel: 508-647-7000
Fax: 508-647-7001
E-mail: info@mathworks.com
Web: www.mathworks.com

1 First-order ordinary differential equations

In the fundamental principles of classical mechanics established by Sir Isaac Newton, known as Newton's laws, the position of an object is expressed as a function of time. By introducing the concepts of differentiation, which yields velocity when applied to position and acceleration when applied to velocity, these laws enable the representation of physical concepts such as force, momentum, and energy through differential equations. Therefore, differential equations serve as the foundation for nearly all engineering and physics disciplines, including electrical and electronic engineering, structural mechanics, dynamics, thermodynamics, fluid mechanics, and aerospace engineering.

1.1 Differential equations and their solutions

1.1.1 Fundamentals of differential equations

When we seek to interpret physical phenomena, we often mathematically formulate these phenomena using various variables and functions. The resulting mathematical expression is referred to as a mathematical model. Furthermore, this process of creating such equations is known as mathematical modeling.

Most mathematical models involving physical concepts incorporate derivatives and are expressed as equations containing derivatives. These equations are called differential equations. By finding solutions that satisfy these differential equations and analyzing the characteristics of these solutions, we gain a deeper understanding of the physical properties involved.

1.1.2 First-order differential equations

In a differential equation, if the highest order of the derivative involved is of the n-th degree, then that differential equation is referred to as an n-th order (order n) differential equation.

In this chapter, we will focus on first-order differential equations. In other words, we will deal with equations that contain with $y'\left(= \dfrac{dy}{dx} \right)$, y, x, etc.

1.1.3 Ordinary differential equations and partial differential equations

An ordinary differential equation (ODE) involves first-order or higher-order derivatives. Examples of ODEs include the following:

$$y' = \sin x + 2, \tag{1.1}$$

$$y'' + 2y' + 9y = e^{-x}, \tag{1.2}$$

$$y''' - 3y' = 0, \tag{1.3}$$

DOI: 10.1201/9781003608912-1

where y is a function of the variable, which means $y = y(x)$. Also, y'' and y''' represent the second and third derivatives of d^2y/dx^2, d^3y/dx^3, respectively.

On the other hand, a *partial differential equation* (PDE) includes partial derivatives with respect to two or more variables. An example of PDEs is as follows:

$$\frac{\partial^2 u}{\partial x^2} + \frac{\partial^2 u}{\partial y^2} = 0, \qquad (1.4)$$

where u is a function with two variables x and y, $u = u(x, y)$.

PDEs are extensively applied in engineering, but we will learn more about them in Chapter 11.

1.1.4 Representation of differential equations

In general, there are two methods for representing equations: *implicit form* and *explicit form*. For example, the equation of a circle with a radius of 2 can be expressed implicitly as $x^2 + y^2 = 4$, while it can be expressed explicitly as $y = \pm\sqrt{4 - x^2}$.

Therefore, first-order ODEs can be represented in two different forms, as follows:

$$F(x, y, y') = 0, \qquad (1.5)$$

$$y' = f(x, y). \qquad (1.6)$$

Here, Eq. (1.5) represents the implicit form, while Eq. (1.6) represents the explicit form. For example, $x^{-2}y' - 3y^2 = 0$ is an implicit form, while $y' = 3x^2y^2$ is an explicit form.

1.1.5 Differential equations and solutions

In a system where a point moves around a circle with a radius of 2, the height y with respect to the angle of rotation x can be represented as follows:

$$y = 2\sin x. \qquad (1.7)$$

Taking the derivative of this expression, we get $y' = 2\cos x$. This derivative satisfies the following differential equation:

$$y'^2 + y^2 = 4. \qquad (1.8)$$

Therefore, Eq. (1.7) is called a solution of the differential equation (1.8).

Example 1.1

a. Prove that $y = \dfrac{2}{x}$ is a solution to the ODE $xy' = -y$.

b. Prove that $y = Ce^{2x}$ (where C is arbitrary) is a solution to the ODE $y' = 2y$.

Solution

a. Given that $y' = -\dfrac{2}{x^2}$,

   Left-hand side (LHS): $xy' = -\dfrac{2}{x}$,

   Right-hand side (RHS): $-y = -\dfrac{2}{x}$.

   Therefore, LHS = RHS, demonstrating that $y = \dfrac{2}{x}$ is indeed a solution to the ODE $xy' = -y$.

b. LHS: $y' = 2Ce^{2x}$,

   RHS: $2y = 2Ce^{2x}$.

   Therefore, LHS = RHS, demonstrating that $y = Ce^{2x}$ is indeed a solution to the ODE $y' = 2y$.

Example 1.2

Solve the following ODEs.

a. $y' = \sin x + 2$

b. $y' = \dfrac{1}{1 + x^2}$

c. $y' = \dfrac{1}{1 - x^2}$

d. $y' = \dfrac{f'(x)}{f(x)}$

Solution

a. $y = \displaystyle\int (\sin x + 2)\,dx = -\cos x + 2x + C$. (Where C is the constant of integration.)

$$\textbf{Answer } y = -\cos x + 2x + C$$

b. Substituting into $x = \tan\theta \left(|\theta| < \frac{\pi}{2} \right)$, we obtain $dx = \sec^2\theta \, d\theta$ and $1 + x^2 = 1 + \tan^2\theta = \sec^2\theta$.

Then, we get

$$y = \int \frac{1}{1+x^2} dx = \int \frac{1}{\sec^2\theta} \sec^2\theta \, d\theta = \theta + C = \arctan x + C,$$

where C is the constant of integration.

Answer $y = \arctan x + C$

c. By performing partial fraction decomposition, we can rewrite the expression as:

$$\frac{1}{1-x^2} = \frac{1}{2}\left(\frac{1}{1+x} + \frac{1}{1-x} \right).$$

Then, we get

$$dy = \frac{1}{2}\left(\frac{1}{1+x} + \frac{1}{1-x} \right) dx.$$

Therefore,

$$y = \frac{1}{2}\left(\ln|1+x| - \ln|1-x| \right) + C$$

$$= \frac{1}{2} \ln\left| \frac{1+x}{1-x} \right| + C. \quad (C \text{ is the constant of integration.})$$

Answer $y = \frac{1}{2} \ln\left| \frac{1+x}{1-x} \right| + C$

d. $dy = \dfrac{f'(x)}{f(x)} dx$

By integrating both sides, we obtain:

$$y = \ln|f(x)| + C. \quad (C \text{ is the constant of integration.})$$

Answer $y = \ln|f(x)| + C$

1.1.6 *Initial value problems*

Generally, the general solution of a differential equation includes arbitrary constants, resulting in infinitely many solutions. To find a specific solution, we use initial conditions. This type of differential equation with the initial conditions is referred to as an initial value problem (IVP). In other words, an IVP is expressed in the following form:

$$y' = f(x, y), \quad y(x_0) = y_0. \tag{1.9}$$

Example 1.3

Solve the following IVP:

$$y' = 2y, \quad y(0) = 3.$$

Solution

In Example 1.1(b), we determined that the general solution for $y' = 2y$ is

$$y = Ce^{2x}. \quad (C \text{ is the constant of integration.})$$

(We will learn the direct method for this in Section 1.2.)
From the solution $y = Ce^{2x}$ and the initial condition $y(0) = 3$, we obtain $C = 3$.

Hence, the IVP has the solution $y = 3e^{2x}$. This is the particular solution.

Figure 1.1 shows the solution in Example 1.3.

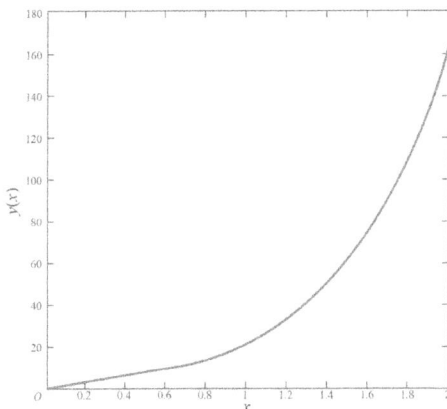

Figure 1.1 The solution in Example 1.3.

1.1.7   *Differentiation*

The following is a summary of fundamental *differentiation*:

- Differentiation of a Rational Function

$$\frac{d}{dx} x^n = nx^{n-1} \tag{a}$$

- Differentiation of a Composite Function

$$\frac{d}{dx}\{f(x)\}^n = n\{f(x)\}^{n-1}f'(x) \tag{b}$$

- Differentiation of Product Functions

$$\{f(x)g(x)\}' = f'(x)g(x)+f(x)g'(x) \tag{c1}$$

$$\{f(x)g(x)h(x)\}' = f'(x)g(x)h(x)+f(x)g'(x)h(x)+f(x)g(x)h'(x) \tag{c2}$$

- Differentiation of a Fractional Function

$$\left\{\frac{f(x)}{g(x)}\right\}' = \frac{f'(x)g(x)-f(x)g'(x)}{\{g(x)\}^2} \tag{d}$$

- Differentiation of Exponential and Logarithmic Functions

$$\left(e^x\right)' = e^x \tag{e1}$$

$$\left(a^x\right)' = a^x \ln a \tag{e2}$$

$$\left(\ln x\right)' = \frac{1}{x} \tag{e3}$$

$$\left\{\ln|f(x)|\right\}' = \frac{f'(x)}{f(x)} \tag{e4}$$

$$\left\{e^{f(x)}\right\}' = e^{f(x)}f'(x) \tag{e5}$$

$$\left\{a^{f(x)}\right\}' = a^{f(x)}f'(x)\ln a \tag{e6}$$

- Differentiation of Trigonometric Functions

$$\frac{d}{dx}\sin x = \cos x \tag{f1}$$

$$\frac{d}{dx}\cos x = -\sin x \tag{f2}$$

$$\frac{d}{dx}\tan x = \sec^2 x \tag{f3}$$

$$\frac{d}{dx}\cot x = -\csc^2 x \qquad\qquad (f4)$$

$$\frac{d}{dx}\sec x = \sec x \tan x \qquad\qquad (f5)$$

$$\frac{d}{dx}\csc x = -\csc x \cot x \qquad\qquad (f6)$$

Check

$$\frac{d}{dx}\tan x = \frac{d}{dx}\left(\frac{\sin x}{\cos x}\right) = \frac{\cos x \cos x - \sin x(-\sin x)}{\cos^2 x} = \frac{1}{\cos^2 x} = \sec^2 x$$

$$\frac{d}{dx}\cot x = \frac{d}{dx}\left(\frac{\cos x}{\sin x}\right) = \frac{(-\sin x)\sin x - \cos x(\cos x)}{\sin^2 x} = \frac{-1}{\sin^2 x} = -\mathrm{cosec}^2 x$$

$$\frac{d}{dx}\sec x = \frac{d}{dx}\cos^{-1} x = -\cos^{-2} x \cdot(-\sin x) = \sec x \tan x$$

$$\frac{d}{dx}\csc x = \frac{d}{dx}\sin^{-1} x = -\sin^{-2} x \cdot \cos x = -\csc x \cot x$$

- Differentiation of Hyperbolic Functions

Remark

$$\sinh x = \frac{e^x - e^{-x}}{2}, \ \cosh x = \frac{e^x + e^{-x}}{2}$$

$$\frac{d}{dx}\sinh x = \cosh x \qquad\qquad (g1)$$

$$\frac{d}{dx}\cosh x = \sinh x \qquad\qquad (g2)$$

$$\frac{d}{dx}\tanh x = \mathrm{sech}^2 x \qquad\qquad (g3)$$

$$\frac{d}{dx}\coth x = -\mathrm{csch}^2 x \qquad\qquad (g4)$$

Check

$$\frac{d}{dx}\tanh x = \frac{d}{dx}\left(\frac{\sinh x}{\cosh x}\right) = \frac{\cosh x \cosh x - \sinh x \sinh x}{\cosh^2 x}$$

$$= \frac{\left(\dfrac{e^x + e^{-x}}{2}\right)^2 - \left(\dfrac{e^x - e^{-x}}{2}\right)^2}{\cosh^2 x}$$

$$= \frac{\left(\dfrac{e^{2x} + 2 + e^{-2x}}{2}\right) - \left(\dfrac{e^{2x} - 2 + e^{-2x}}{2}\right)}{\cosh^2 x}$$

$$= \frac{1}{\cosh^2 x} = \mathrm{sech}^2 x$$

$$\frac{d}{dx}\coth x = \frac{d}{dx}\left(\frac{\cosh x}{\sinh x}\right) = \frac{\sinh x \sinh x - \cosh x \cosh x}{\sinh^2 x}$$

$$= -\frac{1}{\sinh^2 x} = -\mathrm{csch}^2 x$$

1.1.8   Integration

The following is a summary of basic *integration* (C is the constant of integration):

- Integration of Rational Functions

$$\int x^n \, dx = \frac{1}{n+1} x^{n+1} + C \qquad (n \neq -1) \tag{h1}$$

$$\int \frac{1}{x} \, dx = \ln|x| + C \tag{h2}$$

- Integration of Composite Functions

$$\int \{f(x)\}^n f'(x) \, dx = \frac{1}{n+1} \{f(x)\}^{n+1} + C \qquad (n \neq -1) \tag{i1}$$

$$\int \frac{f'(x)}{f(x)} \, dx = \ln|f(x)| + C \tag{i2}$$

- Integration by Parts

$$\int (f \cdot g)\,dx = F \cdot g - \int (F \cdot g')\,dx \qquad \left(F = \int f\,dx \right) \qquad \text{(j)}$$

- Integration of Exponential Functions

$$\int e^x\,dx = e^x + C \qquad (a>0,\ a\neq 1) \qquad \text{(k1)}$$

$$\int e^{f(x)} f'(x)\,dx = e^{f(x)} + C \qquad \text{(k2)}$$

$$\int a^x\,dx = \frac{1}{\ln a} a^x + C \qquad (a>0,\ a\neq 1) \qquad \text{(k3)}$$

$$\int a^{f(x)} f'(x)\,dx = \frac{1}{\ln a} a^{f(x)} + C \qquad (a>0,\ a\neq 1) \qquad \text{(k4)}$$

$$\int x e^x\,dx = x e^x - e^x + C \qquad \text{(k5)}$$

Check Integration by parts

$$\int x e^x\,dx = x e^x - \int 1 \cdot e^x\,dx = x e^x - e^x + C$$

- Integration of Logarithmic Functions

$$\int \frac{1}{x}(\ln x)^n\,dx = \frac{1}{n+1}(\ln x)^{n+1} + C \qquad (n\neq -1) \qquad \text{(l1)}$$

$$\int \frac{1}{x\ln x}\,dx = \ln|\ln x| + C \qquad \text{(l2)}$$

$$\int \ln x\,dx = x\ln x - x + C \qquad \text{(l3)}$$

$$\int \ln(x+1)\,dx = (x+1)\ln(x+1) - x + C \qquad \text{(l4)}$$

$$\int x\ln x\,dx = \frac{x^2}{2}\ln x - \frac{x^2}{4} + C \qquad \text{(l5)}$$

Check   **Integration by parts**

$$\int 1 \cdot \ln x \, dx = x \ln x - \int x \cdot \frac{1}{x} \, dx = x \ln x - x + C$$

$$\int 1 \cdot \ln(x+1) \, dx = (x+1)\ln(x+1) - \int (x+1) \cdot \frac{1}{x+1} \, dx$$

$$= (x+1)\ln(x+1) - x + C$$

$$\int x \cdot \ln x \, dx = \frac{x^2}{2} \ln x - \int \frac{x^2}{2} \cdot \frac{1}{x} \, dx = \frac{x^2}{2} \ln x - \frac{x^2}{4} + C$$

- Integration of Trigonometric Functions

$$\int \sin x \, dx = -\cos x + C \tag{m1}$$

$$\int \cos x \, dx = \sin x + C \tag{m2}$$

$$\int \tan x \, dx = -\ln|\cos x| + C \tag{m3}$$

$$\int \cot x \, dx = \ln|\sin x| + C \tag{m4}$$

Check

$$\int \tan x \, dx = \int \frac{\sin x}{\cos x} \, dx = \int -\frac{(-\sin x)}{\cos x} \, dx = -\ln|\cos x| + C$$

$$\int \cot x \, dx = \int \frac{\cos x}{\sin x} \, dx = \ln|\sin x| + C$$

$$\int x \sin x \, dx = -x \cos x + \sin x + C \tag{m5}$$

$$\int x \cos x \, dx = x \sin x + \cos x + C \tag{m6}$$

Check   **Integration by Parts**

$$\int x \sin x \, dx = x(-\cos x) - \int 1 \cdot (-\cos x) \, dx + C = -x \cos x + \sin x + C$$

$$\int x \cos x \, dx = x \sin x - \int 1 \cdot \sin x \, dx + C = x \sin x + \cos x + C$$

- Integration of Hyperbolic Functions

$$\int \sinh x \, dx = \cosh x + C \tag{n1}$$

$$\int \cosh x \, dx = \sinh x + C \tag{n2}$$

$$\int \operatorname{sech}^2 x \, dx = \tanh x + C \tag{n3}$$

$$\int \operatorname{csch}^2 x \, dx = -\coth x + C \tag{n4}$$

$$\int \tanh x \, dx = \ln(\cosh x) + C \tag{n5}$$

$$\int \coth x \, dx = \ln|\sinh x| + C \tag{n6}$$

Check

$$\int \tanh x \, dx = \int \frac{\sinh x}{\cosh x} \, dx = \ln|\cosh x| + C$$

$$\int \coth x \, dx = \int \frac{\cosh x}{\sinh x} \, dx = \ln|\sinh x| + C$$

- Integration using Partial Fractions

$$\int \frac{1}{(x-a)(x-b)} \, dx = \frac{1}{a-b} \ln\left|\frac{x-a}{x-b}\right| + C \tag{o}$$

Check

Using the formula $\dfrac{1}{(x-a)(x-b)} = \dfrac{1}{a-b}\left(\dfrac{1}{x-a} - \dfrac{1}{x-b}\right)$, then

$$\int \frac{1}{(x-a)(x-b)} \, dx = \frac{1}{a-b} \int \left(\frac{1}{x-a} - \frac{1}{x-b}\right) dx$$

$$= \frac{1}{a-b}\left(\ln|x-a| - \ln|x-b|\right) + C$$

$$= \frac{1}{a-b} \ln\left|\frac{x-a}{x-b}\right| + C.$$

- Integration by Substitution

$$\int \frac{1}{x^2+1}\,dx = \arctan x + C \tag{p}$$

Check

When we substitute $x = \tan\theta \left(|\theta| < \frac{\pi}{2}\right)$ for x,

we have $dx = \sec^2\theta\,d\theta$, and $x^2+1 = \tan^2\theta+1 = \sec^2\theta$.

(Dividing both sides of $\sin^2\theta + \cos^2\theta = 1$ by $\cos^2\theta$ gives $\tan^2\theta + 1 = \sec^2\theta$.)

Therefore, we derive $\displaystyle\int \frac{dx}{x^2+1} = \int \frac{\sec^2\theta\,d\theta}{\sec^2\theta} = \int d\theta = \theta + C = \arctan x + C.$

Problem 1.1

Solve the following ODE by integration or by remembering a differentiation formula. [1 ~ 12]

1. $y' = 2\cos 3x - x$

2. $y' = \dfrac{1}{x} + 3e^{-x}$

3. $y' = 2xe^{x^2}$

4. $y' = \sinh 3x + 2^x$

5. $y' = \dfrac{\ln x}{x} + \sin 2x$

6. $y' = \dfrac{1}{x\ln x} - x^2$

7. $y' = xe^x$

8. $y' = \dfrac{x}{e^x}$

9. $y' = x\sin x$

10. $y' = x\ln x$

11. $y' = e^x \cos x$

12. $y' = e^x \sin x$

Show that the equation y is a solution of the ODE, and determine the particular solution of the IVP. (C arbitrary) [13 ~ 17]

13. $y' + 2y = 4,\ y = Ce^{-2x} + 2,\quad y(0) = 3$

14. $y' + 2xy = 0,\ y = Ce^{-x^2},\quad y(0) = 2$

15. $y' = 2y + e^x,\ y = Ce^{2x} - e^x,\quad y(0) = 1$

16. $yy' = 2x$, $y^2 = 2x^2 + C$, $y(1) = 2$

17. $y' = y - y^2$, $y = \dfrac{1}{1 + Ce^{-x}}$, $y(0) = \dfrac{1}{2}$

1.2   Separable first-order ODEs

1.2.1   *Method of separating variables*

Among the many differential equations, there are *separable ODEs* like Eq. (1.10), where the variables x and y can be separated and expressed independently according to each variable.

$$g(y)y' = f(x). \tag{1.10}$$

Here, since $y' = dy/dx$, it can be expressed as follows:

$$g(y)dy = f(x)dx. \tag{1.11}$$

Integrating both sides, we obtain:

$$\int g(y)dy = \int f(x)dx + C. \tag{1.12}$$

By calculating this, we can find the general solution for the separable ODE (Eq. 1.10). This method of solution is called the *method of separating variables*, and Eq. (1.10) is referred to as a separable equation.

Example 1.4

Solve the following ODE:

a. $y' = 1 + y^2$

b. $y' = (x+1)y$

Solution

a. $\dfrac{dy}{1+y^2} = dx$

By integrating both sides, we obtain:

$$\int \frac{dy}{1+y^2} = \int 1\,dx + C.$$

Then, $\arctan y = x + C$.

Answer $y = \tan(x+C)$

Check

From the answer $y = \tan(x + C)$,

LHS: $y' = \sec^2(x + C)$,

RHS: $1 + y^2 = 1 + \tan^2(x + C) = \sec^2(x + C)$.

Therefore, LHS = RHS, demonstrating that $y = \tan(x + C)$ is indeed a solution to the given ODE $y' = (x + 1)y$.

---

b. $\dfrac{dy}{y} = (x + 1)dx$

Integrating both sides, we obtain: $\displaystyle\int \dfrac{dy}{y} = \int (x + 1)dx$.

Therefore, $\ln|y| = \dfrac{x^2}{2} + x + \ln C$, and

$$y = Ce^{\frac{x^2}{2} + x}.$$

$$\textbf{Answer } y = Ce^{\frac{x^2}{2} + x}$$

Check

From the answer $y = Ce^{\frac{x^2}{2} + x}$,

LHS: $y' = C(x + 1)e^{\frac{x^2}{2} + x}$.

RHS: $(x + 1)y = (x + 1)Ce^{\frac{x^2}{2} + x}$.

Therefore, LHS = RHS, demonstrating that $y = Ce^{\frac{x^2}{2} + x}$ is indeed a solution to the given ODE $\dfrac{dy}{y} = (x + 1)dx$.

Example 1.5

Solve the following IVP.

$$yy' + 2e^x = 0, \ y(0) = 2$$

Solution

By integrating both sides of the given equation $y\,dy = -2e^x dx$, we obtain:

$$\int y\,dy = -\int 2e^x\,dx.$$

Then, $\dfrac{y^2}{2} = -2e^x + \dfrac{C}{2}$,

$$\text{or } y^2 + 4e^x = C.$$

From the solution $y^2 + 4e^x = C$ and the initial condition $y(0) = 2$, we obtain $C = 8$.

Answer $y^2 + 4e^x = 8$

Figure 1.2 shows the solution in Example 1.5.

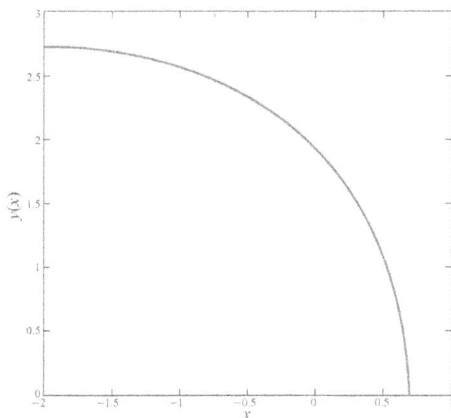

Figure 1.2 The solution in Example 1.5.

Check

From the answer $y^2 + 4e^x = 8$
By differentiating both sides, we obtain: $2yy' + 4e^x = 0$.
Then, we can get the given ODE, $yy' + 2e^x = 0$.

1.2.2 *Extended method of separating variables*

Eq. (1.10) is not in a separable form, but there are simple transformations that can convert it into a separable ODE. An example of this is as follows:

$$y' = f\left(\frac{y}{x}\right).\tag{1.13}$$

In Eq. (1.13), we set $u = y/x$, then $y = ux$.

By product differentiation, $y' = u'x + u$, and Eq. (1.13) becomes

$$u'x + u = f(u).\tag{1.14}$$

Therefore, Eq. (1.14) is converted into the following separable ODE form:

$$\frac{du}{f(u)-u} = \frac{dx}{x}.\tag{1.15}$$

Meanwhile, depending on the form of the differential equation, it can also be converted into a separable ODE by substituting different functions, such as $u = \dfrac{x}{y}$, $u = y - 2x$, etc.

Example 1.6

Solve the following ODE:

a. $2xyy' = x^2 + y^2$

b. $y' = (y + x)^2$

Solution

a. From the given equation $2xyy' = x^2 + y^2$,

$$y' = \frac{x^2 + y^2}{2xy} = \frac{x}{2y} + \frac{y}{2x}.$$

Here, we set $u = \dfrac{y}{x}$, then

$$y = ux.$$

And by product differentiation,

$$y' = u'x + u,$$

then,

$$u'x + u = \frac{1}{2u} + \frac{u}{2}.$$

We see that in the last equation we can now separate the variables,

$$\frac{2u}{u^2 - 1} du = -\frac{dx}{x}.$$

By integration,

$$\ln|u^2 - 1| = -\ln|x| + \ln C,$$

or

$$\ln\left|x\left(u^2 - 1\right)\right| = \ln C.$$

we substitute back in $u = y/x$, then

$$x\left\{\left(\frac{y}{x}\right)^2 - 1\right\} = C.$$

Answer $y^2 - x^2 = Cx$

Check

From the answer $y^2 - x^2 = Cx$,
by differentiating both sides, we obtain:

$$2yy' - 2x = C, \quad \text{or} \quad 2yy' = 2x + C.$$

LHS: $2xyy' = 2x^2 + Cx$.

RHS: $x^2 + y^2 = x^2 + \left(x^2 + Cx\right) = 2x^2 + Cx$.

Therefore, LHS = RHS, demonstrating that $y^2 - x^2 = Cx$ is indeed a solution to the given ODE $2xyy' = x^2 + y^2$.

b. $y' = (y + x)^2$

Let's set $u = y + x$, then $u' = y' + 1$.

When we substitute this into the equation, then

$$u' - 1 = u^2.$$

We see that in the last equation we can now separate the variables,

$$\frac{du}{u^2 + 1} = dx.$$

By integration,

$$\arctan u = x + C,$$

or

$$u = \tan(x + C).$$

Answer $y + x = \tan(x + C)$

Check

From the answer $y = -x + \tan(x + C)$

LHS: $y' = -1 + \sec^2(x + C)$.

RHS: $(y + x)^2 = \tan^2(x + C)$.

Therefore, LHS = RHS, demonstrating that $y = -x + \tan(x + C)$ is indeed a solution to the given ODE $y' = (y + x)^2$.

Problem 1.2

Find the general solution to the following ODEs. And verify the general solution by substituting it into the ODE. [1 ~ 10]

1. $y^2 y' + x^2 = 0$
2. $y' \sin x = y^2 \cos x$
3. $xy' = y^2 + 2y$
4. $y' = e^{x-1} y^2$
5. $xy' = 2y \ln x$
6. $2yy' = (2x + 1) e^{-y^2}$

7. $xy' = y - x^2 e^{y/x}$ (hint: set $u = y/x$)

8. $y' = (y - x + 1)^2$ (hint: set $u = y - x + 1$)

9. $y' = y + 2x$ (hint: set $u = y + 2x$)

10. $xy' = x + y$ $\left(\text{hint: set } u = \dfrac{y}{x} \right)$

Solve the following IVP. [11 ~ 16]

11. $xy' + 2y = 0, \quad y(1) = 2$

12. $yy' = 1 + 2y^2, \quad y(0) = 1$

13. $(x^2 + 1)y' = \tan y, \quad y(0) = \dfrac{\pi}{2}$

14. $y' = 2(\cos x)y, \quad y(0) = 1$

15. $xy' = y + 2x^3 \cos^2(y/x), \quad y(1) = \pi/4$

16. $y' = (x + y + 1)^2, \quad y(0) = 0$

1.3   Exact first-order ODEs and integrating factors

1.3.1   *Exact ODE*

If a function $u = u(x, y)$ has continuous partial derivatives, its differential (also called its total differential) is

$$du = \frac{\partial u}{\partial x} dx + \frac{\partial u}{\partial y} dy. \qquad (1.16)$$

If $u = 2x + xy^2 = C$, for example, then

$$du = (2 + y^2) dx + (2xy) dy = 0,$$

or

$$y' = \frac{dy}{dx} = -\frac{2 + y^2}{2xy}.$$

When this process is reversed, it leads to a new powerful method for obtaining the general solution $2x + xy^2 = C$ from its ODE $y' = -\dfrac{2 + y^2}{2xy}$.

By integrating $du = 0$, which is called an *exact ODE*, we obtain the *general solution* of Eq. (1.16) in the form $u = C$.

In Eq. (1.16), when we put

$$\frac{\partial u}{\partial x} = M(x, y), \tag{1.17a}$$

$$\frac{\partial u}{\partial y} = N(x, y), \tag{1.17b}$$

then it can be expressed as

$$M(x, y)dx + N(x, y)dy = 0. \tag{1.18}$$

Then by partial differentiation of Eq. (1.17),

$$\frac{\partial M}{\partial y} = \frac{\partial^2 u}{\partial y \partial x}, \quad \text{and} \quad \frac{\partial N}{\partial x} = \frac{\partial^2 u}{\partial x \partial y}.$$

Thus,

$$\frac{\partial M}{\partial y} = \frac{\partial N}{\partial x}. \tag{1.19}$$

This condition is not only necessary but also a sufficient condition for Eq. (1.18) to be an exact ODE.

If Eq. (1.18) is exact, we have by integrating with respect to x or y, respectively, from Eq. (1.17) as follows:

$$u = \int M(x, y)dx + f(y), \tag{1.20a}$$

$$u = \int N(x, y)dy + g(x). \tag{1.20b}$$

In Eq. (1.20a), during the integration with respect to x, y should be treated as a constant and $f(y)$ plays a role similar to an integration constant. Similarly, in Eq. (1.20b), during the integration with respect to y, x should be treated as a constant and $g(x)$ plays a role similar to an integration constant. Therefore, by comparing $f(y)$ and $g(x)$, we can obtain the general solution of Eq. (1.16) in the form $u = C$.

Example 1.7

Solve the following ODE:

$$\{x + \sin(x + y)\}dx + \{y^2 + \sin(x + y)\}dy = 0.$$

Solution

First, it needs to be checked whether the given equation is separable. Since the given equation is not separable,

$$M = \frac{\partial u}{\partial x} = x + \sin(x+y), \qquad \text{①}$$

$$N = \frac{\partial u}{\partial y} = y^2 + \sin(x+y). \qquad \text{②}$$

Thus

$$\frac{\partial M}{\partial y} = \frac{\partial}{\partial y}\left(\frac{\partial u}{\partial x}\right) = \cos(x+y),$$

$$\frac{\partial N}{\partial x} = \frac{\partial}{\partial x}\left(\frac{\partial u}{\partial y}\right) = \cos(x+y).$$

From

$$\frac{\partial M}{\partial y} = \frac{\partial N}{\partial x},$$

we see the given equation is exact.

By integrating Eq. ① and Eq. ②, respectively,

$$u = \frac{x^2}{2} - \cos(x+y) + f(y),$$

$$u = \frac{y^3}{3} - \cos(x+y) + g(x).$$

Therefore, by comparing $f(y)$ and $g(x)$, we can obtain the general solution as follows:

$$u = \frac{x^2}{2} - \cos(x+y) + \frac{y^3}{3} = C.$$

$$\textbf{Answer} \quad \frac{x^2}{2} + \frac{y^3}{3} - \cos(x+y) = C$$

Check

By integrating the answer $u = \dfrac{x^2}{2} - \cos(x+y) + \dfrac{y^3}{3} = C,$

we can obtain the given equation,

$$du = \frac{\partial u}{\partial x}dx + \frac{\partial u}{\partial y}dy = \{x + \sin(x+y)\}dx + \{y^2 + \sin(x+y)\}dy = 0.$$

Example 1.8

Solve the following IVP:

$$(y\sinh x + y)dx + (\cosh x + x)dy = 0, \ y(0) = 2.$$

Solution

Here, this section is for practicing the solution of an exact ODE, so let's use the method for solving an exact ODE.

Given equation is with

$$M = \frac{\partial u}{\partial x} = y\sinh x + y, \qquad \qquad ①$$

$$N = \frac{\partial u}{\partial y} = \cosh x + x. \qquad \qquad ②$$

Thus,

$$\frac{\partial M}{\partial y} = \frac{\partial}{\partial y}\left(\frac{\partial u}{\partial x}\right) = \sinh x + 1,$$

$$\frac{\partial N}{\partial x} = \frac{\partial}{\partial x}\left(\frac{\partial u}{\partial y}\right) = \sinh x + 1.$$

From

$$\frac{\partial M}{\partial y} = \frac{\partial N}{\partial x},$$

we see the given equation is exact.

By integrating, respectively,

$$u = y\cosh x + xy + f(y),$$

$$u = y\cosh x + xy + g(x).$$

Therefore, by comparing $f(y)$ and $g(x)$, we can obtain the general solution as follows:

$$u = y\cosh x + xy = C.$$

Using initial conditions $x = 0$, $y = 2$, then $y\cosh x + xy = 2$.

Answer $y\cosh x + xy = 2$

Figure 1.3 shows the solution in Example 1.8.

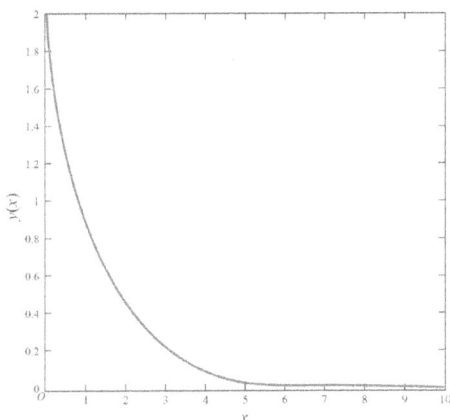

Figure 1.3 The solution in Example 1.8.

Check

By differentiating the answer $u = y \cosh x + xy = C$,
we can obtain the given equation,

$$du = \frac{\partial u}{\partial x} dx + \frac{\partial u}{\partial y} dy = (y \sinh x + y) dx + (\cosh x + x) dx = 0.$$

Another Solution   Using the method of separating variables

By slightly modifying the given equation, we can see that it becomes separable form.

$$y(\sinh x + 1) dx + (\cosh x + x) dy = 0,$$

or

$$\left(\frac{\sinh x + 1}{\cosh x + x} \right) dx + \frac{dy}{y} = 0.$$

Therefore, integrating both sides yields

$$\ln|\cosh x + x| + \ln|y| = \ln C,$$

or

$$(\cosh x + x)y = C.$$

Using initial conditions $x = 0$, $y = 2$, then $y \cosh x + y = 2$.

Answer $y \cosh x + xy = 2$

Example 1.9

Solve the following ODE:

$$-y dx + x dy = 0.$$

Solution

Here, this section is for practicing the solution of an exact ODE, so let's use the method for solving an exact ODE.

In the given equation, we put

$$P = \frac{\partial u}{\partial x} = -y, \quad Q = \frac{\partial u}{\partial y} = x.$$

By differentiating, then

$$\frac{\partial P}{\partial y} = \frac{\partial}{\partial y}\left(\frac{\partial u}{\partial x}\right) = -1, \quad \frac{\partial Q}{\partial x} = \frac{\partial}{\partial x}\left(\frac{\partial u}{\partial y}\right) = 1.$$

From

$$\frac{\partial P}{\partial y} \neq \frac{\partial Q}{\partial x},$$

we see the given equation is not exact.

Therefore, this problem can be solved using the method of separation of variables, discussed in Section 1.2 or by employing a different approach.

When we use the method of separation of variables, then

$$\frac{dy}{y} = \frac{dx}{x}.$$

By integrating, then

$$\ln|y| = \ln|x| + \ln C,$$

or

$$y = Cx.$$

Answer $y = Cx$

Check

By differentiating the answer $y = Cx$, or $dy = Cdx$,
we can obtain the given equation,

$$-ydx + xdy = -(Cx)dx + x(Cdx) = 0.$$

Another Solution   Using the Method of Separating Variables

We see that the given expression is separable as follows:

$$-\frac{dx}{x} + \frac{dy}{y} = 0,$$

or

$$-\ln|x| + \ln|y| = \ln C,$$

$$\therefore \; \frac{y}{x} = C.$$

Answer $y = Cx$

1.3.2   *Exact ODE using integrating factors*

Let's attempt to solve Example 1.9 using the *integrating factor*. Multiplying both sides of the equation by the integrating factor yields the following:

$$-\frac{y}{x^2}dx + \frac{1}{x}dy = 0. \tag{1.21}$$

In the given equation, we put

$$M = \frac{\partial u}{\partial x} = -\frac{y}{x^2}, \tag{1.22a}$$

$$N = \frac{\partial u}{\partial y} = \frac{1}{x}. \tag{1.22b}$$

By integrating, Eq. (1.22) becomes

$$\frac{\partial M}{\partial y} = \frac{\partial}{\partial y}\left(\frac{\partial u}{\partial x}\right) = -\frac{1}{x^2},$$

$$\frac{\partial N}{\partial x} = \frac{\partial}{\partial x}\left(\frac{\partial u}{\partial y}\right) = -\frac{1}{x^2}.$$

Therefore, from $\dfrac{\partial M}{\partial y} = \dfrac{\partial N}{\partial x}$, we see the given equation is exact.

By integrating Eq. (1.22a) and Eq. (1.22b), respectively, then

$$u = \frac{y}{x} + f(y),$$

$$u = \frac{y}{x} + g(x).$$

Therefore, by comparing $f(y)$ and $g(x)$, we can obtain the general solution as follows:

$$u = \frac{y}{x} = C.$$

And Eq. (1.21) also becomes

$$-\frac{y}{x^2}dx + \frac{1}{x}dy = d\left(\frac{y}{x}\right) = 0,$$

or

$$\frac{y}{x} = C.$$

Check

By multiplying both sides of the given equation $-ydx + xdy = 0$ by the integrating factor $1/x^2$, we obtained a first-order ODE.

$$-\frac{y}{x^2}dx + \frac{1}{x}dy = 0$$

I encourage you to practice until you can easily obtain the following equation directly from the exact ODE.

$$d\left(\frac{y}{x}\right) = 0$$

Answer $\dfrac{y}{x} = C$

1.3.3 How to find integrating factors

The method of solving ODEs using an integrating factor can be summarized as follows.

First, consider a differential equation that is in the form of a not-exact ODE as follows:

$$P(x, y)dx + Q(x, y)dy = 0. \tag{1.23}$$

If we multiply this equation by an appropriate integrating factor F, transforming it into an exact form, we should be able to find the solution more easily:

$$FPdx + FQdy = 0. \tag{1.24}$$

The exactness condition for Eq. (1.24) is as follows:

$$\frac{\partial(FP)}{\partial y} = \frac{\partial(FQ)}{\partial x}. \tag{1.25}$$

Using the differentiation of product functions and simplifying, we obtain the following equation:

$$\frac{\partial F}{\partial y}P + F\frac{\partial P}{\partial y} = \frac{\partial F}{\partial x}Q + F\frac{\partial Q}{\partial x}. \tag{1.26}$$

To simplify the expression, if the integrating factor F is a function of x only, that is, $F = F(x)$, or $\dfrac{\partial F}{\partial y} = 0$, then Eq. (1.26) becomes

$$F\frac{\partial P}{\partial y} = \frac{dF}{dx}Q + F\frac{\partial Q}{\partial x}. \tag{1.27}$$

When we divide both sides of Eq. (1.27) by FQ, then

$$\frac{1}{F}\frac{dF}{dx} = R, \quad \text{where} \quad R = \frac{1}{Q}\left(\frac{\partial P}{\partial y} - \frac{\partial Q}{\partial x}\right). \tag{1.28}$$

Here, if the integrating factor F is a function of x only, $F = F(x)$, R should also be a function of x, $R = R(x)$. Therefore, by integrating, we can obtain the integrating factor F as follows:

$$F(x) = e^{\int R(x)\,dx}. \tag{1.29}$$

Similarly, if the integrating factor F^* is a function of y only, $F^* = F^*(y)$, R^* should also be a function of y, $R^* = R^*(y)$.

$$\frac{1}{F^*}\frac{dF^*}{dy} = R^*. \quad \text{where} \quad R^* = \frac{1}{P}\left(\frac{\partial Q}{\partial x} - \frac{\partial P}{\partial y}\right) \tag{1.30}$$

Therefore, by integrating, we can obtain the integrating factor F^* as follows:

$$F^*(y) = e^{\int R^*(y)\,dy}. \tag{1.31}$$

Remark   How to find integrating factors

i. If the integrating factor F is a function of x only, $F = F(x)$, R should also be a function of x as follows:

$$R = \frac{1}{Q}\left(\frac{\partial P}{\partial y} - \frac{\partial Q}{\partial x}\right) \text{ and } F(x) = e^{\int R(x)\,dx}.$$

ii. If the integrating factor F^* is a function of y only, $F^* = F^*(y)$, R^* should also be a function of y as follows:

$$R^* = \frac{1}{P}\left(\frac{\partial Q}{\partial x} - \frac{\partial P}{\partial y}\right) \text{ and } F^*(y) = e^{\int R^*(y)\,dy}.$$

Example 1.10

Find an integrating factor $F(x)$ of the ODE $-y\,dx + x\,dy = 0$.

Solution

From the given equation, $-ydx + xdy = 0$,

$$P = -y, Q = x$$

$$\frac{\partial P}{\partial y} = -1, \frac{\partial Q}{\partial x} = 1$$

For $\dfrac{\partial P}{\partial y} \neq \dfrac{\partial Q}{\partial x}$, a given equation that is not exact.

If the integrating factor F is a function of x only, $F = F(x)$, then

$$R = \frac{1}{Q}\left(\frac{\partial P}{\partial y} - \frac{\partial Q}{\partial x}\right) = \frac{1}{x}(-1-1) = -\frac{2}{x}.$$

Therefore, the integrating factor F becomes

$$F(x) = e^{\int R(x)\,dx} = e^{\int -\frac{2}{x}dx} = e^{-2\ln|x|} = \frac{1}{x^2}.$$

Answer $F(x) = \dfrac{1}{x^2}$

Check

From the given equation $-ydx + xdy = 0$, we get the exact equation multiplying an integrating factor $F(x) = \dfrac{1}{x^2}$ as follows:

$$-\frac{y}{x^2}dx + \frac{1}{x}dy = 0,$$

or

$$d\left(\frac{y}{x}\right) = 0.$$

Then

$$\frac{y}{x} = c, \left(\text{where } c \text{ is an integrating constant}\right)$$

or

$$y = cx.$$

Answer $y = cx$

Another Solution   Integrating factor $F^*(y)$

If the integrating factor F^* is a function of y only, $F^* = F^*(y)$

$$R^* = \frac{1}{P}\left(\frac{\partial Q}{\partial x} - \frac{\partial P}{\partial y}\right) = -\frac{1}{y}\left(1-(-1)\right) = -\frac{2}{y}.$$

Therefore, the integrating factor F^* becomes

$$F^*(y) = e^{\int R^*(y)dy} = e^{-\int \frac{2}{y}dy} = e^{-2\ln|y|} = y^{-2}.$$

Answer $F^*(y) = \dfrac{1}{y^2}$

Check

In the ODE $-ydx + xdy = 0$, we get the exact equation multiplying an integrating factor $F^*(y) = \dfrac{1}{y^2}$ as follows:

$$-\frac{1}{y}dx + \frac{x}{y^2}dy = 0,$$

or

$$d\left(-\frac{x}{y}\right) = 0.$$

Then

$$-\frac{x}{y} = c^*. \left(\text{where } c^* \text{ is an integrating constant}\right)$$

Let $c = -\dfrac{1}{c^*}$, the equation becomes

$$y = cx.$$

Answer $y = cx$

Example 1.11

Find the integrating factor for the following ODE, and then determine the solution to the equation.

$$e^{x-y}dx + \left(1 - e^{-y}\right)dy = 0$$

Solution

From the given equation, $e^{x-y}dx + \left(1 - e^{-y}\right)dy = 0$,

$$P = e^{x-y}, \quad Q = 1 - e^{-y}$$

$$\frac{\partial P}{\partial y} = -e^{x-y}, \quad \frac{\partial Q}{\partial x} = 0.$$

For $\frac{\partial P}{\partial y} \neq \frac{\partial Q}{\partial x}$, a given equation that is not exact.

If the integrating factor F is a function of x only, $F = F(x)$, then

$$R = \frac{1}{Q}\left(\frac{\partial P}{\partial y} - \frac{\partial Q}{\partial x}\right) = \frac{1}{1 - e^{-y}}\left(-e^{x-y} - 0\right).$$

Because R is not a function of x only, let's use the integrating factor $F^* = F^*(y)$. Then

$$R^* = \frac{1}{P}\left(\frac{\partial Q}{\partial x} - \frac{\partial P}{\partial y}\right) = \frac{1}{e^{x-y}}\left(0 + e^{x-y}\right) = 1.$$

Therefore, the integrating factor becomes

$$F^*(y) = e^{\int R^*(y)dy} = e^{\int 1 dy} = e^y.$$

When we multiply the given equation by the integrating factor $F^*(y) = e^y$, it becomes an exact ODE as follows:

$$e^x dx + \left(e^y - 1\right)dy = 0,$$

or

$$d\left(e^x + e^y - y\right) = 0.$$

Answer $e^x + e^y - y = C$

Check

By differentiating the answer $e^x + e^y - y = C$, then

$$e^x dx + (e^y - 1)dy = 0.$$

By multiplying e^{-y}, it becomes the given equation:

$$e^{x-y} dx + (1 - e^{-y})dy = 0.$$

Problem 1.3

Find the general solution to the following ODE. [1 ~ 8]

1. $(x^2 + y)dx + xdy = 0$

2. $e^y dx + (xe^y + 3y^2)dy = 0$

3. $(\cos y + ye^{-x})dx - (x\sin y + e^{-x})dy = 0$

4. $(y\sec^2 x + 2x\ln y)dx + \left(\tan x + \dfrac{x^2}{y}\right)dy = 0$

5. $(x^2 + y^2)dx - 2xydy = 0$

6. $(4xe^y + 3y)dx + (x^2 e^y + x)dy = 0$

7. $(ye^y + e^{x+y})dx + (xe^y - 1)dy = 0$

8. $(y\ln y + 2xy)dx + (x + 2y^2)dy = 0$

Solve the following IVP. [9 ~ 15]

9. $(4x^3 + y^3)dx + 3xy^2 dy = 0, \quad y(1) = 1$

10. $(y - y\cos xy)dx + (x + 2y - x\cos xy)dy = 0, \quad y(0) = 0$

11. $(\sin 2x + xy^2)dx + (1 + x^2)ydy = 0, \quad y(0) = 1$

12. $(3x + y^2)dx + xydy = 0, \quad y(1) = -1$

13. $(2 - y + e^x)dx - dy = 0, \quad y(0) = 1$

14. $3xydx + (2y + 3x^2)dy = 0, \quad y(0) = 1$

15. $y\cos xdx + (y + 1)\sin xdy = 0, \quad y(\pi/2) = 1$

1.4 General solution of the first-order ODE

1.4.1 Standard form of a first-order ODE

When an ODE can be expressed in the following form, it is referred to as the *standard form of a first-order ODE*

$$y' + p(x)y = r(x). \tag{1.32}$$

Furthermore, when the expression on the right-hand side of the equation is zero, $r(x) = 0$, it is called a homogeneous ODE, and when it is not zero, $r(x) \neq 0$, it is referred to as a nonhomogeneous ODE.

The homogeneous ODE becomes

$$y' + p(x)y = 0. \tag{1.33}$$

By the method of separation of variables, it becomes

$$\frac{dy}{y} = -p(x)dx.$$

And by integrating, we obtain

$$\ln|y| = -\int p(x)dx + \ln C.$$

Therefore, we obtain the general solution of the homogeneous ODE as follows:

$$y(x) = Ce^{-\int p(x)dx}. \tag{1.34}$$

And when we multiply both sides of the nonhomogeneous ODE (1.32) by an integrating factor $F(x)$, it becomes

$$Fy' + pFy = rF. \tag{1.35}$$

If the second term on the left-hand side of Eq. (1.35) satisfies the following condition:

$$pFy = F'y, \quad \text{or} \quad pF = F', \tag{1.36}$$

Eq. (1.35) becomes

$$Fy' + F'y = rF,$$

or

$$(Fy)' = rF.$$

Therefore, we obtain

$$y(x) = \frac{1}{F}\left(\int rF\,dx + C \right). \tag{1.37}$$

And $F(x)$ must satisfy the following condition as derived from Eq. (1.36).

$$\frac{dF}{F} = p(x)\,dx.$$

Therefore, the integrating factor $F(x)$ for the first-order ODE (1.32) is given by

$$F(x) = e^{\int p(x)\,dx}. \tag{1.38}$$

By using the integrating factor, Eq. (1.38), let's reorganize the steps from Eq. (1.35) to Eq. (1.37) as follows:

$$e^{\int p(x)\,dx} y' + p(x) e^{\int p(x)\,dx} y = e^{\int p(x)\,dx} r(x). \tag{1.39}$$

Since the left-hand side of the above equation is in the form of the derivative of $e^{\int p(x)\,dx} y$, let's denote it as $F = e^{\int p(x)\,dx}$.

Then, Eq. (1.39) becomes

$$(Fy)' = Fr.$$

By integrating the equation and dividing by F, we obtain the *general solution of the first-order ODE* as follows:

$$y(x) = \frac{1}{F}\left(\int Fr\,dx + C \right). \tag{1.40}$$

Remark   General solution of the first-order ODE

The general solution of the first-order ODE $y' + p(x)y = r(x)$ is as follows:

$$y(x) = \frac{1}{F}\left(\int Fr\,dx + C \right). \tag{1.40}$$

(where $F = e^{\int p(x)\,dx}$ is an integrating factor)

Example 1.12

Obtain the general solution of the first-order ODE.

$$y' - 2y = e^x$$

Solution

Since $p(x) = -2$, $r = e^x$, the integrating factor is

$$F = e^{\int p(x)dx} = e^{-2x}.$$

Therefore, the general solution is

$$y(x) = \frac{1}{F}\left(\int Fr\,dx + C \right) = e^{2x}\left(\int e^{-2x} \cdot e^x\,dx + C \right) = e^{2x} \cdot \left(-e^{-x} + C \right) = Ce^{2x} - e^x.$$

Answer $y(x) = Ce^{2x} - e^x$

Check

Differentiating the answer, $y(x) = Ce^{2x} - e^x$, then

$$y' = 2Ce^{2x} - e^x$$

LHS: $y' - 2y = \left(2Ce^{2x} - e^x \right) - 2\left(Ce^{2x} - e^x \right) = e^x$

RHS: e^x

Therefore, LHS = RHS.

Example 1.13

Solve the following IVP.

$$xy' + y = 2x, \quad y(1) = 2$$

Solution

Since the given equation becomes $y' + \frac{1}{x}y = 2$, then

$$p(x) = \frac{1}{x}, r = 2.$$

The integrating factor is

$$F = e^{\int p(x)\,dx} = e^{\int \frac{1}{x}\,dx} = e^{\ln|x|} = x.$$

And the general solution is

$$y(x) = \frac{1}{F}\left(\int Fr\,dx + C\right) = \frac{1}{x}\left(\int x \cdot 2\,dx + C\right) = \frac{1}{x}\cdot\left(x^2 + C\right) = x + Cx^{-1}.$$

Using initial conditions $y(1) = 2$, then $C = 1$.

Answer $y(x) = x + x^{-1}$

Check

Differentiating the general solution, $y(x) = x + Cx^{-1}$, we obtain

$$y' = 1 - Cx^{-2}.$$

LHS: $xy' + y = x\left(1 - Cx^{-2}\right) + \left(x + Cx^{-1}\right) = 2x$

RHS: $2x$

Therefore, LHS = RHS.

1.4.2*   Bernoulli equation (*optional)

Some nonlinear ODEs can be transformed into linear ODEs through a simple process. Among them, one of the representative nonlinear ODEs is the following *Bernoulli equation*:

$$y' + p(x)y = g(x)y^a. \quad (a: \text{any real number}) \qquad (1.41)$$

Here, if a is equal to 0 or 1, Eq. (1.41) is a linear ODE, but otherwise, it is a nonlinear ODE. In this case, we put

$$u(x) = \left[y(x)\right]^{1-a}, \qquad (1.42)$$

and by differentiating Eq. (1.42), then

$$u' = (1-a)y^{-a}y'. \qquad (1.43)$$

When we substitute Eq. (1.41) into Eq. (1.43), then

$$u' = (1-a)y^{-a}\left(gy^a - py\right)$$
$$= (1-a)\left(g - py^{1-a}\right)$$
$$= (1-a)(g - pu).$$

Therefore, the Bernoulli equation (1.41), which is one of the nonlinear ODE, is transformed into the following linear ODE:

$$u' + (1-a)pu = (1-a)g. \tag{1.44}$$

Example 1.14

Solve the following Bernoulli equation:

$$y' + 2y = 3y^2, \quad y(0) = 0.4.$$

Solution

From the Bernoulli equation (1.40), we see

$$p(x) = 2, \; g(x) = 3, \; a = 2.$$

So, we set

$$u = y^{1-a}, \text{ or } u = y^{-1}.$$

We differentiate this, then

$$u' = -\frac{y'}{y^2} = -u^2 y'.$$

Simplification gives

$$u' = -u^2\left(-2y + 3y^2\right) = -u^2\left(-2u^{-1} + 3u^{-2}\right) = 2u - 3,$$

so that we get the linear ODE

$$u' - 2u = -3.$$

Now, let's find the solution for this. We set

$$h = \int (-2)dx = -2x,$$

and the integrating factor is

$$F = e^h = e^{\int (-2)dx} = e^{-2x}.$$

Then we get

$$u = \frac{1}{F}\left(\int Fr\,dx + C \right) = e^{2x}\left(\int e^{-2x}(-3)dx + C \right) = Ce^{2x} + \frac{3}{2}.$$

Since we substitute $u = y^{-1}$ into the above equation, we get

$$\frac{1}{y} = Ce^{2x} + \frac{3}{2}.$$

Using initial conditions $y(0) = 0.4$, then $C = 1$.

$$\textbf{Answer } y = \frac{1}{e^{2x} + 1.5}$$

Figure 1.4 shows the solution in Example 1.14.

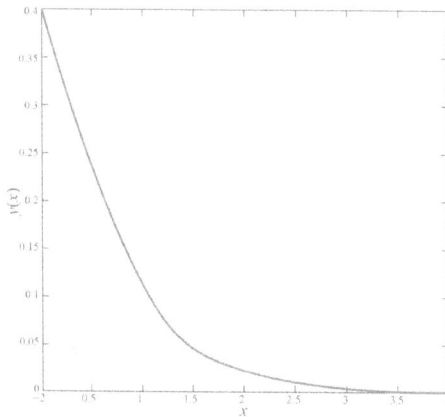

Figure 1.4 The solution in Example 1.14.

Check

Differentiating the answer $y = \dfrac{1}{e^{2x} + 1.5}$, it becomes

$$y' = -\frac{2e^{2x}}{(e^{2x} + 1.5)^2}$$

LHS: $y' + 2y = -\dfrac{2e^{2x}}{(e^{2x} + 1.5)^2} + 2 \cdot \dfrac{1}{e^{2x} + 1.5} = \dfrac{3}{(e^{2x} + 1.5)^2}$

RHS: $3y^2 = \dfrac{3}{(e^{2x} + 1.5)^2}$

Therefore, LHS = RHS.

Problem 1.4

Solve the following ODE or IVP. [1 ~ 12]

1. $y' - y = 2$
2. $y' + 2y = -4x$
3. $y' + y = 2xe^{-x}$
4. $y' + \dfrac{y}{x} = e^x$
5. $y' + 2y = e^{-2x}$
6. $y' + y\tan x = 2e^x \cos x$
7. $xy' + y = 4x^3$
8. $y' + y = 2\cos x$
9. $y' - y\cot x = -\cot x, \quad y(\pi/2) = 2$
10. $y' + y\tan x = \sin x, \quad y(0) = 1$
11. $xy' + 2y = 4x^2, \quad y(1) = 3$
12. $xy' - 2y = x^3 e^x, \quad y(1) = 2e$

Solve the following ODE or IVP. [13 ~ 18]

13. $y' + y = y^2$
14. $y' + xy = xy^{-2}$
15. $y' = 2(y - 3)\tanh x, \ y(0) = 4$
16. $y' = 3y - 2y^3, \ y(0) = 1$
17. (*optional) $yy' + y^2 = -x$
18. (*optional) $xy' - y = -xy^2, \ y(1) = 1$

1.5   Application of first-order ODEs

The first-order ODEs are applied in heat transfer, radioactive decay, electrical circuits, and other areas.

1.5.1   *Heat transfer (Newton's law of cooling)*

The rate of change of the temperature of an object that conducts heat well is known to be proportional to the temperature difference $T - T_\infty$, between the object's temperature T and the ambient temperature T_∞. (*Newton's law of cooling*)

This phenomenon is described mathematically by the following first-order ODE:

$$\frac{dT}{dt} = k(T - T_\infty), \tag{1.45}$$

where k is constant.

If the current temperature T is higher than the ambient temperature T_∞, the temperature T decreases with time (cooling, $\dfrac{dT}{dt} < 0$), and the proportionality constant becomes $k < 0$. Similarly, if the current temperature T is lower than the ambient temperature T_∞, the temperature T increases with time (heating, $\dfrac{dT}{dt} > 0$), and the proportionality constant also becomes $k < 0$.

By using the method of separation of variables, the ODE (1.45) becomes the following:

$$\frac{dT}{T - T_\infty} = kdt$$

or

$$\ln(T - T_\infty) = kt + \ln c.$$

Thus, we can obtain the following general solution:

$$T = ce^{kt} + T_\infty. \tag{1.46}$$

When we apply the initial condition $T(0) = T_0$, we obtain the following particular solution:

$$T(t) = (T_0 - T_\infty)e^{kt} + T_\infty. \tag{1.47}$$

Figure 1.5 shows the sample of the heat transfer.

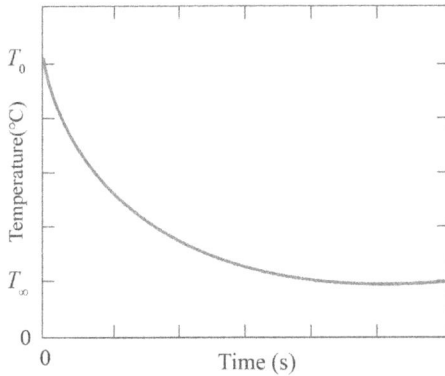

Figure 1.5 Heat transfer.

Example 1.15

A 100°C object was placed at room temperature (20°C) for 5 minutes, resulting in a temperature of 40°C. If it is left for an additional 5 minutes, what will be the temperature of the object?

Solution

According to Newton's law of cooling, the following equation holds for the temperature of the object T, and the ambient temperature T_0:

$$\frac{dT}{dt} = k(T - T_0).$$

The general solution for this is as follows:

$$T = ce^{kt} + T_0.$$

At $t = 0$, from $100 = c + 20$, then $c = 80$.

At $t = 5$, $\quad 40 = 80e^{5k} + 20.$ ①

At $t = 10$, $\quad T = 80e^{10k} + 20.$ ②

From equation ① and ②, then

$$T = 25° \text{ C}.$$

Answer $T = 25°C$

1.5.2 *Decay of a radioactive element*

It is known that the rate of decay $\dfrac{dy}{dt}$ of a *radioactive element* with respect to time is proportional to the amount remaining $y(t)$ as follows:

$$\frac{dy}{dt} = -ky, \tag{1.48}$$

where k is positive constant $(k > 0)$. By using the method of separation of variables, the ODE (1.48) becomes as follows:

$$\frac{dy}{y} = -kdt,$$

or

$$\ln y = -kt + \tilde{c}.$$

The general solution for the Eq. (1.48) is as follows:

$$y = y_0 e^{-kt}. \qquad \left(\text{where } y_0 = e^{\tilde{c}}\right) \tag{1.49}$$

The time it takes for a radioactive element to decrease to half of its initial amount is called the half-life. Figure 1.6 shows the radioactive decay.

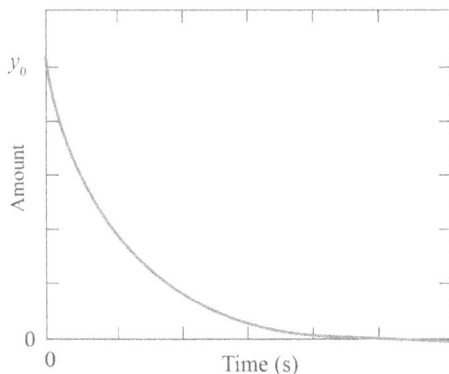

Figure 1.6 Decay of a radioactive element.

Example 1.16

Suppose the half-life of a radioactive element is 400 years. How many years will it take for this substance to decrease to 1/10 of its current amount?

Solution

The rate of decay $\dfrac{dy}{dt}$ of a radioactive element with respect to time is proportional to the amount remaining $y(t)$ as follows:

$$\frac{dy}{dt} = -ky.$$

The general solution for this is as follows:

$$y = y_0 e^{-kt}.$$

Substituting the half-life of 400 years into the above equation, we get:

$$0.5y_0 = y_0 e^{-400k}. \hspace{3cm} ①$$

When the amount remaining decreases to 1/10 of its current amount after time t passes, then

$$0.1y_0 = y_0 e^{-kt}. \hspace{3cm} ②$$

From Eq. ①, then

$$\ln 0.5 = -400k$$

From Eq. ②, then

$$\ln 0.1 = -kt$$

Therefore, we get

$$t = 400\frac{\ln 0.1}{\ln 0.5} = 400\frac{\ln 10}{\ln 2} \cong 1329.$$

Answer after about 1329 years

1.5.3 *Electric circuit*

An *electric circuit* refers to a circuit through which electric current $i(t)$ flows, and it consists of resistor R, inductor L, and capacitor C. A series circuit is supplied with electrical power from a power source, and the direction of current remains constant when power is supplied.

① *RL*-circuit

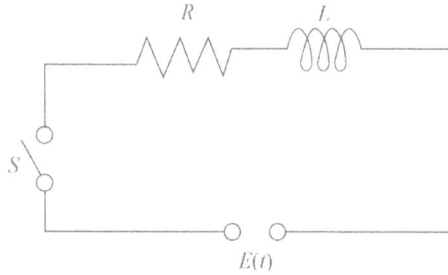

Figure 1.7 RL-circuit.

In a series circuit composed only of a resistor R and an inductor L, as shown in Figure 1.7, when the switch is connected at $t = 0$, the voltage $E(t)$ across the circuit is the sum of the voltage drop $E_R(t)(= Ri)$ across the resistor R and the voltage drop $E_L(t)\left(= L\dfrac{di}{dt}\right)$ across the inductor L as follows:

$$E(t) = E_R(t) + E_L(t) = Ri + L\frac{di}{dt}. \qquad (1.50)$$

If it is assumed that the voltage is constant, it can be represented by the following first-order ODE:

$$L\frac{di}{dt} + Ri = E. \qquad (E \text{ is constant}) \qquad (1.51)$$

By using the method of separation of variables, the ODE (1.51) becomes as follows:

$$\frac{di}{E - Ri} = \frac{1}{L}dt. \qquad (1.52)$$

By integration,

$$-\frac{1}{R}\ln|Ri - E| = \frac{1}{L}t + \tilde{c}.$$

or

$$i(t) = \frac{E}{R}\left(1 + ce^{-\frac{R}{L}t}\right). \qquad \left(\text{where } c = \frac{1}{E}e^{-R\tilde{c}}\right) \qquad (1.53)$$

By substituting the initial condition $i = 0$ at $t = 0$ into Eq. (1.53), we can obtain the following particular solution:

$$i(t) = \frac{E}{R}\left(1 - e^{-\frac{R}{L}t}\right).$$

(1.54)

Figure 1.8 shows the solution of the RL-circuit.

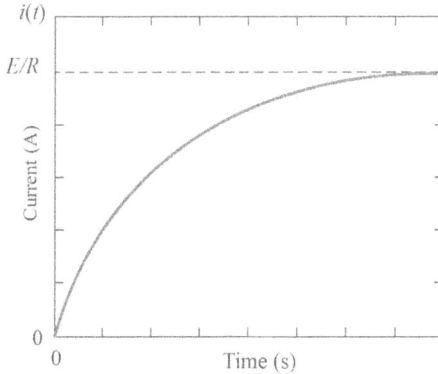

Figure 1.8 The solution of the RL-circuit.

When the switch in this circuit is short-circuited to make the voltage zero, it can be represented by the following first-order ODE:

$$L\frac{di}{dt} + Ri = 0.$$

(1.55)

By using the method of separation of variables, the ODE (1.55) becomes as follows:

$$\frac{di}{i} = -\frac{R}{L}dt.$$

(1.56)

By integration,

$$\ln|i| = -\frac{R}{L}t + \tilde{c}$$

The general solution for the Eq. (1.55) is as follows:

$$i(t) = ce^{-\frac{R}{L}t}. \quad \left(\text{where } c = e^{\tilde{c}}\right)$$

(1.57)

By substituting the initial condition $i = E/R$ at $t = 0$ into Eq. (1.57), we can obtain the following particular solution:

$$i(t) = \frac{E}{R} e^{-\frac{R}{L}t}. \tag{1.58}$$

Figure 1.9 shows the solution (current) of the RL-circuit when the switch is short-circuited.

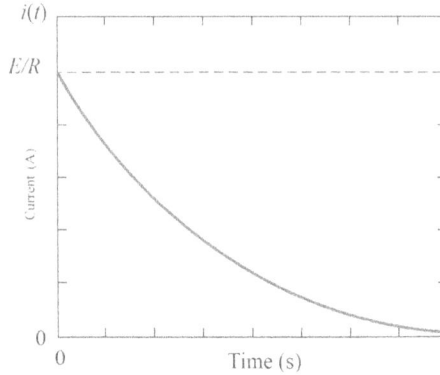

Figure 1.9 RL-circuit when the switch is short-circuited.

② *RC*-circuit

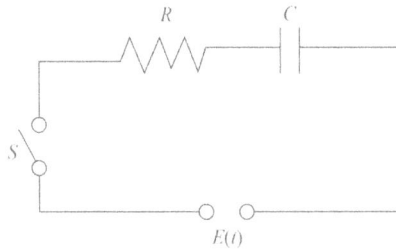

Figure 1.10 RC-circuit.

In a series circuit composed only of a resistor R and a capacity C, as shown in Figure 1.10, when the switch is connected at $t = 0$, the voltage $E(t)$ across the circuit is the sum of the voltage drop $E_R(t)(= Ri)$ across the resistor R and the voltage drop $E_C(t)\left(= \frac{1}{C}\int i\,dt\right)$ across the capacity C as follows:

$$E(t) = E_R(t) + E_C(t) = Ri + \frac{1}{C}\int i\,dt. \tag{1.59}$$

If it is assumed that the voltage is constant, it can be represented by the following integral equation:

$$Ri + \frac{1}{C}\int i\,dt = E. \qquad \text{(where } E \text{ is constant)} \qquad (1.60)$$

When we place a charge $Q = \int i\,dt$, we obtain the following first-order ODE:

$$\frac{dQ}{dt} + \frac{1}{RC}Q = \frac{E}{R}. \qquad (1.61)$$

By using the method of separation of variables, the ODE (1.61) becomes as follows:

$$\frac{dQ}{Q - CE} = -\frac{1}{RC}dt. \qquad (1.62)$$

By integration,

$$\ln|Q - CE| = -\frac{1}{RC}t + \tilde{c}$$

or

$$Q(t) = ce^{-\frac{1}{RC}t} + CE. \qquad \left(\text{where } c = e^{\tilde{c}}\right) \qquad (1.63)$$

By substituting the initial condition $Q = 0$ at $t = 0$ into Eq. (1.63), we can obtain the following particular solution:

$$Q(t) = CE\left(1 - e^{-\frac{1}{RC}t}\right). \qquad (1.64)$$

Since $i = \frac{dQ}{dt}$, the current $i(t)$ is represented as follows:

$$i(t) = \frac{E}{R}e^{-\frac{t}{RC}}. \qquad (1.65)$$

Figure 1.11 shows the solution (current) of the RC-circuit.

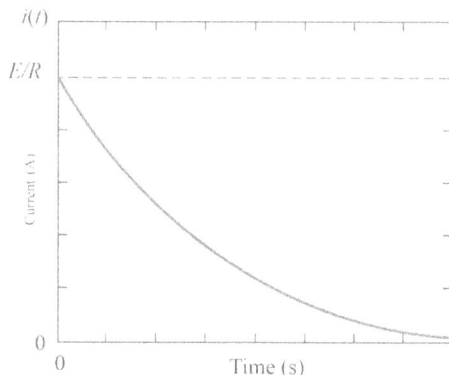

Figure 1.11 The solution of the *RC*-circuit.

When the switch in this circuit is short-circuited to make the voltage zero, it can be represented by the following integral equation:

$$Ri + \frac{1}{C}\int i\,dt = 0. \tag{1.66}$$

When we place a charge $Q = \int i\,dt$, we obtain the following first-order ODE:

$$\frac{dQ}{dt} + \frac{1}{RC}Q = 0. \tag{1.67}$$

By using the method of separation of variables, the ODE (1.67) becomes as follows:

$$\frac{dQ}{Q} = -\frac{1}{RC}dt. \tag{1.68}$$

By integration,

$$\ln|Q| = -\frac{1}{RC}t + \tilde{c}$$

or

$$Q(t) = ce^{-\frac{1}{RC}t}. \quad \left(\text{where } c = e^{\tilde{c}}\right) \tag{1.69}$$

By substituting the initial condition $Q = CE$ at $t = 0$ into Eq. (1.69), we can obtain the following particular solution:

$$Q(t) = CEe^{-\frac{1}{RC}t}. \tag{1.70}$$

Since $i = \dfrac{dQ}{dt}$, the current $i(t)$ is represented as follows:

$$i(t) = -\frac{E}{R}e^{-\frac{1}{RC}t}. \tag{1.71}$$

Here, if the current value is negative (–), it signifies that the current is flowing in the opposite direction to the reference direction.

Figure 1.12 shows the solution (current) of the RC-circuit when the switch is short-circuited.

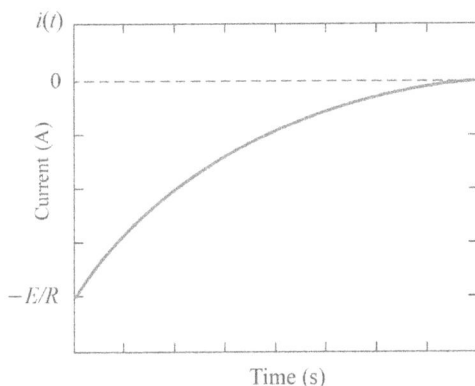

Figure 1.12 RC-circuit when the switch is short-circuited.

Problem 1.5

1. **Newton's law of cooling** If an object heated to 50°C is placed in a refrigerator at a constant temperature of 4°C, and after 10 minutes, the temperature of the object has become 30°C, what can be predicted as the temperature of the object after 1 hour?
2. **Newton's law of cooling** In an environment where the ambient temperature is maintained at 25°C, a frozen object at –5°C was left for 2 hours, resulting in the temperature of the object changing to 3°C. Predict when the temperature of the object will reach 18°C.
3. **Decay of a radioactive element** The half-life of the radioactive isotope cobalt $60(Co_{60})$ is 5.27 years. How many years will it take for the current 4 grams of cobalt to decrease to less than 0.1 grams?
4. **Increase of the number of bacteria** If a certain bacterium takes 2 days to double in number, predict how many days it will take for the population to increase 100-fold.
5. **Electric circuit** Find the current $i(t)$ flowing through the circuit when the switch S is connected at $t = 0$, in a series circuit composed of a resistor R and an inductor L as shown in Figure 1.13. Assume $E(t) = 10$ V, $R = 1\,\Omega$, and $L = 0.1$ H. Also, assume $i(0) = 0$.

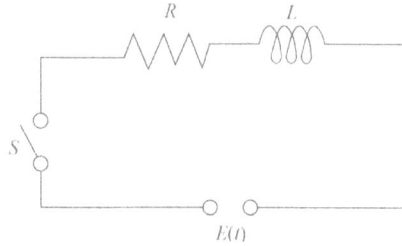

Figure 1.13 RL-circuit.

6. **Electric circuit**   Find the current $i(t)$ flowing through the circuit when the switch S is connected during $0 \le t \le 0.1$ s and short-circuited when $t > 0.1$ s in a series circuit composed of a resistor R and an inductor L as shown in Figure 1.14. Assume $E(t) = 10$ V, $R = 1\,\Omega$, and $L = 0.1$ H. Also, assume $i(0) = 0$.

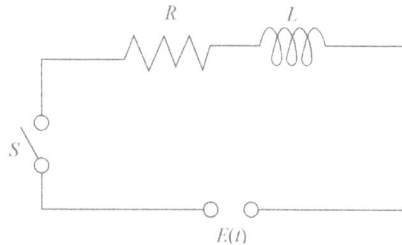

Figure 1.14 RL-circuit.

7. **Electric circuit**   Find the current $i(t)$ flowing through the circuit when the switch S is connected at $t = 0$ in a series circuit composed of a resistor R and a capacitor C, as shown in Figure 1.15. Assume $E(t) = E = 10$ V, $R = 1\,\Omega$, and $C = 0.1$ C. Also, assume $Q(0) = 0$.

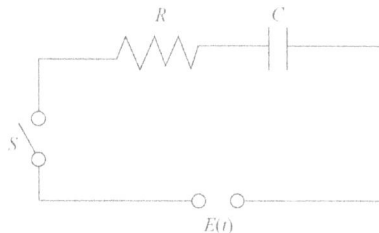

Figure 1.15 RC-circuit.

8. **Electric circuit**   Find the current $i(t)$ flowing through the circuit when the switch S is connected during $0 \le t \le 0.1$ s and short-circuited when $t > 0.1$ s in a series circuit composed of a resistor R and a capacitor C, as shown in Figure 1.16. Assume $E(t) = E = 10$ V, $R = 1\,\Omega$, and $C = 0.1$ C. Also, assume $Q(0) = 0$.

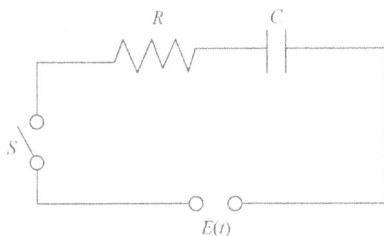

Figure 1.16 RC-circuit.

1.6* Numerical methods for ODEs (*optional)

Up to this point, we have learned methods to find solutions to ODEs analytically. However, many general ODEs may have solutions that cannot be
obtained analytically. In such cases, there are *numerical methods* for approx-
imating solutions.

Representative numerical methods include the Euler method and the
Runge-Kutta method.

1.6.1 *Euler method*

Let's assume that the following first-order IVP has a solution:

$$y' = f(x, y), \ y(x_0) = y_0. \tag{1.72}$$

As seen in Figure 1.17, we can obtain an approximation of y at by drawing
a tangent line at the initial value as follows:

$$y_1 = y_0 + hf(x_0, y_0), \tag{1.73}$$

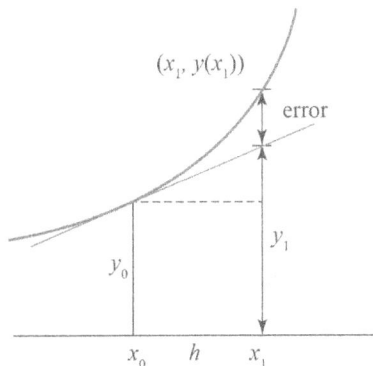

Figure 1.17 An approximation of $y(x_1)$ using the tangent line.

where h is the positive step size of x. Using this obtained value of (x_1, y_1) as a new starting point, we can also determine the next point.

$$y_2 = y_1 + hf(x_1, y_1) \tag{1.74}$$

Therefore, by repeating the same process, the *Euler method* can be summarized as follows.

Remark   Euler method

The approximation of y_{n+1} in the ODE $y' = f(x, y)$ is obtained as follows:

$$y_{n+1} = y_n + hf(x_n, y_n), \qquad (n = 0, 1, 2, \cdots) \tag{1.75}$$

where $x_{n+1} = x_0 + nh$, and h is the positive step size of x.

Example 1.17

Using the Euler method, find the approximation value of $y(0.2)$ at $x = 0.2$ for the following ODE. Given that the initial value is $(0, 1)$ and the positive step size of x is 0.1.

$$y' = 2y$$

Solution

Since the initial value is $(0, 1)$, let $x_0 = 0$, $y_0 = 1$.
Because slope $y'|_{x_0=0} = 2y|_{x_0=0} = 2 \cdot 1 = 2$ is at $x_0 = 0$, then

$$x_1 = x_0 + h = 0 + 0.1 = 0.1$$

$$y_1 = y_0 + h \cdot 2 = 1 + 0.2 = 1.2$$

Similarly, the next slope $y'|_{x_1=0.1} = 2y|_{x_1=0.1} = 2 \cdot 1.2 = 2.4$ is at $x_1 = 0.1$, then

$$y_2 = y_1 + h \cdot 1.2 = 1.2 + 0.1 \cdot 2.4 = 1.44$$

Answer $y_2 = 1.44$

Remark

The solution of the IVP $y' = 2y$, $y(0) = 1$ is

$$y = e^{2x}.$$

Therefore, the approximation value by the Euler method is as follows:

	Approximation value by the Euler method	True value of $y = e^{2x}$
$x_0 = 0$	$y_0 = 1$	$y(x_0) = 1$
$x_1 = 0.1$	$y_1 = 1.2$	$y(x_1) = 1.2214$
$x_2 = 0.2$	$y_2 = 1.44$	$y(x_2) = 1.4918$
$x_3 = 0.3$	$y_3 = 1.728$	$y(x_3) = 1.8221$

As the step size of x becomes smaller, we can obtain an approximation value closer to the true value.

Figure 1.18 shows the solution in Example 1.17.

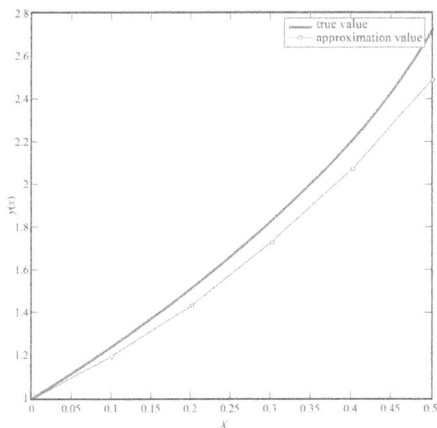

Figure 1.18 The true value and the approximation value in Example 1.17.

1.6.2 *Runge-Kutta method*

The *Runge-Kutta method* is a generalization of the Euler method explained in the previous section. In this method, the slope function $f(x_n, y_n)$ is replaced by the average slope measured over the interval $x_n < x < x_{n+1}$, as shown in the following equation.

Remark   Runge-Kutta method

i. Runge-Kutta method of the second-order

$$y_{n+1} = y_n + h\frac{k_1 + k_2}{2}, \quad (n = 0, 1, 2, \cdots) \qquad (1.76)$$

where $k_1 = f(x_n, y_n)$, $k_2 = f(x_n + h, y_n + hk_1)$.

ii. Runge-Kutta method of the fourth-order

$$y_{n+1} = y_n + h\frac{k_1 + 2k_2 + 2k_3 + k_4}{6}, \quad (n = 0, 1, 2, \cdots) \qquad (1.77)$$

where $k_1 = f(x_n, y_n)$, $k_2 = f\left(x_n + \dfrac{h}{2}, y_n + \dfrac{h}{2}k_1\right)$,

$$k_3 = f\left(x_n + \frac{h}{2}, y_n + \frac{h}{2}k_2\right), \quad k_4 = f(x_n + h, y_n + hk_3).$$

1.7   Utilizing MATLAB®

Please refer to Appendix B for instructions on how to use MATLAB®.

1.7.1   *Starting with MATLAB®*

1. On the main screen of the computer where the program is installed, double-click the MATLAB® icon with the left part of the mouse.
2. Start your work in the Command Window.

MATLAB® Toolbox

To perform symbolic processing in MATLAB®, you need the 'Symbolic Math' Toolbox.

Let's familiarize ourselves with the basic functions of MATLAB® while performing the following exercises.

M_Example 1.1

Practice the following:
```
>> x=1:10                % Automatic array from 1 to 10
>> x=1:0.5:10            % Array from 1 to 10 with a step size of 0.5

>> c=1:3; d=2:4;         % The semicolon suppresses screen output
>> x=c.*d    % Multiplication of each element with its corresponding
             element
>> x=c./d    % Division of each element by its corresponding element
>> x=c.^2             % Squares of each element in matrix c
```

M_Example 1.2

$$\text{Enter matrix } A = \begin{bmatrix} 1 & 1 & 1 \\ 1 & 2 & 3 \\ 1 & 3 & 6 \end{bmatrix} \text{ in MATLAB.}$$

Solution

```
>> A=[1 1 1; 1 2 3; 1 3 6] % the semicolon (;) represents a new row
```

M_Example 1.3

$$\text{Calculate the determinant of the matrix } A = \begin{bmatrix} 1 & 2 & -3 \\ -1 & 1 & 2 \\ 1 & 1 & 0 \end{bmatrix}. \text{ (det.m)}$$

Solution % det.m

```
>> A=[ 1 2 -3; -1 1 2; 1 1 0]
>> det(A)
```

Answer 8

M_Example 1.4 Calculate the inverse of the matrix $A = \begin{bmatrix} 1 & 2 & -3 \\ -1 & 1 & 2 \\ 1 & 1 & 0 \end{bmatrix}$.
(inv.m)

Solution % inv.m

```
>> A=[ 1 2 -3; -1 1 2 ; 1 1 0]
>> inv(A)
```

Answer
$$\begin{matrix} -0.25 & -0.375 & 0.875 \\ 0.25 & 0.375 & 0.125 \\ -0.25 & 0.125 & 0.375 \end{matrix}$$

1.7.2   To create an m-file in MATLAB®

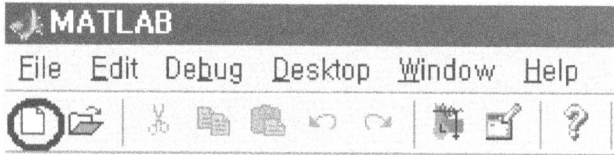

Click on ☐ (new m-file) at the top left corner of MATLAB® to open a new file window, enter the content, and save it with a new file name (💾, save).

Check   Notification when Naming the File

Do not use spaces, or special characters in the file name, and it should not consist only of numbers. (Using an underscore '_' is allowed. For example, M1_5.m.)

1.7.3   Plotting the graph of functions

M_Example 1.5

Plot the graph of $y(x) = e^{-2x}$.

Solution

```
% main file mex1_5.m
close all; clear all;
x=0:0.01:2;
y=exp(-2*x);
plot(x, y, 'linewidth', 3)
xlabel('x', 'fontsize', 14)
ylabel('y(x)', 'fontsize', 14)
title('by analytic solution', 'fontsize', 14)
```

Figure 1.19 shows the solution in M-Example 1.5.

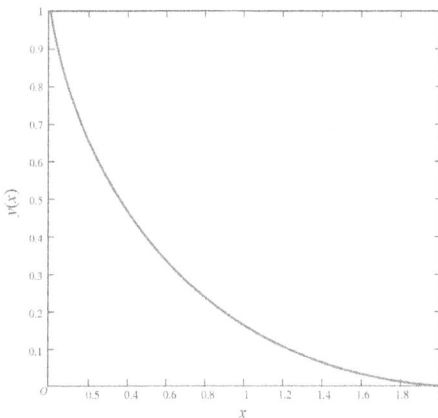

Figure 1.19 The solution in M-Example 1.5.

1.7.4 *Plotting the solution graph of the ODEs by using Runge-Kutta method*

Using the Runge-Kutta method in MATLAB® allows you to obtain the graph of a solution without finding an analytic solution for linear ODEs.

It is also valuable for cases where analytic solutions are not readily available, such as nonlinear ODEs.

M_Example 1.6

Plot the solution graph of the linear first-order ODE $y' + 2y = 0$, with the initial condition $y(0) = 1$.

Solution

When we put $f = y'$, then

$$f = -2y.$$

First, create a function file that defines the function. Let's name the file "song1_6."

```
% song1_6.m
function f=song1_6(x, y)
f=zeros(1,1);
f=-2*y;
```

Additionally, create a main file as follows and name it "mex1_6."

```
% main file mex1_6.m
close all; clear all;
x=0:0.01:2;
y_initial=1;                                    % initial condition
[x, y]=ode23('song1_6', x, y_initial);          % ode (ordinary
differential equation)
plot(x, y, 'linewidth', 3)
xlabel('x', 'fontsize', 15)
ylabel('y(x)', 'fontsize', 15)
```

Figure 1.20 shows the solution in M-Example 1.6.

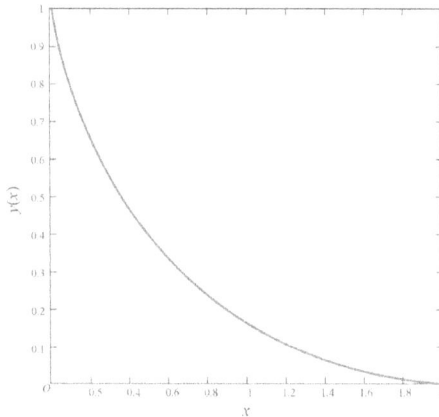

Figure 1.20 The solution in M-Example 1.6.

M_Example 1.7

Plot the solution graph of the nonlinear first-order ODE $y' + 2\sin y = 1$, with the initial condition $y(0) = 0$.

```
% song1_7.m                    % function file
function f=song1_7(x,y)
f=zeros(1,1);
f=1-2*sin(y);
```

```
% main file mex1_7.m           % main file
clear all; close all;
x=0:0.01:3;
y00=0;                         % initial condition
[x, y]=ode23('song1_7', x, y00);   % ode (ordinary differential
                               equation)

plot(x, y, 'linewidth', 3)
xlabel('x', 'fontsize', 15)
ylabel('y(x)', 'fontsize', 15)
```

Figure 1.21 shows the solution in M-Example 1.7.

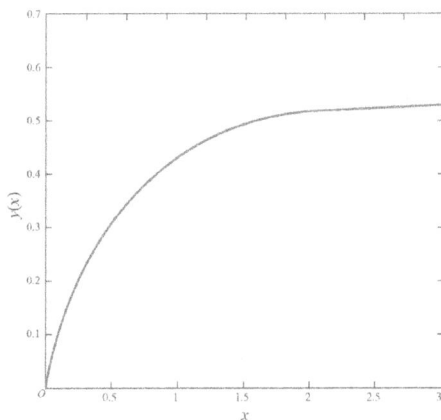

Figure 1.21 The solution in M-Example 1.7.

1.7.5   *Solve the first-order ODE by using MATLAB®*

Using the MATLAB® command dsolve.m, you can directly obtain solutions to ODEs with the initial conditions. (Requires the 'Symbolic Math' Toolbox)

M_Example 1.8

Solve the following first-order ODE by using MATLAB®.

a. $\dot{y} + 2y = 2, \quad y(0) = 0$
b. $\dot{y} + ay = \sin t, \quad y(0) = 1$
c. $t\dot{y} + y = t^2, \quad y(0) = 1$
d. $\dot{y} + y \tan t = \sin t, \quad y(0) = 1$

Solution

a. The given ODE can be rewritten in an explicit form as follows:

$$\frac{dy}{dt} = -2y + 2, \quad y(0) = 0.$$

First, let's solve the ODE $\frac{dy}{dt} = -2y + 2.$

```
>> syms y(t)
>> Dy=diff(y);                % diff(y): the first-order differential
                               equation
>> dsolve('Dy=-2*y+2')         % dsolve.m
```

---

ans =
        C1*exp(-2*t) + 1

Answer $y = 1 - C_1 e^{-2t}$

Now, let's find the particular solution by substituting the initial values $y(0) = 0.$

```
>> dsolve('Dy=-2*y+2', 'y(0)=0')          % paticular solution
```

---

ans =
        1 - exp(-2*t)

Answer $y = 1 - e^{-2t}$

b. The given ODE can be rewritten in an explicit form as follows:

$$\frac{dy}{dt} = -ay + \sin t, \quad y(0) = 0.$$

First, solve the ODE $\frac{dy}{dt} = -ay + \sin t$.

>> dsolve('Dy=−a*y+sin(t)')   % general solution

---

ans =
     C2*exp(−a*t) − (cos(t) − a*sin(t))/(a^2 + 1)

$$\textbf{Answer } y = C_2 e^{-at} + \frac{1}{a^2+1}\left(-\cos t + a\sin t\right)$$

Now, let's find the particular solution by substituting the initial values $y(0) = 0$.

>> dsolve('Dy=−a*y+sin(t)', 'y(0)=0')   % particular solution

---

ans =
   exp(−a*t)/(a^2 + 1) − (cos(t) − a*sin(t))/(a^2 + 1)

$$\textbf{Answer } y = \frac{1}{a^2+1}\left(e^{-at} - \cos t + a\sin t\right)$$

c. The given differential equation can be rewritten in an explicit form as follows:

$$\frac{dy}{dt} = -\frac{1}{t}y + t, \quad y(0) = 0.$$

>> dsolve('Dy=−1/t*y+t', 'y(0)=0')

---

ans =
     t^2/3

$$\textbf{Answer } y = \frac{t^2}{3}$$

d. The given differential equation can be rewritten in an explicit form as follows:

$$\frac{dy}{dt} = -y\tan t + \sin t, \quad y(0) = 1.$$

> \>> dsolve('Dy=−tan(t)*y+sin(t)', 'y(0)=1')

ans =
cos(t) − log(cos(t))*cos(t)

Answer $y = \cos t \cdot (1 - \ln|\cos t|)$

Problem 1.7

1. Calculate the determinant of the matrix $A = \begin{bmatrix} 1 & 2 & -3 & 1 \\ -1 & 1 & 2 & -2 \\ 1 & 1 & 0 & 2 \\ 2 & -1 & 2 & 3 \end{bmatrix}$ by using MATLAB®.

2. Calculate the inverse of the matrix $A = \begin{bmatrix} 1 & 3 & -1 \\ -2 & 1 & 2 \\ 0 & 1 & 3 \end{bmatrix}$ by using MATLAB®.

3. Plot the graph of $y = e^{-0.1x} \sin x$.
4. Plot the solution graph of the linear ODE $y' + y = 0$, with the initial condition $y(0) = 2$.
5. Plot the solution graph of the nonlinear ODE $y' - y\cos x = \sin x$, with the initial condition $y(0) = 1$.
6. Plot the solution graph of the nonlinear ODE $y' = 3y - 2y^3$, with the initial condition $y(0) = 0.1$.
7. Solve the following ODE by using MATLAB®.

   a. $\dot{y} + 2y = -3t, \quad y(0) = 0$
   b. $\dot{y} = (y - 1)\cot t, \quad y(\pi/2) = 2$
   c. $t\dot{y} + 2y = 4t^2, \quad y(1) = 3$
   d. $\dot{y} + 2y = 3y^2, \quad y(0) = 0.4$
   e. $\dot{y} = 2(y - 3)\tanh t, \quad y(0) = 2$
   f. $y\dot{y} + 2e^t = 0, \quad y(0) = 2$

Answer

Problem 1.1

1. $y = \dfrac{2}{3}\sin 3x - \dfrac{1}{2}x^2 + C$

2. $y = \ln|x| - 3e^{-x} + C$

3. $y = e^{x^2} + C$

4. $y = \dfrac{1}{3}\cosh 3x + \dfrac{1}{\ln 2}2^x$

5. $y = \dfrac{1}{2}(\ln x)^2 - \dfrac{1}{2}\cos 2x + C$

6. $y = \ln|\ln x| - \dfrac{1}{3}x^3 + C$

7. $y = (x-1)e^x + C$

8. $y = -(x+1)e^{-x} + C$

9. $y = -x\cos x + \sin x + C$

10. $y = \dfrac{x^2}{2}\ln x - \dfrac{x^2}{4} + C$

11. $y = \dfrac{1}{2}e^x(\sin x + \cos x) + C$

12. $y = \dfrac{1}{2}e^x(-\cos x + \sin x) + C$

13. $y = e^{-2x} + 2$

14. $y = 2e^{-x^2}$

15. $y = 2e^{2x} - e^x$

16. $y^2 = 2x^2 + 2$

17. $y = \dfrac{1}{1+e^{-x}}$

Problem 1.2

1. $x^3 + y^3 = C$

2. $y = -\dfrac{1}{\ln|\sin x| + C}$

3. $y = \dfrac{2x^2}{C - x^2}$

4. $y = \dfrac{1}{C - e^{x-1}}$

5. $y = Ce^{(\ln x)^2}$

6. $y^2 = \ln|x^2 + 2x + C|$

7. $y = -x \ln|x + C|$

8. $y = x - \dfrac{2}{Ce^{2x} - 1}$

9. $y = -2x - 2 + Ce^x$

10. $y = x(\ln x + C)$

11. $y = \dfrac{2}{x^2}$

12. $y^2 = \dfrac{3}{2}e^{4x} - \dfrac{1}{2}$

13. $\sin y = e^{\arctan x}$

14. $y = e^{2\sin x}$

15. $y = x \arctan x^2$

16. $y = -x - 1 + \tan\left(x + \dfrac{\pi}{4}\right)$

Problem 1.3

1. $\dfrac{x^3}{3} + xy^2 = C$

2. $xe^y + y^3 = C$

3. $x \cos y - ye^{-x} = C$

4. $y \tan x + x^2 \ln y = C$

5. $x^2 - y^2 = Cx$

6. $x^4 e^y + x^3 y = C$

7. $xy + e^x + e^{-y} = C$

8. $x \ln y + x^2 + y^2 = C$

9. $x^4 + xy^3 = 2$

10. $xy + y^2 - \sin xy = 0$

11. $-\cos 2x + x^2 y^2 + y^2 = 0$

12. $2x^3 + x^2 y^2 = 3$

13. $2(2 - y)e^x + e^{2x} = 3$

14. $9x^2 y^2 + 4y^3 = 4$

15. $ye^y \sin x = e$

Problem 1.4

1. $y = Ce^x - 2$

2. $y = e^{-2x} - 2x + 1$

3. $y = e^{-x}(x^2 + C)$

4. $y = \dfrac{(x-1)e^x}{x} + \dfrac{C}{x}$

5. $y = e^{-2x}(x+C)$

6. $y = 2e^x \cos x + C \cos x$

7. $y = x^3 + \dfrac{C}{x}$

8. $y = \sin x + \cos x + Ce^{-x}$

9. $y = \sin x + 1$

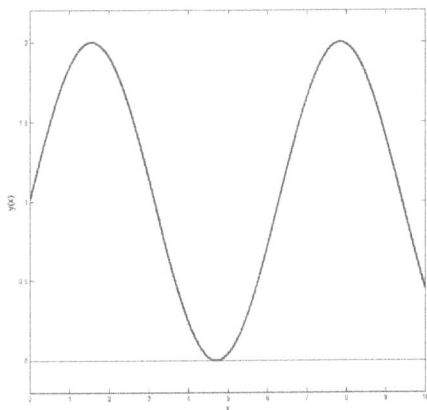

10. $y(x) = \cos x \cdot \left(-\ln|\cos x| + 1\right)$

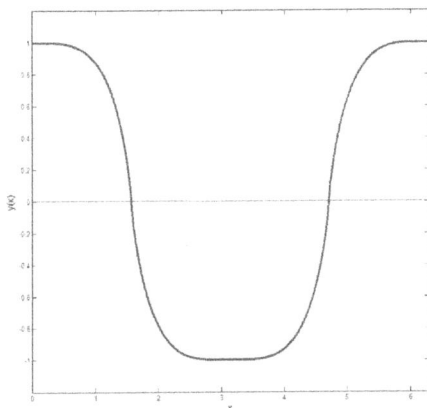

11. $y = x^2 + \dfrac{2}{x^2}$

12. $y = x^2 e^x + ex^2$

13. $y = \dfrac{1}{1+Ce^x}$

14. $y^3 = 1 + Ce^{-\frac{3}{2}x^2}$

15. $y = \cosh^2 x + 3$

16. $y^2 = \dfrac{3}{2 + e^{-6x}}$

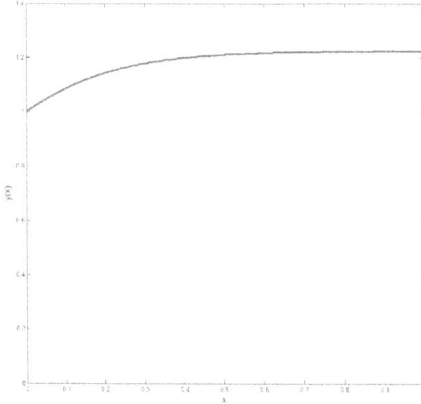

17. (*optional) $y^2 = -x + \dfrac{1}{2} + Ce^{-2x}$

18. (*optional) $y = \dfrac{2x}{x^2 + 1}$

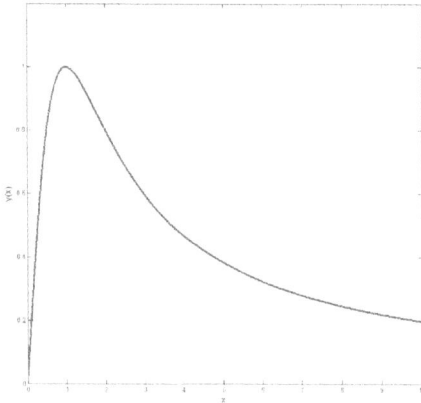

Problem 1.5

1. $T = 5.5°C$
2. 9.38 hours
3. about 28.05 years
4. about 13.3 days

5. $i(t) = 10\left(1 - e^{-10t}\right)[\text{A}]$

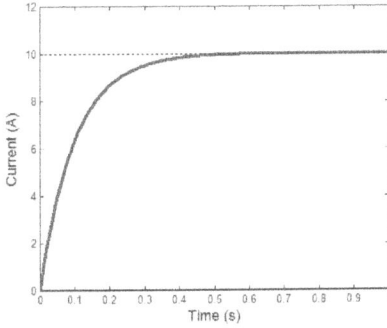

6. $i(t) = \begin{cases} 10\left(1 - e^{-10t}\right)[\text{A}], & 0 \le t \le 0.1 \text{ s} \\ 10(e-1)e^{-10t}[\text{A}], & t > 0.1 \text{ s} \end{cases}$

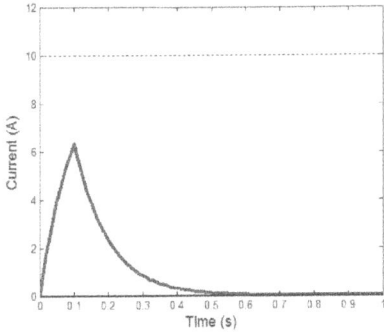

7. $i(t) = 10e^{-10t}[\text{A}]$ (Figure (a) shows the current, and Figure (b) shows the capacity.)

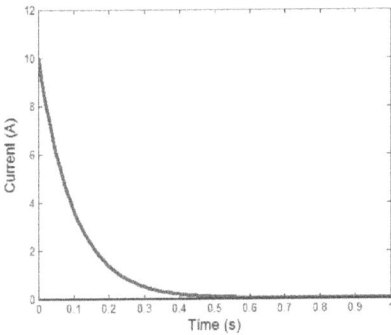

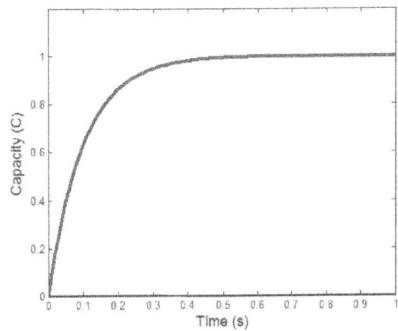

(a)   (b)

8. $i(t) = \begin{cases} 10e^{-10t} [\text{A}], & 0 \le t \le 0.1 \text{ s} \\ -10(e-1)e^{-10t} [\text{A}], & t > 0.1 \text{ s} \end{cases}$ (Figure (a) shows the current, and Figure (b) shows the capacity.)

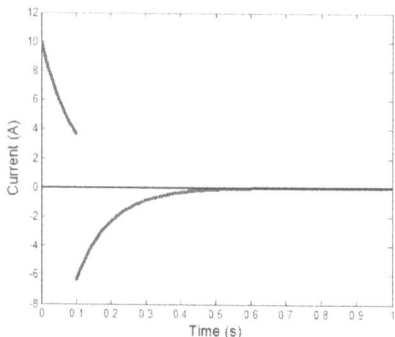

(a)

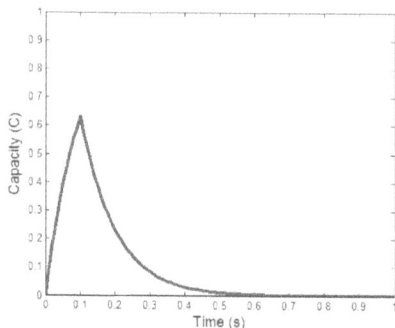

(b)

Problem 1.7

1. −18
2. 0.0476   −0.4762   0.3333
    0.2857   0.1429   0
    −0.0952   −0.0476   0.3333

3.

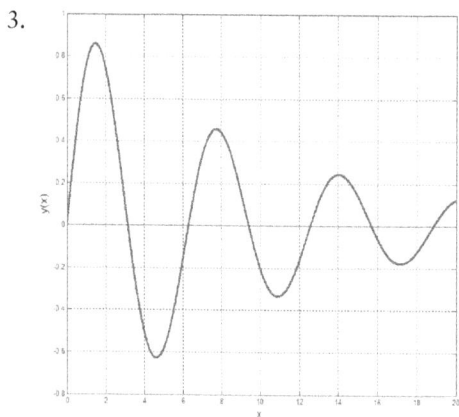

4.

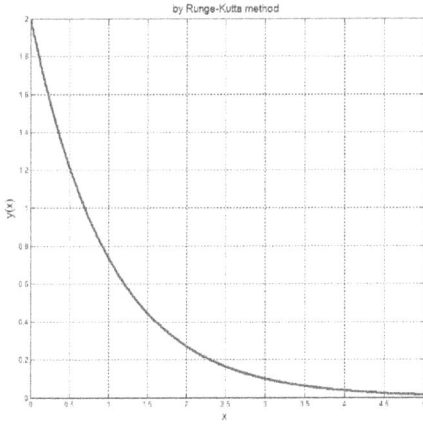
by Runge-Kutta method

5.

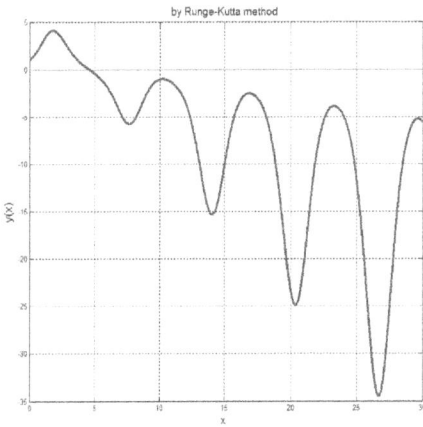
by Runge-Kutta method

6.

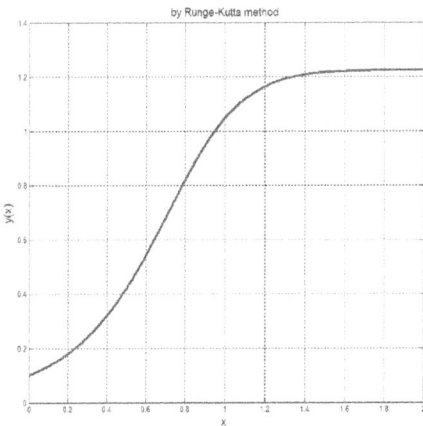
by Runge-Kutta method

7. a. $y = \dfrac{3}{4} - \dfrac{3}{4}e^{-2t} - \dfrac{3}{2}t$

b. $y = \sin t + 1$

c. $y = t^2 + \dfrac{2}{t^2}$

d. $y = \dfrac{2}{2e^{2t} + 3}$

e. $y(t) = 3 - \cosh^2 t$

f. $y^2 + 4e^t = 8$

2 Second-order ordinary differential equations

Differential equations with the highest order of the derivative being 2 are called second-order differential equations. Particularly, the damped vibration system and the RLC-circuit, often seen in fields like mechanical engineering and electrical engineering, are typically represented by second-order ODEs.

2.1 Homogeneous ODEs

2.1.1 Standard form of the second-order ODEs

When a differential equation is represented in the following form, it is referred to as the *standard form of a second-order ODE*,

$$y'' + p(x)y' + q(x)y = r(x). \tag{2.1}$$

When the right-hand side of Eq. (2.1) is zero, $r(x) = 0$, it is called a second-order homogeneous ODE, and it is as follows:

$$y'' + p(x)y' + q(x)y = 0. \tag{2.2}$$

2.1.2 Linear combination, basis, and general solution

If two independent solutions, y_1 and y_2, satisfy the second-order homogeneous ODE (2.2), then the *general solution* is a *linear combination* of these solutions, as given by $y = C_1 y_1 + C_2 y_2$, where C_1 and C_2 are arbitrary constants.

The proof of this statement is as follows.

Since y_1 and y_2, respectively, are *basis* solutions to Eq. (2.2), they satisfy the following equations:

$$y_1'' + py_1' + qy_1 = 0,$$
$$y_2'' + py_2' + qy_2 = 0.$$

Therefore, by substituting the general solution $y = C_1 y_1 + C_2 y_2$ into Eq. (2.2), we have

$$C_1 y_1'' + C_2 y_2'' + p(C_1 y_1' + C_2 y_2') + q(C_1 y_1 + C_2 y_2)$$
$$= C_1 (y_1'' + py_1' + qy_1) + C_2 (y_2'' + py_2' + qy_2)$$
$$= 0.$$

DOI: 10.1201/9781003608912-2

2.1.3   Initial value problem (IVP)

The unknown constants, C_1 and C_2 in the general solution $y = C_1 y_1 + C_2 y_2$ of the second-order homogeneous ODE $y'' + p(x)y' + q(x)y = 0$ can be determined by the initial conditions, given as $y(x_0)$ and $y'(x_0)$.

2.1.4   Find the second solution if one solution is known (method of variation of parameters)

The method for finding the second independent solution y_2, when one solution y_1 satisfying the homogeneous ODE (2.2) $y'' + p(x)y' + q(x)y = 0$ is known, is called the *method of variation of parameters* (or "*method of reduction of order*").

Following is the procedure.

If we substitute a function u for the coefficient C in the general solution $y(x) = Cy_1$ (reminding "method of variation of parameters"), we can assume the second independent solution y_2 as follows:

$$y_2 = uy_1. \tag{2.3}$$

To get y_2, we substitute it into Eq. (2.2). Then

$$
\begin{aligned}
(uy_1)'' &+ p(uy_1)' + q(uy_1) \\
&= (u''y_1 + 2u'y_1 + uy_1'') + p(u'y_1 + uy_1') + q(uy_1) \\
&= u''y_1 + u'(2y_1' + py_1) + u(y_1'' + py_1' + qy_1) \\
&= 0.
\end{aligned}
$$

Since y_1 is a solution of Eq. (2.2), when we apply $y_1'' + py_1' + qy_1 = 0$, the above equation becomes as follows:

$$u''y_1 + u'(2y_1' + py_1) = 0,$$

or

$$u'' + u'\left(2\frac{y_1'}{y_1} + p\right) = 0.$$

We set $v = u'$, then

$$v' + v\left(2\frac{y_1'}{y_1} + p\right) = 0.$$

This is the desired first-order ODE, the reduced-order ODE (reminding "Method of Reduction of Order"). Separation of variables and integration gives

$$\frac{dv}{v} = -\left(2\frac{y_1'}{y_1} + p\right)dx,$$

or

$$\ln|v| = -2\ln|y_1| - \int p\,dx.$$

We finally obtain

$$v = \frac{1}{y_1^2}e^{-\int p\,dx}. \tag{2.4}$$

Since $v = u'$, then $u = \int v\,dx$. Hence, the desired second solution is as follows:

$$y_2 = y_1 u = y_1 \int v\,dx. \tag{2.5}$$

From Eq. (2.4), since $v > 0$, $\int v\,dx$ cannot be constant. Therefore, y_1 and y_2 form a basis of solutions.

Remark   Method of variation of parameters (or Method of reduction of order)

How to find another basis $y_2(x)$ when one basis $y_1(x)$ is known in the homogeneous ODE $y'' + p(x)y' + q(x)y = 0$.

$$y_2 = y_1 \int v\,dx \qquad \left(\text{where } v = \frac{1}{y_1^2}e^{-\int p(x)\,dx}\right) \tag{2.5}$$

Example 2.1

When $e^{\lambda x}$ is one solution that satisfies the following homogeneous ODE, find the second solution.

$$y'' - 2\lambda y' + \lambda^2 y = 0$$

Solution

Since one solution is $y_1 = e^{\lambda x}$, we set $y_2 = uy_1 = ue^{\lambda x}$.

To get y_2, we substitute it into the given equation. This gives

$$\left(ue^{\lambda x}\right)'' - 2\lambda\left(ue^{\lambda x}\right)' + \lambda^2\left(ue^{\lambda x}\right)$$

$$= \left(u'' + 2\lambda u' + \lambda^2 u\right)e^{\lambda x} - 2\lambda\left(u' + \lambda u\right)e^{\lambda x} + \lambda^2 ue^{\lambda x}$$

$$= u''e^{\lambda x} = 0.$$

Then

$$u' = C_2, \text{ and } u = C_2 x + C_1.$$

Therefore,

$$y_2 = uy_1 = \left(C_2 x + C_1\right)e^{\lambda x}. \quad (C_1, C_2 \text{ arbitrary})$$

Here, since $C_1 e^x$ has the first basis $y_1 = e^{\lambda x}$, the second basis is $y_2 = xe^{\lambda x}$.

Answer $y_2 = xe^{\lambda x}$

Another Solution   Use method of variation of parameters (or Method of reduction of order)

Since $y_1 = e^{\lambda x}$, $p(x) = -2\lambda$ in the ODE $y'' + p(x)y' + q(x)y = 0$, then

$$v = \frac{1}{y_1^2}e^{-\int p\,dx} = \frac{1}{\left(e^{\lambda x}\right)^2}e^{-\int -2\lambda\,dx} = \frac{1}{\left(e^{\lambda x}\right)^2}e^{2\lambda x} = 1,$$

and

$$y_2 = y_1 u = y_1 \int v\,dx = e^{\lambda x}\int 1\,dx = xe^{\lambda x}.$$

Answer $y_2 = xe^{\lambda x}$

Remark

In the ODE $y'' - 2\lambda y' + \lambda^2 y = 0$, bases are $y_1 = e^{\lambda x}$ and $y_2 = xe^{\lambda x}$

Example 2.2

When x^m is one solution that satisfies the following homogeneous ODE, find the second solution.

$$x^2 y'' - (2m-1)xy' + m^2 y = 0$$

Solution

Since one solution is $y_1 = x^m$, we set $y_2 = uy_1 = ux^m$.

To get y_2, we substitute it into the given equation. This gives

$$x^2 \left(ux^m \right)'' - (2m-1)x\left(ux^m \right)' + m^2 \left(ux^m \right)$$

$$= x^2 \left(u''x^m + 2mu'x^{m-1} + m(m-1)ux^{m-2} \right) - (2m-1)x\left(u'x^m + mux^{m-1} \right)$$

$$+ m^2 \left(ux^m \right)$$

$$= x^m \left(x^2 u'' + xu' \right) = 0,$$

or

$$xu'' + u' = 0.$$

Here, we set $v = u'$, then

$$xv' + v = 0.$$

By using the separation of variables, then

$$\frac{dv}{v} = -\frac{dx}{x} \quad \text{and} \quad \ln|v| = -\ln|x| + \ln C_1$$

or

$$v = \frac{C_1}{x}.$$

Since we already set $v = u'$, then $u' = \frac{C_1}{x}$.

We finally obtain $u = C_1 \ln|x| + C_2$.

Therefore,

$$y_2 = uy_1 = \left(C_1 \ln|x| + C_2 \right)x^m \quad (C_1, C_2 \text{ arbitrary})$$

Here, since $C_1 x^m$ has the first basis $y_1 = x^m$, the second basis is $y_2 = x^m \ln|x|$.

Answer $y_2 = x^m \ln|x|$

Remark

In the ODE $x^2y'' - (2m-1)xy' + m^2y = 0$, bases are $y_1 = x^m$ and $y_2 = x^m \ln|x|$.

Problem 2.1

Given one solution that satisfies the following homogeneous ODE, find the second solution. [1 ~ 8]

1. e^{-x}, $y'' - 2y' - 3y = 0$
2. e^x, $y'' - 2y' + y = 0$
3. e^{-2x}, $y'' + 4y' + 4y = 0$
4. x, $x^2y'' + 2xy' - 2y = 0$
5. x^2, $x^2y'' - 3xy' + 4y = 0$
6. $\cos 2x$, $y'' + 4y = 0$
7. $y = e^{\lambda x}$, $y'' - 2\lambda y' + \lambda^2 y = 0$
8. $y = x^a$, $x^2y'' + (-2a+1)xy' + a^2y = 0$

2.2   Second-order homogeneous ODEs with constant coefficients

2.2.1   *Characteristic equation*

Let's consider the second-order homogeneous ODE (2.2) with constant coefficients a and b.

$$y'' + ay' + by = 0 \tag{2.6}$$

We assume the solution $y(x)$ to this ODE is as follows:

$$y = Ce^{\lambda x}. \tag{2.7}$$

Here, the coefficient C is a nonzero constant. When we substitute Eq. (2.7) into the ODE (2.6), we obtain the following *characteristic equation*:

$$\lambda^2 + a\lambda + b = 0. \tag{2.8}$$

2.2.2   *General solution*

According to the two roots λ_1, λ_2 of the characteristic equation (2.8), it is summarized as follows.

(Case I) If the discriminant $a^2 - 4b > 0$, the characteristic equation (2.8) has distinct real roots $\lambda_{1,2} = \dfrac{-a \pm \sqrt{a^2 - 4b}}{2}$.

So, the bases of the ODE (2.6) are

$$y_1 = e^{\lambda_1 x} \text{ and } y_2 = e^{\lambda_2 x}.$$

Therefore, the *general solution* in (Case I) is as follows:

$$y(x) = C_1 e^{\lambda_1 x} + C_2 e^{\lambda_2 x} \quad (C_1, C_2 \text{ arbitrary}) \tag{2.9}$$

(Case II) If the discriminant $a^2 - 4b = 0$, the characteristic equation (2.8) has a real double root $\lambda (= \lambda_1 = \lambda_2) = -\dfrac{a}{2}$.

One solution of the ODE (2.6) is

$$y_1 = e^{\lambda x}.$$

As derived in **Example 2.1**, another solution of the ODE (2.6) is

$$y_2 = xe^{\lambda x}.$$

Therefore, the *general solution* in (Case II) is as follows:

$$y(x) = (C_1 + C_2 x)e^{\lambda x} \quad (C_1, C_2 \text{ arbitrary}) \tag{2.10}$$

(Case III) If the discriminant $a^2 - 4b < 0$, the characteristic equation (2.8) has complex conjugate roots $\lambda_{1,2} = -\dfrac{a}{2} \pm \omega i$ (where $\omega = \dfrac{\sqrt{4b - a^2}}{2}$).

Therefore, the *general solution* in (Case III) is as follows:

$$y = C_1 e^{\lambda_1 x} + C_2 e^{\lambda_2 x}$$

$$= e^{-\frac{a}{2}x} \left(C_1 e^{\omega x i} + C_2 e^{-\omega x i} \right)$$

$$= e^{-\frac{a}{2}x} \left(C_1 (\cos \omega x + i \sin \omega x) + C_2 (\cos \omega x - i \sin \omega x) \right)$$

$$= e^{-\frac{a}{2}x} (A \cos \omega x + B \sin \omega x).$$

Here, $A = C_1 + C_2$, $B = i(C_1 - C_2)$. Therefore, the general solution in (Case III) is as follows:

$$y(x) = e^{-\frac{a}{2}x} (A \cos \omega x + B \sin \omega x) \quad (A, B \text{ arbitrary}) \tag{2.11}$$

Remark   General solution of the second-order homogeneous ODE

The characteristic equation of the second-order homogeneous ODE $y'' + ay' + by = 0$ has

(Case I) distinct real roots λ_1, λ_2, the general solution of the ODE is

$$y(x) = C_1 e^{\lambda_1 x} + C_2 e^{\lambda_2 x}. \tag{2.9}$$

(Case II) a real double root $\lambda_1 (= \lambda_2)$, the general solution of the ODE is

$$y(x) = (C_1 + C_2 x) e^{\lambda_1 x}. \tag{2.10}$$

(Case III) complex conjugate roots $\lambda_{1,2} = p \pm qi$, the general solution of the ODE is

$$y(x) = e^{px} (A \cos qx + B \sin qx). \tag{2.11}$$

Example 2.3

Solve the second-order homogeneous ODE.

a. $y'' - 3y' + 2y = 0$
b. $y'' + 6y' + 9y = 0$
c. $y'' + 2y' + 5y = 0$

Solution

a. We set $y(x) = Ce^{\lambda x}$, then the characteristic equation is

$$\lambda^2 - 3\lambda + 2 = 0, \text{ or } (\lambda - 1)(\lambda - 2) = 0.$$

Two real roots are $\lambda = 1, \lambda = 2$.
Therefore, we obtain the general solution

$$y(x) = C_1 e^x + C_2 e^{2x}.$$

Answer $y(x) = C_1 e^x + C_2 e^{2x}$ (C_1, C_2 arbitrary)

b. We set $y(x) = Ce^{\lambda x}$, then

$$\lambda^2 + 6\lambda + 9 = 0, \text{ or } (\lambda + 3)^2 = 0.$$

A real double root is $\lambda = -3$.
Therefore, we obtain the general solution

$$y(x) = (C_1 + C_2 x)e^{-3x}.$$

Answer $y(x) = (C_1 + C_2 x)e^{-3x}$ (C_1, C_2 arbitrary)

c. We set $y(x) = Ce^{\lambda x}$, then

$$\lambda^2 + 2\lambda + 5 = 0.$$

Complex conjugate roots are $= -1 \pm 2i$.
Therefore, we obtain the general solution

$$y(x) = e^{-x}(A \cos 2x + B \sin 2x)$$

Answer $y(x) = e^{-x}(A \cos 2x + B \sin 2x)$ (C_1, C_2 arbitrary)

2.2.3 Differential operators

If the *differential operator* D is defined as $\dfrac{d}{dx}$, we can write

$$Dy = y' = \frac{dy}{dx}. \tag{2.12}$$

And for a homogeneous ODE $y'' + ay' + by = 0$ with constant coefficients, we can write that ODE as

$$\left(D^2 + aD + bI\right)y = 0. \tag{2.13}$$

where I is the *identity operator* defined by $Iy = y$.
For example, because $I^2 = I$, a homogeneous ODE $\left(D^2 - D - 6I\right)y = 0$ becomes

$$(D - 3I)(D + 2I)y = 0.$$

Now, $(D - 3I)y = y' - 3y = 0$ has the solution $y_1 = e^{3x}$. Similarly, the solution of $(D + 2I)y = y' + 2y = 0$ is $y_2 = e^{-2x}$.
Therefore, the general solution of $\left(D^2 - D - 6I\right)y = 0$ is

$$y = C_1 e^{3x} + C_2 e^{-2x} \quad (C_1, C_2 \text{ arbitrary})$$

Problem 2.2

Solve the following homogeneous ODEs. [1 ~ 6]

1. $y'' - 4y = 0$
2. $3y'' + 10y' + 3y = 0$
3. $y'' - 4y' + 4y = 0$
4. $9y'' + 6y' + y = 0$
5. $y'' + 3y = 0$
6. $y'' + y' + y = 0$

Solve the following IVPs. [7 ~ 9]

7. $y'' - y' - 2y = 0$, $y(0) = 3, \; y'(0) = 0$
8. $y'' + 3y' + 2.25y = 0$, $y(1) = 2e^{-1.5}, \; y'(1) = 0$
9. $y'' - 2y' + 2y = 0$, $y(0) = 0, \; y'(0) = 1$

Solve the following homogeneous ODEs. [10 ~ 13]

10. $\left(9D^2 - I\right)y = 0$
11. $\left(D^2 + 2D + I\right)y = 0$
12. $\left(D^2 - 3D\right)y = 0$
13. $\left(D^2 + 2I\right)y = 0$

2.3   Euler-Cauchy equations

2.3.1   *Characteristic equation*

Euler-Cauchy equations are ODEs of the form

$$x^2 y'' + axy' + by = 0 \tag{2.14}$$

with given constants a, b, and unknown function $y(x)$. We set

$$y = Cx^m, \tag{2.15a}$$

and then

$$y' = Cmx^{m-1}, \quad y'' = Cm(m-1)x^{m-2}. \tag{2.15b}$$

We substitute Eq. (2.15a) and Eq. (2.15b) into Eq. (2.14). This gives

$$x^2 \cdot m(m-1)x^{m-2} + ax \cdot mx^{m-1} + bx^m = 0.$$

Therefore, we can get the following characteristic equation:

$$m(m-1) + am + b = 0. \tag{2.16}$$

2.3.2 General solution

The roots of the characteristic equation (2.16) are

$$m_{1,2} = \frac{-(a-1) \pm \sqrt{(a-1)^2 - 4b}}{2}.$$

(Case I) When the discriminant $(a-1)^2 - 4b > 0$, the characteristic equation
(2.16) has distinct real roots m_1, m_2.
   They give two real solutions $y_1 = x^{m_1}$ and $y_2 = x^{m_2}$.
   The corresponding *general solution* for all these x is

$$y(x) = C_1 x^{m_1} + C_2 x^{m_2} \quad (C_1, C_2 \text{ arbitrary}) \tag{2.17}$$

(Case II) When the discriminant $(a-1)^2 - 4b = 0$, the characteristic equation
(2.16) has a double real root $m(= m_1 = m_2) = -\dfrac{a-1}{2}$.
   It gives a double real solution $y_1 = x^m$.
   As derived in **Example 2.2** by the method of variation of parameters (or
the method of reduction of order), another solution is

$$y_2 = x^m \ln|x|.$$

   The corresponding *general solution* for all these x is

$$y(x) = x^m \left(C_1 + C_2 \ln|x| \right) \quad (C_1, C_2 \text{ arbitrary}) \tag{2.18}$$

(Case III) When the discriminant $(a-1)^2 - 4b < 0$, the characteristic equation (2.16) has complex conjugate roots $m_{1,2} = -\dfrac{a-1}{2} \pm \omega i$ (where
$\omega = \dfrac{\sqrt{4b - (a-1)^2}}{2}$).
   The general solution for all these x is

$$y = C_1 x^{m_1} + C_2 x^{m_2}$$

$$= x^{-\frac{a-1}{2}} \left(C_1 x^{\omega i} + C_2 x^{-\omega i} \right)$$

$$= x^{-\frac{a-1}{2}} \left(C_1 (e^{\ln|x|})^{\omega i} + C_2 (e^{\ln|x|})^{-\omega i} \right)$$

$$= x^{-\frac{a-1}{2}} \left(C_1 e^{(\omega \ln|x|)i} + C_2 e^{-(\omega \ln|x|)i} \right)$$

$$= x^{-\frac{a-1}{2}} \left(A \cos(\omega \ln|x|) + B \sin(\omega \ln|x|) \right),$$

where $A = C_1 + C_2$, $B = i(C_1 - C_2)$.

The corresponding *general solution* for all these x is

$$y(x) = x^{-\frac{a-1}{2}} \left(A\cos(\omega \ln|x|) + B\sin(\omega \ln|x|) \right). \quad (A, B \text{ arbitrary}) \qquad (2.19)$$

Remark **General solution of the second-order homogeneous Euler-Cauchy equation**

The characteristic equation of the second-order homogeneous ODE $x^2 y'' + axy' + by = 0$ has

(Case I) two real roots m_1, m_2, the general solution of the ODE is

$$y(x) = C_1 x^{m_1} + C_2 x^{m_2} \quad (C_1, C_2 \text{ arbitrary}) \qquad (2.17)$$

(Case II) a real double root $m_1 (= m_2)$, the general solution of the ODE is

$$y(x) = x_1^m \left(C_1 + C_2 \ln|x| \right) \quad (C_1, C_2 \text{ arbitrary}) \qquad (2.18)$$

(Case III) complex conjugate roots $\lambda_{1,2} = p \pm qi$, the general solution of the ODE is

$$y(x) = x^p \left\{ A\cos(q \ln|x|) + B\sin(q \ln|x|) \right\} \quad (A, B \text{ arbitrary}) \qquad (2.19)$$

Example 2.4

Solve the following homogeneous ODE:

a. $x^2 y'' - 2xy' + 2y = 0$
b. $x^2 y'' + 7xy' + 9y = 0$
c. $x^2 y'' + 3xy' + 5y = 0$

Solution

a. We set $y(x) = Cx^m$.
   This gives the characteristic equation

$$m(m-1) - 2m + 2 = 0, \quad \text{or} \quad (m-1)(m-2) = 0.$$

Two real roots of the characteristic equation are

$$m = 1, \ m = 2,$$

and give the general solution

$$y(x) = C_1 x + C_2 x^2.$$

Answer $y(x) = C_1 x + C_2 x^2$ (C_1, C_2 arbitrary)

b. We set $y(x) = C x^m$.
   This gives the characteristic equation

$$m(m-1) + 7m + 9 = 0, \text{ or } (m+3)^2 = 0.$$

A real double root of the characteristic equation is

$$m = -3,$$

and gives the general solution

$$y(x) = x^{-3} \left(C_1 + C_2 \ln|x| \right).$$

Answer $y(x) = x^{-3} \left(C_1 + C_2 \ln|x| \right)$ (C_1, C_2 arbitrary)

c. We set $y(x) = C x^m$.
   This gives the characteristic equation

$$m(m-1) + 3m + 5 = 0.$$

Complex conjugate roots of the characteristic equation are

$$m = -1 \pm 2i$$

and gives the general solution

$$y = C_1 x^{-1+2i} + C_2 x^{-1-2i}$$

$$= x^{-1} \left(C_1 x^{2i} + C_2 x^{-2i} \right)$$

$$= x^{-1} \left(C_1 (e^{\ln|x|})^{2i} + C_2 (e^{\ln|x|})^{-2i} \right)$$

$$= x^{-1} \left(C_1 e^{(2\ln|x|)i} + C_2 e^{-(2\ln|x|)i} \right)$$

$$= x^{-1} \left(A\cos(2\ln|x|) + B\sin(2\ln|x|) \right). \text{ (where } A = C_1 + C_2, \ B = i(C_1 - C_2))$$

Answer $x^{-1} \left(A\cos(2\ln|x|) + B\sin(2\ln|x|) \right)$ (A, B arbitrary)

Problem 2.3

Solve the following homogeneous ODEs. [1 ~ 6]

1. $x^2y'' + xy' - 4y = 0$
2. $3x^2y'' + 5xy' + y = 0$
3. $x^2y'' - 3xy' + 4y = 0$
4. $4x^2y'' + y = 0$
5. $x^2y'' + xy' + 3y = 0$
6. $x^2y'' + 5xy' + 5y = 0$

Solve the following homogeneous IVPs. [7 ~ 9]

7. $2x^2y'' - xy' + y = 0,$ $\quad y(1) = 1,\ y'(1) = 0$
8. $x^2y'' - 3xy' + 4y = 0,$ $\quad y(1) = 2,\ y'(1) = 1$
9. $x^2y'' - xy' + 3y = 0,$ $\quad y(1) = 2,\ y'(1) = 2 + \sqrt{2}$

2.4 Second-order nonhomogeneous ODEs

A *general solution of the second-order nonhomogeneous ODE* (2.1) is a solution of the form

$$y(x) = y_h(x) + y_p(x) \tag{2.20}$$

where $y_h(x)$ is a *homogeneous solution of the homogeneous ODE* (2.2) and $y_p(x)$ is any *particular solution of the nonhomogeneous ODE* (2.1) that has the nonzero term $r(x)$.

There are typically two methods to find the particular solution $y_p(x)$ of the second-order nonhomogeneous ODE (2.1) as follows:

- Method of undetermined coefficients
- Method of variation of parameters.

2.4.1 Method of undetermined coefficients

We choose a form for $y_p(x)$ similar to $r(x)$, but with unknown coefficients A, B, $\cdots$ to be determined by substituting that $y_p(x)$ and its derivatives into the second-order nonhomogeneous ODE (2.1).

Table 2.1 shows the choice of $y_p(x)$ for practically important forms of $r(x)$.

Table 2.1 Method of undetermined coefficients

$r(x)$	$y_p(x)$
x^m	$Ax^m + Bx^{m-1} + Cx^{m-2} + \dots.$
$\cos px$ $\sin px$	$A\cos px + B\sin px$
e^{px}	Ce^{px}

Example 2.5

Solve the following nonhomogeneous ODEs:

a. $y'' + 2y' + 3y = 3x^2$
b. $y'' - 6y' + 9y = 2e^{-3x}$
c. $y'' + 3y' + 2y = 10\sin x$
d. $y'' + y = 2\cos x$

Solution

a. General solution of the given nonhomogeneous ODE $y'' + 2y' + 3y = 3x^2$ is

$$y(x) = y_h(x) + y_p(x).$$

First, we can get the homogeneous solution of the ODE

$$y_h(x) = e^{-x}\left(C_1 \cos\sqrt{2}x + C_2 \sin\sqrt{2}x\right).$$

Now, we set the particular solution as follows:

$$y_p = Ax^2 + Bx + C.$$

Its derivatives are

$$y_p' = 2Ax + B$$

$$y_p'' = 2A.$$

We substitute these expressions into the given nonhomogeneous ODE. This yields

$$2A + 2(2Ax + B) + 3(Ax^2 + Bx + C) = 3x^2.$$

Comparing the coefficients of x^2, x^1, x^0 gives

$$A = 1, \ B = -\frac{4}{3}, \ C = \frac{2}{9}.$$

This gives the particular solution

$$y_p = x^2 - \frac{4}{3}x + \frac{2}{9}.$$

Hence, the given ODE has the general solution

$$y(x) = y_h + y_p = e^{-x}\left(C_1 \cos\sqrt{2}x + C_2 \sin\sqrt{2}x\right) + x^2 - \frac{4}{3}x + \frac{2}{9}.$$

Answer $y(x) = e^{-x}\left(C_1 \cos\sqrt{2}x + C_2 \sin\sqrt{2}x\right) + x^2 - \frac{4}{3}x + \frac{2}{9}$

(C_1, C_2 arbitrary)

b. General solution of the nonhomogeneous ODE $y'' - 6y' + 9y = 2e^{-3x}$ is

$$y(x) = y_h(x) + y_p(x).$$

First, we can get the homogeneous solution of the ODE

$$y_h(x) = (C_1 + C_2 x)e^{3x}.$$

Now, we set the particular solution as follows:

$$y_p = Ae^{-3x}.$$

Its derivatives are

$$y_p' = -3Ae^{-3x}$$

$$y_p'' = 9Ae^{-3x}.$$

We substitute these expressions into the given nonhomogeneous ODE and omit the factor e^{-3x}. This yields

$$9A - 6\cdot(-3A) + 9A = 2,$$

$$\therefore A = \frac{1}{18}.$$

This gives the particular solution

$$y_p = \frac{1}{18}e^{-3x}.$$

Hence, the given ODE has the general solution

$$y(x) = y_h + y_p = (C_1 + C_2 x)e^{3x} + \frac{1}{18}e^{-3x}.$$

Answer $y(x) = (C_1 + C_2 x)e^{3x} + \frac{1}{18}e^{-3x}$ (C_1, C_2 arbitrary)

Remark   Duplication

When the given nonhomogeneous ODE is $y'' - 6y' + 9y = 2e^{3x}$, we can get the homogeneous solution of the ODE

$$y_h(x) = (C_1 + C_2 x)e^{3x}.$$

Because the component e^{3x} in the homogeneous solution is duplicated twice with $r(x) = 2e^{3x}$, we should set the particular solution as follows:

$$y_p = Ax^2 e^{3x}.$$

c. General solution of the nonhomogeneous ODE $y'' + 3y' + 2y = 10\sin x$ is

$$y(x) = y_h(x) + y_p(x).$$

First, we can get the homogeneous solution of the ODE

$$y_h = C_1 e^{-x} + C_2 e^{-2x}.$$

Now, we set the particular solution as follows:

$$y_p = A\cos x + B\sin x.$$

Its derivatives are

$$y_p' = B\cos x - A\sin x,$$
$$y_p'' = -A\cos x - B\sin x.$$

We substitute these expressions into the given nonhomogeneous ODE. This yields

$$(-A + 3B + 2)\cos x + (-B - 3A + 2B)\sin x = 10\sin x.$$

Comparing the coefficients of $\cos x$, $\sin x$ gives

$$A + 3B = 0, \quad -3A + B = 10,$$

hence

$$A = -3, \quad B = 1.$$

This gives the particular solution

$$y_p = -3\cos x + \sin x.$$

Hence, the given ODE has the general solution

$$y(x) = y_h + y_p = C_1 e^{-x} + C_2 e^{-2x} - 3\cos x + \sin x.$$

Answer $y(x) = C_1 e^{-x} + C_2 e^{-2x} - 3\cos x + \sin x$ (C_1, C_2 arbitrary)

d. General solution of the nonhomogeneous ODE $y'' + y = 2\cos x$ is

$$y(x) = y_h(x) + y_p(x).$$

First, we can get the homogeneous solution of the ODE

$$y_h = C_1 \cos x + C_2 \sin x.$$

Because the homogeneous solution $y_h = C_1 \cos x + C_2 \sin x$ is duplicated with $r(x) = 2\cos x$, we should set the particular solution as follows:

$$y_p = x(A\cos x + B\sin x).$$

Its derivatives are

$$y_p' = (A\cos x + B\sin x) + x(B\cos x - A\sin x),$$
$$y_p'' = 2(B\cos x - A\sin x) + x(-A\cos x - B\sin x).$$

We substitute these expressions into the given nonhomogeneous ODE. This yields

$$(2B - Ax + Ax)\cos x + (-2A - Bx + Bx)\sin x = 2\cos x.$$

Comparing the coefficients of $\cos x$, $\sin x$ gives

$$A = 0, \ B = 1.$$

This gives the particular solution

$$y_p = x\sin x.$$

Hence, the given ODE has the general solution

$$y(x) = y_h + y_p = C_1 \cos x + C_2 \sin x + x\sin x.$$

Answer $y(x) = C_1 \cos x + C_2 \sin x + x\sin x$ (C_1, C_2 arbitrary)

2.4.2 *Method of variation of parameters*

The *method of variation of parameters* is a technique for finding the particular solution $y_p(x)$ using the bases $y_1(x)$, $y_2(x)$, $\cdots$, $y_n(x)$ of a n-th-order homogeneous ODE, sometimes referred to as the *Wronskian's method*.

In the homogeneous solution $y_h(x) = C_1 y_1 + C_2 y_2$ of the second-order homogeneous ODE (2.2), we replace the constants C_1, C_2 by functions $u(x)$, $v(x)$, respectively. This process is similar to the method of variation of parameters (or the method of reduction of order) for finding second solution if one solution is known in Section 2.1.4.

We shall determine u and v so that the function

$$y_p(x) = u(x)y_1(x) + v(x)y_2(x) \tag{2.21}$$

is a particular solution of the second-order nonhomogeneous ODE (2.1). We determine u and v by substituting Eq. (2.21) and its derivatives into Eq. (2.1). Differentiating Eq. (2.21), we obtain

$$y_p' = u'y_1 + uy_1' + v'y_2 + vy_2'. \tag{2.22}$$

When in the Eq. (2.22) we set

$$u'y_1 + v'y_2 = 0, \tag{2.23}$$

Eq. (2.22) becomes the first derivative to the simpler form as follows:

$$y_p' = uy_1' + vy_2'. \tag{2.24}$$

Differentiating Eq. (2.24), we obtain

$$y_p'' = u'y_1' + uy_1'' + v'y_2' + vy_2''. \tag{2.25}$$

We substitute Eq. (2.21), Eq. (2.24), and Eq. (2.25) into the ODE $y_p'' + py_p' + qy_p = r$. Then we obtain

$$\left(u'y_1' + uy_1'' + v'y_2' + vy_2''\right) + p\left(uy_1' + vy_2'\right) + q\left(uy_1 + vy_2\right) = r$$

or

$$u\left(y_1'' + py_1' + qy_1\right) + v\left(y_2'' + py_2' + qy_2\right) + u'y_1' + v'y_2' = r.$$

Since $y_1'' + py_1' + qy_1 = 0$ and $y_2'' + py_2' + qy_2 = 0$, this reduces to

$$u'y_1' + v'y_2' = r. \tag{2.26}$$

Then, Eq. (2.23) and Eq. (2.26) are two algebraic equations for unknown functions u', v'. So we can obtain

$$u' = -\frac{y_2 r}{W}, \quad v' = \frac{y_1 r}{W}. \qquad \left(\text{where Wronskian } W = y_1 y_2' - y_2 y_1'\right)$$

By integration,

$$u = -\int \frac{y_2 r}{W} dx, \quad v = \int \frac{y_1 r}{W} dx.$$

Therefore, the particular solution $y_p(x)$ of the second-order nonhomogeneous ODE (2.1) is as follows:

$$y_p(x) = -y_1 \int \frac{y_2 r}{W} dx + y_2 \int \frac{y_1 r}{W} dx. \qquad (2.27)$$

Remark   Wronskian method

When the homogeneous solution is expressed as $y_h(x) = C_1 y_1 + C_2 y_2$ with two bases y_1 and y_2 in the second-order nonhomogeneous ODE, the particular solution $y_p(x)$ can be obtained as

$$y_p(x) = -y_1 \int \frac{y_2 r}{W} dx + y_2 \int \frac{y_1 r}{W} dx \qquad (2.27)$$

where Wronskian $W = \begin{vmatrix} y_1 & y_2 \\ y_1' & y_2' \end{vmatrix} = y_1 y_2' - y_2 y_1'.$

Example 2.6

Solve the following nonhomogeneous ODE:

a. $y'' + y = 1/\sin x$
b. $x^2 y'' - 2xy' - 4y = x^2$

Solution

a. From the given ODE $y'' + y = 1/\sin x$, we obtain the homogeneous solution

$$y_h = C_1 \cos x + C_2 \sin x.$$

So two bases of the homogeneous ODE are

$$y_1 = \cos x, \quad y_2 = \sin x.$$

The Wronskian of y_1, y_2 is

$$W = \begin{vmatrix} \cos x & \sin x \\ -\sin x & \cos x \end{vmatrix} = 1, \quad \text{and} \quad r(x) = \frac{1}{\sin x}.$$

And we obtain the particular solution as follows:

$$y_p(x) = -y_1 \int \frac{y_2 r}{W} dx + y_2 \int \frac{y_1 r}{W} dx$$

$$= -\cos x \int \sin x \cdot \frac{1}{\sin x} dx + \sin x \int \cos x \cdot \frac{1}{\sin x} dx$$

$$= -\cos x \cdot x + \sin x \cdot \ln|\sin x|.$$

Hence, the given ODE has the general solution

$$y(x) = y_h + y_p = C_1 \cos x + C_2 \sin x - x\cos x + \sin x \cdot \ln|\sin x|.$$

$$\textbf{Answer } y(x) = C_1 \cos x + C_2 \sin x - x\cos x + \sin x \cdot \ln|\sin x|$$
$$(C_1, C_2 \text{ arbitrary})$$

b. From the homogeneous ODE $x^2 y'' - 2xy' - 4y = 0$, we set $y(x) = Cx^m$. The characteristic equation of the homogeneous ODE is

$$m(m-1) - 2m - 4 = 0, \quad (m+1)(m-4) = 0.$$

It has distinct real roots

$$m = -1, \quad m = 4.$$

So, we obtain the homogeneous solution

$$y_h = C_1 x^{-1} + C_2 x^4.$$

And two bases of the homogeneous ODE are

$$y_1 = x^{-1}, \, y_2 = x^4.$$

The Wronskian of y_1, y_2 is

$$W = \begin{vmatrix} x^{-1} & x^4 \\ -x^{-2} & 4x^3 \end{vmatrix} = 5x^2 \text{ and } r(x) = 1.$$

And we obtain the particular solution as follows:

$$y_p(x) = -y_1 \int \frac{y_2 r}{W} dx + y_2 \int \frac{y_1 r}{W} dx$$

$$= -x^{-1} \int \frac{x^4 \cdot 1}{5x^2} dx + x^4 \int \frac{x^{-1} \cdot 1}{5x^2} dx$$

$$= -\frac{1}{6} x^2.$$

Hence, the given ODE has the general solution

$$y(x) = y_h + y_p = C_1 x^{-1} + C_2 x^4 - \frac{1}{6} x^2.$$

Answer $y(x) = C_1 x^{-1} + C_2 x^4 - \frac{1}{6} x^2$ (C_1, C_2 arbitrary)

Remark   Finding $r(x)$

When we divide both sides of the given ODE $x^2 y'' - 2xy' - 4y = x^2$ by x^2 to make the coefficient of y'' equal to 1, the right-hand side of the given ODE becomes $r(x)$.

$$r(x) = 1. \quad \left(\text{caution: } r(x) \neq x^2 \right)$$

Problem 2.4

Solve the following nonhomogeneous ODE. [1 ~ 6]

1. $y'' + 4y = 3\sin x$
2. $y'' - 2y' - 3y = -3x^2 - x + 1$
3. $y'' + 2y' + 2y = 2e^{-2x}$
4. $y'' + 4y' + 4y = 2e^{-2x}$
5. $y'' - 2y' + y = 2e^x$
6. $y'' - y' - 2y = 3e^{2x}$

Solve the following nonhomogeneous ODE. [7 ~ 12]

7. $y'' + y = 2\cos x$
8. $y'' + y = 1/\cos x$

9. $y'' + 4y = 4/\sin 2x$
10. $x^2 y'' - 4xy' + 6y = 12/x$
11. $x^2 y'' + xy' - 4y = 4x^2$
12. $x^2 y'' - 2y = 2x$

Solve the following nonhomogeneous IVP. [13 ~ 18]

13. $y'' + y = 1$, $\qquad\qquad\quad y(0) = 0,\ y'(0) = 1$
14. $y'' - 3y' + 2y = 2x^2$, $\qquad y(0) = 7/2,\ y'(0) = 4$
15. $y'' - 2y' + y = 2\sin x$, $\qquad y(0) = 2,\ y'(0) = 0$
16. $x^2 y'' - 2y = x^2$, $\qquad\qquad y(1) = 2,\ y'(1) = \dfrac{4}{3}$
17. $x^2 y'' + xy' - y = 8x^3$, $\qquad y(1) = 2,\ y'(1) = 0$
18. $y'' - y = 8\cosh x$, $\qquad\quad y(0) = 2,\ y'(0) = 0$

2.5 Application of second-order ODEs

Second-order ODEs are commonly applied in fields such as vibration and electrical circuits.

2.5.1 Vibration

1. Undamped vibration

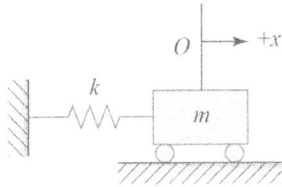

Figure 2.1 Undamped vibration system.

When we ignore resistance in the mass-spring system in Figure 2.1, there would be no element dissipating energy, thus the motion of this system would sustain a constant amplitude. Such vibration is termed as undamped vibration. When considering motion displacement x from the equilibrium position, the spring's restorative force is represented by $-kx$ (where k is called spring constant or stiffness), and applying Newton's second law $\sum F = m\ddot{x}$ yields

$$-kx = m\ddot{x}. \tag{2.28}$$

or,

$$m\ddot{x} + kx = 0. \tag{2.29}$$

The roots of the characteristic equation $m\lambda^2 + k = 0$ obtained from the ODE (2.29) are $\lambda = \pm\sqrt{\dfrac{k}{m}}i$. Therefore, the general solution is as follows:

$$x(t) = A\cos\sqrt{\frac{k}{m}}t + B\sin\sqrt{\frac{k}{m}}t. \qquad (2.30)$$

If we denote the initial displacement and initial velocity at time $t = 0$ as x_0 and $\dot{x}_0$ respectively, we can obtain the following solution:

$$x(t) = x_0\cos\omega_n t + \frac{\dot{x}_0}{\omega_n}\sin\omega_n t. \qquad (2.31)$$

where $\omega_n = \sqrt{\dfrac{k}{m}}$ is called by the natural frequency of the system.

Example 2.7

Given a mass-spring system with a spring constant of $k = 50$ N/m and a mass of 2 kg, in the undamped syatem, with initial conditions as $x(0) = 2$ cm and $\dot{x}(0) = 0$, shown in Figure 2.2, find the displacement.

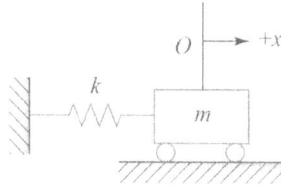

Figure 2.2 Undamped vibration system.

Solution

The natural frequency of the system is

$$\omega_n = \sqrt{\frac{k}{m}} = \sqrt{\frac{50}{2}} = 5 \text{ rad/s.}$$

The general solution for the equation of motion of undamped vibration $m\ddot{x} + kx = 0$ is as follows:

$$x(t) = A\cos 5t + B\sin 5t.$$

The particular solution obtained by using initial conditions x_0, $\dot{x}_0$ to determine the coefficients A, B is as follows:

$$x(t) = 2\cos 5t.$$

Answer $x(t) = 2\cos 5t$

2. Damped vibration

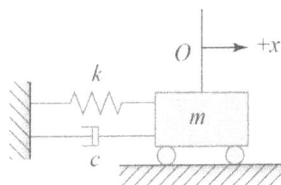

Figure 2.3 Damped vibration system.

In the mass-damping-spring system in Figure 2.3, damping dissipates energy, thus the motion of this system will lead to a gradual decrease in amplitude. Such vibration is termed as damped vibration. When considering motion displacement x from the equilibrium position, the damping force is represented by $-c\dot{x}$ (where c is referred to as the damping constant), and applying Newton's second law $\Sigma F = m\ddot{x}$ yields

$$-c\dot{x} - kx = m\ddot{x}. \tag{2.32}$$

Therefore,

$$m\ddot{x} + c\dot{x} + kx = 0. \tag{2.33}$$

The roots of the characteristic equation $m\lambda^2 + c\lambda + k = 0$ obtained from the ODE (2.33) are

$$\lambda_{1,2} = \frac{-c \pm \sqrt{c^2 - 4mk}}{2m}. \tag{2.34}$$

Therefore, the general solution of the ODE (2.33) is

$$x(t) = C_1 e^{\frac{-c+\sqrt{c^2-4mk}}{2m}t} + C_2 e^{\frac{-c-\sqrt{c^2-4mk}}{2m}t}. \tag{2.35}$$

Therefore, the general solution (2.35) appears in different forms depending on the sign of the discriminant $c^2 - 4mk$. In the case of damping $c = 0$, it becomes the undamped vibration already mentioned in Eq. (2.30). Let's consider the case of $c \neq 0$ in the following three cases.

(Case I) $c^2 - 4mk > 0$, (overdamped)

Because the roots of the characteristic equation $\lambda_{1,2} = \dfrac{-c \pm \sqrt{c^2 - 4mk}}{2m}(< 0)$

become distinct negative real numbers, the general solution is:

$$x(t) = C_1 e^{\lambda_1 t} + C_2 e^{\lambda_2 t}.$$

If we denote the initial displacement and initial velocity at $t = 0$ as x_0, $\dot{x}_0$ respectively, then we obtain the following constants:

$$C_1 = \frac{\lambda_2 x_0 - \dot{x}_0}{\lambda_2 - \lambda_1}, \quad C_2 = \frac{-\lambda_1 x_0 + \dot{x}_0}{\lambda_2 - \lambda_1}. \tag{2.36}$$

Since both roots of the characteristic equation are negative, it indicates that in overdamped motion, the displacement decreases rapidly over time.

(Case II) $c^2 - 4mk = 0$, (critical damped)

The roots of the characteristic equation $m\lambda^2 + c\lambda + k = 0$ obtained from the ODE (2.33) become a real double root as follows:

$$\lambda_{1,2} = -\sqrt{\frac{k}{m}}. \tag{2.37}$$

Therefore, the general solution of the ODE (2.33) can be expressed as follows:

$$x(t) = (C_0 + C_1 t)e^{-\sqrt{\frac{k}{m}}t}. \tag{2.38}$$

If we denote the initial displacement and initial velocity at $t = 0$ as x_0, $\dot{x}_0$ respectively, then we obtain the following constants:

$$C_1 = x_0, \quad C_2 = \dot{x}_0 + \sqrt{\frac{k}{m}}x_0. \tag{2.39}$$

Therefore, as time progresses, the solution of the equation also decreases rapidly, indicating a rapid decay in the system's behavior.

(Case III) $c^2 - 4mk < 0$, (underdamped)

The roots of the characteristic equation, Eq. (2.34), can be changed to the following form:

$$\lambda_{1,2} = \frac{-c \pm i\sqrt{4mk - c^2}}{2m}. \tag{2.40}$$

Then, the general solution of the ODE (2.33) can be expressed as follows:

$$x(t) = C_1 e^{\frac{-c+i\sqrt{4mk-c^2}}{2m}t} + C_2 e^{\frac{-c-i\sqrt{4mk-c^2}}{2m}t} \qquad (2.41)$$

or

$$x(t) = e^{-\frac{c}{2m}t}\left(A\cos\frac{\sqrt{4mk-c^2}}{2m}t + B\sin\frac{\sqrt{4mk-c^2}}{2m}t\right). \qquad (2.42)$$

If we denote the initial displacement and initial velocity at $t = 0$ as x_0, $\dot{x}_0$ respectively, then we obtain the following constants:

$$A = x_0, \quad B = \frac{2m}{\sqrt{4mk-c^2}}\left(\dot{x}_0 + \frac{c}{2m}x_0\right). \qquad (2.43)$$

Therefore, as observed in Figure 2.4, the solution of the equation contains exponential terms, $e^{-\frac{c}{2m}t}$, resulting in a gradual decrease in amplitude over time, converging to zero. Additionally, due to the influence of the harmonic function, it forms an oscillatory pattern over time.

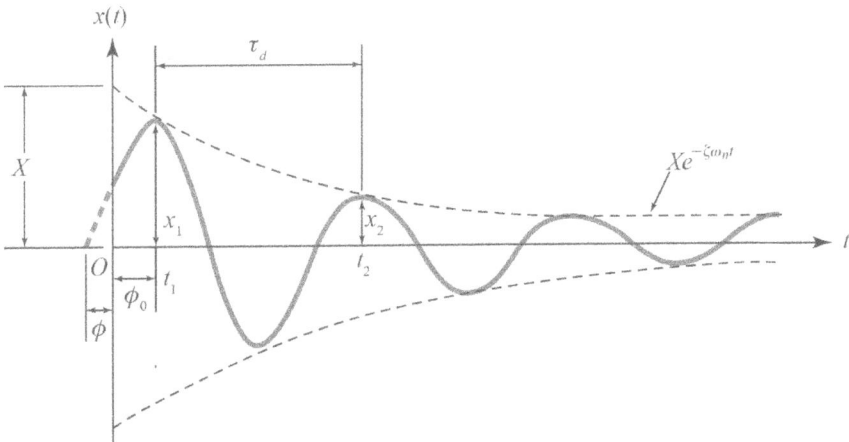

Figure 2.4 Time function of underdamped system.

Figure 2.5 compares the motion over time of overdamped, critical damped, and underdamped systems, respectively.

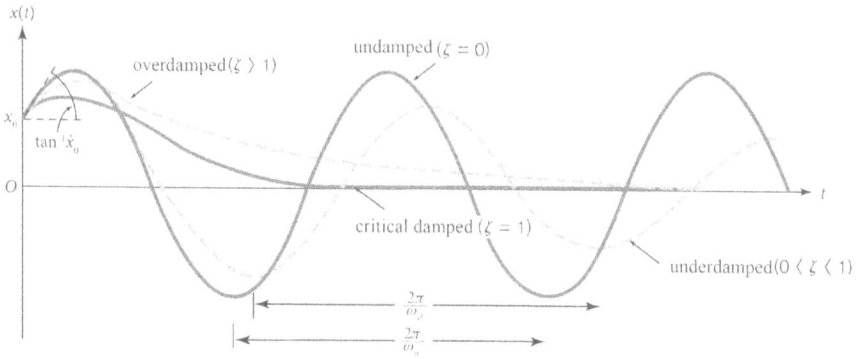

Figure 2.5 Vibration systems.

Example 2.8

Compare the motion when damping is added to the mass-spring system with $m = 2$ kg, $k = 800$ N/m as shown in Figure 2.6. Given initial conditions are $x(0) = 3$ cm and $\dot{x}(0) = 0$, respectively.

a. $c = 100$ Ns/m,
b. $c = 80$ Ns/m,
c. $c = 40$ Ns/m

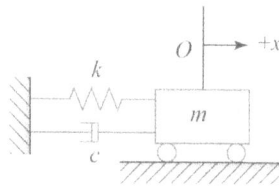

Figure 2.6 Damped vibration system.

Solution

a. The equation of motion of the vibration system with $m = 2$ kg, $k = 800$ N/m, and $c = 100$ Ns/m is:

$$\ddot{x} + 50\dot{x} + 400x = 0.$$

The roots of the characteristic equation $\lambda^2 + 50\lambda + 400 = 0$ are as follows:

$$\lambda_1 = -10, \quad \lambda_2 = -40.$$

Therefore, the general solution is

$$x(t) = C_1 e^{-10t} + C_2 e^{-40t}.$$

When applying the initial conditions, $x(0) = 3$ cm, $\dot{x}(0) = 0$ to solve the differential equation, the solution is as follows:

$$x(t) = 4e^{-10t} - e^{-40t}.$$

Answer $x(t) = 4e^{-10t} - e^{-40t}$ [cm]

b. The equation of motion of the vibration system with $m = 2$ kg, $k = 800$ N/m, and $c = 80$ Ns/m is:

$$\ddot{x} + 40\dot{x} + 400x = 0.$$

The roots of the characteristic equation $\lambda^2 + 40\lambda + 400 = 0$ are as follows:

$$\lambda = -20 \ (\text{real double root}).$$

Therefore, the homogeneous solution is

$$x(t) = (C_1 + C_2 t)e^{-20t}.$$

When applying the initial conditions, $x(0) = 3$ cm, $\dot{x}(0) = 0$ to solve the differential equation, the solution is as follows:

$$x(t) = (3 + 60t)e^{-20t}.$$

Answer $x(t) = (3 + 60t)e^{-20t}$ [cm]

c. The equation of motion of the vibration system with $m = 2$ kg, $k = 800$ N/m, and $c = 40$ Ns/m is:

$$\ddot{x} + 20\dot{x} + 400x = 0.$$

The roots of the characteristic equation $\lambda^2 + 20\lambda + 400 = 0$ are as follows:

$$\lambda_{1,2} = -10 \pm 10\sqrt{3}i.$$

Therefore, the general solution is

$$x(t) = e^{-10t}\left(A\cos 10\sqrt{3}t + B\sin 10\sqrt{3}t\right).$$

When applying the initial conditions, $x(0) = 3$ cm, $\dot{x}(0) = 0$ to solve the differential equation, the solution is as follows:

$$x(t) = e^{-10t}\left(3\cos 10\sqrt{3}t + \sqrt{3}\sin 10\sqrt{3}t\right).$$

Answer $x(t) = e^{-10t}\left(3\cos 10\sqrt{3}t + \sqrt{3}\sin 10\sqrt{3}t\right)$ [cm]

Figure 2.7 shows the solutions in Example 2.8.

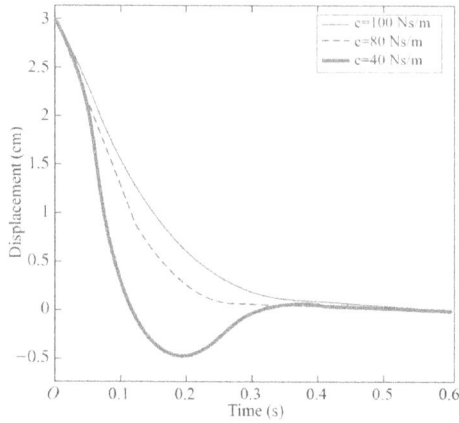

Figure 2.7 The solutions in Example 2.8.

Example 2.9

In the mass-damping-spring system of $m = 4$ kg, $k = 40$ N/m, $c = 8$ Ns/m as shown in Figure 2.8 find the response displacement function $x(t)$ when a constant force $F = 80$ N is applied. Given initial conditions are $x(0) = 1$ m and $\dot{x}(0) = 0$, respectively.

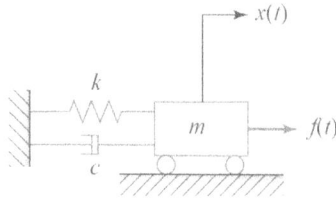

Figure 2.8 Damped vibration system with an external force.

Solution

The equation of motion is:

$$4\ddot{x} + 8\dot{x} + 40x = 80,$$

or

$$\ddot{x} + 2\dot{x} + 10x = 20.$$ ①

Therefore, the homogeneous solution of Eq. ① is:

$$x_h(t) = e^{-t}\left(C_1 \cos 3t + C_2 \sin 3t\right). \quad \left(C_1,\ C_2\ \text{arbitrary}\right)$$ ②

And the particular solution of Eq. ① is:

$$x_p(t) = 2$$ ③

Then the particular solution of Eq. ① is

$$x(t) = x_h(t) + x_p(t) = e^{-t}\left(C_1 \cos 3t + C_2 \sin 3t\right) + 2$$ ④

When applying the initial conditions, $x(0) = 1$ and $\dot{x}(0) = 0$ to solve the differential equation, the solution is

$$x(t) = 2 - e^{-t}\left(\cos 3t + \frac{1}{3}\sin 3t\right).$$

$$\textbf{Answer } x(t) = 2 - e^{-t}\left(\cos 3t + \frac{1}{3}\sin 3t\right)$$

Figure 2.9 shows the solutions in Example 2.9.

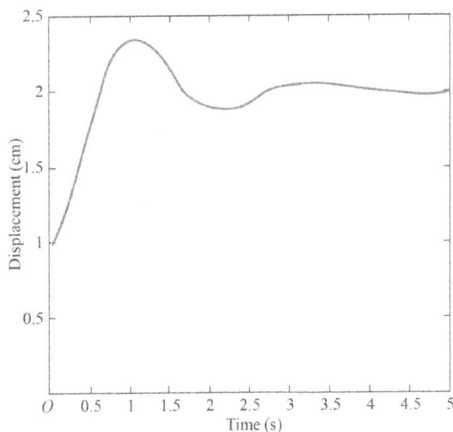

Figure 2.9 The solution in Example 2.9.

2.5.2   Electrical circuit

a. *LC*-circuit

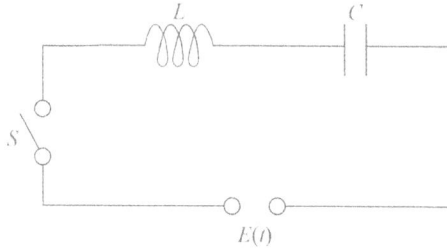

Figure 2.10 LC-circuit.

In an *LC*-circuit composed of an inductor *L* and a capacitor *C* as shown in Figure 2.10, when the switch is closed at $t = 0$, the voltage $E(t)$ becomes the sum of the voltage drop $E_L(t) \left(= L\dfrac{di}{dt} \right)$ across the inductor *L* and the voltage drop $E_C(t) \left(= \dfrac{1}{C}\int i\,dt \right)$ across the capacitor *C*. Then,

$$E(t) = E_L(t) + E_C(t) = L\frac{di}{dt} + \frac{1}{C}\int i\,dt. \tag{2.44}$$

If we denote charge $Q(t)$ as $\int i\,dt$, it can be expressed by the following second-order ODE in terms of charge $Q(t)$.

$$L\frac{d^2 Q}{dt} + \frac{1}{C}Q = E \tag{2.45}$$

To find the general solution of the ODE (2.45), let's first find the homogeneous solution $Q_h(t)$.

The roots of the characteristic equation $\lambda^2 + \dfrac{1}{LC} = 0$ are as follows:

$$\lambda_{1,2} = \pm\sqrt{\frac{1}{LC}}i.$$

Therefore, the homogeneous solution of the ODE (2.45) is:

$$Q_h(t) = A\cos\sqrt{\frac{1}{LC}}t + B\sin\sqrt{\frac{1}{LC}}t. \tag{2.46}$$

And if the voltage *E* is constant, the particular solution of Eq. (2.45) is:

$$Q_p(t) = qE \quad (q \text{ arbitrary}). \tag{2.47}$$

Then the general solution of the ODE (2.45) is as follows:

$$Q(t) = Q_h(t) + Q_p(t) = qE + A\cos\sqrt{\frac{1}{LC}}t + B\sin\sqrt{\frac{1}{LC}}t \qquad (2.48)$$

The current $i(t)$ is obtained by differentiating the charge $Q(t)$ with respect to time as follows:

$$i(t) = -\frac{A}{\sqrt{LC}}\sin\sqrt{\frac{1}{LC}}t + \frac{B}{\sqrt{LC}}\cos\sqrt{\frac{1}{LC}}t. \qquad (2.49)$$

If we apply the initial current $i(0) = 0$ at $t = 0$, we obtain the constant $B = 0$.

Similarly, if we apply the initial charge $Q(0) = 0$ at $t = 0$, we obtain the constant $A = -qE$.

Therefore, the current $i(t)$ is:

$$i(t) = \frac{qE}{\sqrt{LC}}\sin\sqrt{\frac{1}{LC}}t. \qquad (2.50)$$

b. *RLC-circuit*

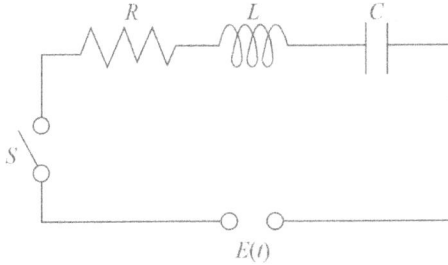

Figure 2.11 RLC-circuit.

In an *RLC*-circuit composed of a resister R, an inductor L, and a capacitor C as shown in Figure 2.11, when the switch is closed at $t = 0$, the voltage $E(t)$ becomes the sum of the voltage drop $E_R(t)(= Ri)$ across the resister R, the voltage drop $E_L(t)\left(= L\frac{di}{dt}\right)$ across the inductor L, and the voltage drop $E_C(t)\left(= \frac{1}{C}\int i\,dt\right)$ across the capacitor C. Then,

$$E(t) = E_R(t) + E_L(t) + E_C(t) = Ri + L\frac{di}{dt} + \frac{1}{C}\int i\,dt. \qquad (2.51)$$

If we denote charge $Q(t)$ as $\int i\,dt$, it can be expressed by the following second-order ODE in terms of charge $Q(t)$.

$$L\frac{d^2Q}{dt} + R\frac{dQ}{dt} + \frac{1}{C}Q = E \qquad (2.52)$$

To find the general solution of the ODE (2.52), let's first find the homogeneous solution.

The roots of the characteristic equation $\lambda^2 + \frac{R}{L}\lambda + \frac{1}{LC} = 0$ are

$$\lambda_{1,2} = -\frac{R}{2L} \pm \frac{1}{2L}\sqrt{R^2 - \frac{4L}{C}}. \qquad (2.53)$$

From algebra, we further know that the characteristic equation may have three cases of roots, depending on the sign of the discriminant $D = R^2 - \frac{4L}{C}$, namely,

(Case I) $D = R^2 - \dfrac{4L}{C} > 0$ overdamped

(Case II) $D = R^2 - \dfrac{4L}{C} = 0$ critial damped

(Case III) $D = R^2 - \dfrac{4L}{C} < 0$ underdamped

In all three cases, the solutions include the form $e^{-\frac{R}{2L}t}$, causing the amplitude to gradually decrease and approach zero as time progresses. Particularly, in Case III of underdamping, due to the influence of harmonic functions, it forms an oscillatory pattern. In other words, as time passes, it oscillates, repeating the cycle of charging and discharging, converging to zero in magnitude.

Example 2.10

In the *RLC*-circuit as shown in Figure 2.12, with $R = 20\ \Omega$, $L = 0.5$ H, $C = 0.001$ F, and $E = 0$ V given, find the charge $Q(t)$ in the capacitor. Given initial charges and initial current are $Q(0) = 1$ C and $i(0) = 0$ A, respectively.

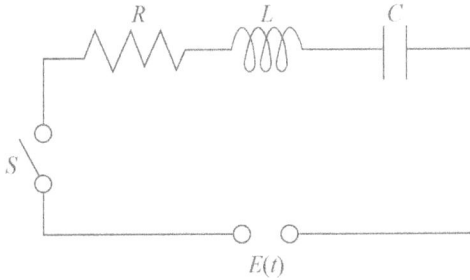

Figure 2.12 RLC-circuit.

Solution

The differential equation $L\dfrac{d^2Q}{dt} + R\dfrac{dQ}{dt} + \dfrac{1}{C}Q = E$ becomes

$$0.5\frac{d^2Q}{dt} + 20\frac{dQ}{dt} + \frac{1}{0.001}Q = 0,$$

or, $\dfrac{d^2Q}{dt} + 40\dfrac{dQ}{dt} + 2000Q = 0.$

The roots of the characteristic equation $\lambda^2 + 40\lambda + 2000 = 0$ are as follows.

$$\lambda_{1,2} = -20 \pm 40i.$$

Then the general solution is

$$Q(t) = e^{-20t}\left(A\cos 40t + B\sin 40t\right).$$

By applying the initial conditions $Q(0) = 1$, $\dot{Q}(0) = i(0) = 0$,

$$Q(0) = A = 1 \qquad\qquad ①$$
$$\dot{Q}(0) = -20A + 40B = 0 \qquad\qquad ②$$

From Eq. ① and Eq. ②,

$$A = 1,\ B = 0.5.$$

Therefore, the solution of the ODE is

$$Q(t) = e^{-20t}\left(\cos 40t + 0.5\sin 40t\right).$$

Answer $Q(t) = e^{-20t}\left(\cos 40t + 0.5\sin 40t\right)\left[C\right]$

Figure 2.13 shows the solutions in Example 2.10.

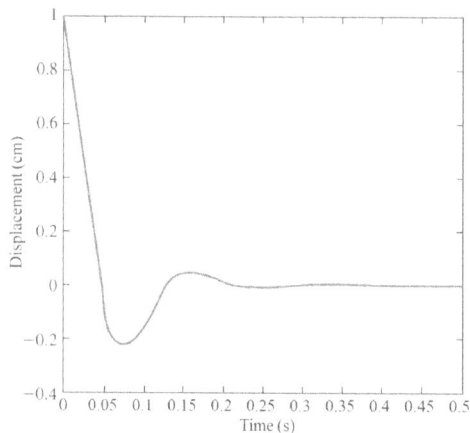

Figure 2.13 The solution in Example 2.10.

Problem 2.5

Solve the following second-order ODE. [1 ~ 8]

1. **Vibration** When modeling a mass-damping-spring system as shown in Figure 2.14, design the damping coefficient c when attaching the spring of $k = 50$ kN/m to the mass of 400 kg to be a critical damping.

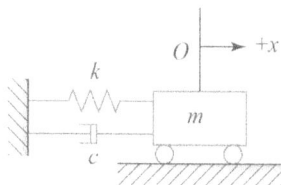

Figure 2.14 Damped vibration system.

2. **Vibration** In the mass-damping-spring system of $m = 1$ kg, $k = 101$ N/m, $c = 2$ Ns/m without external force as shown in Figure 2.15, find the response displacement function $x(t)$. Given initial conditions are $x_0 = 0.1$ m and $\dot{x}_0 = 0$.

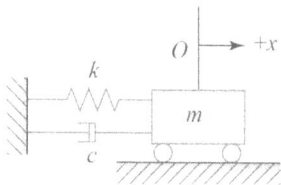

Figure 2.15 Damped vibration system.

3. **Vibration** In the mass-damping-spring system of $m = 1$ kg, $k = 68$ N/m, $c = 4$ Ns/m as shown in Figure 2.16 find the response displacement function $x(t)$ when a constant force $F = 68$ N is applied. Given initial conditions are $x(0) = 0$ m and $\dot{x}(0) = 0.4$ m/s.

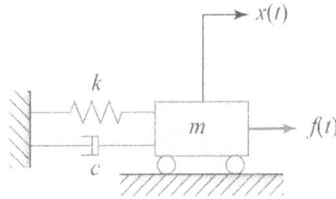

Figure 2.16 Damped vibration system with an external force.

4. **Vibration** In the mass-damping-spring system of $m = 5$ kg, $k = 105$ N/m, $c = 10$ Ns/m as shown in Figure 2.17 find the response displacement function $x(t)$ when the force $F = 101 \sin t$ [N] is applied. Given initial conditions are $x(0) = 1$ m and $\dot{x}(0) = 0$.

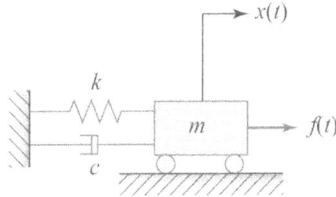

Figure 2.17 Damped vibration system with an external force.

5. **Electric circuit** In the *LC*-circuit as shown in Figure 2.18, with $L = 0.5$ H, $C = 0.02$ F, and $E = 0$ V given, find the current $i(t)$ flowing through the circuit. Given initial charges and initial current are $Q(0) = 1$ C and $i(0) = 0$ A, respectively.

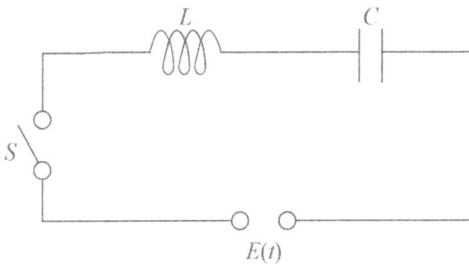

Figure 2.18 LC-circuit.

6. **Electric circuit** In the LC-circuit as shown in Figure 2.19, with $L = 0.5$ H, $C = 0.02$ F, and $E = 5$ V given, find the charge $Q(t)$ in the capacitor. Given initial charges and initial current are $Q(0) = 0$ C and $i(0) = 0$ A, respectively.

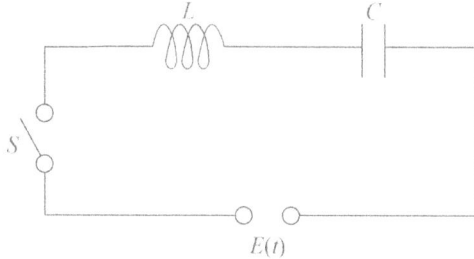

Figure 2.19 LC-circuit.

7. **Electric circuit** In the RLC-circuit as shown in Figure 2.20, with $R = 20$ Ω, $L = 0.5$ H, $C = 0.005$ F, and $E = 0$ V given, find the current $i(t)$ flowing through the circuit. Given initial charges and initial current are $Q(0) = 1$ C and $i(0) = 0$ A, respectively.

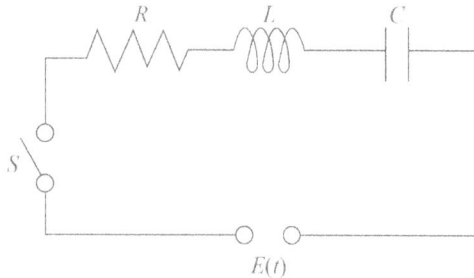

Figure 2.20 RLC-circuit.

8. **Electric circuit** In the RLC-circuit as shown in Figure 2.21, with $R = 20$ Ω, $L = 0.5$ H, $C = 0.005$ F, and $E = 50$ V given, find the current $i(t)$ flowing through the circuit. Given initial charges and initial current are $Q(0) = 0$ C and $i(0) = 0$ A, respectively.

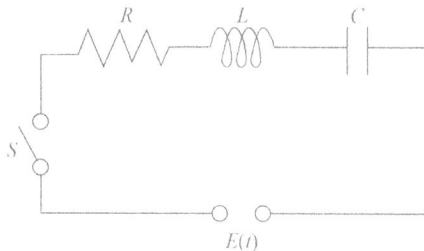

Figure 2.21 RLC-circuit.

2.6 Utilizing MATLAB®

The Runge-Kutta method is a technique for transforming an n-th-order ODE into a system of first-order ODE to find the solution of the given ODE (we will learn more about it in Section 9.1).

Let's use the Runge-Kutta method to plot the solution graph of the second-order ODEs.

M_Example 2.1

Plot the solution graph of the second-order ODE $y'' + 0.2y' + 50y = 20$ with initial conditions $y(0) = 0.1$, $y'(0) = 0$.

Solution

When we define the state $y = \left\{ \begin{array}{c} y \\ y' \end{array} \right\}$, the function $f = \left\{ \begin{array}{c} y' \\ y'' \end{array} \right\}$ becomes

$$f(1) = y(2)$$
$$f(2) = -0.2y(2) - 50y(1) + 20.$$

Then, let's create a function file named 'song2_1', and the main file named 'mex2_1' as follows.

```
% function f=song2_1(x,y)
f= zeros(2,1);
f(1)=y(2);
f(2)=-0.2*y(2)-50*y(1)+20;
```

```
% main file mex2_1.m
close all; clear all;
x=0:0.01:20;
y00=[0.1; 0];                          % initial condition
[x, y]= ode23('song2_1', x, y00);   % ode (ordinary differential equation)
plot(x, y(:,1), 'linewidth', 3); hold on
plot(x, 0*x, 'black')
xlabel('x', 'fontsize', 14);
ylabel('y(x)', 'fontsize', 14); grid
```

Figure 2.22 shows the solution in M-Example 2.1.

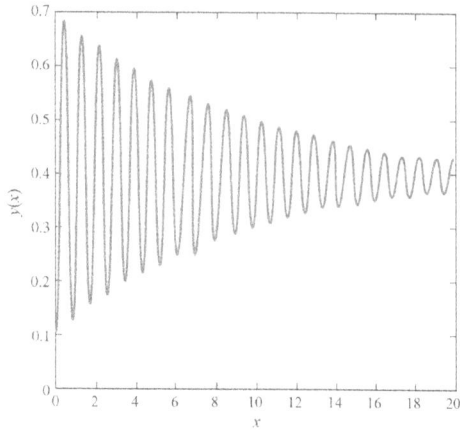

Figure 2.22 The solution in M-Example 2.1.

M_Example 2.2

Plot the solution graph of the second-order ODE $y'' + (\sin x)y' + 50y = 4e^{-0.05x}\cos 2x$ with initial conditions $y(0) = 0.1$, $y'(0) = 0.2$.

Solution

When we define the state $y = \begin{Bmatrix} y \\ y' \end{Bmatrix}$, the function $f = \begin{Bmatrix} y' \\ y'' \end{Bmatrix}$ becomes

$$f(1) = y(2)$$
$$f(2) = -(\sin x)y(2) - 50y(1) + 4e^{-0.05x}\cos 2x.$$

Then, let's create a function file named 'song2_2', and the main file named 'mex2_2' as follows.

```
% song2_2.m
function f=song2_2(x,y)
f=zeros(2,1);
f(1)=y(2);
f(2)=-sin(x).*y(2)-50*y(1)+4*exp(-0.05 *x).*cos(2*x);
```

```
% main file mex2_2.m
close all; clear all;
x=0:0.01:30;
y00=[0.1; 0.2];                        % initial condition
[x, y]=ode23('song2_2', x, y00);
plot(x, y(:,1), 'linewidth', 3); hold on
plot(x, 0*x, 'black')
xlabel('x', 'fontsize', 14);
ylabel('y(x)', 'fontsize', 14); grid
```

Figure 2.23 shows the solution in M-Example 2.2.

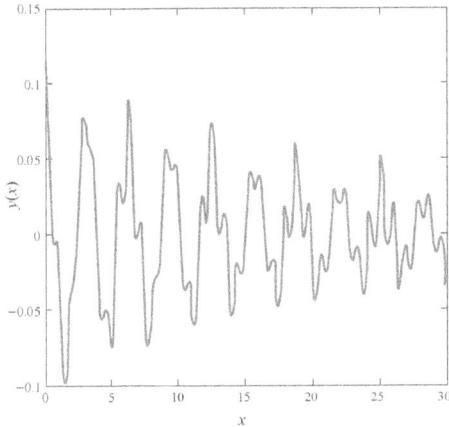

Figure 2.23 The solution in M-Example 2.2.

M_Example 2.3

Solve the following ODEs by using MATLAB®:

a. $\ddot{y}+4y=2,\quad y(0)=1,\quad \dot{y}(0)=0$
b. $\ddot{y}-\dot{y}-2y=0,\quad y(0)=3,\quad \dot{y}(0)=0$
c. $2t^2\ddot{y}-t\dot{y}+y=0,\quad y(1)=1,\quad \dot{y}(1)=0$
d. $t^2\ddot{y}+t\dot{y}-y=8t^3,\quad y(1)=2,\quad \dot{y}(1)=0$
e. $\ddot{y}+y=1/\sin t,\quad y(\pi/2)=0,\quad \dot{y}(\pi/2)=0$

Solution

a. Given the ODE, we can rewrite it in explicit form as follows:

$$\ddot{y} = -4y + 2, \quad y(0) = 1, \quad \dot{y}(0) = 0.$$

Now, using dsolve.m in MATLAB®, find the general solution for the given ODE $\ddot{y} = -4y + 2$.

```
>> syms y(t)
>> D2y = diff(y, 2);      % diff(y,2 ): 2nd-order differential equation
>> dsolve('D2y=–4*y+2')       % solution without initial conditions
```

ans =

C1*cos(2*t) + C2*sin(2*t) + 1/2

Answer $y = C_1 \cos 2t + C_2 \sin 2t + \dfrac{1}{2}$ (C_1, C_2 are constant)

Let's substitute the initial conditions, $y(0) = 1$ and $\dot{y}(0) = 0$, and find the particular solution.

```
>> dsolve('D2y=–4*y+2', 'y(0)=1', 'Dy(0)=0')      % solution with
initial conditions
```

ans =

cos(2*t)/2 + 1/2

Answer $y = \dfrac{1}{2}(\cos 2t + 1)$

b. Given the differential equation, we can rewrite it in explicit form as follows:
$\ddot{y} = \dot{y} + 2y$,

$$y(0) = 3, \quad \dot{y}(0) = 0$$

Now, using dsolve.m in MATLAB®, find the general solution for the given differential equation, $\ddot{y} = \dot{y} + 2y$.

```
>> dsolve('D2y=Dy+2*y')     % solution without initial conditions
```

ans =

C3*exp(2*t) + C4*exp(–t)

Answer $y(x) = C_3 e^{2t} + C_4 e^{-t}$ (C_3, C_4 arbitrary)

Let's substitute the initial conditions, $y(0) = 3$ and $\dot{y}(0) = 0$, and find the particular solution.

```
>> dsolve('D2y=Dy+2*y', 'y(0)=3', 'Dy(0)=0')        % solution with
initial conditions
```

ans =
    2*exp(−t) + exp(2*t)

Answer $y(x) = 2e^{-t} + e^{2t}$

c. Given the differential equation, we can rewrite it in explicit form as follows:

$$\ddot{y} = \frac{1}{2t}\dot{y} - \frac{1}{2t^2}y, \quad y(1) = 1, \quad \dot{y}(1) = 0.$$

Now, use dsolve.m in MATLAB®.

```
dsolve('D2y=1/2/t*Dy−1/2/t^2*y', 'y(1)=1', 'Dy(1)=0')
```

ans =
    2*t^(1/2) − t

Answer $y(t) = 2\sqrt{t} - t$

d. Given the differential equation, we can rewrite it in explicit form as follows:

$$\ddot{y} = -\frac{1}{t}\dot{y} + \frac{1}{t^2}y + 8t, \quad y(1) = 2, \quad \dot{y}(1) = 0.$$

Now, use dsolve.m in MATLAB®.

```
dsolve('D2y=−1/t*Dy+1/t^2*y+8*t', 'y(1)=2', 'Dy(1)=0')
```

ans =
    2/t − (− t^4 + t^2)/t

Answer $y(t) = 2t^{-1} - t + t^3$

e. Given the differential equation, we can rewrite it in explicit form as follows:

$$\ddot{y} = -y + 1/\sin t, \quad y(\pi/2) = 0, \quad \dot{y}(\pi/2) = 0.$$

Now, use dsolve.m in MATLAB®.

>> dsolve('D2y=−y+1/sin(t)', 'y(pi/2)=0', 'Dy(pi/2)=0')

ans =

(pi*cos(t))/2 + log(sin(t))*sin(t) − t*cos(t)

Answer $y(t) = \dfrac{\pi}{2}\cos t - t\cos t + \sin t \cdot \ln|\sin x|$

Problem 2.6

1. Plot the solution graph of the second-order ODE $y'' + 4y = e^{-0.2x}$ with initial conditions $y(0) = 0.2$, $y'(0) = 0$.
2. Plot the solution graph of the second-order ODE $y'' + 0.3y' \operatorname{sign}(y) + y^2 = 1$ with initial conditions $y(0) = 0$, $y'(0) = 0$.
3. Solve the following IVPs by using MATLAB®:

   a. $\ddot{y} - 2\dot{y} + 2y = 0$, $\quad y(0) = 0$, $\quad \dot{y}(0) = 1$
   b. $t^2\ddot{y} - t\dot{y} + 3y = 0$, $\quad y(1) = 2$, $\quad \dot{y}(1) = 2 + \sqrt{2}$
   c. $\ddot{y} + y = \cos t$, $\quad y(0) = 0$, $\quad \dot{y}(0) = 0$
   d. $\ddot{y} - 2\dot{y} + y = 2\sin t$, $\quad y(0) = 2$, $\quad \dot{y}(0) = 0$
   e. $t^2\ddot{y} - 2t\dot{y} - 4y = t^2$, $\quad y(1) = 0$, $\quad \dot{y}(1) = 2$

Answer

Problem 2.1

1. e^{3x}, e^{-x}
2. e^x, xe^x
3. e^{-2x}, xe^{-2x}
4. x, x^{-2}
5. $x^2, x^2\ln|x|$
6. $\cos 2x, \sin 2x$
7. e^{3x}, e^{-x}
8. $x^a, x^a\ln|x|$

Problem 2.2

1. $y(x) = C_1 e^{2x} + C_2 e^{-2x}$ (C_1, C_2 arbitrary)

2. $y(x) = C_1 e^{-3x} + C_2 e^{-\frac{1}{3}x}$ (C_1, C_2 arbitrary)

3. $y(x) = (C_1 + C_2 x)e^{2x}$ (C_1, C_2 arbitrary)

4. $y(x) = (C_1 + C_2 x)e^{-\frac{1}{3}x}$ (C_1, C_2 arbitrary)

5. $y(x) = A\cos\sqrt{3}x + B\sin\sqrt{3}x$ (A, B arbitrary)

6. $y(x) = e^{-\frac{1}{2}x}\left(A\cos\frac{\sqrt{3}}{2}x + B\sin\frac{\sqrt{3}}{2}x\right)$ (A, B arbitrary)

7. $y(x) = e^{2x} + 2e^{-x}$

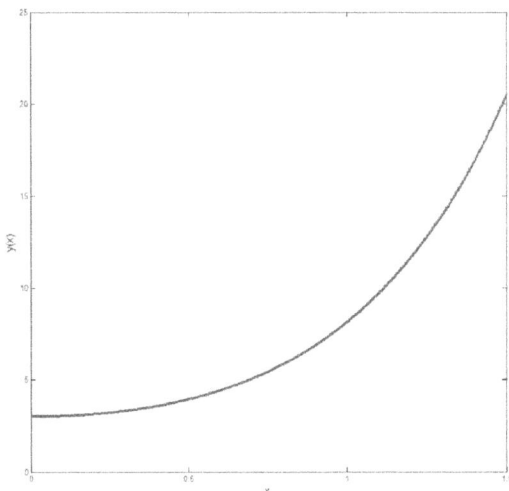

8. $y(x) = (-1 + 3x)e^{-1.5x}$

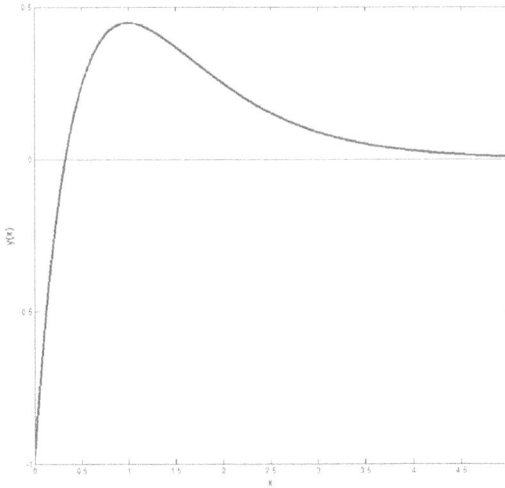

9. $y(x) = e^x \sin x$

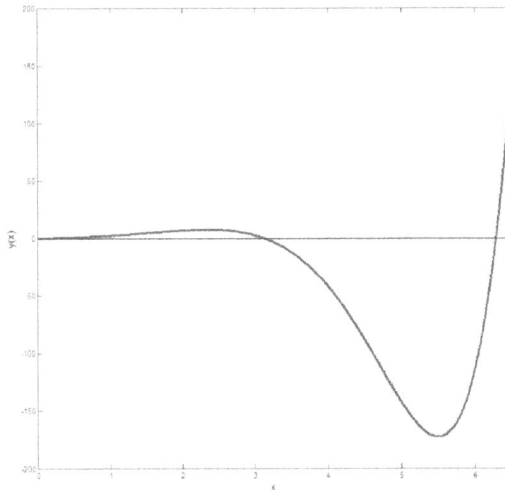

10. $y = C_1 e^{\frac{1}{3}x} + C_2 e^{-\frac{1}{3}x}$ (C_1, C_2 arbitrary)

11. $y = (C_1 + C_2 x)e^x$ (C_1, C_2 arbitrary)

12. $y = C_1 + C_2 e^{3x}$ (C_1, C_2 arbitrary)

13. $y = A\cos\sqrt{2}x + B\sin\sqrt{2}x$ (A, B arbitrary)

Problem 2.3

1. $y(x) = C_1 x^2 + C_2 x^{-2}$ (C_1, C_2 arbitrary)

2. $y(x) = C_1 x^{\frac{1}{3}} + C_2 x^{-1}$ (C_1, C_2 arbitrary)

3. $y(x) = (C_1 + C_2 \ln|x|) x^2$ (C_1, C_2 arbitrary)

4. $y(x) = (C_1 + C_2 \ln|x|) x^{\frac{1}{2}}$ (C_1, C_2 arbitrary)

5. $y = A\cos(\sqrt{3}\ln|x|) + B\sin(\sqrt{3}\ln|x|)$ (A, B arbitrary)

6. $y = x^{-2}\left[A\cos(\ln|x|) + B\sin(\ln|x|)\right]$ (A, B arbitrary)

7. $y(x) = 2\sqrt{x} - x$

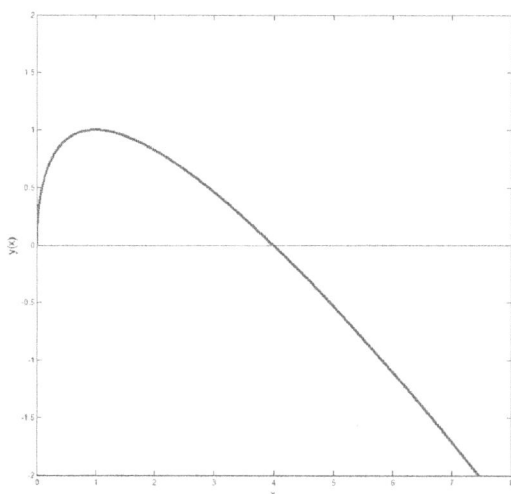

8. $y(x) = x^2(2 - 3\ln|x|)$

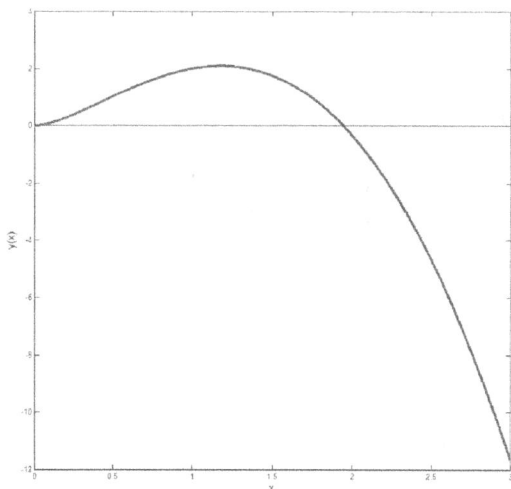

9. $y(x) = x\left\{2\cos\left(\sqrt{2}\ln|x|\right) + \sin\left(\sqrt{2}\ln|x|\right)\right\}$

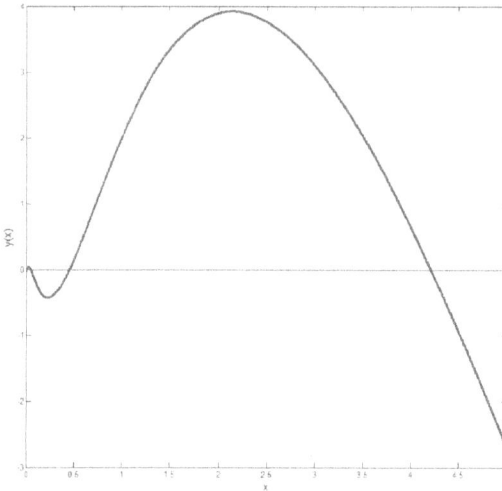

Problem 2.4

1. $y(x) = C_1\cos 2x + C_2\sin 2x + \sin x$ (C_1, C_2 arbitrary)
2. $y(x) = C_1 e^{3x} + C_2 e^{-x} + x^2 - x + 1$ (C_1, C_2 arbitrary)
3. $y(x) = e^{-x}\left(C_1\cos x + C_2\sin x\right) + e^{-2x}$ (C_1, C_2 arbitrary)
4. $y(x) = \left(C_1 + C_2 x\right)e^{-2x} + x^2 e^{-2x}$ (C_1, C_2 arbitrary)
5. $y(x) = \left(C_1 + C_2 x\right)e^{x} + x^2 e^{x}$ (C_1, C_2 arbitrary)
6. $y(x) = C_1 e^{-x} + C_2 e^{2x} + x e^{2x}$ (C_1, C_2 arbitrary)
7. $y(x) = C_1\cos x + C_2\sin x + x\sin x$ (C_1, C_2 arbitrary)
8. $y(x) = \cos x + C_2\sin x + x\sin x + \cos x \cdot \ln|\cos x|$ (C_1, C_2 arbitrary)
9. $y(x) = C_1\cos 2x + C_2\sin 2x - 2x\cos 2x + \sin 2x \cdot \ln|\sin 2x|$ (C_1, C_2 arbitrary)
10. $y(x) = C_1 x^2 + C_2 x^3 + \dfrac{1}{x}$ (C_1, C_2 arbitrary)
11. $y(x) = C_1 x^{-2} + C_2 x^2 + x^2\ln|x|$ (C_1, C_2 arbitrary)
12. $y(x) = C_1 x^{-1} + C_2 x^2 - x$ (C_1, C_2 arbitrary)
13. $y(x) = -\cos x + \sin x + 1$

14. $y(x) = -e^x + e^{2x} + x^2 + 3x + \dfrac{7}{2}$

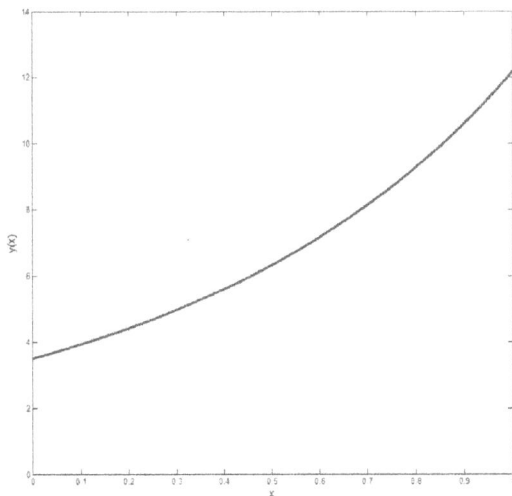

15. $y(x) = (1-x)e^x + \cos x$

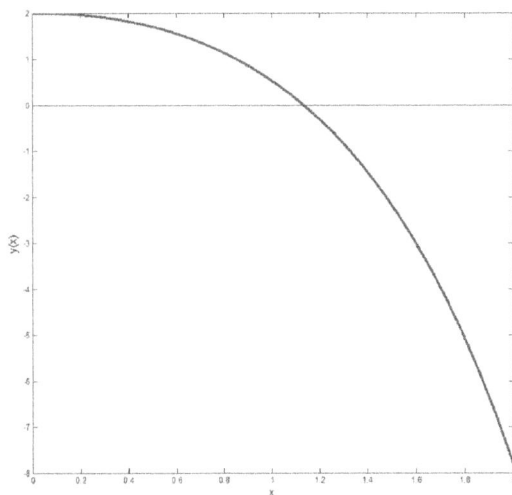

16. $y(x) = \dfrac{1}{x} + x^2 + \dfrac{1}{3}x^2 \ln|x|$

17. $y(x) = \dfrac{2}{x} - x + x^3$

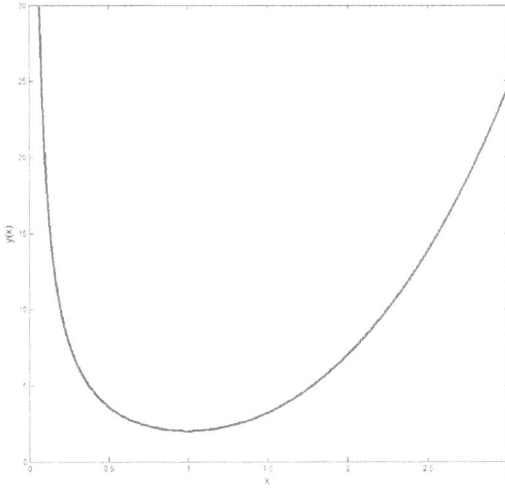

18. $y(x) = (1 - 2x)e^{-x} + (1 + 2x)e^x$

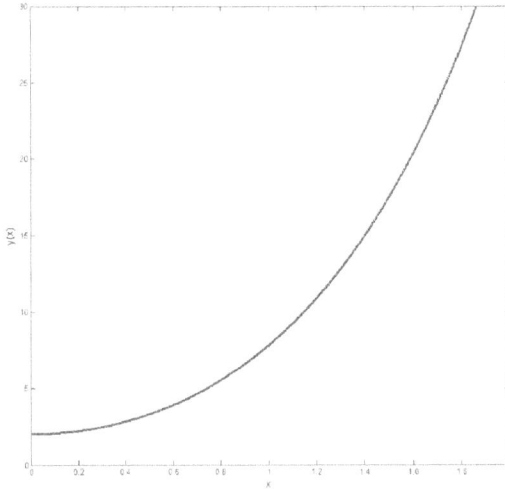

Problem 2.5

1. 8944 N s/m
2. $x(t) = e^{-t}(0.1\cos 10t + 0.01\sin 10t)$ $[m]$

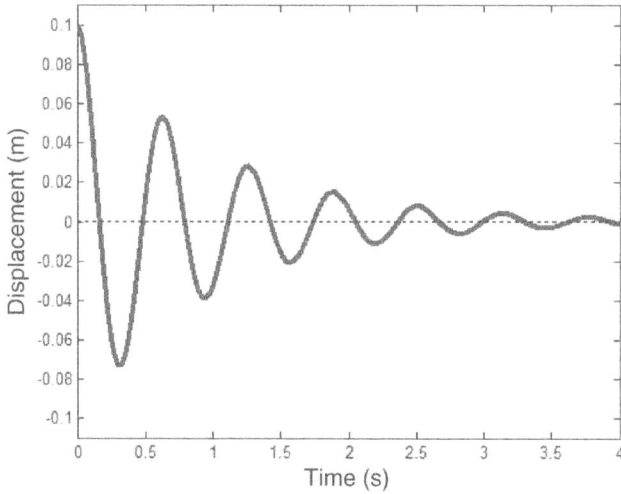

3. $x(t) = 1 - e^{-2t}(\cos 8t + 0.2\sin 8t)$ $[m]$

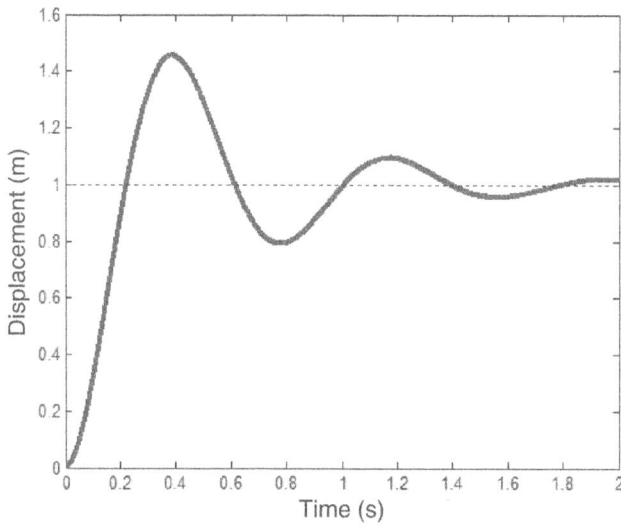

4. $x(t) = e^{-t}\left(1.1\cos\sqrt{20}t + 0.0224\sin\sqrt{20}t\right) - 0.1\cos t + \sin t \;\; [\text{m}]$

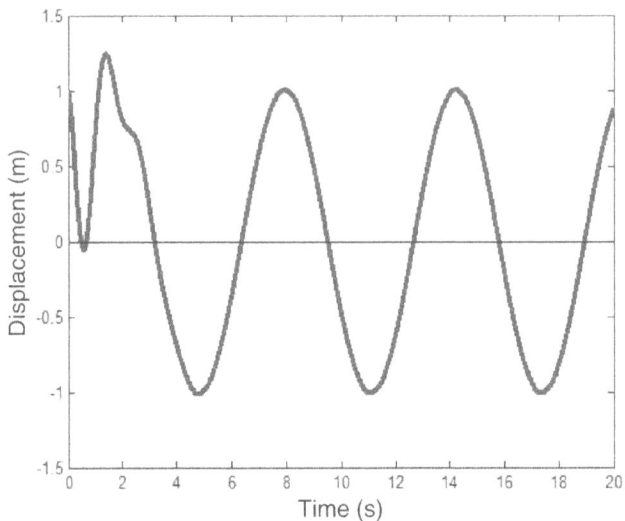

5. $i(t) = -10\sin 10t \;\; [\text{A}]$

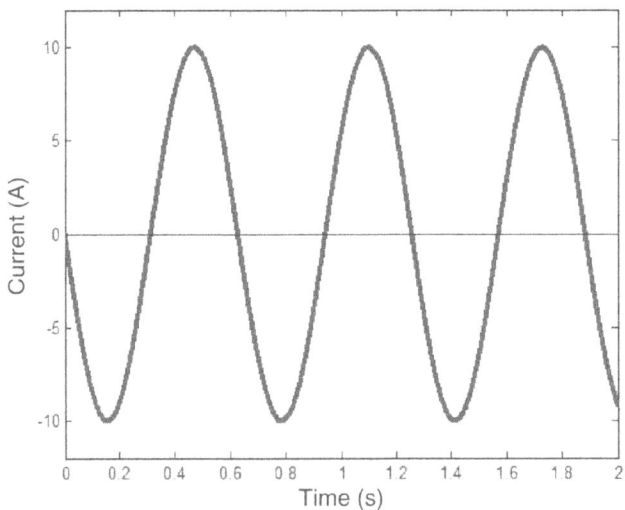

6. $Q(t) = 0.1(1 - \cos 10t)$ [C]

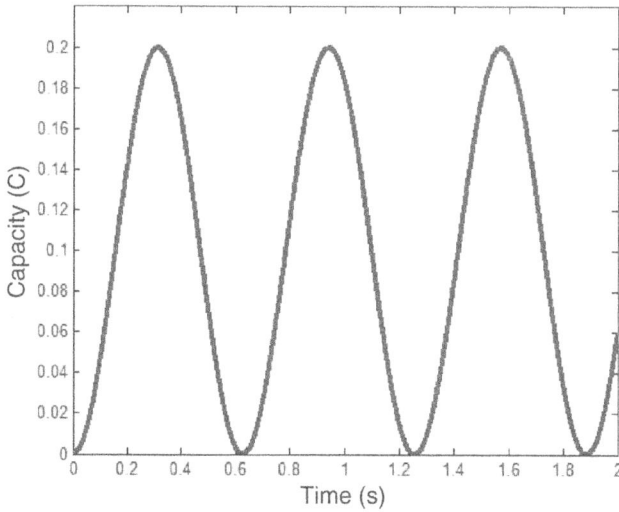

7. $i(t) = -400te^{-20t}$ [A]

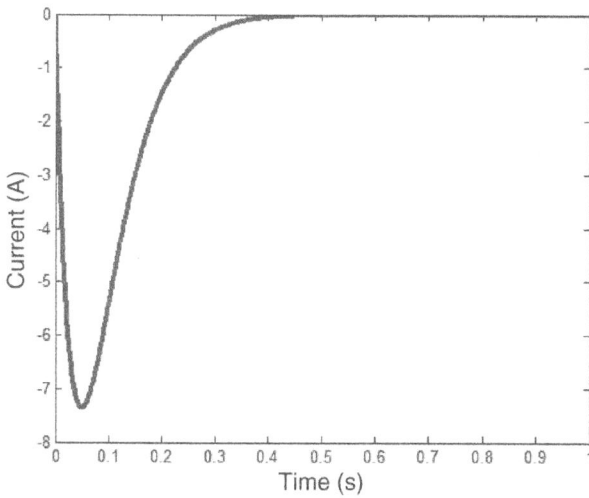

8. $i(t) = 100te^{-20t}$ $[\text{A}]$

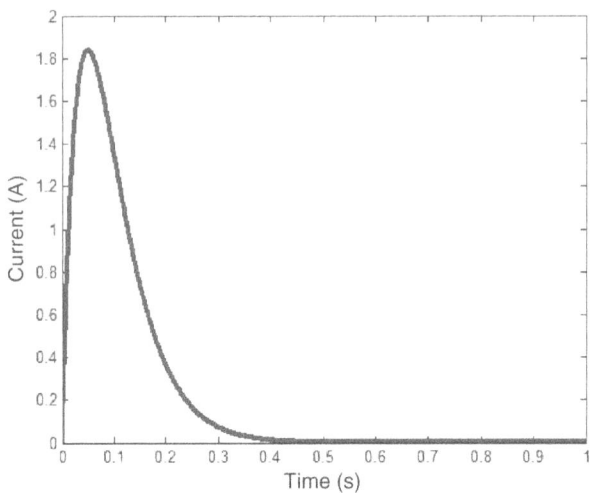

Problem 2.6

1.

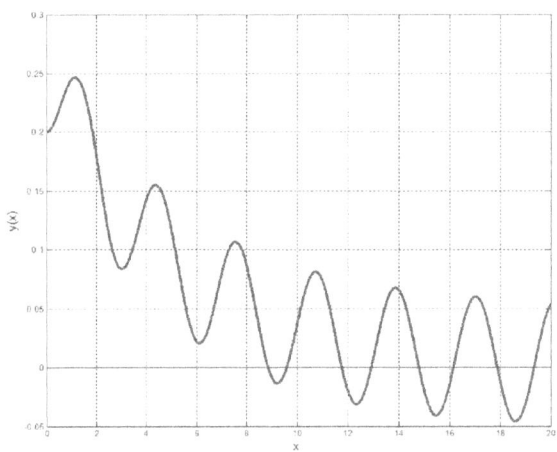

2.

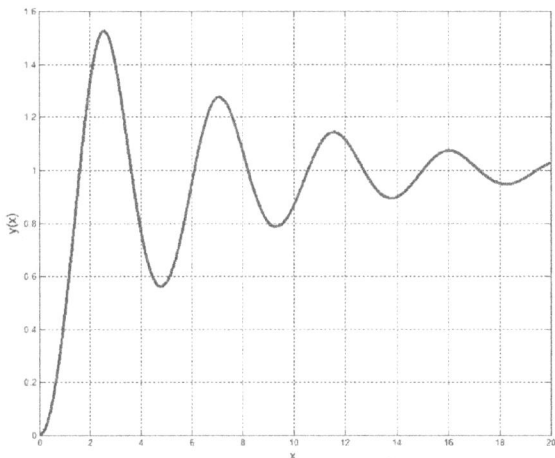

3. a. $y(t) = e^t \sin t$

b. $y(t) = t \left\{ 2 \cos\left(\sqrt{2} \ln|t| \right) + \sin\left(\sqrt{2} \ln|t| \right) \right\}$

c. $y(t) = \dfrac{1}{2} t \sin t$

d. $y(t) = \cos t + (1-t) e^t$

e. $y(t) = \dfrac{1}{2} t^4 - \dfrac{1}{6} t^2 - \dfrac{1}{3t}$

3 Higher-order ordinary differential equations

The solution methods that we have learned for second-order differential equations, such as linear combination of solutions, initial value problems, characteristic equations, the method of undetermined coefficients, and the method of variation of parameters, can be extended directly to higher-order ODEs.

3.1 Homogeneous ODE

3.1.1 Standard form of a homogeneous ODE

When a higher-order ODE is expressed in the following form, it is referred to as the *standard form* of a higher-order ODE.

$$y^{(n)} + p_{n-1}(x)y^{(n-1)} + \cdots + p_1(x)y' + p_0(x)y = r(x) \tag{3.1}$$

When the expression on the right-hand side of the Eq. (3.1) is zero, it is referred to as a homogeneous ODE of n-th-order and is defined as follows:

$$y^{(n)} + p_{n-1}(x)y^{(n-1)} + \cdots + p_1(x)y' + p_0(x)y = 0. \tag{3.2}$$

On the other hand, when the expression on the right-hand side of the Eq. (3.1) is nonzero, $r(x) \neq 0$, it is referred to as a nonhomogeneous ODE.

To simplify notation, the differential operator L is defined using $D = \frac{d}{dx}$ as follows:

$$L = \frac{d^n}{dx^n} + p_{n-1}(x)\frac{d^{n-1}}{dx^{n-1}} + \cdots + p_1(x)\frac{d}{dx} + p_0(x)$$

$$= D^n + p_{n-1}(x)D^{n-1} + \cdots + p_1(x)D + p_0(x). \tag{3.3}$$

The standard form Eq. (3.1) of a higher-order ODE can be expressed using the differential operator L.

$$Ly = \left\{ D^n + p_{n-1}(x)D^{n-1} + \cdots + p_1(x)D + p_0(x) \right\}y = r(x) \tag{3.4}$$

DOI: 10.1201/9781003608912-3

3.1.2 Bases and a general solution

If we denote each independent basis satisfying the homogeneous ODE (3.2) as $y_1, \cdots, y_n$, then the *general solution* of Eq. (3.2) can be expressed as a *linear combination* of these bases as follows:

$$y(x) = C_1 y_1(x) + \cdots + C_n y_n(x) \tag{3.5}$$

where $C_1, C_2, \cdots, C_n$ are arbitrary.

In the general solution of a second-order ODE $y = C_1 y_1 + C_2 y_2$, the constants C_1 and C_2 can be determined from the initial conditions $y(x_0)$ and $y'(x_0)$.

Similarly, in the general solution $y(x) = C_1 y_1(x) + \cdots + C_n y_n(x)$ of a homogeneous ODE (3.2), the n constants can be determined from n initial conditions.

3.1.3 Third-order linear ODE

The *characteristic equation* of a third-order homogeneous linear ODE $y''' + ay'' + by' + cy = 0$ with constant coefficients is as follows:

$$\lambda^3 + a\lambda^2 + b\lambda + c = 0. \tag{3.6}$$

Depending on the types of roots of the characteristic equation (3.6), the homogeneous solution can be summarized as follows.

(Case I) When having distinct real roots $\lambda_1, \lambda_2, \lambda_3$, the *homogeneous solution* is

$$y_h(x) = C_1 e^{\lambda_1 x} + C_2 e^{\lambda_2 x} + C_3 e^{\lambda_3 x}. \tag{3.7a}$$

(Case II) When having one real root λ_1 and one real double root $\lambda_2 (= \lambda_3)$, the *homogeneous solution* is

$$y_h(x) = C_1 e^{\lambda_1 x} + (C_2 + C_3 x) e^{\lambda_2 x}. \tag{3.7b}$$

(Case III) When having a triple real root $\lambda_1 (= \lambda_2 = \lambda_3)$, the *homogeneous solution* is

$$y_h(x) = e^{\lambda_1 x} (C_1 + C_2 x + C_3 x^2). \tag{3.7c}$$

(Case IV) When having a real root λ_1 and complex conjugate roots $\lambda_{2,3} (= p \pm qi)$, the *homogeneous solution* is

$$y_h(x) = C_1 e^{\lambda_1 x} + e^{px} (C_2 \cos qx + C_3 \sin qx). \tag{3.7d}$$

3.1.4 *Fourth-order or higher-order linear ODE*

The characteristic equation of an n-th-order linear homogeneous ODE $y^{(n)} + a_{n-1}y^{(n-1)} + \cdots + a_1 y' + a_0 y = 0$ with constant coefficients is

$$\lambda^n + a_{n-1}\lambda^{n-1} + \cdots + a_1\lambda + a_0 = 0. \tag{3.8}$$

Therefore, depending on the types of roots of the characteristic equation (3.8), the n-th-order solution can be summarized as follows.

(Case I) When having distinct real roots $\lambda_1, \lambda_2, \lambda_3, \cdots$, the homogeneous solution is

$$y_h = C_1 e^{\lambda_1 x} + C_2 e^{\lambda_2 x} + C_3 e^{\lambda_3 x} + \cdots. \tag{3.9a}$$

(Case II) When having multiple real roots $\lambda_1 (= \lambda_2 = \cdots = \lambda_m)$ $(m \leq n)$, the homogeneous solution includes

$$y_h = e^{\lambda_1 x} \left(C_1 + C_2 x + \cdots + C_m x^{m-1} \right). \tag{3.9b}$$

For example, the homogeneous solution in the case of a double root includes

$$\left(C_1 + C_2 x \right) e^{\lambda x},$$

the homogeneous solution in the case of a triple root includes

$$\left(C_1 + C_2 x + C_3 x^2 \right) e^{\lambda x},$$

and the homogeneous solution in the case of a quadruple root includes

$$\left(C_1 + C_2 x + C_3 x^2 + C_4 x^3 \right) e^{\lambda x}.$$

(Case III) When having double complex conjugate roots $\lambda_1 (= \lambda_2 = p \pm qi)$, the homogeneous solution includes

$$y_h = \left(A_1 + A_2 x \right) e^{px} \cos qx + \left(B_1 + B_2 x \right) e^{px} \sin qx. \tag{3.9c}$$

While rarely encountered in practical problems, it is possible to extend this to complex conjugate roots of order three or more.

Remark   Homogeneous Solution of an *n*-th-order Linear ODE

The characteristic equation of an *n*-th-order linear homogeneous ODE $y^{(n)} + a_{n-1}y^{(n-1)} + \cdots + a_1 y' + a_0 y = 0$ with constant coefficients is

$$\lambda^n + a_{n-1}\lambda^{n-1} + \cdots + a_1\lambda + a_0 = 0.$$

(Case I) When the characteristic equation has distinct real roots $\lambda_1, \lambda_2, \lambda_3, \cdots$, the homogeneous solution includes

$$C_1 e^{\lambda_1 x} + C_2 e^{\lambda_2 x} + C_3 e^{\lambda_3 x} + \cdots.$$

(Case II) When the characteristic equation has multiple real roots $\lambda_1 (= \lambda_2 = \cdots = \lambda_m)$, the homogeneous solution includes

$$e^{\lambda_1 x}\left(C_1 + C_2 x + \cdots + C_m x^{m-1}\right).$$

(Case III) When the characteristic equation has multiple complex conjugate roots $\lambda_1 (= \lambda_2 = \cdots = \lambda_m = p \pm qi)$, the homogeneous solution includes

$$\left(A_1 + A_2 x + A_3 x^2 + \cdots\right)e^{px}\cos qx + \left(B_1 + B_2 x + B_3 x^2 + \cdots\right)e^{px}\sin qx.$$

Example 3.1

Find the general solution to the following homogeneous ODE:

a. $y''' - 2y'' - y' + 2y = 0$
b. $y''' - 3y' - 2y = 0$
c. $y^{(5)} + 2y''' + y' = 0$

Solution

a. We set $y(x) = Ce^{\lambda x}$, then the characteristic equation is

$$\lambda^3 - 2\lambda^2 - \lambda + 2 = 0,$$

or

$$(\lambda + 1)(\lambda - 1)(\lambda - 2) = 0.$$

Three real roots are

$$\lambda = -1, \ \lambda = 1, \ \lambda = 2.$$

Therefore, the homogeneous solution is

$$y(x) = C_1 e^{-x} + C_2 e^x + C_3 e^{2x}.$$

Answer $y(x) = C_1 e^{-x} + C_2 e^x + C_3 e^{2x}$ (C_1, C_2, C_3 arbitrary).

Check

When reviewing the method of partial fraction decomposition, it can be summarized as follows:

Only the coefficients in the equation $\lambda^3 - 2\lambda^2 - \lambda + 2 = 0$ are written in order.

$$
\begin{array}{r|rrrr}
1) & 1 & -2 & -1 & 2 \\
   &   & 1 & -1 & -2 \\ \hline
-1) & 1 & -1 & -2 & \boxed{0} \\
    &   & -1 & 2 & \\ \hline
2) & 1 & -2 & \boxed{0} & \\
   &   & 2 & & \\ \hline
   & 1 & \boxed{0} & &
\end{array}
$$

The factors of the expression, which are represented by 1, –1, and 2 written vertically on the left side, lead to the factored form of the expression as:

$$(\lambda + 1)(\lambda - 1)(\lambda - 2) = 0.$$

---

b. We set $y(x) = Ce^{\lambda x}$, then the characteristic equation is

$$\lambda^3 - 3\lambda - 2 = 0.$$

By using the method of partial fraction decomposition, it becomes

$$
\begin{array}{r|rrrr}
-1) & 1 & 0 & -3 & -2 \\
    &   & -1 & 1 & 2 \\ \hline
-1) & 1 & -1 & -2 & \boxed{0} \\
    &   & -1 & 2 & \\ \hline
2) & 1 & -2 & \boxed{0} & \\
   &   & 2 & & \\ \hline
   & 1 & \boxed{0} & &
\end{array}
$$

or

$$(\lambda + 1)^2 (\lambda - 2) = 0.$$

Three real roots are

$$\lambda = -1 \text{ (double real root)}, \ \lambda = 2.$$

Therefore, the homogeneous solution is

$$y(x) = e^{-x}(C_1 + C_2 x) + C_3 e^{2x}.$$

Answer $y(x) = e^{-x}(C_1 + C_2 x) + C_3 e^{2x}$ (C_1, C_2, C_3 arbitrary)

c. We set $y(x) = Ce^{\lambda x}$, then the characteristic equation is

$$\lambda^5 + 2\lambda^3 + \lambda = 0.$$

or

$$\lambda(\lambda^2 + 1)^2 = 0.$$

Five roots are

$$\lambda = 0, \ \lambda = \pm i. \text{ (double complex conjugate root)}$$

Therefore, the homogeneous solution is

$$y(x) = C + (A_1 + A_2 x)\cos x + (B_1 + B_2 x)\sin x.$$

Answer $y(x) = C + (A_1 + A_2 x)\cos x + (B_1 + B_2 x)\sin x.$
$(A_1, A_2, B_1, B_2, C \text{ arbitrary}).$

3.1.5 Third-order or higher-order Euler-Cauchy equation

The third-order *Euler-Cauchy equations* are ODEs of the form

$$x^3 y''' + ax^2 y'' + bxy' + cy = 0 \tag{3.10}$$

with given constants a and b and unknown function $y(x)$.
We set $y(x) = Cx^m$, then the characteristic equation is

$$m(m-1)(m-2) + am(m-1) + bm + c = 0. \tag{3.11}$$

According to the roots of the characteristic equation (3.11), the general solution of the Euler-Cauchy equation is summarized as follows. (C_1, C_2, C_3 arbitrary)

(Case I) When the characteristic equation has different real roots $m_1, m_2, m_3, \cdots$, the homogeneous solution is

$$y_h(x) = C_1 x^{m_1} + C_2 x^{m_2} + C_3 x^{m_3}. \tag{3.12a}$$

(Case II) When the characteristic equation has a real root m_1 and a double real root $m_2 (= m_3)$, the homogeneous solution is

$$y_h(x) = C_1 x^{m_1} + x^{m_2}\left(C_2 + C_3 \ln|x|\right). \tag{3.12b}$$

(Case III) When the characteristic equation has a triple real root $m_1 (= m_2 = m_3)$, the homogeneous solution is

$$y_h(x) = x^{m_1}\left\{C_1 + C_2 \ln|x| + C_3 (\ln|x|)^2\right\}. \tag{3.12c}$$

When the characteristic equation has a real root m_1 and a complex conjugate root $m_{2,3} (= p \pm qi)$, the homogeneous solution is

$$y_h(x) = C_1 x^{m_1} + x^p\left\{C_2 \cos(q\ln|x|) + C_3 \sin(q\ln|x|)\right\}. \tag{3.12d}$$

The four-order or higher Euler-Cauchy equation seldom arises in practical problems, but it can be summarized as follows:

Remark   Euler-Cauchy equation

The characteristic equation of the homogeneous ODE with constant coefficients, $x^n y^{(n)} + a_{n-1}x^{n-1}y^{(n-1)} + \cdots + a_1 xy' + a_0 y = 0$, is as follows:

(Case I) When it has district real roots $m_1, m_2, m_3, \cdots$, the homogeneous solution includes

$$C_1 x^{m_1} + C_2 x^{m_2} + C_3 x^{m_3} + \cdots$$

(Case II) When it has a double real root $m_1 (= m_2)$, the homogeneous solution includes

$$x^{m_1}\left(C_1 + C_2 \ln|x|\right)$$

(Case III) When it has complex conjugate roots $p \pm qi$, the homogeneous solution includes

$$x^p\left\{C_1 \cos(q\ln|x|) + C_2 \sin(q\ln|x|)\right\}$$

Example 3.2

Solve the following third-order homogeneous ODE:

a. $x^3y''' - 3xy' + 3y = 0$
b. $x^3y''' - x^2y'' - 2xy' + 6y = 0$
c. $4x^2y''' + 8xy'' + y' = 0$

Solution

a. We set $y(x) = Cx^m$, then

$$y' = Cmx^{m-1},$$

$$y'' = Cm(m-1)x^{m-2},$$

$$y''' = Cm(m-1)(m-2)x^{m-3}.$$

Then, the characteristic equation is

$$m(m-1)(m-2) - 3m + 3 = 0,$$

or

$$(m+1)(m-1)(m-3) = 0.$$

The roots of the characteristic equation are

$$m = -1,\ 1,\ 3.$$

Therefore, the homogeneous solution is

$$y(x) = C_1x^{-1} + C_2x + C_3x^3.$$

Answer $y(x) = C_1x^{-1} + C_2x + C_3x^3$ (C_1, C_2, C_3 arbitrary)

b. $x^3y''' - x^2y'' - 2xy' + 6y = 0$
We set $y(x) = Cx^m$, then

$$y' = Cmx^{m-1},$$

$$y'' = Cm(m-1)x^{m-2},$$

$$y''' = Cm(m-1)(m-2)x^{m-3}.$$

Substituting these into the given expression, the characteristic equation is

$$m(m-1)(m-2) - m(m-1) - 2m + 6 = 0,$$

or

$$m^3 - 4m^2 + m + 6 = 0.$$

By using the method of partial fraction decomposition, it becomes

$$
\begin{array}{r}
-1\overline{)}\begin{array}{rrrr} 1 & -4 & 1 & 6 \\ & -1 & 5 & -6 \end{array} \\
 \begin{array}{rrrr} & & & \underline{0} \end{array}
\end{array}
$$

$$
\begin{array}{r}
2\overline{)}\begin{array}{rrr} 1 & -5 & 6 \\ & 2 & -6 \end{array} \\
\begin{array}{rrr} & & \underline{0}\end{array}
\end{array}
$$

$$
\begin{array}{r}
3\overline{)}\begin{array}{rr} 1 & -3 \\ & 3 \end{array} \\
\begin{array}{rr} 1 & \underline{0} \end{array}
\end{array}
$$

or

$$(m + 1)(m - 2)(m - 3) = 0.$$

The roots of the characteristic equation are

$$m = -1,\ 2,\ 3.$$

Therefore, the homogeneous solution is

$$y(x) = C_1 x^{-1} + C_2 x^2 + C_3 x^3.$$

Answer $y(x) = C_1 x^{-1} + C_2 x^2 + C_3 x^3$ (C_1, C_2, C_3 arbitrary)

c. By multiplying x into the given equation, then

$$4x^3 y''' + 8x^2 y'' + xy' = 0.$$

We set $y(x) = Cx^m$, then

$$y' = Cmx^{m-1},$$
$$y'' = Cm(m - 1)x^{m-2},$$
$$y''' = Cm(m - 1)(m - 2)x^{m-3}.$$

Substituting these into the given expression, the characteristic equation is

$$4m(m - 1)(m - 2) + 8m(m - 1) + m = 0,$$

or

$$m(2m - 1)^2 = 0.$$

The roots of the characteristic equation are

$$m = 0,\ m = 1/2\ \text{(double real root)}.$$

Therefore, the homogeneous solution is

$$y(x) = C_1 + x^{1/2}\left(C_2 + C_3 \ln|x|\right).$$

Answer $y(x) = C_1 + x^{\frac{1}{2}}\left(C_2 + C_3 \ln|x|\right)$ $(C_1, C_2, C_3$ arbitrary$)$

Problem 3.1

Solve the following homogeneous ODEs. [1 ~ 7]

1. $y^{(4)} - 9y'' = 0$
2. $y^{(4)} + 3y''' + y'' - 3y' - 2y = 0$
3. $\left(D^4 + 8D^2 + 16I\right)y = 0$
4. $\left(D^5 - 3D^4 + 3D^3 - D^2\right)y = 0$
5. $x^2y''' + 3xy'' = 0$
6. $x^3y''' + 2x^2y'' - xy' + y = 0$
7. $x^3y''' + 2x^2y'' + 2y = 0$

Solve the following IVPs. [8 ~ 11]

8. $y^{(4)} - 2y'' + y = 0,$ $y(0) = 0,$ $y'(0) = 1,$ $y''(0) = 6,$ $y'''(0) = -1$
9. $y''' - y'' + 2y' - 2y = 0,$ $y(0) = 2,$ $y'(0) = 1,$ $y''(0) = -1$
10. $x^3y''' + x^2y'' - 2xy' + 2y = 0,$ $y(1) = 2,$ $y'(1) = 1,$ $y''(1) = 4$
11. $x^3y''' + 4x^2y'' + xy' - y = 0,$ $y(1) = 1,$ $y'(1) = 1,$ $y''(1) = -4$

3.2   Nonhomogeneous ODE

The *general solution of a higher-order nonhomogeneous ODE* is

$$y(x) = y_h(x) + y_p(x). \tag{3.13}$$

where $y_h(x)$ is the homogeneous solution of the homogeneous ODE (3.2), and $y_p(x)$ is any *particular solution* of the nonhomogeneous ODE (3.1) which has the nonzero term $r(x)$.

There are typically two methods to find the particular solution $y_p(x)$ of the nonhomogeneous ODE (3.1) as follows:

- Method of undetermined coefficients
- Method of variation of parameters.

3.2.1   Method of undetermined coefficients

Similar to the *method of undetermined coefficients* for a second-order non-homogeneous ODE discussed in Section 2.4.1, we choose a form for $y_p(x)$ similar to $r(x)$ using Table 2.1, but with unknown coefficients $A, B, \cdots$ to be determined by substituting that $y_p(x)$ and its derivatives into the nonhomogeneous ODE (3.1).

Example 3.3

Solve the following nonhomogeneous ODEs:

a. $y''' - 2y'' - y' + 2y = 2x$
b. $y''' + y'' + y' + y = e^{-x}$

Solution

a. General solution of the given nonhomogeneous ODE $y''' - 2y'' - y' + 2y = 2x$ is

$$y(x) = y_h(x) + y_p(x).$$

First, we let the homogeneous solution of the ODE $y_h = Ce^{\lambda x}$. Then, the characteristic equation is

$$\lambda^3 - 2\lambda^2 - \lambda + 2 = (\lambda + 1)(\lambda - 1)(\lambda - 2) = 0,$$

or

$$\lambda = -1,\ 1,\ 2.$$

Therefore, the homogeneous solution of the given ODE is

$$y_h(x) = C_1 e^{-x} + C_2 e^{x} + C_3 e^{2x}.$$

Now, we set the particular solution as follows:

$$y_p = Ax + B.$$

Its derivatives are

$$y_p' = A,\ y_p'' = 0,\ y_p''' = 0.$$

We substitute these expressions into the given ODE $y_p''' - 2y_p'' - y_p' + 2y_p = 2x$. This yields

$$-A + 2(Ax + B) = 2x.$$

Comparing the coefficients of x^1, x^0 gives

$$A = 1,\quad B = \frac{1}{2}.$$

This gives the particular solution

$$y_p = x + \frac{1}{2}.$$

Therefore, the general solution of the given ODE is

$$y(x) = y_h + y_p = C_1 e^{-x} + C_2 e^x + C_3 e^{2x} + x + \frac{1}{2}.$$

Answer $y(x) = C_1 e^{-x} + C_2 e^x + C_3 e^{2x} + x + \frac{1}{2}$ (C_1, C_2, C_3 arbitrary)

Check

By differentiating the particular solution $y_p(x) = x + \frac{1}{2}$, we obtain

$$y_p' = 1, \; y_p'' = 0, \; y_p''' = 0.$$

LHS: $y''' - 2y'' - y' + 2y = -1 + 2\left(x + \frac{1}{2}\right) = 2x$

RHS: $2x$

Therefore, LHS = RHS.

---

b. General solution of the given nonhomogeneous ODE $y''' + y'' + y' + y = e^{-x}$ is

$$y(x) = y_h(x) + y_p(x).$$

First, we let the homogeneous solution of the ODE $y_h = Ce^{\lambda x}$. Then, the characteristic equation is

$$\lambda^3 + \lambda^2 + \lambda + 1 = (\lambda + 1)(\lambda^2 + 1) = 0,$$

or

$$\lambda = -1, \; \lambda = \pm i.$$

Therefore, the homogeneous solution of the given ODE is

$$y_h(x) = C_1 e^{-x} + C_2 \cos x + C_3 \sin x.$$

Since the component e^{-x} in the homogeneous solution is duplicated twice with $r(x)$, e^{-x} is duplicate.
So, we should set the particular solution $y_p(x)$ as follows:

$$y_p = Axe^{-x}.$$

Its derivatives are

$$y_p' = A(-x+1)e^{-x},$$
$$y_p'' = A(x-2)e^{-x},$$
$$y_p''' = A(-x+3)e^{-x}.$$

We substitute these expressions into the given nonhomogeneous ODE. This yields

$$2Ae^{-x} = e^{-x}.$$

Comparing the coefficients of e^{-x} gives

$$A = \frac{1}{2}.$$

This gives the particular solution

$$y_p = \frac{1}{2}xe^{-x}.$$

Therefore, the general solution of the given ODE is

$$y(x) = y_b + y_p = C_1e^{-x} + C_2\cos x + C_3\sin x + \frac{1}{2}xe^{-x}.$$

Answer $y(x) = C_1e^{-x} + C_2\cos x + C_3\sin x + \frac{1}{2}xe^{-x}$ (C_1, C_2, C_3 arbitrary)

Check

By differentiating the particular solution $y_p(x) = \frac{1}{2}xe^{-x}$, we obtain

$$y_p' = \frac{1}{2}e^{-x} - \frac{1}{2}xe^{-x},$$

$$y_p'' = -e^{-x} + \frac{1}{2}xe^{-x},$$

$$y_p''' = \frac{3}{2}e^{-x} - \frac{1}{2}xe^{-x}.$$

LHS: $y''' + y'' + y' + y = \frac{3}{2}e^{-x} - e^{-x} + \frac{1}{2}e^{-x} = e^{-x}$

RHS: e^{-x}

Therefore, LHS = RHS.

3.3.2 Method of variation of parameters (Wronskian's method)

Expanding the *method of variation of parameters* from Section 2.4 yields the following.

$$y_p(x) = y_1 \int \frac{W_1 \cdot r}{W} dx + y_2 \int \frac{W_2 \cdot r}{W} dx + \cdots + y_n \int \frac{W_n \cdot r}{W} dx \qquad (3.14)$$

where Wronskian W is a function defined from the bases $y_1, y_2, \cdots, y_n$ of the homogeneous solution $y_h(x) = C_1 y_1(x) + \cdots + C_n y_n(x)$

$$W(y_1, y_2, \cdots, y_n) = \begin{vmatrix} y_1 & y_2 & \cdots & y_n \\ y_1' & y_2' & \cdots & y_n' \\ . & . & \cdots & . \\ y_1^{(n-1)} & y_2^{(n-1)} & \cdots & y_n^{(n-1)} \end{vmatrix} \qquad (3.15)$$

And $W_j (j = 1, 2, \cdots, n)$ is obtained from W by replacing the j-th column of W by the column $[0\ 0\ \cdots\ 0\ 1]^T$.

$$W = \begin{vmatrix} y_1 & y_2 \\ y_1' & y_2' \end{vmatrix},$$

$$W_1 = \begin{vmatrix} 0 & y_2 \\ 1 & y_2' \end{vmatrix} = -y_2, \quad W_2 = \begin{vmatrix} y_1 & 0 \\ y_1' & 1 \end{vmatrix} = y_1.$$

Thus, when $n = 2$, this becomes with Eq. (2.26) in Section 2.4 as follows:

$$y_p(x) = -y_1 \int \frac{y_2\, r}{W} dx + y_2 \int \frac{y_1\, r}{W} dx. \qquad (2.27)$$

Therefore, we obtain the third-order Wronskian as follows:

$$W = \begin{vmatrix} y_1 & y_2 & y_3 \\ y_1' & y_2' & y_3' \\ y_1'' & y_2'' & y_3'' \end{vmatrix},$$

$$W_1 = \begin{vmatrix} 0 & y_2 & y_3 \\ 0 & y_2' & y_3' \\ 1 & y_2'' & y_3'' \end{vmatrix},$$

$$W_2 = \begin{vmatrix} y_1 & 0 & y_3 \\ y_1' & 0 & y_3' \\ y_1'' & 1 & y_3'' \end{vmatrix},$$

$$W_3 = \begin{vmatrix} y_1 & y_2 & 0 \\ y_1' & y_2' & 0 \\ y_1'' & y_2'' & 1 \end{vmatrix}.$$

Example 3.4

Solve the following nonhomogeneous ODE:

a. $x^2 y''' + xy'' - 4y' = x$
b. $x^3 y''' - 3x^2 y'' + 6xy' - 6y = x^3 \ln x$

Solution

a. Multiplying both sides of the given ODE by x yields

$$x^3 y''' + x^2 y'' - 4xy' = x^2.$$

We set $y_h(x) = Cx^m$, and then

$$y_h' = Cmx^{m-1},$$
$$y_h'' = Cm(m-1)x^{m-2},$$
$$y_h''' = Cm(m-1)(m-2)x^{m-3}.$$

We substitute them into the given ODE. And we obtain the following characteristic equation:

$$m(m-1)(m-2) + m(m-1) - 4m = 0,$$

or

$$m(m+1)(m-3) = 0.$$

Three real roots of the characteristic equation are

$$m = -1,\ 0,\ 3,$$

and give the homogeneous solution of

$$y_h(x) = C_1 x^{-1} + C_2 + C_3 x^3.$$

Then three bases are

$$y_1 = x^{-1}, \; y_2 = 1, \; y_3 = x^3.$$

Then, the Wronskians are as follows:

$$W = \begin{vmatrix} x^{-1} & 1 & x^3 \\ -x^{-2} & 0 & 3x^2 \\ 2x^{-3} & 0 & 6x \end{vmatrix} = 12x^{-1},$$

$$W_1 = \begin{vmatrix} 0 & 1 & x^3 \\ 0 & 0 & 3x^2 \\ 1 & 0 & 6x \end{vmatrix} = 3x^2,$$

$$W_2 = \begin{vmatrix} x^{-1} & 0 & x^3 \\ -x^{-2} & 0 & 3x^2 \\ 2x^{-3} & 1 & 6x \end{vmatrix} = -4x,$$

$$W_3 = \begin{vmatrix} x^{-1} & 1 & 0 \\ -x^{-2} & 0 & 0 \\ 2x^{-3} & 0 & 1 \end{vmatrix} = x^{-2}.$$

On the other hand, dividing both sides of the given ODE by x^2 so that the coefficient of y''' becomes 1 yields

$$r(x) = x^{-1}.$$

Then, the particular solution is

$$y_p(x) = y_1 \int \frac{W_1 \cdot r}{W} dx + y_2 \int \frac{W_2 \cdot r}{W} dx + \cdots + y_n \int \frac{W_n \cdot r}{W} dx$$

$$= x^{-1} \int \frac{3x^2 \cdot x^{-1}}{12x^{-1}} dx + \int \frac{-4x \cdot x^{-1}}{12x^{-1}} dx + x^3 \int \frac{x^{-2} \cdot x^{-1}}{12x^{-1}} dx$$

$$= x^{-1} \int \frac{x^2}{4} dx - \int \frac{x}{3} dx + x^3 \int \frac{x^{-2}}{12} dx$$

$$= x^{-1} \cdot \frac{x^3}{12} - \frac{x^2}{6} + x^3 \cdot \left(-\frac{x^{-1}}{12} \right)$$

$$= -\frac{x^2}{6}$$

or, $y_p = -\dfrac{x^2}{6}$.

Therefore, the general solution of the given nonhomogeneous ODE is

$$y(x) = y_h + y_p = C_1 x^{-1} + C_2 + C_3 x^3 - \frac{x^2}{6}.$$

Answer $y(x) = C_1 x^{-1} + C_2 + C_3 x^3 - \dfrac{x^2}{6}$ (C_1, C_2, C_3 arbitarary)

Check

By differentiating the particular solution $y_p = -\dfrac{x^2}{6}$, we obtain

$$y_p' = -\frac{x}{3}, \quad y_p'' = -\frac{1}{3}, \quad y_p''' = 0.$$

LHS: $x^2 y_p''' + x y_p'' - 4 y_p' = x\left(-\dfrac{1}{3}\right) - 4\left(-\dfrac{x}{3}\right) = x$

RHS: x

Therefore, LHS = RHS.

---

b. We set $y_h(x) = Cx^m$, and then

$$y_h' = Cmx^{m-1},$$
$$y_h'' = Cm(m-1)x^{m-2},$$
$$y_h''' = Cm(m-1)(m-2)x^{m-3}.$$

We substitute them into the given homogeneous ODE. And we obtain the following characteristic equation:

$$m(m-1)(m-2) - 3m(m-1) + 6m - 6 = 0,$$

or

$$(m-1)(m-2)(m-3) = 0.$$

So, three real roots of the characteristic equation are

$$m = 1, \ 2, \ 3,$$

and give the homogeneous solution of

$$y_h(x) = C_1 x + C_2 x^2 + C_3 x^3.$$

Then three bases are

$$y_1 = x, \quad y_2 = x^2, \quad y_3 = x^3.$$

Then, the Wronskians are as follows:

$$W = \begin{vmatrix} x & x^2 & x^3 \\ 1 & 2x & 3x^2 \\ 0 & 2 & 6x \end{vmatrix} = 2x^3,$$

$$W_1 = \begin{vmatrix} 0 & x^2 & x^3 \\ 0 & 2x & 3x^2 \\ 1 & 2 & 6x \end{vmatrix} = x^4,$$

$$W_2 = \begin{vmatrix} x & 0 & x^3 \\ 1 & 0 & 3x^2 \\ 0 & 1 & 6x \end{vmatrix} = -2x^3,$$

$$W_3 = \begin{vmatrix} x & x^2 & 0 \\ 1 & 2x & 0 \\ 0 & 2 & 1 \end{vmatrix} = x^2.$$

On the other hand, dividing both sides of the given ODE by x^3 so that the coefficient of y''' becomes 1 yields

$$r(x) = \ln x.$$

Then, the particular solution is

$$y_p(x) = y_1 \int \frac{W_1 \cdot r}{W} dx + y_2 \int \frac{W_2 \cdot r}{W} dx + \cdots + y_n \int \frac{W_n \cdot r}{W} dx$$

$$= x \int \frac{x^4 \cdot \ln x}{2x^3} dx + x^2 \int \frac{-2x^3 \cdot \ln x}{2x^3} dx + x^3 \int \frac{x^2 \cdot \ln x}{2x^3} dx$$

$$= \frac{x}{2} \cdot \left(\frac{x^2}{2} \ln x - \frac{x^2}{4} \right) - x^2 \cdot (x \ln x - x) + \frac{x^3}{2} \cdot \frac{(\ln x)^2}{2}$$

$$= \frac{1}{4} x^3 (\ln x)^2 - \frac{3}{4} x^3 \ln x + \frac{7}{8} x^3.$$

But since $\dfrac{7}{8}x^3$ can be included in $y_h(x)$, the particular solution becomes

$$y_p = \frac{1}{4}x^3(\ln x)^2 - \frac{3}{4}x^3\ln x.$$

Therefore, the general solution of the given nonhomogeneous ODE is

$$y(x) = y_h + y_p = C_1x + C_2x^2 + C_3x^3 + \frac{1}{4}x^3(\ln x)^2 - \frac{3}{4}x^3\ln x.$$

Answer $y(x) = C_1x + C_2x^2 + C_3x^3 + \dfrac{1}{4}x^3\left\{(\ln x)^2 - 3\ln x\right\}$

(C₁, C₂, C₃ arbitarary)

Check

By differentiating the particular solution $y_p = \dfrac{1}{4}x^3\left\{(\ln x)^2 - 3\ln x\right\}$, we obtain

$$y_p' = \frac{1}{4}x^2\left\{3(\ln x)^2 - 7\ln x - 3\right\},$$

$$y_p'' = \frac{1}{4}x\left\{6(\ln x)^2 - 8\ln x - 13\right\},$$

$$y_p''' = \frac{1}{4}\left\{6(\ln x)^2 + 4\ln x - 21\right\}.$$

LHS: $x^3y''' - 3x^2y'' + 6xy' - 6y = x^3\ln x$
RHS: $x^3\ln x$

Therefore, LHS = RHS.

Problem 3.2

Solve the following nonhomogeneous ODE. [1 ~ 8]

1. $y''' + 4y' = 3\sin x$
2. $y''' - y' = 1$
3. $y''' + 2y'' - y' - 2y = e^x$
4. $y''' - 3y'' + 3y' - y = 2e^x$
5. $\left(D^3 + D\right)y = 2\cos x$
6. $\left(D^3 + D^2 - 2D\right)y = 18xe^x$
7. $x^3y''' + x^2y'' - 2xy' + 2y = 8x^3$
8. $x^3y''' + 2x^2y'' = x$

Solve the following nonhomogeneous IVP. [9 ~ 12]

9. $y''' + 2y'' + y' = 4e^x$, $y(0) = 3$, $y'(0) = 2$, $y''(0) = -1$
10. $y''' + y' = 2e^x$, $y(0) = 3$, $y'(0) = 2$, $y''(0) = 0$
11. $xy''' + 3y'' = 1$, $y(1) = 2$, $y'(1) = 1$, $y''(1) = 1$
12. $x^3 y''' + 2x^2 y'' - xy' + y = 3x^2$, $y(1) = 2$, $y'(1) = 2$, $y''(1) = 5$

3.3   Application of the higher-order ODEs

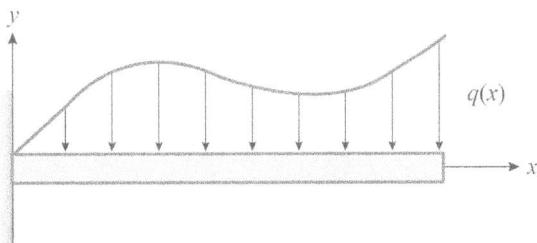

Figure 3.1 Elastic beam.

A classic example utilizing higher-order ODEs is an *elastic beam* subjected to a distributed load $q(x)$ per unit length, as illustrated in Figure 3.1. Typically, the deflection $y(x)$ of the beam is determined by the applied distributed load $q(x)$. The relationship between the bending moment $M(x)$ and the distributed load $q(x)$ per unit length is as follows:

$$\frac{d^2 M(x)}{dx^2} = q(x). \tag{3.16}$$

In addition, the bending moment $M(x)$ is proportional to the curvature of the elastic curve, represented by ρ, which denotes the radius of curvature. That is,

$$M(x) = EI \frac{1}{\rho} \tag{3.17}$$

where E is the Young's modulus, I is the area moment of inertia about the centroidal axis of the beam, and EI is sometimes referred to as the flexural rigidity of the elastic beam. The curvature $1/\rho$ is determined by the following equation, dictated by the geometric shape.

$$\frac{1}{\rho} = \frac{y''}{(1 + y'^2)^{3/2}} \tag{3.18}$$

If the slope of deflection is very small, that is, $y' = 0$, Eq. (3.18) simplifies to

$$\frac{1}{\rho} \cong y''.$$

Substituting this into Eq. (3.17) yields the following equation:

$$M(x) = EIy''. \tag{3.19}$$

Substituting Eq. (3.19) into Eq. (3.16) yields the following fourth-order ODE:

$$EI \frac{d^{(4)}y}{dx^4} = q(x). \tag{3.20}$$

The boundary conditions for the ODE (3.20) are classified as follows:

i. clamped: displacement $y = 0$,   slope $\dfrac{dy}{dx} = 0$

ii. simply supported: displacement $y = 0$,   bending moment
$$EI \frac{d^2 y}{dx^2} = 0$$

iii. free: bending moment $EI \dfrac{d^2 y}{dx^2} = 0$,   shear force $EI \dfrac{d^3 y}{dx^3} = 0$

Example 3.5

Find the deflection $y(x)$ of the cantilever beam subjected to a uniform load per unit length as shown in Figure 3.2.

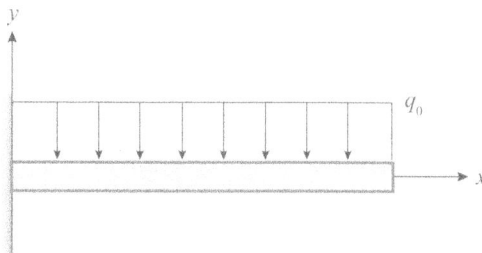

Figure 3.2 Cantilever beam with a uniform load.

Solution

By integrating the equation of motion $EI \dfrac{d^{(4)}y}{dx^4} = -q_0$, we obtain

$$EI \frac{d^{(3)}y}{dx^3} = -q_0 x + C_1 \qquad \qquad ①$$

$$EI \frac{d^2 y}{dx^2} = -\frac{q_0}{2} x^2 + C_1 x + C_2 \qquad \qquad ②$$

$$EI \frac{dy}{dx} = -\frac{q_0}{6} x^3 + \frac{C_1}{2} x^2 + C_2 x + C_3 \qquad \qquad ③$$

$$EI\,y = -\frac{q_0}{24} x^4 + \frac{C_1}{6} x^3 + \frac{C_2}{2} x^2 + C_3 x + C_4 \qquad \qquad ④$$

By applying the boundary conditions $y(0) = 0$ and $\left.\dfrac{dy}{dx}\right|_{x=0} = 0$ at $x = 0$, and $\left.\dfrac{d^3 y}{dx^3}\right|_{x=L} = 0$ and $\left.\dfrac{d^2 y}{dx^2}\right|_{x=L} = 0$ at $x = L$, we can determine the coefficients.

$$C_1 = q_0 L, \quad C_2 = -\frac{1}{2} q_0 L^2, \quad C_3 = 0, \quad C_4 = 0$$

Therefore, the deflection is

$$EI\,y(x) = -\frac{1}{24} q_0 x^4 + \frac{1}{6} q_0 L x^3 - \frac{1}{4} q_0 L^2 x^2$$

$$\textbf{Answer } y(x) = -\frac{q_0 L^4}{24 EI} \left(\frac{x}{L}\right)^2 \left\{\left(\frac{x}{L}\right)^2 + 4\left(\frac{x}{L}\right) - 6\right\}$$

Figure 3.3 shows the solution in Example 3.5.

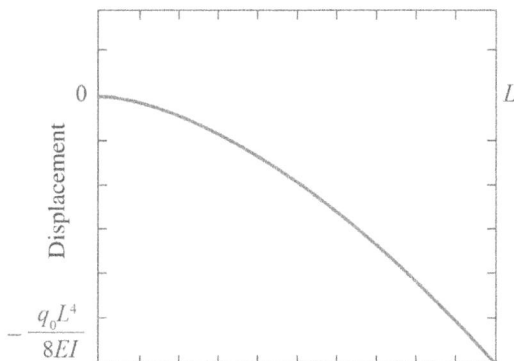

Figure 3.3 The solution in Example 3.5.

Problem 3.3

Solve the following second-order ODE. [1 ~ 3]

1. **Elastic beam**   Find the deflection of the clamped beam subjected to a uniform load per unit length as shown in Figure 3.4.

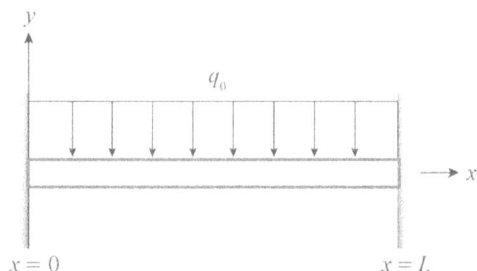

Figure 3.4 Clamped beam with a uniform load.

2. **Elastic beam**   Find the deflection of the simply supported beam subjected to a uniform load per unit length as shown in Figure 3.5.

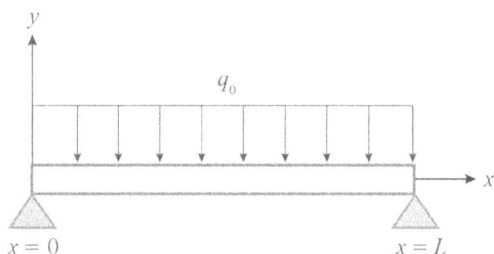

Figure 3.5 Simply supported beam with a uniform load.

3. **Elastic beam**   Find the deflection of the cantilever beam subjected to a load $q(x) = \dfrac{q_0}{L} x$ per unit length as shown in Figure 3.6.

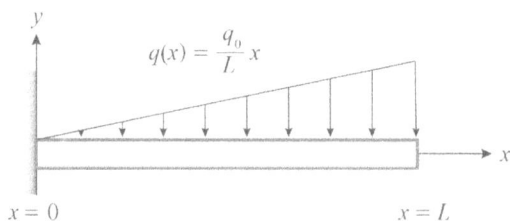

Figure 3.6 Cantilever beam.

3.4 Utilizing MATLAB®

Let's learn how to transform a higher-order ODE into a system of first-order ODEs and then apply the Runge-Kutta method to plot the solution. We will delve into this in detail in Section 7.1.

Furthermore, using the MATLAB® command dsolve.m, we can directly obtain the solution of an ODE with initial conditions.

M-Example 3.1

Plot the solution graph of the ODE $y''' + 2y'' + y' = 0.5$ with initial conditions $y(0) = 0.1$, $y'(0) = 0$, $y''(0) = 0$.

Solution

When we define the state $y = \begin{Bmatrix} y \\ y' \\ y'' \end{Bmatrix}$, the function $f = \begin{Bmatrix} y' \\ y'' \\ y''' \end{Bmatrix}$ becomes

$$f(1) = y(2)$$
$$f(2) = y(3)$$
$$f(3) = -2y(3) - y(2) + 0.5.$$

Then, let's create a function file named 'song3_1', and the main file named 'mex3_1' as follows.

```
% song3_1.m
function f=song3_1(x,y)
f=zeros(3,1);
f(1)=y(2);
f(2)=y(3);
f(3)=-2*y(3)-y(2)+0.5;
```

```
% main file mex3_1.m
close all; clear all;
x=0:0.01:5;
y00=[1; 0; 0];                  % initial condition
[x, y]=ode23('song3_1', x, y00);  % ode (ordinary differential equation)
```

```
plot(x, y(:,1), 'linewidth', 3); hold on
plot(x, 0*x, 'black')
xlabel('x', 'fontsize', 14);
ylabel('y(x)', 'fontsize', 14);
```

Figure 3.7 shows the solution in M-Example 3.1.

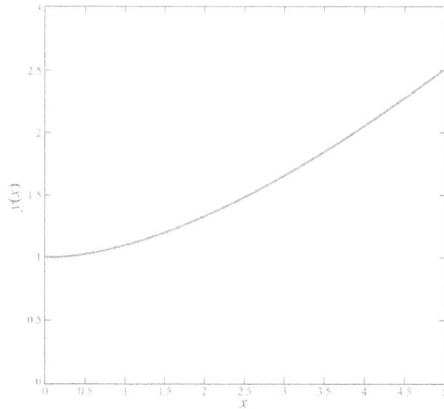

Figure 3.7 The solution in M-Example 3.1.

M-Example 3.2

Plot the solution graph of the ODE $y^{(4)} + 4y''' + 6y'' + 4y' + y = 4e^{-0.1x}\sin x$ with initial conditions $y(0) = 1$, $y'(0) = 0$, $y''(0) = 0$, $y'''(0) = 0$.

Solution

When we define the state $\mathbf{y} = \begin{Bmatrix} y \\ y' \\ y'' \\ y''' \end{Bmatrix}$, the function $\mathbf{f} = \begin{Bmatrix} y' \\ y'' \\ y''' \\ y^{(4)} \end{Bmatrix}$ becomes

$$f(1) = y(2)$$
$$f(2) = y(3)$$
$$f(3) = y(4)$$
$$f(4) = -4y(4) - 6y(3) - 4y(2) - y(1) + 4e^{-0.1x}\sin x.$$

Then, let's create a function file named 'song3_2', and the main file named 'mex3_2' as follows:

```
% song3_2.m
function f=song3_2(x,y)
f=zeros(3,1);
f(1)=y(2); f(2)=y(3); f(3)=y(4);
f(4)=-4*y(4)-6*y(3)-4*y(2)-y(1)+4*exp(-0.1 *x).*sin(x);
```

```
% main file mex3_2.m
close all; clear all;
x=0:0.01:5;
y00=[1; 0; 0; 0];                    % initial condition
[x, y]=ode23('song3_2', x, y00);     % ode (ordinary differential equation)
plot(x, y(:,1), 'linewidth', 3); hold on
plot(x, 0*x, 'black')
xlabel('x', 'fontsize', 14);
ylabel('y(x)', 'fontsize', 14);
```

Figure 3.8 shows the solution in M-Example 3.2.

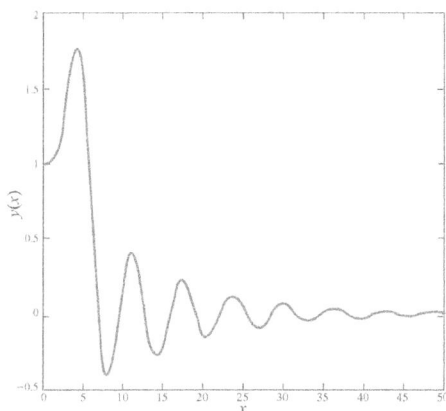

Figure 3.8 The solution in M-Example 3.2.

M-Example 3.3

Solve the ODEs using MATLAB®.

a. $y^{(3)} - \ddot{y} + 2\dot{y} - 2y = 0$, $\quad\quad y(0) = 2, \; \dot{y}(0) = 1, \; \ddot{y}(0) = -1$
b. $t^3 y^{(3)} + t^2\ddot{y} - 2t\dot{y} + 2y = 0$, $\quad y(1) = 2, \; \dot{y}(1) = 1, \; \ddot{y}(1) = 4$
c. $y^{(3)} + 2\ddot{y} + \dot{y} = 4e^t$, $\quad\quad y(0) = 3, \; \dot{y}(0) = 2, \; \ddot{y}(0) = -1$
d. $t^3 y^{(3)} + 2t^2\ddot{y} - t\dot{y} + y = 3t^2$, $\quad y(1) = 2, \; \dot{y}(1) = 2, \; \ddot{y}(1) = 5$

Solution

a. Given the ODE, we can rewrite it in explicit form as follows:

$$y^{(3)} = \ddot{y} - 2\dot{y} + 2y, \; y(0) = 2, \; \dot{y}(0) = 1, \; \ddot{y}(0) = -1.$$

Now, using dsolve.m in MATLAB®, find the general solution for the given ODE $y^{(3)} - \ddot{y} + 2\dot{y} - 2y = 0$.

```
syms y(t)
D3y=diff(y,3);      % diff(y,3): the 3rd-order differential equation
dsolve('D3y=D2y-2*Dy+2*y', 'y(0)=2', 'Dy(0)=1', 'D2y(0)=-1')
```

Remark

It's fine to omit the first two lines of content, syms y(t) and D3y = diff(y,3).

```
ans =
      cos(2^(1/2)*t) + exp(t)
```

Answer $y(t) = \cos\sqrt{2}t + e^t$

b. Given the ODE, we can rewrite it in explicit form as follows:

$$y^{(3)} = -\frac{1}{t}\ddot{y} + \frac{2}{t^2}\dot{y} - \frac{2}{t^3}y, \; y(1) = 2, \; \dot{y}(1) = 1, \; \ddot{y}(1) = 4.$$

Now, using dsolve.m in MATLAB®, find the general solution for the given ODE $t^3 y^{(3)} + t^2\ddot{y} - 2t\dot{y} + 2y = 0$.

> \>\> dsolve('D3y=−1/t*D2y+2/t^2*Dy−2/t^3*y', 'y(1)=2', 'Dy(1)=1',
> 'D2y(1)=4')

ans =
    1/t + t^2

Answer $y(t) = t^{-1} + t^2$

c. Given the ODE, we can rewrite it in explicit form as follows:

$$y^{(3)} = -2\ddot{y} - \dot{y} + 4e^t,\ y(0) = 3,\ \dot{y}(0) = 2,\ \ddot{y}(0) = -1.$$

Now, using dsolve.m in MATLAB®, find the general solution for the given ODE $y^{(3)} + 2\ddot{y} + \dot{y} = 4e^t$.

> \>\> dsolve('D3y=−2*D2y−Dy+4*exp(t)', 'y(0)=3', 'Dy(0)=2', 'D2y(0)=−1')

ans =
    exp(t) − 2*t + t*exp(−t) + t*(2*exp(t) + 2) − 2*t*exp(t) + 2

> \>\> simple(ans)

ans =
    exp(t) + t*exp(−t) + 2

Answer $y(t) = e^t + te^{-t} + 2$

d. Given the ODE, we can rewrite it in explicit form as follows:

$$y^{(3)} = -\frac{2}{t}\ddot{y} + \frac{1}{t^2}\dot{y} - \frac{1}{t^3}y + \frac{3}{t},\ y(1) = 2,\ \dot{y}(1) = 2,\ \ddot{y}(1) = 5.$$

Now, using dsolve.m in MATLAB®, find the general solution for the given ODE $t^3 y^{(3)} + 2t^2\ddot{y} - t\dot{y} + y = 3t^2$.

> \>\> dsolve('D3y=−2/t*D2y+1/t^2*Dy−1/t^3*y+3/t', 'y(1)=2', 'Dy(1)=2',
> 'D2y(1)=5')

ans =
    (3*t^2*log(t))/2 + t*log(t) − (3*t^2*(2*log(t) − 1))/4 + 1/t + t^2/4

>> simple(ans)

ans =
    t*(t + log(t)) + 1/t

Answer $y(t) = t^2 + t \ln|t| + t^{-1}$

Problem 3.4

1. Plot the solution graph of the ODE $y''' + 3y'' + 3y' + y = e^{-0.2x}$ with initial conditions $y(0) = 0.2$, $y'(0) = 0$, $y''(0) = 0$.
2. Plot the solution graph of the ODE $y^{(4)} + 6y''' + 11y'' + 6y' = e^{-x}$ with initial conditions $y(0) = 0$, $y'(0) = 1$, $y''(0) = 0$, $y'''(0) = 0$.
3. Solve the following ODEs using MATLAB®:

   a. $y^{(4)} - 2\ddot{y} + y = 0$, $y(0) = 0$, $\dot{y}(0) = 1$, $\ddot{y}(0) = 2$, $y^{(3)}(0) = 3$

   b. $t^3 y^{(3)} + 4t^2 \ddot{y} + t\dot{y} - y = 0$, $y(1) = 1$, $\dot{y}(1) = 1$, $\ddot{y}(1) = -4$

   c. $y^{(3)} + \ddot{y} + \dot{y} + y = e^{-t}$, $y(0) = 0$, $\dot{y}(0) = 0$, $\ddot{y}(0) = 1$

   d. $y^{(3)} + 3\ddot{y} + 3\dot{y} + y = e^{-t}$, $y(0) = 1$, $\dot{y}(0) = 0$, $\ddot{y}(0) = 0$

Answer

Problem 3.1

1. $y(x) = C_1 + C_2 x + C_3 e^{-3x} + C_4 e^{3x}$ (C_1, C_2, C_3 arbitrary)
2. $y(x) = (C_1 + C_2 x)e^{-x} + C_3 e^{x} + C_4 e^{-2x}$ (C_1, C_2, C_3, C_4 arbitrary)
3. $y(x) = (C_1 + C_2 x)\cos 2x + (C_3 + C_4 x)\sin 2x$ (C_1, C_2, C_3, C_4 arbitrary)
4. $y(x) = C_1 + C_2 x + e^{x}(C_3 + C_4 x + C_5 x^2)$ (C_1, C_2, C_3, C_4, C_5 arbitrary)
5. $y(x) = C_1 x^{-1} + C_2 + C_3 x$ (C_1, C_2, C_3 arbitrary)
6. $y(x) = C_1 x^{-1} + x\{C_2 + C_3 \ln|x|\}$ (C_1, C_2, C_3 arbitrary)
7. $y(x) = C_1 x^{-1} + x\{C_2 \cos(\ln|x|) + C_3 \sin(\ln|x|)\}$.   ($C_1$, C_2, C_3 arbitrary)
8. $y(x) = e^{x}(1+x) - e^{-x}(1+2x)$
9. $y(x) = e^{x} + \cos\sqrt{2}x$

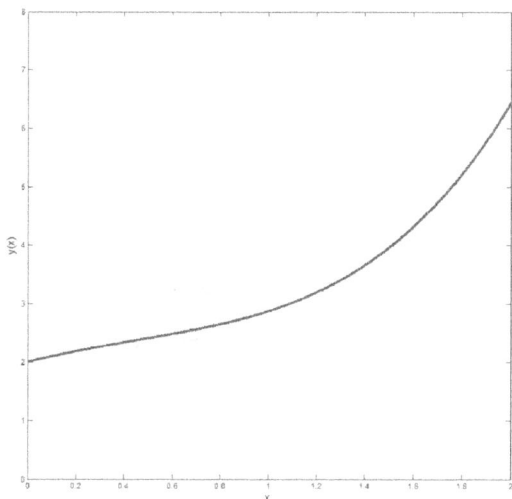

10. $y(x) = x^{-1} + x^2$

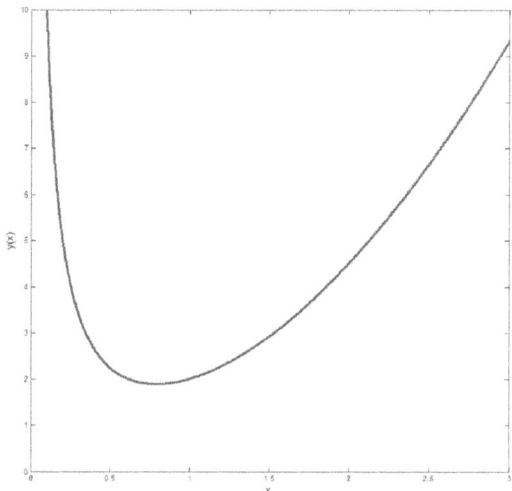

11. $y(x) = x^{-1}(1 + 2\ln|x|)$

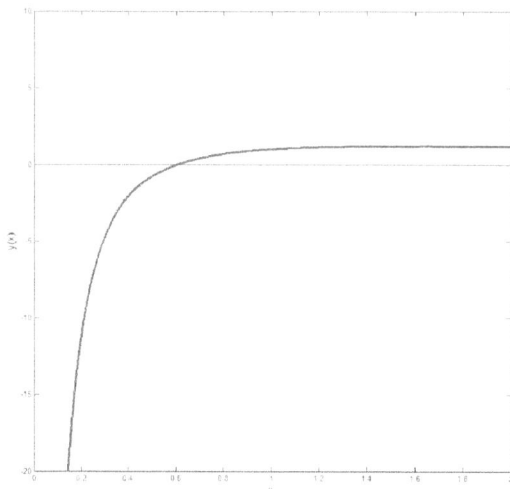

Problem 3.2

1. $y(x) = C_1 + C_2 \cos 2x + C_3 \sin 2x - \cos x$ (C_1, C_2, C_3 arbitrary)

2. $y(x) = C_1 e^{-x} + C_2 + C_3 e^x - x$ (C_1, C_2, C_3 arbitrary)

3. $y(x) = C_1 e^x + C_2 e^{-x} + C_3 e^{-2x} + \dfrac{1}{6} x e^x$ (C_1, C_2, C_3 arbitrary)

4. $y(x) = e^x (C_1 + C_2 x + C_3 x^2) + \dfrac{1}{3} x^3 e^x$ (C_1, C_2, C_3 arbitrary)

5. $y(x) = C_1 + C_2 \cos x + C_3 \sin x - x \cos x + x \sin x$ (C_1, C_2, C_3 arbitrary)

6. $y(x) = y_h + y_p = C_1 + C_2 e^x + C_3 e^{-2x} + 3x^2 e^x - 8x e^x$ (C_1, C_2, C_3 arbitrary)

7. $y(x) = C_1 x^{-1} + C_2 x + C_3 x^2 + x^3$ (C_1, C_2, C_3 arbitrary)

8. $y(x) = C_1 + C_2 \ln|x| + C_3 x + x \ln|x|$ (C_1, C_2, C_3 arbitrary)

9. $y(x) = 2 + x e^{-x} + e^x$

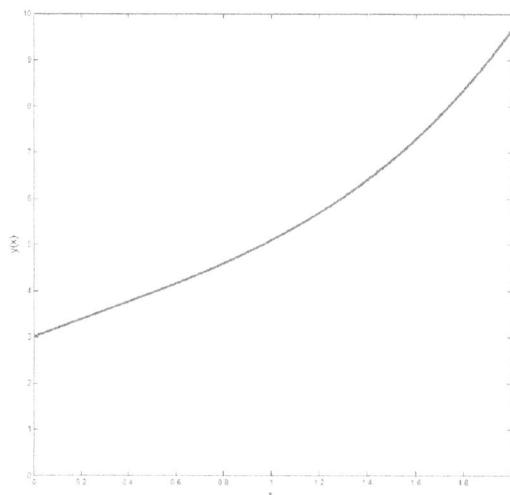

10. $y(x) = 1 + \cos x + \sin x + e^x$

11. $y(x) = \dfrac{1}{3x} + \dfrac{1}{2} + x + \dfrac{x^2}{6}$

12. $y(x) = x \ln|x| + x^{-1} + x^2$

Problem 3.3

1. $y(x) = -\dfrac{q_0 L^4}{24 EI} \left(\dfrac{x}{L}\right)^2 \left(1 - \dfrac{x}{L}\right)^2$

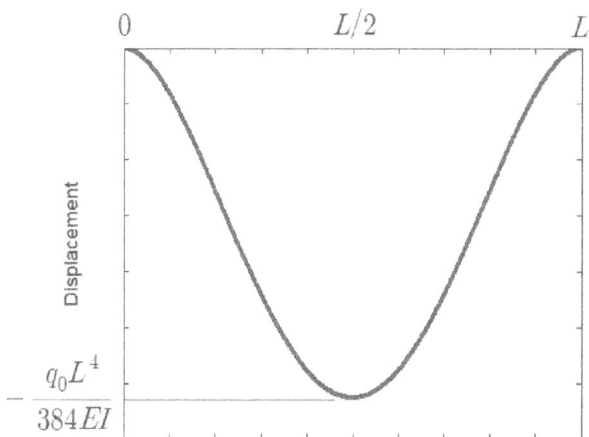

2. $y(x) = -\dfrac{q_0 L^4}{24 EI} \left(\dfrac{x}{L}\right) \left(1 - \dfrac{x}{L}\right) \left\{1 + \dfrac{x}{L} - \left(\dfrac{x}{L}\right)^2\right\}$

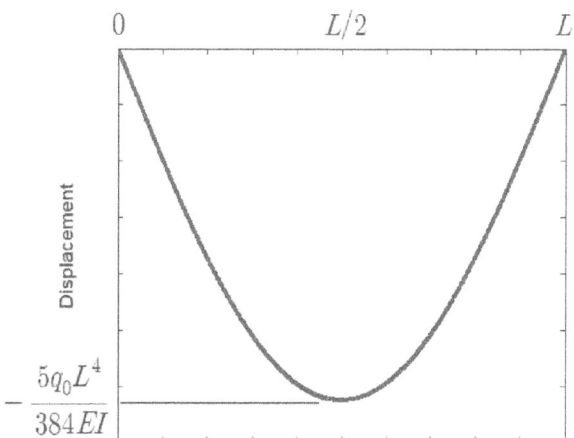

3. $y(x) = -\dfrac{q_0 L^4}{120EI}\left(\dfrac{x}{L}\right)^2\left\{\left(\dfrac{x}{L}\right)^3 - 10\left(\dfrac{x}{L}\right) + 20\right\}$

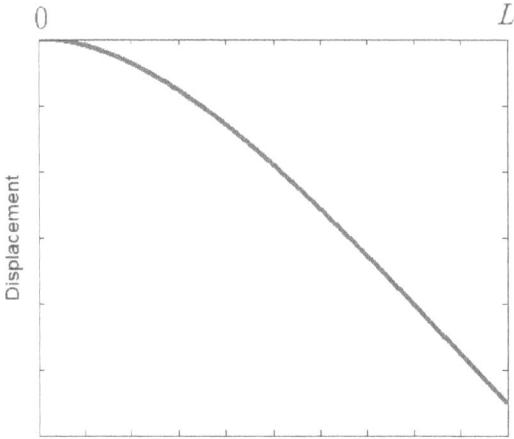

Problem 3.4

1.

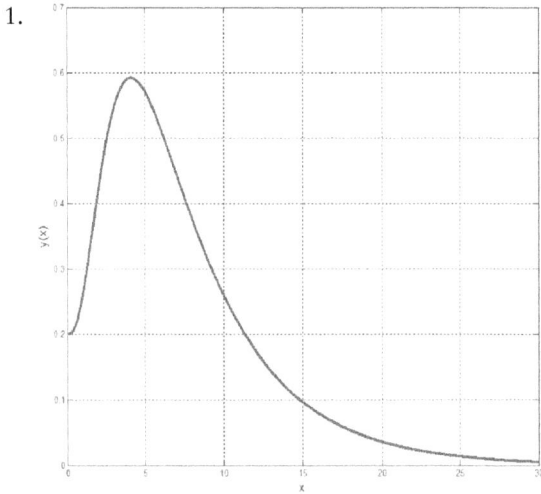

2.

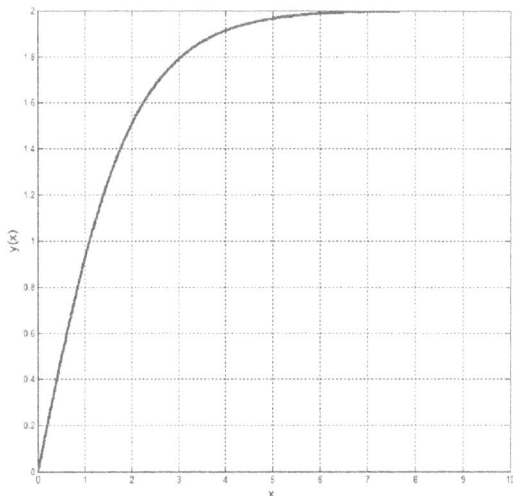

3. a. $y(t) = te^t$

   b. $y(t) = \dfrac{2\ln t + 1}{t}$

   c. $y(t) = e^{-t} - \cos t + \dfrac{1}{2}\sin t + \dfrac{1}{2}te^{-t}$

   d. $y(t) = \left(1 + t + \dfrac{t^2}{2} + \dfrac{t^3}{6}\right)e^{-t}$

4 Laplace transforms

The Laplace transforms are useful methods for finding the solutions of linear ODEs that include initial conditions. It is particularly handy when dealing with ODEs containing inputs such as delta functions and step functions, which are challenging to solve using conventional analytical methods. Typically, this method involves transforming the ODE into the Laplace domain first, obtaining a function of 's-domain', and then applying the inverse Laplace transform to find the solution with a function of 'time domain'. The Laplace transform finds extensive applications in engineering such as control theory, mechanical vibration, signal processing, and physics.

4.1 Definition of the Laplace transform

> **Remark   Laplace transform**
>
> If $f(t)$ is a function defined for all $t \geq 0$, Laplace transform $F(s)$ of the function $f(t)$ is defined as follows:
>
> $$F(s) = \mathcal{L}\{f(t)\} = \int_0^\infty f(t)e^{-st}\, dt. \quad (s > 0) \qquad (4.1)$$

In the *Laplace transform*, a function in the t-domain is transformed into a function in another domain, namely the s-domain, through a one-to-one correspondence. This transformation enables the possibility of performing an *inverse Laplace transform*, which allows us to convert a function from the s-domain back to its original space in the t-domain.

Furthermore, the given function $f(t)$ is called the inverse Laplace of $F(s)$, and is denoted by $\mathcal{L}^{-1}(F)$. That is

$$f(t) = \mathcal{L}^{-1}(F). \qquad (4.2)$$

4.1.1   Notation convention of Laplace transforms

When representing the original t-domain functions and their corresponding Laplace-transformed functions, we adhere to the following convention:

$$f(t) \xrightarrow{\mathcal{L}} F(s), \quad x(t) \xrightarrow{\mathcal{L}} X(s), \quad y(t) \xrightarrow{\mathcal{L}} Y(s), \quad r(t) \xrightarrow{\mathcal{L}} R(s)$$

$$F(s) \xrightarrow{\mathcal{L}^{-1}} f(t), \quad X(s) \xrightarrow{\mathcal{L}^{-1}} x(t), \quad Y(s) \xrightarrow{\mathcal{L}^{-1}} y(t), \quad R(s) \xrightarrow{\mathcal{L}^{-1}} r(t).$$

DOI: 10.1201/9781003608912-4

4.1.2 Linearity of the Laplace transform

The Laplace transform is a linear operation. For any functions $f(t)$ and $g(t)$ with corresponding Laplace transforms $F(s)$ and $G(s)$ that exist, and for any constants a and b, the Laplace transform of $af(t) + bg(t)$ exists, and

$$af(t) + bg(t) \xrightarrow{\mathcal{L}} aF(s) + bG(s)$$

or

$$\mathcal{L}\left[af(t) + bg(t)\right] = \int_0^\infty \left\{af(t) + b(t)\right\}e^{-st}\, dt$$

$$= a\int_0^\infty f(t)e^{-st}\, dt + b\int_0^\infty g(t)e^{-st}\, dt$$

$$= aF(s) + bG(s). \tag{4.3}$$

This property is essential in simplifying the analysis of ODEs using the Laplace transform, as it allows the transform to be applied to individual terms in a linear combination separately.

4.1.3 Laplace transform of 1 and t^n

Let's first take the Laplace transform of $f(t) = 1$ for $t \geq 0$. It is as follows:

$$F(s) = \mathcal{L}(1) = \int_0^\infty e^{-st}\, dt = \left.\frac{e^{-st}}{-s}\right|_0^\infty = \frac{1}{s}. \qquad (s > 0) \tag{4.4}$$

Moreover, when $f(t) = t$, its Laplace transform is as follows:

$$F(s) = \mathcal{L}(t) = \int_0^\infty te^{-st}\, dt = \left.t\frac{e^{-st}}{-s}\right|_0^\infty - \int_0^\infty 1 \cdot \frac{e^{-st}}{-s}\, dt = \frac{1}{s^2}. \tag{4.5}$$

Similarly, when $(t) = t^n$ $(n = 2, 3, \cdots)$, its Laplace transform is as follows:

$$\mathcal{L}(t^n) = \int_0^\infty t^n e^{-st}\, dt = \left.t^n\frac{e^{-st}}{-s}\right|_0^\infty - \int_0^\infty nt^{n-1} \cdot \frac{e^{-st}}{-s}\, dt = \frac{n}{s}\mathcal{L}(t^{n-1}). \tag{4.6}$$

Consequently, summarizing this, it can be represented as shown in Table 4.1.

Table 4.1 Laplace transforms of the
$t^n (n = 0, 1, 2, \cdots)$

$f(t)$	$F(s)$
1	$\dfrac{1}{s}$
t	$\dfrac{1}{s^2}$
t^2	$\dfrac{2}{s^3}$
t^3	$\dfrac{3!}{s^4}$
$t^n (n = 0, 1, 2, \cdots)$	$\dfrac{n!}{s^{n+1}}$

Example 4.1

Find the Laplace transform of the function $f(t) = t^2 - 2t + 3$.

Solution

By utilizing the linearity property, the Laplace transform of the function $f(t) = t^2 - 2t + 3$ is

$$F(s) = \mathcal{L}(t^2 - 2t + 3) = \mathcal{L}(t^2) - 2\mathcal{L}(t) + 3 = \frac{2}{s^3} - 2 \cdot \frac{1}{s^2} + 3 \cdot \frac{1}{s}.$$

$$\text{Answer } F(s) = \frac{2 - 2s + 3s^2}{s^3}$$

4.1.4   Laplace transforms of exponential function e^{at} and trigonometric functions

The Laplace transform of $f(t) = e^{at}$ for $t \geq 0$ is as follows:

$$F(s) = \mathcal{L}(e^{at}) = \int_0^\infty e^{at} e^{-st} \, dt = \frac{e^{-(s-a)t}}{a-s}\bigg|_0^\infty = \frac{1}{s-a}. \qquad (s-a>0) \qquad (4.7)$$

In Eq. (4.7), setting $a = i\omega$, we have $f(t) = e^{i\omega t}$, and its Laplace transform is as follows:

$$F(s) = \mathcal{L}(e^{i\omega t}) = \frac{1}{s - i\omega}. \qquad (4.8)$$

Applying Euler's formula, $e^{i\omega t} = \cos\omega t + i\sin\omega t$, utilizing the linearity property of the Laplace transform, we can express the left-hand side of Eq. (4.8) as follows:

$$F(s) = L\left(e^{i\omega t}\right) = L\left(\cos\omega t + i\sin\omega t\right) = L\left(\cos\omega t\right) + i\, L\left(\sin\omega t\right). \quad (4.9a)$$

Also, rationalizing the right-hand side of Eq. (4.8), we obtain:

$$\frac{1}{s - i\omega} = \frac{s}{s^2 + \omega^2} + i\frac{\omega}{s^2 + \omega^2}. \quad (4.9b)$$

Separating the real and imaginary parts from both the left-hand side and the right-hand side terms, we have:

$$L\left(\cos\omega t\right) = \frac{s}{s^2 + \omega^2}, \quad (4.10a)$$

$$L\left(\sin\omega t\right) = \frac{\omega}{s^2 + \omega^2}. \quad (4.10b)$$

When summarizing the Laplace transforms of exponential and trigonometric functions, it can be represented as shown in Table 4.2.

Table 4.2 Laplace transforms of exponential and trigonometric functions

$f(t)$	$F(s)$
e^{at}	$\dfrac{1}{s - a}$
$\cos\omega t$	$\dfrac{s}{s^2 + \omega^2}$
$\sin\omega t$	$\dfrac{\omega}{s^2 + \omega^2}$
$\cosh at$	$\dfrac{s}{s^2 - a^2}$
$\sinh at$	$\dfrac{a}{s^2 - a^2}$

Example 4.2

Find the Laplace transform of the function

$$f(t) = \cosh at.$$

Solution

By utilizing the linearity property, the Laplace transform of the function is

$$\mathcal{L}(\cosh at) = \mathcal{L}\left(\frac{e^{at} + e^{-at}}{2}\right)$$

$$= \frac{1}{2}\left\{\mathcal{L}(e^{at}) + \mathcal{L}(e^{-at})\right\}$$

$$= \frac{1}{2}\left(\frac{1}{s-a} + \frac{1}{s+a}\right) = \frac{s}{s^2 - a^2}.$$

Answer $\dfrac{s}{s^2 - a^2}$

4.1.5   s-shifting: Substituting s with s − a in the Laplace transform

---

Remark   Laplace transform of the function $e^{at}f(t)$

When the Laplace transform of a function $f(t)$ is $F(s)$, the Laplace transform of $e^{at}f(t)$ becomes $F(s-a)$ as follows:

$$e^{at}f(t) \quad \overset{\mathcal{L}}{\rightarrow} \quad F(s-a), \qquad s-a>0$$

$$\mathcal{L}\left\{e^{at}f(t)\right\} = F(s-a). \qquad\qquad (4.11)$$

Substituting s with $s-a$ in the Laplace transform $F(s) = \displaystyle\int_0^\infty f(t)e^{-st}\,dt$, then

$$F(s-a) = \int_0^\infty f(t)e^{-(s-a)t}\,dt$$

$$= \int_0^\infty \left\{e^{at}f(t)\right\}e^{-st}\,dt$$

$$= \mathcal{L}\left\{e^{at}f(t)\right\}. \qquad\qquad \text{[Proof of Eq. (4.11)]}$$

Table 4.3 summarizes the Laplace transforms of the product of the exponential function e^{at} and an arbitrary function $f(t)$, where s is replaced by $s-a$.

Table 4.3 Laplace transforms of the product of the exponential function e^{at} and an arbitrary function $f(t)$

$e^{at} f(t)$	$F(s-a)$
$e^{at} \cdot 1$	$\dfrac{1}{s-a}$
$e^{at} t$	$\dfrac{1}{(s-a)^2}$
$e^{at} t^2$	$\dfrac{2}{(s-a)^3}$
$e^{at} t^n$	$\dfrac{n!}{(s-a)^{n+1}}$
$e^{at} \cos \omega t$	$\dfrac{s-a}{(s-a)^2 + \omega^2}$
$e^{at} \sin \omega t$	$\dfrac{\omega}{(s-a)^2 + \omega^2}$

Example 4.3

Find the Laplace transforms of the following functions:

a. $f(t) = e^{at} \cos \omega t$
b. $f(t) = e^{at} \sin \omega t$.

Solution

Using $\mathcal{L}\left(e^{at} \cdot 1\right) = \dfrac{1}{s-a}$, we obtain

$$\mathcal{L}\left(e^{at} e^{i\omega t}\right) = \mathcal{L}\left[e^{(a+i\omega)t}\right] = \frac{1}{s-(a+i\omega)}.$$

Then,

$$\text{LHS: } \mathcal{L}\left(e^{at} e^{i\omega t}\right) = \mathcal{L}\left\{e^{at}\left(\cos \omega t + i \sin \omega t\right)\right\}$$

$$= \mathcal{L}\left(e^{at} \cos \omega t\right) + i\mathcal{L}\left(e^{at} \sin \omega t\right)$$

$$\text{RHS: } \frac{1}{(s-a) - i\omega} = \frac{s-a}{(s-a)^2 + \omega^2} + i\frac{\omega}{(s-a)^2 + \omega^2}$$

Therefore, organizing the real and imaginary parts, we derive the following equations:

$$\mathcal{L}\left(e^{at}\cos\omega t\right) = \frac{s-a}{(s-a)^2 + \omega^2},$$

$$\mathcal{L}\left(e^{at}\sin\omega t\right) = \frac{\omega}{(s-a)^2 + \omega^2}.$$

Answer a. $\mathcal{L}\left(e^{at}\cos\omega t\right) = \dfrac{s-a}{(s-a)^2 + \omega^2}$, b. $\mathcal{L}\left(e^{at}\sin\omega t\right) = \dfrac{\omega}{(s-a)^2 + \omega^2}$

4.1.6   Inverse Laplace transform

Since the Laplace transform is a one-to-one mapping, it is possible to perform the *inverse Laplace transform* from the transformed s-domain function $F(s)$ back to the original t-domain function $f(t)$, or, $f(t) \overset{\mathcal{L}}{\rightarrow} F(s) \overset{\mathcal{L}^{-1}}{\rightarrow} f(t)$.

Example 4.4

Find the inverse Laplace transforms of the following function:

$$F(s) = \frac{s+2}{s^2 - 2s + 10}.$$

Solution

$$F(s) = \frac{s+2}{s^2 - 2s + 10} = \frac{(s-1)+3}{(s-1)^2 + 3^2} \overset{\mathcal{L}^{-1}}{\rightarrow} \quad f(t) = e^t\left(\cos 3t + \sin 3t\right)$$

$$\textbf{Answer } f(t) = e^t\left(\cos 3t + \sin 3t\right)$$

Figure 4.1 shows the solution in Example 4.4.

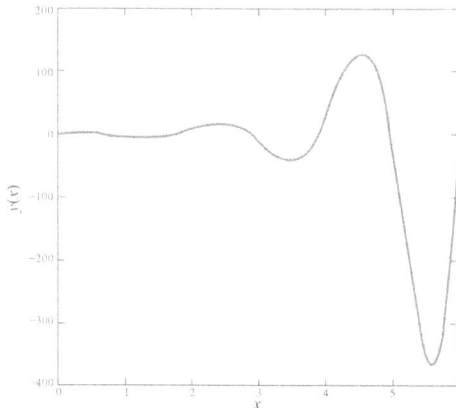

Figure 4.1 The solution in Example 4.4.

Problem 4.1

Find the Laplace transform. [1 ~ 6]

1. $f(t) = t^3 - 3t^2 + 6t + 6$
2. $f(t) = e^{2t} + 2\sin 2t$
3. $f(t) = e^{2t}\cos 3t - 1$
4. $f(t) = 4\cos^2 3t + t^2$
5. $f(t) = (t^2 + 2t)e^{-3t}$
6. $f(t) = e^{-2t}\cosh 3t$

Find the inverse Laplace transform. [7 ~ 12]

7. $F(s) = \dfrac{s^2 - 2s + 2}{s^3}$

8. $F(s) = \dfrac{2s}{(s-3)^2}$

9. $F(s) = \dfrac{-3s + 3}{s^2 + 2s + 5}$

10. $F(s) = \dfrac{5s + 4}{s^2 + 2s}$

11. $F(s) = \dfrac{2s + 2}{(s-2)^3}$

12. $F(s) = \dfrac{s^2 - 10s + 20}{s(s^2 - 6s + 10)}$

4.2 Laplace transforms of derivatives and integrals

The Laplace transform is a valuable method for solving linear ODEs with initial values. Therefore, the Laplace transform of derivatives and integrals involved in ODEs becomes an important process.

4.2.1 Laplace transform of derivatives

Remark Laplace transform of derivatives

Here are the Laplace transforms of the first, second, and third derivatives of a function:

i. First derivative:
   The Laplace transform of the first derivative $f'(t)$ of a function $f(t)$ is given by:

$$\mathcal{L}(f') = s\mathcal{L}(f) - f(0). \qquad (4.12)$$

ii. Second derivative:
The Laplace transform of the second derivative $f''(t)$ of a function $f(t)$ is given by:

$$\mathcal{L}(f'') = s^2\mathcal{L}(f) - sf(0) - f'(0). \tag{4.13}$$

iii. Third derivative:
The Laplace transform of the third derivative $f'''(t)$ of a function $f(t)$ is given by:

$$\mathcal{L}(f''') = s^3\mathcal{L}(f) - s^2f(0) - sf'(0) - f''(0). \tag{4.14}$$

Expanding $\mathcal{L}(f') = \int_0^\infty f'(t)e^{-st}\,dt$ using the method of integration by parts, we obtain

$$\mathcal{L}(f') = \int_0^\infty f'(t)e^{-st}\,dt = f(t)e^{-st}\Big|_0^\infty + s\int_0^\infty f(t)e^{-st}\,dt, \quad (s>0)$$

or

$$\mathcal{L}(f') = s\mathcal{L}(f) - f(0). \tag{4.12}$$

Now the proof of Eq. (4.13) follows by applying Eq. (4.12) to f'' and then substituting Eq. (4.12), that is,

$$\mathcal{L}(f'') = s\mathcal{L}(f') - f'(0)$$
$$= s\{s\mathcal{L}(f) - f(0)\} - f'(0)$$
$$= s^2\mathcal{L}(f) - sf(0) - f'(0). \tag{4.13}$$

Continuing by substitution as in the proof of Eq. (4.13) and using induction, we obtain the following extension:

$$\mathcal{L}(f^{(n)}) = s^n\mathcal{L}(f) - s^{(n-1)}f(0) - s^{(n-2)}f'(0) - \cdots - f^{(n-1)}(0). \tag{4.15}$$

Example 4.5

Find the Laplace form.

a. $f(t) = te^{at}$
b. $f(t) = t\sin t$

Solution

a. Let $f(t) = te^{at}$.

Then $f(0) = 0$,

$$f'(t) = e^{at} + ate^{at}.$$

That expression, when Laplace transformed, becomes

$$sF(s) - f(0) = \frac{1}{s-a} + aF(s).$$

Therefore, we obtain

$$F(s) = \frac{1}{(s-a)^2}.$$

Answer $\mathcal{L}(te^{at}) = \dfrac{1}{(s-a)^2}$

Another Solution

In Table 4.3, it can be noted that just like the Laplace transform of the product of the exponential function e^{at} and the function t, the Laplace transform of the function t is replaced by $s - a$ instead of s in the expression $\dfrac{1}{s^2}$.

---

b. Let $f(t) = t\sin t$.

Then $f(0) = 0$,

$$f'(t) = \sin t + t\cos t.$$

And then $f'(0) = 0$,

$$f''(t) = 2\cos t - t\sin t,$$

or $f''(t) = 2\cos t - f(t)$.

That expression, when Laplace transformed, becomes

$$s^2 F(s) - sf(0) - f'(0) = \frac{2s}{s^2+1} - F(s)$$

or

$$\left(s^2 + 1\right)F(s) = \frac{2s}{s^2 + 1}.$$

Therefore, we obtain

$$F(s) = \frac{2s}{\left(s^2 + 1\right)^2}.$$

Answer $\mathcal{L}(t\sin t) = \dfrac{2s}{\left(s^2 + 1\right)^2}$

4.2.2   Laplace transform of integral

Remark   Laplace transform of integral

Let $F(s)$ denote the Laplace transform of a function $f(t)$ for $t \geq 0$. Then, for $s > 0$ and $t > 0$, we obtain

$$\mathcal{L}\left\{\int_0^t f(\tau)d\tau\right\} = \frac{1}{s}F(s). \qquad (s > 0) \qquad (4.16)$$

Let $g(t) = \displaystyle\int_0^t f(\tau)d\tau.$

   Then $g(0) = 0$ and $g'(t) = f(t)$,

$$F(s) = \mathcal{L}(f) = \mathcal{L}(g') = s\mathcal{L}(g) - g(0), \qquad (s > 0)$$

or $F(s) = s\mathcal{L}(g).$
   Therefore, we obtain

$$\frac{1}{s}F(s) = \mathcal{L}\left\{\int_0^t f(\tau)d\tau\right\}. \qquad (4.16)$$

And the inverse Laplace transform of Eq. (4.16) is

$$\mathcal{L}^{-1}\left\{\frac{1}{s}F(s)\right\} = \int_0^t f(\tau)d\tau. \qquad (4.17)$$

Example 4.6

Find the inverse Laplace transform.

$$\frac{1}{s\left(s^2 + \omega^2\right)}$$

Solution

Let $\dfrac{1}{s}F(s) = \dfrac{1}{s\left(s^2 + \omega^2\right)}$, then

$$F(s) = \frac{1}{s^2 + \omega^2} = \mathcal{L}\left\{\frac{\sin \omega t}{\omega}\right\}$$

where

$$f(t) = \frac{\sin \omega t}{\omega}.$$

From $\mathcal{L}^{-1}\left\{\dfrac{1}{s}F(s)\right\} = \displaystyle\int_0^t f(\tau)d\tau$, we obtain

$$\mathcal{L}^{-1}\left\{\frac{1}{s\left(s^2 + \omega^2\right)}\right\} = \int_0^t \frac{\sin \omega \tau}{\omega}d\tau = -\left.\frac{\cos \omega \tau}{\omega^2}\right|_0^t = \frac{1 - \cos \omega t}{\omega^2}.$$

$$\textbf{Answer } \mathcal{L}^{-1}\left\{\frac{1}{s\left(s^2 + \omega^2\right)}\right\} = \frac{1 - \cos \omega t}{\omega^2}$$

Another Solution

Given equation can be expanded into partial fractions:

$$\frac{1}{s\left(s^2 + \omega^2\right)} = \frac{a}{s} + \frac{bs + c}{s^2 + \omega^2},$$

or $1 = a\left(s^2 + \omega^2\right) + (bs + c)s$.
   When comparing coefficients, we get

$$a = \frac{1}{\omega^2}, \quad b = -\frac{1}{\omega^2}, \quad c = 0.$$

Then

$$\frac{1}{s(s^2+\omega^2)} = \frac{1}{\omega^2}\left(\frac{1}{s}-\frac{s}{s^2+\omega^2}\right).$$

And the inverse Laplace transform of the above equation is

$$\mathcal{L}^{-1}\left\{\frac{1}{\omega^2}\left(\frac{1}{s}-\frac{s}{s^2+\omega^2}\right)\right\} = \frac{1-\cos\omega t}{\omega^2}.$$

4.2.3   Solution of the ODE with initial conditions

Let's find the solution to the second-order ODE with initial values using the Laplace transform.

$$y'' + ay' + by = r(t), \quad y(0) = y_0, \quad y'(0) = y_0', \qquad (4.18)$$

where a and b are constants.

Here $r(t)$ is the given input, or the external force applied to the system, and $y(t)$ is the output or response to the $r(t)$ to be obtained.

Using the Laplace transform, solutions to initial value problems in ODEs can be relatively easily obtained, as shown in Figure 4.2, by sequentially performing three steps. Particularly, problems that couldn't be solved from Chapter 1 to Chapter 3 in ODEs, such as unit step functions, delta functions, time-shifted functions, and periodic signal functions, can be solved.

i. First step: Apply the Laplace transform to both sides of the given ODE.
ii. Second step: Solve the algebraic equation to find the Laplace transform $Y(s)$ in the s-domain.
iii. Third step: Inverse Transform: Apply the inverse Laplace transform to $Y(s)$ to obtain the solution $y(t)$ in the t-domain.

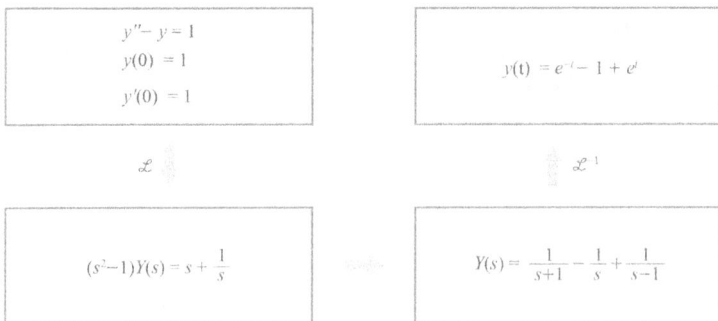

Figure 4.2 The process of finding the solution to an ODE.

Example 4.7

Solve the following ODE using the Laplace transform:

$$y'' - y = 1, \quad y(0) = 1, \quad y'(0) = 0.$$

Solution

Let's transform the given ODE $y'' - y = 1$, then

$$\left[s^2 Y(s) - sy(0) - y'(0) \right] - Y(s) = \frac{1}{s}.$$

Collecting the $Y(s)$-terms, we have the subsidiary equation

$$\left(s^2 - 1 \right) Y(s) = s + \frac{1}{s}.$$

Then this gives the solution

$$Y(s) = \frac{s^2 + 1}{s\left(s^2 - 1 \right)} = \frac{1}{s+1} - \frac{1}{s} + \frac{1}{s-1}.$$

From this expression for $Y(s)$, we obtain the solution

$$y(t) = e^{-t} - 1 + e^{t}.$$

Answer $y(t) = e^{t} + e^{-t} - 1$

Check

From $y(t) = e^{t} + e^{-t} - 1$, we obtain

$$y' = e^{t} - e^{-t}, \quad \text{and} \quad y'' = e^{t} + e^{-t}.$$

Substituting these expressions into the given equation yields:

$$y'' - y = \left(e^{t} + e^{-t} \right) - \left(e^{t} + e^{-t} - 1 \right) = 1.$$

Figure 4.3 shows the solution in Example 4.7.

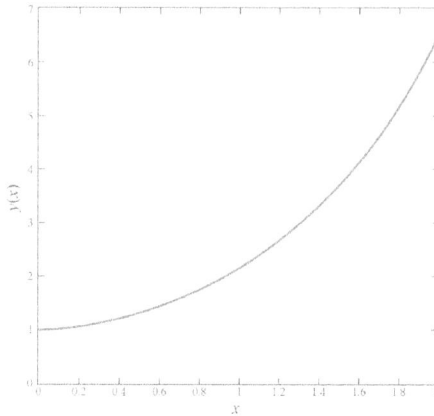

Figure 4.3 The solution in Example 4.7.

Example 4.8

Solve the initial value problem using the Laplace transform.

$$y'' + 2y' + 10y = 10t, \quad y(0) = 0, \quad y'(0) = 1$$

Solution

Let's transform the given ODE $y'' + 2y' + 10y = 10t$, then

$$\left[s^2 Y(s) - sy(0) - y'(0)\right] + 2\left[sY(s) - y(0)\right] + 10Y(s) = \frac{10}{s^2}.$$

Collecting the $Y(s)$-terms, we have the subsidiary equation

$$\left(s^2 + 2s + 10\right)Y(s) = \frac{s^2 + 10}{s^2}.$$

We divide by $\left(s^2 + 2s + 10\right)$ and obtain

$$Y(s) = \frac{s^2 + 10}{s^2\left(s^2 + 2s + 10\right)} = \frac{a}{s} + \frac{b}{s^2} + \frac{cs + d}{s^2 + 2s + 10}.$$

By comparing coefficients, you can achieve

$$a = -\frac{1}{5}, \ b = 1, \ c = \frac{1}{5}, \ d = \frac{2}{5}.$$

Then this gives the solution

$$Y(s) = -\frac{1}{5s} + \frac{1}{s^2} + \frac{(s+1)+\frac{1}{3}\cdot 3}{5\{(s+1)^2 + 3^2\}}.$$

From this expression for $Y(s)$, we obtain the solution

$$y(t) = -\frac{1}{5} + t + \frac{1}{5}e^{-t}\left(\cos 3t + \frac{1}{3}\sin 3t\right).$$

Answer $y(t) = -\frac{1}{5} + t + \frac{1}{5}e^{-t}\left(\cos 3t + \frac{1}{3}\sin 3t\right)$

Check

Let $y(t) = -\frac{1}{5} + t + \frac{1}{5}e^{-t}\left(\cos 3t + \frac{1}{3}\sin 3t\right).$

Then $y(0) = 0,$

$$y' = 1 - \frac{2}{3}e^{-t}\sin 3t.$$

And then $y'(0) = 1,$

$$y'' = \frac{2}{3}e^{-t}\sin 3t - 2e^{-t}\cos 3t$$

LHS: $y'' + 2y' + 10y$

$$= \left(\frac{2}{3}e^{-t}\sin 3t - 2e^{-t}\cos 3t\right) + 2\left(1 - \frac{2}{3}e^{-t}\sin 3t\right)$$

$$+10\left\{-\frac{1}{5} + t + \frac{1}{5}e^{-t}\left(\cos 3t + \frac{1}{3}\sin 3t\right)\right\} = 10t$$

RHS: $10t$

Therefore, LHS = RHS.

Figure 4.4 shows the solution in Example 4.8.

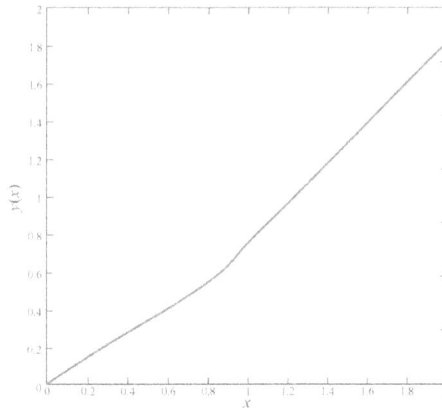

Figure 4.4 The solution in Example 4.8.

4.2.4 *System analysis*

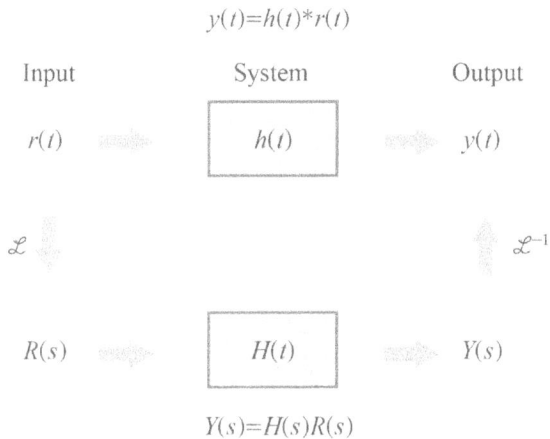

$$y(t)=h(t)*r(t)$$

Input	System	Output
$r(t)$	$h(t)$	$y(t)$

$\mathcal{L}$ $\mathcal{L}^{-1}$

$R(s)$	$H(t)$	$Y(s)$

$$Y(s)=H(s)R(s)$$

Figure 4.5 System analysis.

In a general linear system, the input-system-output relationship can be represented as shown in Figure 4.5. The relationship between input $r(t)$, system $h(t)$, and output $y(t)$ in the t-domain is expressed as $y(t) = h(t)*r(t)$, where $*$ denotes *convolution*.

$$y(t) = h(t)*r(t) = \int_0^t h(\tau)r(t-\tau)d\tau = \int_0^t h(t-\tau)r(\tau)d\tau \qquad (4.19)$$

It would be very challenging to derive the system function $h(t)$ from Eq. (4.19) in t-domain.

On the other hand, in the s-domain, the relationship between the input $R(s)$, *transfer function $H(s)$*, and output $Y(s)$ is as follows:

$$Y(s) = H(s)R(s) \tag{4.20a}$$

or

$$H(s) = \frac{R(s)}{Y(s)}. \tag{4.20b}$$

Therefore, analyzing system characteristics in the s-domain is indeed much easier than in the t-domain, highlighting the advantage of the Laplace transform.

4.2.5 (*optional) Solution of an ODE with t-shifted initial values

In Section 4.2.3, we examined the ODEs with the initial values $y(0) = y_0$, $y'(0) = y_0'$. However, in practical problems, there may be the ODEs with the initial values $y(t_0) = y_0, y'(t_0) = y_0'$ for $t_0 > 0$.

$$y'' + ay' + by = r(t), \quad y(t_0) = y_0, \quad y'(t_0) = y_0' \tag{4.21}$$

When we let $\tau = t - t_0(>0)$ in Eq. (4.21), then $t = t_0$ corresponds to $\tau = 0$. Furthermore, if we denote the t-shifted function by $\tilde{y}(\tau)$ in the direction of t by t_0 in $y(t)$, we get

$$y(t) = y(\tau + t_0) = \tilde{y}(\tau). \tag{4.22}$$

And then Eq. (4.21) becomes

$$\tilde{y}'' + a\tilde{y}' + b\tilde{y} = r(\tau + t_0), \quad \tilde{y}(0) = y_0, \quad \tilde{y}'(0) = y_0'. \tag{4.23}$$

If we denote the Laplace transform of $\tilde{y}(\tau)$ as $\tilde{Y}(s)$, we get

$$\mathcal{L}\{y(t)\} = \mathcal{L}\{\tilde{y}(\tau)\} = \tilde{Y}(s). \tag{4.24}$$

Then the Laplace transform of Eq. (4.23) becomes

$$\left[s^2\tilde{Y} - sy_0 - y_0'\right] + a\left[s\tilde{Y} - y_0\right] + b\tilde{Y} = \mathcal{L}\{r(\tau + t_0)\}. \tag{4.25}$$

Example 4.9

Solve the following ODE using the Laplace transform.

$$y'' - 2y' - 3y = 3t, \quad y(1) = 0, \quad y'(1) = -2$$

Solution

Let $\tau = t - 1 (> 0)$, then $t = \tau + 1$.

If we denote the t-shifted function by $\tilde{y}(\tau)$ in the direction of t by 1 in $y(t)$, we get

$$y(t) = y(\tau + 1) = \tilde{y}(\tau).$$

And the given ODE becomes

$$\tilde{y}'' - 2\tilde{y}' - 3\tilde{y} = 3(\tau + 1), \quad \tilde{y}(0) = 0, \quad \tilde{y}'(0) = -2.$$

Then the Laplace transform of the given ODE becomes

$$\left[s^2 \tilde{Y} - s \cdot 0 - (-2) \right] - 2 \left[s\tilde{Y} - 0 \right] - 3\tilde{Y} = \frac{3}{s^2} + \frac{3}{s}.$$

Collecting the $\tilde{Y}$-terms, we have the subsidiary equation

$$\left(s^2 - 2s - 3 \right) \tilde{Y} = \frac{3}{s^2} + \frac{3}{s} - 2.$$

Then this gives the solution

$$\tilde{Y} = \frac{-2s^2 + 3s + 3}{s^2 (s+1)(s-3)},$$

or

$$\tilde{Y}(s) = -\frac{1}{3s} - \frac{1}{s^2} + \frac{1}{2(s+1)} - \frac{1}{6(s-3)}.$$

And the inverse Laplace transform is

$$\tilde{y}(\tau) = -\frac{1}{3} - \tau + \frac{1}{2}e^{-\tau} - \frac{1}{6}e^{3\tau}.$$

Substituting $\tau = t - 1$ into this equation yields

$$y(t) = -\frac{1}{3} - (t-1) + \frac{1}{2}e^{-(t-1)} - \frac{1}{6}e^{3(t-1)}.$$

$$\textbf{Answer } y(t) = \frac{2}{3} - t + \frac{1}{2}e^{-(t-1)} - \frac{1}{6}e^{3(t-1)}$$

Check

Let
$$y(t) = \frac{2}{3} - t + \frac{1}{2}e^{-(t-1)} - \frac{1}{6}e^{3(t-1)}.$$

Then $y(1) = 0$,
$$y'(t) = -1 - \frac{1}{2}e^{-(t-1)} - \frac{1}{2}e^{3(t-1)}.$$

And then $y'(1) = -2$,
$$y''(t) = \frac{1}{2}e^{-(t-1)} - \frac{3}{2}e^{3(t-1)}.$$

LHS: $y'' - 2y' - 3y$

$$= \left(\frac{1}{2}e^{-(t-1)} - \frac{3}{2}e^{3(t-1)} \right) - 2\left(-1 - \frac{1}{2}e^{-(t-1)} - \frac{1}{2}e^{3(t-1)} \right)$$

$$-3\left\{ \frac{2}{3} - t + \frac{1}{2}e^{-(t-1)} - \frac{1}{6}e^{3(t-1)} \right\}$$

$$= 3t$$

RHS: $3t$

Therefore, LHS = RHS.

Figure 4.6 shows the solution in Example 4.9.

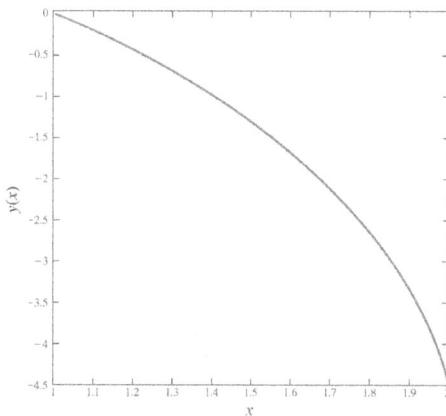

Figure 4.6 The solution in Example 4.9.

Problem 4.2

Find the Laplace transform. [1 ~ 6]

1. $f(t) = t\sin 3t$
2. $f(t) = te^{3t}$
3. $f(t) = t\cosh\omega t$
4. $f(t) = \cos^2\omega t$
5. $f(t) = \sin^2\omega t$
6. $f(t) = \sinh^2\omega t$

Find the inverse Laplace transform. [7 ~ 12]

7. $F(s) = \dfrac{3}{s^2 + 3s}$

8. $F(s) = \dfrac{1}{s^3 - s^2}$

9. $F(s) = \dfrac{9}{s(s^2 + 9)}$

10. $F(s) = \dfrac{4}{s(s^2 - 4)}$

11. $F(s) = \dfrac{5}{s(s^2 + 2s + 5)}$

12. $F(s) = \dfrac{4s + 8}{s^2(s^2 + 4)}$

Solve the IVP using the Laplace transform. [13 ~ 20]

13. $y' + 3y = 0$, $y(0) = 1$
14. $y' - y = \sin t$, $y(0) = 2$
15. $y'' - y' - 6y = 0$, $y(0) = 0$, $y'(0) = 1$
16. $y'' + 4y = 5e^{-t}$, $y(0) = 0$, $y'(0) = 0$
17. $y'' - 4y = 16\cos 2t$, $y(0) = 1$, $y'(0) = 0$
18. $y'' - 4y' + 4y = 0$, $y(0) = 2$, $y'(0) = 0$
19. $y'' + 3y' + 2y = 2t + 1$, $y(0) = 0$, $y'(0) = 0$
20. $y'' + 2y' + y = 4e^{-t}$, $y(0) = 1$, $y'(0) = 0$

Solve the IVP using the Laplace transform. [21 ~ 24]

21. (*optional) $y' - 3y = 0$, $y(1) = 2$
22. (*optional) $y'' + 2y' - 3y = 0$, $y(1) = 1$, $y'(1) = 2$
23. (*optional) $y'' + 2y' + 5y = 5t$, $y(2) = 0$, $y'(2) = 0$
24. (*optional) $y'' - 3y' - 4y = 2e^{t-2}$, $y(2) = 1$, $y'(2) = -1$

4.3 Unit step function and Dirac Delta function

Let's study the Laplace transform of the ODEs containing unit step functions and delta functions, which are difficult to find solutions using conventional methods.

4.3.1 Unit step function

Remark Unit step function

Unit step function (or Heaviside function) $u(t - a)$, $(a \geq 0)$ is defined as follows:

$$u(t - a) = \begin{cases} 1 & (t > a) \\ 0 & (t < a) \end{cases} \tag{4.26}$$

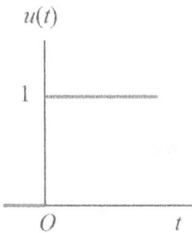

Figure 4.7 Unit step function $u(t)$.

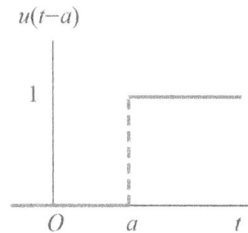

Figure 4.8 Unit step function $u(t - a)$.

Figure 4.7 shows the special step function $u(t)$, which has its jump at $a = 0$, Figure 4.8 shows the general step function $u(t - a)$ for an arbitrary positive a.

The unit step function is a highly useful function in engineering application, particularly in mechanical forces and electrical signals. Utilizing the unit step function, various signals can be manipulated as shown in Figure 4.9.

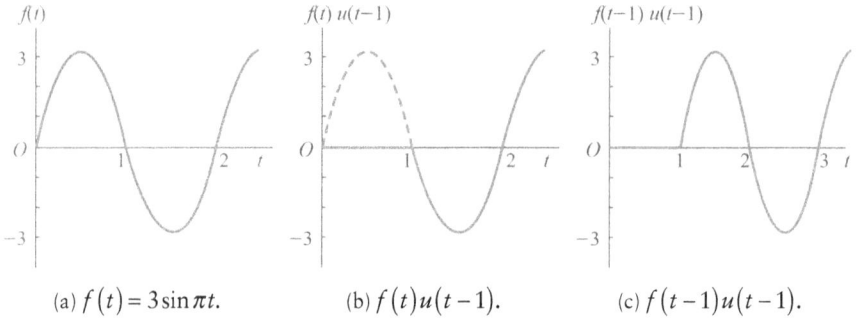

(a) $f(t) = 3\sin \pi t$.   (b) $f(t)u(t-1)$.   (c) $f(t-1)u(t-1)$.

Figure 4.9 Application I of the unit step function

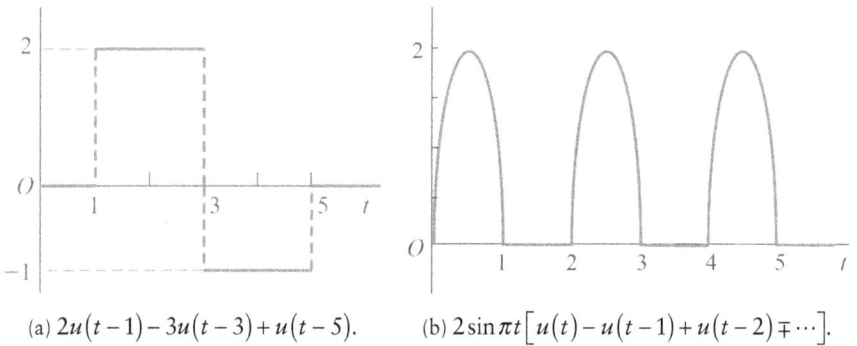

(a) $2u(t-1) - 3u(t-3) + u(t-5)$.   (b) $2\sin \pi t\left[u(t) - u(t-1) + u(t-2) \mp \cdots\right]$.

Figure 4.10 Application II of the unit step function.

Figure 4.9(a) shows a typical harmonic signal $f(t) = 3\sin \pi t$, while Figure 4.9(b) illustrates the selection of time intervals from the original function by multiplying it with a unit step function $u(t-1)$. Here, $u(t-1)$ equals 0 if $t < 1$ and 1 if $t > 1$. Figure 4.9(c) shows a signal shifted in time to the right by an amount of $t = 1$, with interval selection facilitated by the unit step function.

Figure 4.10 shows another application of the unit step function. In Figure 4.10(a), combinations of step functions are displayed, while Figure 4.10(b) shows a half-sine wave signal commonly encountered in electrical engineering.

The Laplace transform of the unit step function $u(t-a)$ is as follows:

$$\mathcal{L}\{u(t-a)\} = \int_0^\infty u(t-a)e^{-st}\, dt$$

$$= \int_0^a 0 \cdot e^{-st}\, dt + \int_a^\infty 1 \cdot e^{-st}\, dt = -\frac{e^{-st}}{s}\bigg|_a^\infty = \frac{e^{-as}}{s}. \qquad (s > 0) \qquad (4.27)$$

4.3.2   t-shifting: Replacing t by t − a in f(t)

In Section 4.1.5, we learned about the *s*-shifting by replacing *s* with $s - a$. Here, let's summarize the *t*-shifting by replacing *t* with $t - a$.

Remark   *t*-shifting

If $f(t)$ has the Laplace transform $F(s)$, then

$$\mathcal{L}\{f(t-a)u(t-a)\} = e^{-as}F(s) \qquad (4.28)$$

where $\tilde{f}(t) = f(t-a)u(t-a) = \begin{cases} 0 & (t < a) \\ f(t-a) & (t > a) \end{cases}$.

Or, if we take the inverse Laplace transform on both sides, we obtain

$$f(t-a)u(t-a) = \mathcal{L}^{-1}\{e^{-as}F(s)\}. \qquad (4.29)$$

Let's use the definition of the Laplace transform, writing τ for t,

$$F(s) = \int_0^\infty f(\tau)e^{-s\tau}\, d\tau.$$

Multiplying $F(s)$ by e^{-as} becomes

$$e^{-as}F(s) = e^{-as}\int_0^\infty f(\tau)e^{-s\tau}\, d\tau = \int_0^\infty f(\tau)e^{-s(\tau+a)}\, d\tau.$$

Substituting $t = \tau + a$, then $\tau = t - a$ and $d\tau = dt$ in the integral, we obtain

$$e^{-as}F(s) = \int_a^\infty f(t-a)e^{-st}\, dt$$

$$= \int_0^\infty f(t-a)u(t-a)e^{-st}\, dt = \mathcal{L}\{f(t-a)u(t-a)\}. \qquad (4.29)$$

Example 4.10

Find the Laplace transform of the function $f(t-1)u(t-1)$, where $f(t) = 3\sin \pi t$.

Solution

The Laplace transform of $f(t) = 3\sin \pi t$ is

$$\frac{3\pi}{s^2 + \pi^2}.$$

Therefore, we obtain

$$\mathcal{L}\{f(t-1)u(t-1)\} = e^{-s}\frac{3\pi}{s^2 + \pi^2}.$$

$$\textbf{Answer } \frac{3\pi e^{-s}}{s^2 + \pi^2}$$

Example 4.11

Write the given function $f(t)$ – (refer to Figure 4.11) into an expression using unit step functions and find its Laplace transform.

$$f(t) = \begin{cases} 1 & (0 < t < 1) \\ -t + 2 & (1 < t < 3). \\ \cos\pi t & (t > 3) \end{cases}$$

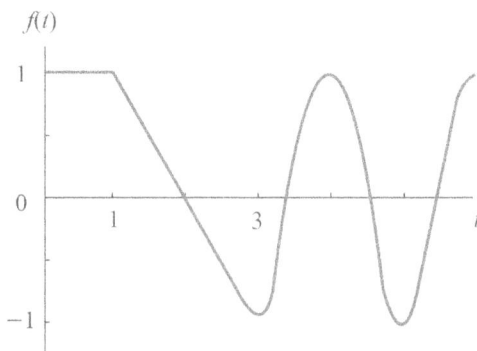

Figure 4.11 The function in Example 4.11.

Solution

In terms of unit step functions, we obtain

$$f(t) = \{1 - u(t-1)\} - (t-2)\{u(t-1) - u(t-3)\} + \cos \pi t \cdot u(t-3)$$

$$= 1 - (t-1)u(t-1) + (t-2+\cos \pi t)u(t-3)$$

$$= 1 - (t-1)u(t-1) + \{(t-3) + 1 - \cos \pi (t-3)\} u(t-3).$$

Its Laplace transform is

$$F(s) = \frac{1}{s} - \frac{1}{s^2} e^{-s} + \left(\frac{1}{s^2} + \frac{1}{s} - \frac{s}{s^2 + \pi^2} \right) e^{-3s}.$$

$$\textbf{Answer } F(s) = \frac{1}{s} - \frac{1}{s^2} e^{-s} + \left(\frac{1}{s^2} + \frac{1}{s} - \frac{s}{s^2 + \pi^2} \right) e^{-3s}$$

Example 4.12

Find the inverse Laplace transform.

$$F(s) = \left(\frac{1}{s^2 + \pi^2} \right) e^{-s} + \left(\frac{1}{s^2 + \pi^2} \right) e^{-2s} + \frac{1}{(s+1)^2} e^{-3s}$$

Solution

Since $\mathcal{L}^{-1} \left(\frac{1}{s^2 + \pi^2} \right) = \frac{1}{\pi} \sin \pi t$ and $\mathcal{L}^{-1} \left\{ \frac{1}{(s+1)^2} \right\} = te^{-t}$, we obtain

$$f(t) = \frac{1}{\pi} \sin \pi (t-1) \cdot u(t-1) + \frac{1}{\pi} \sin \pi (t-2) \cdot u(t-2) + (t-3)e^{-(t-3)} \cdot u(t-3)$$

$$\textbf{Answer } f(t) = \begin{cases} 0 & (0 < t < 1) \\ -\dfrac{1}{\pi} \sin \pi t & (1 < t < 2) \\ 0 & (2 < t < 3) \\ (t-3)e^{-t+3} & (t > 3) \end{cases}$$

Figure 4.12 shows the solution in Example 4.12.

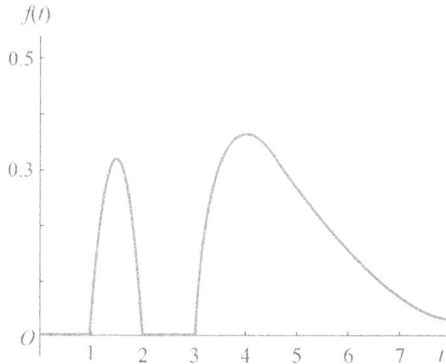

Figure 4.12 The solution in Example 4.12.

Example 4.13

Solve the following ODE:

$$y'' + 4y' + 3y = 3\{u(t-1) - u(t-2)\}, \quad y(0) = 0, \quad y'(0) = 0.$$

Solution

The Laplace transform of the given ODE $y'' + 4y' + 3y = 3\{u(t-1) - u(t-2)\}$ is

$$\{s^2 Y(s) - sy(0) - y'(0)\} + 4\{sY(s) - y(0)\} + 3Y(s) = \frac{3}{s}\left(e^{-s} - e^{-2s}\right).$$

Collecting the $Y(s)$-terms, we have the subsidiary equation

$$Y(s) = \frac{3}{s(s+1)(s+3)}\left(e^{-s} - e^{-2s}\right).$$

The following equation can be expanded into partial fractions:

$$\frac{3}{s(s+1)(s+3)} = \frac{a}{s} + \frac{b}{s+1} + \frac{c}{s+3}.$$

By comparing coefficients, you can achieve

$$a = 1, \quad b = -\frac{3}{2}, \quad c = \frac{1}{2}.$$

Then this gives the solution

$$Y(s) = \left\{ \frac{1}{s} - \frac{3}{2(s+1)} + \frac{1}{2(s+3)} \right\} \left(e^{-s} - e^{-2s} \right).$$

And the inverse Laplace transform of the above equation is

$$\frac{1}{s} - \frac{3}{2(s+1)} + \frac{1}{2(s+3)} \xrightarrow{\mathcal{L}^{-1}} 1 - \frac{3}{2}e^{-t} + \frac{1}{2}e^{-3t}.$$

Therefore, we obtain

$$y(t) = \left\{ u(t-1) - u(t-2) \right\} - \frac{3}{2}\left\{ e^{-(t-1)}u(t-1) - e^{-(t-2)}u(t-2) \right\}$$

$$+ \frac{1}{2}\left\{ e^{-3(t-1)}u(t-1) - e^{-3(t-2)}u(t-2) \right\}.$$

$$\textbf{Answer } y(t) = \begin{cases} 0 & (0 < t < 1) \\ 1 - \frac{3}{2}e^{-(t-1)} + \frac{1}{2}e^{-3(t-1)} & (1 < t < 2) \\ -\frac{3}{2}\left[e^{-(t-1)} - e^{-(t-2)} \right] + \frac{1}{2}\left[e^{-3(t-1)} - e^{-3(t-2)} \right] & (t > 2) \end{cases}$$

Figure 4.13 shows the solution in Example 4.13.

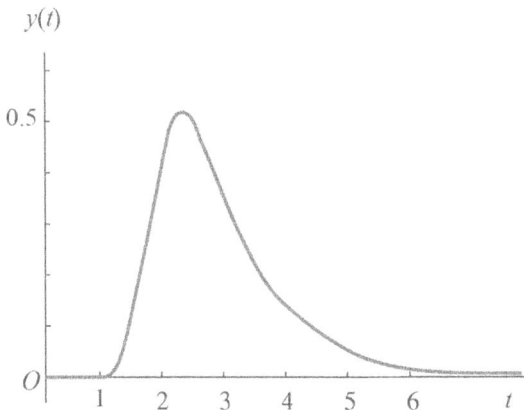

Figure 4.13 The solution in Example 4.13.

4.3.3   *Dirac delta function*

> **Remark   Dirac delta function**
>
> The function defined as follows is called the *Dirac delta function* or unit impulse function.
>
> $$\delta(t-a) = \lim_{\varepsilon \to 0} \frac{1}{\varepsilon}\left[u(t-a)-u\{t-(a+\varepsilon)\}\right] \qquad (4.30)$$

Figure 4.14 Unit step function $\dfrac{1}{\varepsilon}\left[u(t-a)-u\{t-(a+\varepsilon)\}\right]$.

Figure 4.14 shows the shape of the Dirac delta function, where its magnitude at $t = a$ (where a is arbitrary) is infinite, yet it forms a rectangular shape with an overall area of 1. Therefore, the Dirac delta function $\delta(t-a)$, defined as Eq. (4.31), is not a function in the ordinary sense as used in calculus, but so-called generalized function.

$$\delta(t-a) = \begin{cases} \infty & (t=a) \\ 0 & \text{otherwise} \end{cases} \quad \text{and} \quad \int_0^\infty \delta(t-a)\,dt = 1 \qquad (4.31)$$

Specifically, the Dirac delta function $\delta(t)$ when $a = 0$ is expressed as follows.

$$\delta(t) = \begin{cases} \lim_{\varepsilon \to 0} \dfrac{1}{\varepsilon} & (0 < t < \varepsilon) \\ 0 & \text{otherwise} \end{cases} \quad \text{and} \quad \int_0^\infty \delta(t)\,dt = 1 \qquad (4.32)$$

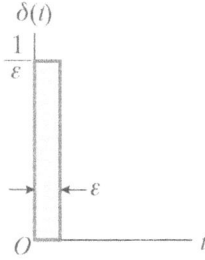

Figure 4.15 Dirac delta function $\delta(t)$.

Figure 4.15 shows the Dirac delta function $\delta(t-a)$ with $a = 0$. From the definition of the Dirac delta function

$$\delta(t-a) = \lim_{\varepsilon \to 0} \frac{1}{\varepsilon}\left[u(t-a) - u\{t-(a+\varepsilon)\}\right], \qquad (4.30)$$

we obtain

$$\int_0^\infty \delta(t-a)g(t)dt = \lim_{\varepsilon \to 0} \frac{1}{\varepsilon}\int_0^\infty \left[u(t-a)g(t) - u\{t-(a+\varepsilon)\}g(t)\right]dt$$

$$= \lim_{\varepsilon \to 0} \frac{1}{\varepsilon}\int_a^{a+\varepsilon} g(t)dt$$

$$= g(a).$$

Therefore, for a continuous function $g(t)$, the Dirac delta function $\delta(t-a)$ has the following property:

$$\int_0^\infty \delta(t-a)g(t)dt = g(a). \qquad (4.33)$$

Remark Laplace Transform of the Dirac Delta Function

The Laplace transform of the Dirac delta function is as follows:

$$\mathcal{L}\{\delta(t)\} = 1 \qquad (4.34a)$$

$$\mathcal{L}\{\delta(t-a)\} = e^{-as}. \qquad (4.34b)$$

By the definition of the Laplace transform, we obtain

$$L\{\delta(t-a)\} = \int_0^\infty \delta(t-a)e^{-st}\,dt.$$

Substituting $g(t) = e^{-st}$ into Eq. (4.33), we obtain

$$L\{\delta(t-a)\} = g(a) = e^{-as}. \tag{4.34}$$

Example 4.14

Solve the following ODE:

$$y'' + 4y' + 3y = \delta(t-1), \quad y(0) = 0, \quad y'(0) = 0.$$

Solution

The Laplace transform of the given ODE $y'' + 4y' + 3y = \delta(t-1)$ is

$$\left[s^2Y(s) - sy(0) - y'(0)\right] + 4\left[sY(s) - y(0)\right] + 3Y(s) = e^{-s}.$$

Collecting the $Y(s)$-terms, we have the subsidiary equation

$$Y(s) = \frac{e^{-s}}{(s+1)(s+3)}.$$

The following equation can be expanded into partial fractions:

$$\frac{1}{(s+1)(s+3)} = \frac{1}{2}\left(\frac{1}{s+1} - \frac{1}{s+3}\right).$$

Then we obtain

$$Y(s) = \frac{1}{2}\left(\frac{1}{s+1} - \frac{1}{s+3}\right)e^{-s}.$$

And the inverse Laplace transform of the above equation is

$$\frac{1}{2}\left(\frac{1}{s+1} - \frac{1}{s+3}\right) \xrightarrow{\;L^{-1}\;} \frac{1}{2}\left(e^{-t} - e^{-3t}\right).$$

Therefore, we obtain the following answer:

$$\text{Answer } y(t) = \frac{1}{2}\left\{e^{-(t-1)} - e^{-3(t-1)}\right\}u(t-1)$$

$$= \begin{cases} 0 & (0 < t < 1) \\ \frac{1}{2}\left\{e^{-(t-1)} - e^{-3(t-1)}\right\} & (t > 1) \end{cases}$$

Check

i. $0 < t < 1$, $\quad y(t) = 0$

ii. $t > 1$

The answer is $y(t) = \frac{1}{2}e^{-(t-1)} - \frac{1}{2}e^{-3(t-1)}$.

Its derivatives are

$$y' = -\frac{1}{2}e^{-(t-1)} + \frac{3}{2}e^{-3(t-1)},$$

$$y'' = \frac{1}{2}e^{-(t-1)} - \frac{9}{2}e^{-3(t-1)}.$$

Inserting y, y', and y'' into the given ODE, we obtain

$$y'' + 4y' + 3y = 0.$$

Upon inspection, it is noted that the term $\delta(t-1)$ cannot be found.

Figure 4.16 shows the solution in Example 4.14.

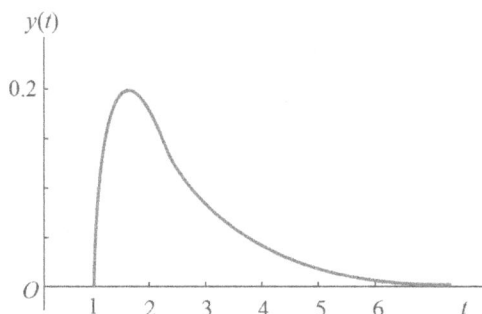

Figure 4.16 The solution in Example 4.14.

Table 4.4 summarizes the Laplace transforms of the step function and the delta function.

Table 4.4 Laplace transforms of the step function and the delta function

$f(t)$	$F(s)$
$u(t)$	$\dfrac{1}{s}$
$u(t-a)$	$\dfrac{e^{-as}}{s}$
$f(t-a)u(t-a)$	$F(s)e^{-as}$
$\delta(t)$	1
$\delta(t-a)$	e^{-as}

Problem 4.3

Find the Laplace transform of the following equation. [1 ~ 6]

1. $f(t) = \begin{cases} t & (0 < t < 2) \\ 0 & (t > 2) \end{cases}$

2. $f(t) = \begin{cases} 0 & (0 < t < 1) \\ t-1 & (t > 1) \end{cases}$

3. $f(t) = \begin{cases} \sin t & (0 < t < \pi) \\ 0 & (t > \pi) \end{cases}$

4. $f(t) = \begin{cases} e^{-t} & (0 < t < 1) \\ 0 & (t > 1) \end{cases}$

5. $f(t) = \begin{cases} 0 & (0 < t < 1) \\ t^2 & (1 < t < 2) \\ 0 & (t > 2) \end{cases}$

6. $f(t) = \begin{cases} t^2 & (0 < t < 1) \\ 0 & (1 < t < 2) \\ t & (t > 2) \end{cases}$

Find the inverse Laplace transform of the following equation. [7 ~ 12]

7. $F(s) = \dfrac{e^{-2s}}{(s-1)^2}$

8. $F(s) = \dfrac{\pi(1+e^{-s})}{s^2+\pi^2}$

9. $F(s) = \dfrac{e^{-2s}-e^{-3s}}{s}$

10. $F(s) = \dfrac{e^{-2s}}{s^4}$

11. $F(s) = \dfrac{e^{-s}-e^{-2s}}{s^2-4}$

12. $F(s) = \dfrac{2}{(s+1)^2+4}\{1+e^{-\pi s}\}$

Find the solution to the ODE using the Laplace transform. [13 ~ 20]

13. $y'' + 4y = \begin{cases} 8t & (0 < t < 1) \\ 8 & (t > 1) \end{cases}$ $y(0) = 0, \quad y'(0) = 0$

14. $y'' + 4y = \begin{cases} 8\cos t & (0 < t < \pi) \\ 0 & (t > \pi) \end{cases}$ $y(0) = 0, \quad y'(0) = 0$

15. $y'' + 3y' + 2y = \begin{cases} 1 & (0 < t < 1) \\ 0 & (t > 1) \end{cases}$ $y(0) = 0, \quad y'(0) = 0$

16. $y'' + 3y' + 2y = \begin{cases} 2t & (0 < t < 1) \\ 4 & (t > 1) \end{cases}$ $y(0) = 0, \quad y'(0) = 0$

17. $y'' - y' - 2y = \begin{cases} 1 & (0 < t < 2) \\ 0 & (t > 2) \end{cases}$ $y(0) = 0, \quad y'(0) = 1$

18. $y'' - y' - 2y = \begin{cases} 2\sin t & (0 < t < 2\pi) \\ 0 & (t > 2\pi) \end{cases}$ $y(0) = 0, \quad y'(0) = 0$

19. (*optional) $y'' + 4y = \begin{cases} 2t & (0 < t < 2) \\ 0 & (t > 2) \end{cases}$ $y(1) = 0, \quad y'(1) = 1$

20. (*optional) $y'' + 2y' + 2y = \begin{cases} 10\sin t & (0 < t < 2\pi) \\ 0 & (t > 2\pi) \end{cases}$ $y(\pi) = 0, \quad y'(\pi) = 0$

Find the solution to the ODE using the Laplace transform. [21 ~ 28]

21. $y'' + 2y' + 10y = \delta(t)$ $y(0) = 0,$ $y'(0) = 0$

22. $y'' + 4y = \delta(t - \pi)$ $y(0) = 0,$ $y'(0) = 0$

23. $y'' + 16y = 4\delta(t - 2\pi)$ $y(0) = 1,$ $y'(0) = 0$

24. $y'' + 2y' + 5y = \delta(t - 1)$ $y(0) = 0,$ $y'(0) = 1$

25. $y'' + 2y' + 2y = 5\sin t + 8\delta(t - \pi)$ $y(0) = 1,$ $y'(0) = 1$

26. $y'' + 3y' + 2y = \delta(t - \pi) + u(t - 2\pi)$ $y(0) = 0,$ $y'(0) = 0$

27. $y'' + 4y' + 5y = 2e^{-t} + \delta(t - 1)$ $y(0) = 0,$ $y'(0) = 0$

28. $y'' + 3y' + 2y = 4t - 6\delta(t - 1)$ $y(0) = 0,$ $y'(0) = 1$

4.4 Convolution, integral equations, and differentiation and integral of transforms

4.4.1 *Convolution*

Remark   Convolution

If two functions $f(t)$ and $g(t)$ are continuous, the *convolution* of $f(t)$ and $g(t)$, as defined in Eq. (4.19), is as follows:

$$h(t) = f(t) * g(t) = \int_0^t f(\tau)g(t - \tau)d\tau \qquad (4.35a)$$

or

$$h(t) = f(t) * g(t) = \int_0^\infty f(t - \tau)g(\tau)d\tau \qquad (4.35b)$$

If the Laplace transforms of the functions $h(t)$, $f(t)$, and $g(t)$ are $H(s)$, $F(s)$, and $G(s)$, respectively, we obtain

$$H(s) = F(s)G(s). \qquad (4.36)$$

Denoting by τ and u in the Laplace forms $F(s)$ and $G(s)$, respectively, then

$$F(s) = \int_0^\infty e^{-s\tau} f(\tau) d\tau,$$

$$G(s) = \int_0^\infty e^{-su} g(u) du.$$

We now set $t = \tau + u$, where τ is at first constant. Then $u = t - \tau$, and t varies from τ to ∞. Then

$$G(s) = \int_\tau^\infty e^{-s(t-\tau)} g(t-\tau) dt = e^{s\tau} \int_\tau^\infty e^{-st} g(t-\tau) dt.$$

Since τ in $F(s)$ and t in $G(s)$ vary independently, multiplying $F(s)$ by $G(s)$ is

$$F(s)G(s) = \int_0^\infty e^{-s\tau} f(\tau) d\tau \, e^{s\tau} \int_\tau^\infty e^{-st} g(t-\tau) dt$$

$$= \int_0^\infty f(\tau) \int_\tau^\infty e^{-st} g(t-\tau) dt \, d\tau.$$

Here we integrate for fixed τ over t from τ to ∞ and then over τ from 0 to ∞. Then we integrate first over τ from 0 to ∞ and then over t from 0 to ∞, that is,

$$F(s)G(s) = \int_0^\infty e^{-st} \int_0^t f(\tau) g(t-\tau) d\tau \, dt$$

$$= \int_0^\infty e^{-st} \{f(t) * g(t)\} dt$$

$$= \int_0^\infty e^{-st} h(t) dt = H(s). \tag{4.36}$$

From the definition, convolution has the following properties:

$$
\begin{array}{lll}
f * g = g * f & \text{(commutative law)} & (4.37) \\
(f * g) * r = f * (g * r) & \text{(associative law)} & (4.38) \\
(f_1 + f_2) * g = f_1 * g + f_2 * g & \text{(distributive law)} & (4.39) \\
f * 0 = 0 * f = 0. & & (4.40)
\end{array}
$$

Example 4.15

Find the inverse Laplace transform of the following equation:

$$H(s) = \frac{s^2}{(s^2 + \omega^2)^2}.$$

Solution

Since

$$\frac{s^2}{(s^2 + \omega^2)^2} = \frac{s}{s^2 + \omega^2} \frac{s}{s^2 + \omega^2}$$

and

$$f(t) = \mathcal{L}^{-1}\left\{ \frac{s}{s^2 + \omega^2} \right\} = \cos\omega t,$$

we obtain

$$h(t) = \cos\omega t * \cos\omega t = \int_0^t \cos\omega\tau \cos\omega(t - \tau)d\tau$$

$$= \int_0^t \cos\omega\tau (\cos\omega t \cos\omega\tau + \sin\omega t \sin\omega\tau)d\tau$$

$$= \cos\omega t \int_0^t \cos^2\omega\tau\, d\tau + \sin\omega t \int_0^t \cos\omega\tau \sin\omega\tau\, d\tau$$

$$= \cos\omega t \int_0^t \frac{1 + \cos 2\omega\tau}{2}d\tau + \sin\omega t \int_0^t \frac{\sin 2\omega\tau}{2}d\tau$$

$$= \cos\omega t \left(\frac{t}{2} + \frac{\sin 2\omega t}{4\omega} \right) + \sin\omega t \left(\frac{1 - \cos 2\omega t}{4\omega} \right)$$

$$= \frac{t}{2}\cos\omega t + \frac{1}{4\omega}\sin\omega t + \frac{1}{4\omega}(\sin 2\omega t \cos\omega t - \cos 2\omega t \sin\omega t)$$

$$= \frac{t}{2}\cos\omega t + \frac{1}{2\omega}\sin\omega t.$$

Answer $h(t) = \frac{t}{2}\cos\omega t + \frac{1}{2\omega}\sin\omega t$

Table 4.5 summarizes the inverse Laplace transforms using convolution.

Table 4.5 Inverse Laplace transforms using convolution

$F(s)$	$f(t)$
$\dfrac{1}{(s^2+\omega^2)^2}$	$\dfrac{1}{2\omega^3}(-\omega t\cos\omega t+\sin\omega t)$
$\dfrac{s}{(s^2+\omega^2)^2}$	$\dfrac{t}{2\omega}\sin\omega t$
$\dfrac{s^2}{(s^2+\omega^2)^2}$	$\dfrac{1}{2\omega}(\omega t\cos\omega t+\sin\omega t)$

4.4.2 Integral equations

In solving *integral equations* with integrals, convolution is sometimes useful for finding solutions more easily.

Example 4.16

Solve the following integral equation:

$$y(t)-\int_0^t y(\tau)\sin(t-\tau)d\tau=1.$$

Solution

The given integral equation can be written as a convolution

$$y(t)-y(t)*\sin t=1.$$

and its Laplace transform is

$$Y(s)-Y(s)\frac{1}{s^2+1}=\frac{1}{s}.$$

Collecting the $Y(s)$-terms, we have the subsidiary equation

$$Y(s)=\frac{s^2+1}{s^3}=\frac{1}{s}+\frac{1}{s^3}.$$

Therefore, the inverse Laplace transform of the above equation is

$$y(t)=1+\frac{t^2}{2}.$$

$$\textbf{Answer } y(t)=1+\frac{t^2}{2}$$

Another Solution

Since $\sin(t-\tau) = \sin t \cos \tau - \cos t \sin \tau$, the given equation is

$$y(t) - \sin t \int_0^t y(\tau) \cos \tau \, d\tau + \cos t \int_0^t y(\tau) \sin \tau \, d\tau = 1. \qquad ①$$

The differentiation of Eq. ① yields

$$y' - \cos t \int_0^t y(\tau) \cos \tau \, d\tau - \sin t \cdot y \cos t - \sin t \int_0^t y(\tau) \sin \tau \, d\tau + \cos t \cdot y \sin t = 0,$$

or

$$y' - \cos t \int_0^t y(\tau) \cos \tau \, d\tau - \sin t \int_0^t y(\tau) \sin \tau \, d\tau = 0. \qquad ②$$

And the differentiation of Eq. ② yields

$$y'' - y + \sin t \int_0^t y(\tau) \cos \tau \, d\tau - \cos t \int_0^t y(\tau) \sin \tau \, d\tau = 0. \qquad ③$$

Adding equations ① and ③ yields

$$y'' = 1 \qquad ④$$

Then we obtain

$$y' = t + C_1$$

and

$$y(t) = \frac{t^2}{2} + C_1 t + C_2.$$

From equations ① and ②, we get

$$y(0) = 1, \quad y'(0) = 0,$$

and then

$$C_1 = 0, \quad C_2 = 1.$$

Therefore, we obtain

$$y(t) = \frac{t^2}{2} + 1.$$

4.4.3   Differentiation of Laplace transforms

Remark   Differentiation of Laplace transforms

Laplace transform $F(s)$ of Eq. (4.1) differentiated with respect to s yields

$$F'(s) = \frac{dF(s)}{ds} = -\int_0^\infty \{tf(t)\} e^{-st}\, dt. \tag{4.41}$$

If $\mathcal{L}\{f(t)\} = F(s)$, then

$$\mathcal{L}\{tf(t)\} = -F'(s) \tag{4.42a}$$

$$\mathcal{L}^{-1}\{F'(s)\} = -tf(t) \tag{4.42b}$$

Example 4.17

Find the Laplace transform of the following equation:

$$f(t) = t\sin\omega t.$$

Solution

Since $\mathcal{L}(\sin\omega t) = \dfrac{\omega}{s^2 + \omega^2}$, then

$$\mathcal{L}(t\sin\omega t) = -\frac{d}{ds}\left(\frac{\omega}{s^2+\omega^2}\right) = \frac{2\omega s}{(s^2+\omega^2)^2}.$$

$$\text{Answer } F(s) = \frac{2\omega s}{(s^2+\omega^2)^2}$$

Check

Since $\mathcal{L}(t\sin\omega t) = \dfrac{2\omega s}{(s^2+\omega^2)^2}$, then

$$\mathcal{L}^{-1}\left\{\frac{s}{(s^2+\omega^2)^2}\right\} = \frac{1}{2\omega} t\sin\omega t.$$

4.4   Integration of Laplace transforms

Remark   Integration of Laplace transforms

When $\lim_{t \to +0} \dfrac{f(t)}{t}$ exists, then for $s > 0$,

$$\mathcal{L}\left\{\frac{f(t)}{t}\right\} = \int_s^\infty F(\tilde{s})\, d\tilde{s} \qquad\qquad (4.43\text{a})$$

or

$$\mathcal{L}^{-1}\left\{\int_s^\infty F(\tilde{s})\, d\tilde{s}\right\} = \frac{f(t)}{t} \qquad\qquad (4.43\text{b})$$

From the definition of Laplace transform $F(\tilde{s}) = \displaystyle\int_0^\infty e^{-\tilde{s}t} f(t)\, dt$, we obtain

$$\int_s^\infty F(\tilde{s})\, d\tilde{s} = \int_s^\infty \left[\int_0^\infty e^{-\tilde{s}t} f(t)\, dt\right] d\tilde{s} = \int_0^\infty f(t)\left[\int_s^\infty e^{-\tilde{s}t}\, d\tilde{s}\right] dt.$$

Integrating $e^{-\tilde{s}t}$ with respect to $\tilde{s}$ gives $\dfrac{e^{-\tilde{s}t}}{-t}$, then

$$\int_s^\infty F(\tilde{s})\, d\tilde{s} = \int_0^\infty f(t)\left[\frac{e^{-\tilde{s}t}}{-t}\right]_s^\infty dt = \int_0^\infty e^{-st}\frac{f(t)}{t}\, dt = \mathcal{L}\left\{\frac{f(t)}{t}\right\}. \qquad (4.43)$$

Example 4.18

Find the inverse Laplace transform of the following equation:

$$F(s) = \ln\frac{s}{s+1}.$$

Solution

The given equation becomes

$$F(s) = \ln\frac{s}{s+1} = \ln s - \ln(s+1).$$

Its derivative is

$$F'(s) = \frac{1}{s} - \frac{1}{s+1}.$$

And the inverse Laplace transform of the above equation is

$$\mathcal{L}^{-1}\{F'(s)\} = 1 - e^{-t}.$$

Since $\mathcal{L}^{-1}\{F'(s)\} = -tf(t)$ in Eq. (4.43b), we get

$$-tf(t) = 1 - e^{-t}.$$

Therefore,

$$f(t) = \frac{e^{-t} - 1}{t}.$$

$$\textbf{Answer } f(t) = \frac{e^{-t} - 1}{t}$$

Problem 4.4

Find the following convolution. [1 ~ 6]

1. $2 * t$
2. $t * t$
3. $1 * \cos \omega t$
4. $t * \sin \omega t$
5. $t * e^t$
6. $\cos t * \sin t$

Find the inverse Laplace transform of the following $F(s)$ using convolution. [7 ~ 12]

7. $\dfrac{1}{s(s+1)}$

8. $\dfrac{1}{(s-a)(s-b)}$

9. $\dfrac{s}{(s^2 + \omega^2)^2}$

10. $\dfrac{1}{(s^2+\omega^2)^2}$

11. $\dfrac{1}{s(s^2-1)}$

12. $\dfrac{e^{-as}}{s(s-1)}$

Solve the integral equation. [13 ~ 20]

13. $y(t)+\displaystyle\int_0^t y(\tau)d\tau = 2$

14. $y(t)+\displaystyle\int_0^t (t-\tau)y(\tau)d\tau = 1$

15. $y(t)+\displaystyle\int_0^t y(\tau)e^{t-\tau}\,d\tau = t$

16. $y(t)+\displaystyle\int_0^t (t-\tau)y(\tau)d\tau = \sin t$

17. $y(t)+\displaystyle\int_0^t y(\tau)\cos(t-\tau)d\tau = 1$

18. $y(t)-2\displaystyle\int_0^t y(\tau)\cos(t-\tau)d\tau = \sin 2t$

19. $y(t)-\displaystyle\int_0^t (1+\tau)y(t-\tau)d\tau = 1-\sinh t$

20. $y(t)+\displaystyle\int_0^t e^{2(t-\tau)}y(\tau)d\tau = \left(t^2-3t+\dfrac{3}{2}\right)u(t-1)+\dfrac{1}{2}e^{2(t-1)}u(t-1)$

Find the inverse Laplace transform. [21 ~ 30]

21. $F(s)=\ln\left(\dfrac{s+1}{s-1}\right)$

22. $F(s)=\ln\left(s+\dfrac{\omega^2}{s}\right)$

23. $F(s)=\ln\left(1+\dfrac{4}{s}+\dfrac{5}{s^2}\right)$

24. $F(s)=\ln\left(\dfrac{s^2+1}{(s-1)^2}\right)$

25. $F(s)=\ln\left(\dfrac{s+3}{(s^2+2s+5)^2}\right)$

26. $F(s) = \ln\left(\dfrac{s^2 + 2s + 3}{(s^2 + 2s + 5)^2}\right)$

27. $\dfrac{s}{(s^2 + \omega^2)^2}$

28. $\dfrac{1}{(s^2 + \omega^2)^2}$

29. $\dfrac{s^2}{(s^2 + \omega^2)^2}$

30. $\dfrac{6s}{(s^2 - 9)^2}$

Find the Laplace transform. [31 ~ 36]

31. $t \sin 2t$
32. $t \sinh 2t$
33. $te^{-t} \sin t$
34. $te^{-2t} \cos 3t$
35. $t^2 e^{3t}$
36. $t^2 \cos 3t$

4.5 Application of Laplace transforms

Laplace transform is extensively utilized in solving the ODEs derived from various application fields such as electrical engineering, mechanical engineering, control engineering, and more.

Example 4.19

The mechanical system consists of a mass of 1 kg, damping of 4 Ns/m, and a spring constant of 104 N/m, as shown in Figure 4.17. Find the output displacement when an impulse $F = \delta(t) [\text{N}]$ is applied instantaneously. Initial conditions are $x(0) = 0$ and $\dot{x}(0) = 0$.

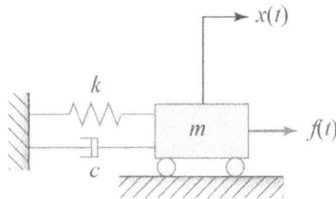

Figure 4.17 Damped vibration system with an external force.

Solution

In equation of motion $m\ddot{x} + c\dot{x} + kx = f(t)$, we get

$$\ddot{x} + 4\dot{x} + 104x = \delta(t). \qquad \text{①}$$

Obtaining the general solution to this equation without using the Laplace transform might not be easy.

Using $\mathcal{L}\{\delta(t)\} = 1$, the Laplace transform of Eq. ① is

$$\{s^2 X(s) - sx(0) - \dot{x}(0)\} + 4\{sX(s) - x(0)\} + 104X(s) = 1. \qquad \text{②}$$

Substituting initial conditions $x(0) = 0$ and $\dot{x}(0) = 0$ into Eq. ②, we obtain

$$X(s) = \frac{1}{s^2 + 4s + 104} = \frac{1}{(s+2)^2 + 10^2}. \qquad \text{③}$$

Therefore, the inverse Laplace transform of Eq. ③ is

$$x(t) = 0.1e^{-2t} \sin 10t.$$

$$\textbf{Answer } x(t) = 0.1e^{-2t} \sin 10t \ [\text{m}]$$

Figure 4.18 shows the solution in Example 4.19.

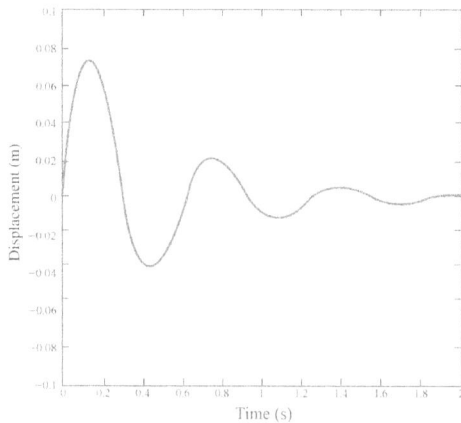

Figure 4.18 The solution in Example 4.19.

Example 4.20

Find the current $i(t)$ in the LC-circuit with $E(t) = 4$ V, $L = 1$ H, and $C = 0.04$ F, as shown in Figure 4.19. Initial charge is $Q(0) = 0$ C and initial current is $i(0) = \dfrac{dQ(0)}{dt} = 0$ A.

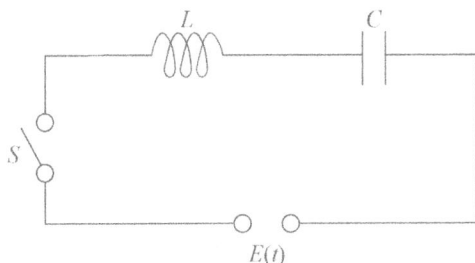

Figure 4.19 LC-circuit.

Solution

The voltage $E(t)$ is the sum of the voltage drops across the inductor L, denoted as $E_L(t)\left(= L\dfrac{di}{dt}\right)$, and the capacitor C, denoted as $E_C(t)\left(= \dfrac{1}{C}\int i\,dt\right)$

$$E(t) = E_L(t) + E_C(t) = L\frac{di}{dt} + \frac{1}{C}\int i\,dt.$$

Then we obtain

$$\frac{di}{dt} + \frac{1}{0.04}\int i\,dt = 4. \qquad \qquad ①$$

Its Laplace transform is

$$sI(s) - i(0) + 25\frac{I(s)}{s} = \frac{4}{s}. \qquad \qquad ②$$

Substituting initial conditions $i(0) = 0$ into Eq. ②, we obtain

$$I(s) = \frac{4}{s^2 + 25}. \qquad \qquad ③$$

Therefore, the inverse Laplace transforms of Eq. ③ is

$$i(t) = 0.8 \sin 5t.$$

Answer $i(t) = 0.8 \sin 5t \ [\text{A}]$

Figure 4.20 shows the solution in Example 4.20.

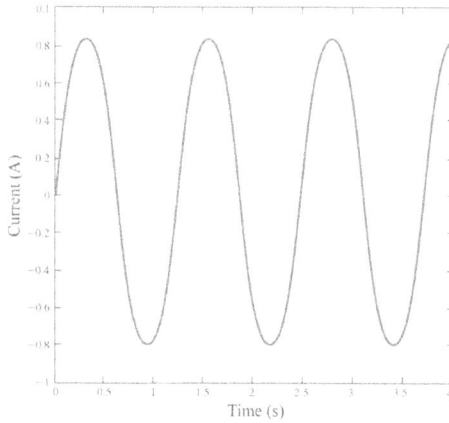

Figure 4.20 The solution in Example 4.20.

Example 4.21

Find the current $i(t)$ in the RLC-circuit with $R = 1 \ \Omega$, $L = 2$ H, $C = 1$ F, and $E(t) = \cos 2t$ V, as shown in Figure 4.21. Initial charge is $Q(0) = 0$ C and initial current is $i(0) = \dfrac{dQ(0)}{dt} = 0$ A.

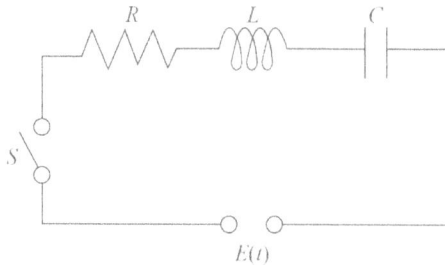

Figure 4.21 RLC-circuit.

Solution

The voltage $E(t)$ is the sum of the voltage drops across the inductor L, denoted as $E_L(t)\left(=L\dfrac{di}{dt}\right)$, the voltage drops across the resistor R, denoted as $E_R(t)(=Ri)$, and the capacitor C, denoted as $E_C(t)\left(=\dfrac{1}{C}\int i\,dt\right)$

$$E(t) = L\frac{di}{dt} + Ri + \frac{1}{C}\int i\,dt.$$

Then we obtain

$$\frac{di}{dt} + 2i + \int i\,dt = \cos 2t. \qquad \text{①}$$

Its Laplace transform is

$$\{sI(s) - i(0)\} + 2I(s) + \frac{I(s)}{s} = \frac{s}{s^2+4}. \qquad \text{②}$$

Substituting initial condition $i(0) = 0$ into Eq. ②, we obtain

$$I(s) = \frac{s^2}{(s+1)^2(s^2+4)}, \qquad \text{③}$$

or

$$I(s) = -\frac{8}{25(s+1)} + \frac{1}{5(s+1)^2} + \frac{8s+12}{25(s^2+4)}. \qquad \text{④}$$

Therefore, the inverse Laplace transforms of Eq. ④ is

$$i(t) = -\frac{8}{25}e^{-t} + \frac{1}{5}te^{-t} + \frac{2}{25}(4\cos 2t + 3\sin 2t).$$

Answer $i(t) = -\dfrac{8}{25}e^{-t} + \dfrac{1}{5}te^{-t} + \dfrac{2}{25}(4\cos 2t + 3\sin 2t)\,[\text{A}]$

Figure 4.22 shows the solution in Example 4.21.

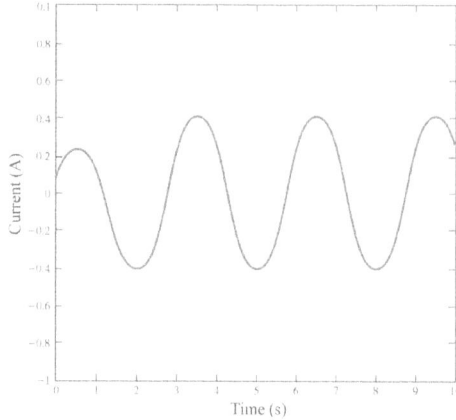

Figure 4.22 The solution in Example 4.21.

Problem 4.5

Solve the following applications. [1 ~ 6]

1. **Vibration**   The mechanical system consists of a mass of 1 kg, damping of 2 Ns/m, and a spring constant of 101 N/m, as shown in Figure 4.23. Find the output displacement when a harmonic external force $F = 5002 \sin t$ [mN] is applied. Initial conditions are $x(0) = 0$ and $\dot{x}(0) = 0$.

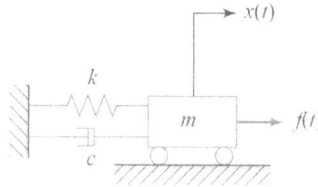

Figure 4.23 Damped vibration system with an external force.

2. **Vibration**   The mechanical system consists of a mass of 1 kg, damping of 4 Ns/m, and a spring constant of 104 N/m, as shown in Figure 4.24. Find the output displacement when a step function force $F = 104\, u(t-1)$ N is applied. Initial conditions are $x(0) = 0$ and $\dot{x}(0) = 0$.

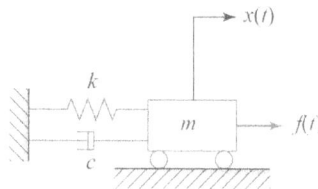

Figure 4.24 Damped vibration system with an external force.

3. **Electric circuit** Find the current $i(t)$ in the LC-circuit with $E(t)=10$ V, $L=1$ H, and $C=0.01$ F, as shown in Figure 4.25. Initial charge is $Q(0)=0$ C and initial current is $i(0)=\dfrac{dQ(0)}{dt}=0$ A.

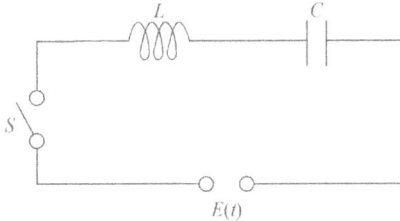

Figure 4.25 LC-circuit.

4. **Electric circuit** Find the current $i(t)$ in the LC-circuit with $L=1$ H, and $C=0.01$ F, as shown in Figure 4.26. The voltage applied to the circuit is $E(t)=-9900$ V for $\pi<t<3\pi$ and $E(t)=0$ V otherwise. Initial charge is $Q(0)=0$ C and initial current is $i(0)=\dfrac{dQ(0)}{dt}=0$ A.

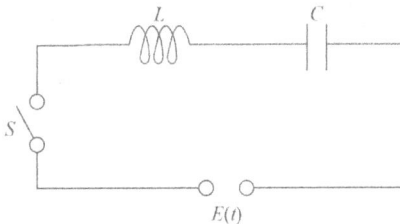

Figure 4.26 LC-circuit.

5. **Electric circuit** Find the current $i(t)$ in the RLC-circuit with $R=160$ Ω, $L=20$ H, $C=0.002$ F, and $E(t)=37\sin 10t$ V, as shown in Figure 4.27. Initial charge is $Q(0)=0$ C and initial current is $i(0)=\dfrac{dQ(0)}{dt}=0$ A.

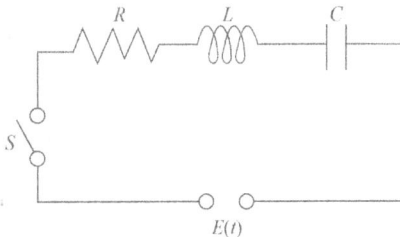

Figure 4.27 RLC-circuit.

6. **Electric circuit**   Find the current $i(t)$ in the *RLC*-circuit with $R = 2\,\Omega$, $L = 1$ H, and $C = 0.5$ F, as shown in Figure 4.28. The voltage applied to the circuit is $E(t) = 1$ kV for $0 < t < 2$ and $E(t) = 0$ V for $t > 2$. Initial charge is $Q(0) = 0$ C and initial current is $i(0) = \dfrac{dQ(0)}{dt} = 0$ A.

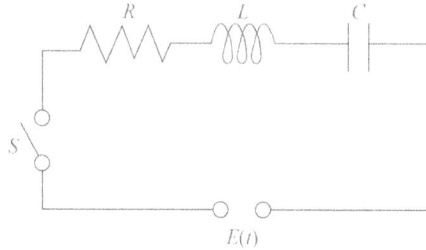

Figure 4.28 RLC-circuit.

Table 4.6 summarizes the Laplace transforms.

Table 4.6 Table of Laplace transforms

	$F(s) = \mathcal{L}\{f(t)\}$	$f(t)$
1	1	$\delta(t)$
2	$\dfrac{1}{s}$	1
3	$\dfrac{1}{s^2}$	t
4	$\dfrac{1}{s^n}$ $(n = 1, 2, \cdots)$	$\dfrac{t^{n-1}}{(n-1)!}$
5	$\dfrac{1}{s-a}$	e^{at}
6	$\dfrac{1}{(s-a)^2}$	te^{at}
7	$\dfrac{1}{(s-a)^n}$ $(n = 1, 2, \cdots)$	$\dfrac{t^{n-1}e^{at}}{(n-1)!}$
8	$\dfrac{1}{s^2+\omega^2}$	$\dfrac{1}{\omega}\sin\omega t$
9	$\dfrac{s}{s^2+\omega^2}$	$\cos\omega t$
10	$\dfrac{1}{s^2-\omega^2}$	$\dfrac{1}{\omega}\sinh\omega t$

(Continued)

Table 4.6 *(Continued)*

	$F(s) = \mathcal{L}\{f(t)\}$	$f(t)$
11	$\dfrac{s}{s^2 - \omega^2}$	$\cosh \omega t$
12	$\dfrac{1}{(s-a)^2 + \omega^2}$	$\dfrac{1}{\omega} e^{at} \sin \omega t$
13	$\dfrac{s}{(s-a)^2 + \omega^2}$	$e^{at} \cos \omega t$
14	$\dfrac{1}{s\left(s^2 + \omega^2\right)}$	$\dfrac{1}{\omega^2}(1 - \cos \omega t)$
15	$\dfrac{1}{s^2\left(s^2 + \omega^2\right)}$	$\dfrac{1}{\omega^3}(\omega t - \sin \omega t)$
16	$\dfrac{1}{(s^2 + \omega^2)^2}$	$\dfrac{1}{2\omega^3}(-\omega t \cos \omega t + \sin \omega t)$
17	$\dfrac{s}{(s^2 + \omega^2)^2}$	$\dfrac{t}{2\omega} \sin \omega t$
18	$\dfrac{s^2}{(s^2 + \omega^2)^2}$	$\dfrac{1}{2\omega}(\omega t \cos \omega t + \sin \omega t)$
19	$\dfrac{e^{-as}}{s}$	$u(t-a)$
20	e^{-as}	$\delta(t-a)$

4.6 Utilizing MATLAB®

Using the MATLAB® command 'laplace.m', we can obtain the Laplace transform of $f(t)$, and with 'ilaplace.m', we can obtain the inverse Laplace transform of $F(s)$.

M-Example 4.1

Find the Laplace transform using the MATLAB® command.

a. $f(t) = t + \sin at + e^{bt}$

b. $f(t) = t^2 e^{at} + e^{bt} \cos \omega t$

c. $f(t) = 2 + t \sin \omega t$

d. $f(t) = te^{-t} \cos t$

Solution

a. $f(t) = t + \sin at + e^{bt}$

```
>> syms a b t
>> laplace(t+sin(a*t)+exp(b*t))
```

ans =

    a/(a^2 + s^2) + 1/s^2 – 1/(b – s)

$$\text{Answer } F(s) = \frac{1}{s^2} + \frac{a}{s^2 + a^2} + \frac{1}{s - b}$$

b. $f(t) = t^2 e^{at} + e^{bt} \cos \omega t$

```
>> syms a b omega t
>> laplace(t^2 *exp(a*t)+exp(b*t)*cos(omega*t))
```

ans =

    – (b – s)/(omega^2 + (b – s)^2) – 2/(a – s)^3

$$\text{Answer } F(s) = \frac{2}{(s - a)^3} + \frac{s - b}{(s - b)^2 + \omega^2}$$

c. $f(t) = 2 + t \sin \omega t$

```
>> syms omega t
>> laplace(2+t*sin(omega*t))
```

ans =

    2/s + (2*omega*s)/(omega^2 + s^2)^2

$$\text{Answer } F(s) = \frac{2}{s} + \frac{2\omega s}{(s^2 + \omega^2)^2}$$

d. $f(t) = te^{-t} \cos t$

```
>> laplace(t*exp(-t)*cos(t))
```

ans =

    ((2*s + 2)*(s + 1))/((s + 1)^2 + 1)^2 – 1/((s + 1)^2 + 1)

```
>> simple(ans)
```

ans =
   (s*(s + 2))/(s^2 + 2*s + 2)^2

$$\textbf{Answer } F(s) = \frac{s(s+2)}{\left(s^2 + 2s + 2\right)^2}$$

M-Example 4.2

Find the inverse Laplace transform using the MATLAB® command.

a. $F(s) = \dfrac{3}{s\left(s^2 + 9\right)}$

b. $F(s) = \dfrac{1}{s\left(s^2 - 4\right)}$

c. $F(s) = \dfrac{1}{\left(s^2 + \omega^2\right)^2}$

d. $F(s) = \dfrac{s^2}{\left(s^2 + \omega^2\right)^2}$

Solution

a. $F(s) = \dfrac{3}{s\left(s^2 + 9\right)}$

```
syms s
ilaplace(3/s/(s^2+9))
```

ans =
   1/3 – cos(3*t)/3

$$\textbf{Answer } f(t) = \frac{1 - \cos 3t}{3}$$

b. $F(s) = \dfrac{1}{s(s^2 - 4)}$

---

syms s
ilaplace(1/s/(s^2−4))

---

ans =

    exp(−2*t)/8 + exp(2*t)/8 − 1/4

Answer $f(t) = \dfrac{1}{8}\left(e^{-2t} + e^{2t}\right) - \dfrac{1}{4}$

c. $F(s) = \dfrac{1}{(s^2 + \omega^2)^2}$

---

syms s omega
ilaplace(1/(s^2 +omega^2)^2)

---

ans =

    (sin(omega*t) − omega*t*cos(omega*t))/(2*omega^3)

Answer $f(t) = \dfrac{1}{2\omega^3}\left(\sin \omega t - \omega t \cos \omega t\right)$

d. $F(s) = \dfrac{s^2}{(s^2 + \omega^2)^2}$

>> syms s omega
>> ilaplace(s^2/(s^2 +omega^2)^2)

---

ans =

    sin(omega*t)/omega − (sin(omega*t) − omega*t*cos(omega*t))/(2*omega)

>> simple(ans)

---

ans =

    (t*cos(omega*t))/2 + sin(omega*t)/(2*omega)

Answer $f(t) = \dfrac{1}{2\omega}\left(\omega t \cos \omega t + \sin \omega t\right)$

Problem 4.6

Find the Laplace transform using the MATLAB® command. [1 ~ 4]

1. $f(t) = te^{at}$
2. $f(t) = t\cos\omega t$
3. $f(t) = \sinh\omega t$
4. $f(t) = t^2\cos t$

Find the inverse Laplace transform using the MATLAB® command. [5 ~ 8]

5. $F(s) = \dfrac{1}{(s-a)^2}$

6. $F(s) = \dfrac{1}{(s-a)^2 + \omega^2}$

7. $F(s) = \dfrac{1}{s^2(s^2 + \omega^2)}$

8. $F(s) = \dfrac{s}{(s^2 + \omega^2)^2}$

Answer

Problem 4.1

1. $F(s) = \dfrac{6\left(1 - s + s^2 + s^3\right)}{s^4}$

2. $F(s) = \dfrac{1}{s-2} + \dfrac{4}{s^2 + 2^2}$

3. $F(s) = \dfrac{s-2}{(s-2)^2 + 3^2} - \dfrac{1}{s}$

4. $F(s) = 2 \cdot \left(\dfrac{1}{s} + \dfrac{s}{s^2 + 6^2} + \dfrac{1}{s^3} \right)$

5. $F(s) = \dfrac{2(s+4)}{(s+3)^3}$

6. $F(s) = \dfrac{s+2}{s^2 + 4s - 5}$

7. $f(t) = (t-1)^2$

8. $f(t) = 2(1 + 3t)e^{3t}$

9. $f(t) = 3e^{-t}(-\cos 2t + \sin 2t)$

10. $f(t) = 2 + 3e^{-2t}$

11. $f(t) = e^{2t}\left(2t + 3t^2\right)$

12. $f(t) = 2 - e^{3t}(\cos t + \sin t)$

Problem 4.2

1. $F(s) = \dfrac{6s}{(s^2 + 9)^2}$

2. $F(s) = \dfrac{1}{(s-3)^2}$

3. $F(s) = \dfrac{s^2 + \omega^2}{(s^2 - \omega^2)^2}$

4. $F(s) = \dfrac{s^2 + 2\omega^2}{s\left(s^2 + 4\omega^2\right)}$

5. $F(s) = \dfrac{2\omega^2}{s\left(s^2 + 4\omega^2\right)}$

6. $F(s) = \dfrac{2\omega^2}{s\left(s^2 - 4\omega^2\right)}$

7. $f(t) = 1 - e^{-3t}$

8. $f(t) = -1 - t + e^t$

9. $f(t) = 1 - \cos 3t$

10. $f(t) = \cosh 2t - 1$

11. $f(t) = 1 - e^{-t}\left(\cos 2t + \frac{1}{2}\sin t\right)$

12. $f(t) = 1 + 2t - (\cos 2t + \sin 2t)$

13. $y(t) = e^{-3t}$

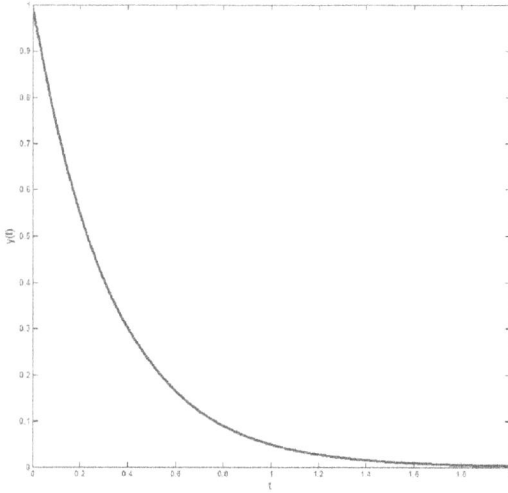

14. $y(t) = \frac{5}{2}e^t - \frac{1}{2}(\cos t + \sin t)$

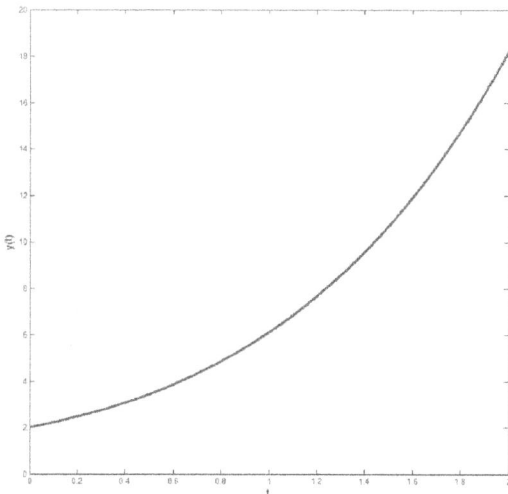

15. $y(t) = -\dfrac{1}{5}\left(e^{-2t} - e^{3t}\right)$

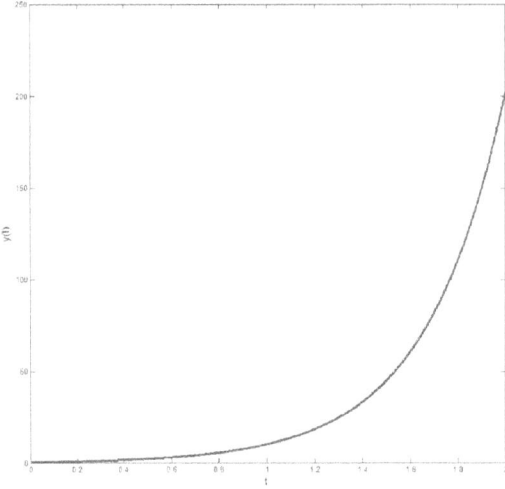

16. $y(t) = e^{-t} - \cos 2t + \dfrac{1}{2}\sin 2t$

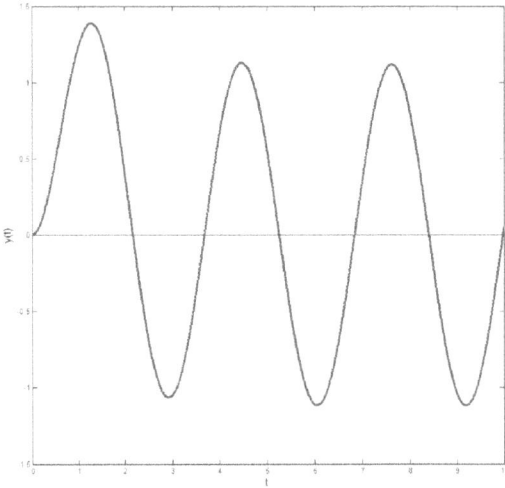

17. $y(t) = 3\cosh 2t - 2\cos 2t$

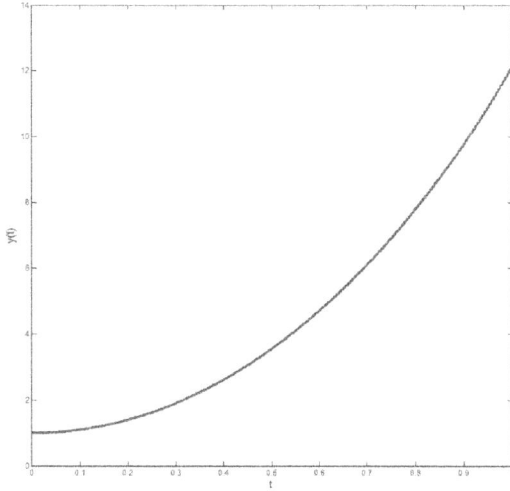

18. $y(t) = 2e^{2t} - 4te^{2t}$

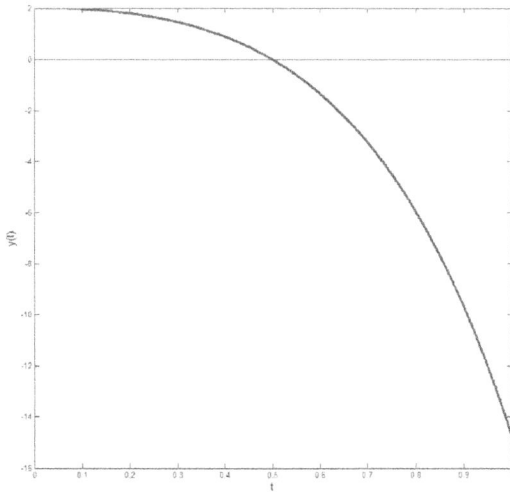

19. $y(t) = -1 + t + e^{-t}$

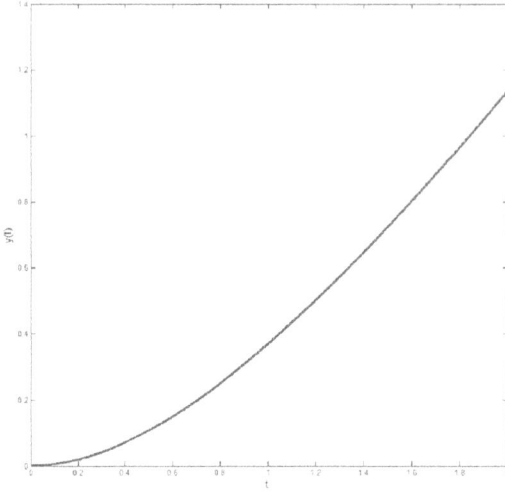

20. $y(t) = (2t^2 + t + 1)e^{-t}$

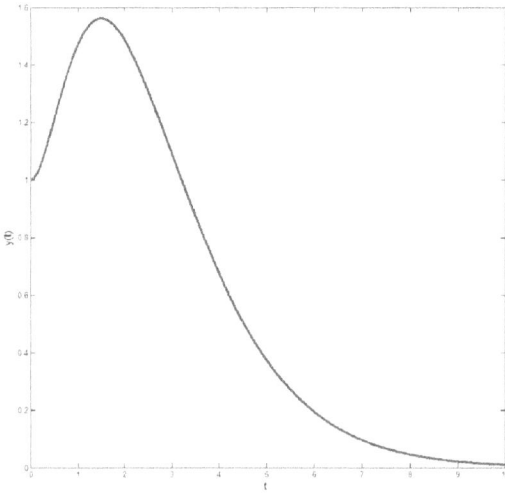

21. $y(t) = 2e^{3(t-1)}$

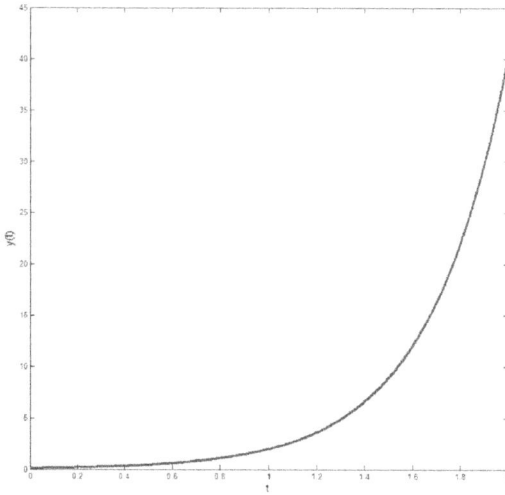

22. $y(t) = \frac{5}{4}e^{t-1} - \frac{1}{4}e^{-3(t-1)}$

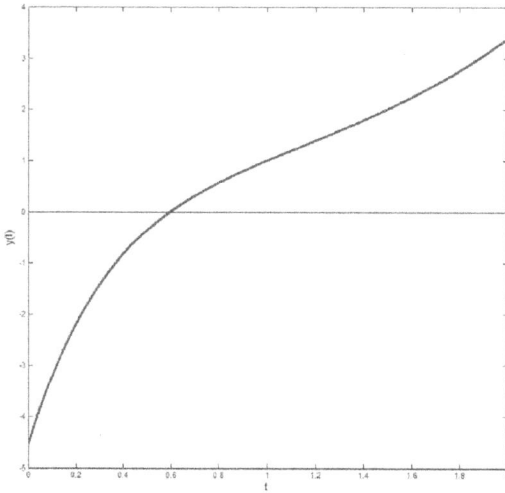

23. $y(t) = t - \dfrac{2}{5} - \dfrac{1}{5}e^{-(t-2)}\left\{8\cos 2(t-2) + \dfrac{13}{2}\sin 2(t-2)\right\}$

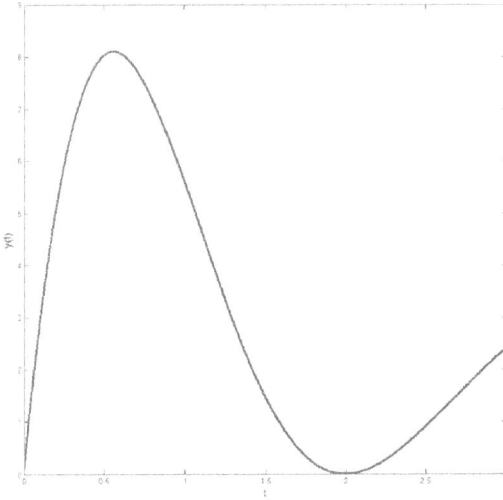

24. $y(t) = \dfrac{6}{5}e^{-(t-2)} - \dfrac{1}{3}e^{t-2} + \dfrac{2}{15}e^{4(t-2)}$

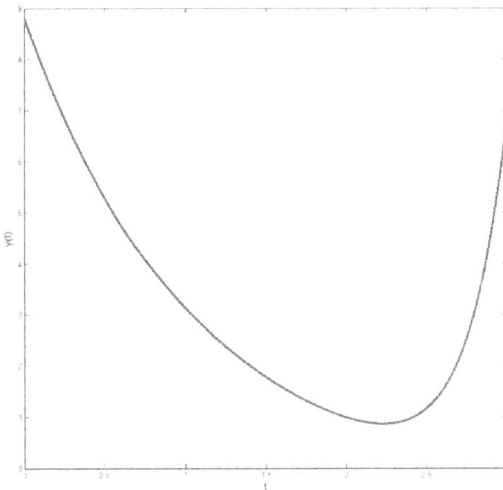

Problem 4.3

1. $F(s) = \dfrac{1}{s^2} - \left(\dfrac{1}{s^2} + \dfrac{2}{s}\right)e^{-2s}$

2. $F(s) = \dfrac{1}{s^2}e^{-s}$

3. $F(s) = \dfrac{s}{s^2+1}\left(1 + e^{-\pi s}\right)$

4. $F(s) = \dfrac{1}{s+1}\left\{1 - e^{-(s+1)}\right\}$

5. $F(s) = \left\{\dfrac{2}{s^3} + \dfrac{2}{s^2} + \dfrac{1}{s}\right\}e^{-s} - \left\{\dfrac{2}{s^3} + \dfrac{4}{s^2} + \dfrac{4}{s}\right\}e^{-2s}$

6. $F(s) = \dfrac{2}{s^3} - \left\{\dfrac{2}{s^3} + \dfrac{2}{s^2} + \dfrac{1}{s}\right\}e^{-s} + \left\{\dfrac{1}{s^2} + \dfrac{1}{s}\right\}e^{-2s}$

7. $f(t) = \begin{cases} 0 & (0 < t < 2) \\ (t-2)e^{t-2} & (t > 2) \end{cases}$

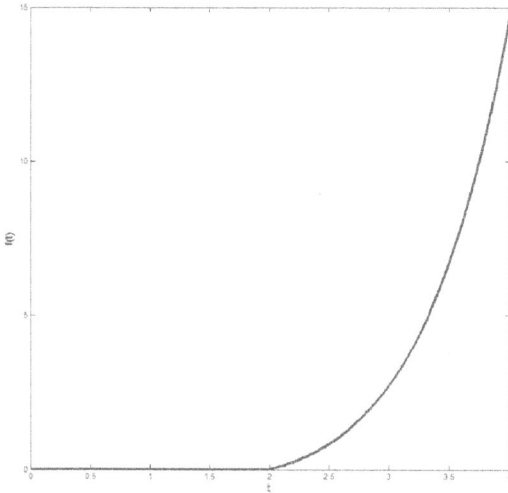

8. $f(t) = \begin{cases} \sin \pi t & (0 < t < 1) \\ 0 & (t > 1) \end{cases}$

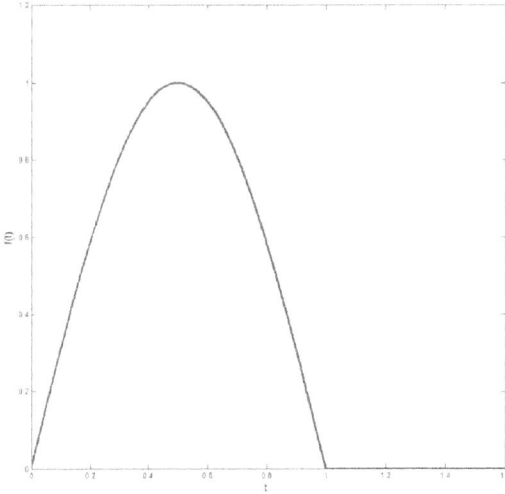

9. $f(t) = \begin{cases} 0 & (0 < t < 2) \\ 1 & (2 < t < 3) \\ 0 & (t > 3) \end{cases}$

10. $f(t) = \begin{cases} 0 & (0 < t < 2) \\ \dfrac{1}{6}(t-2)^3 & (t > 2) \end{cases}$

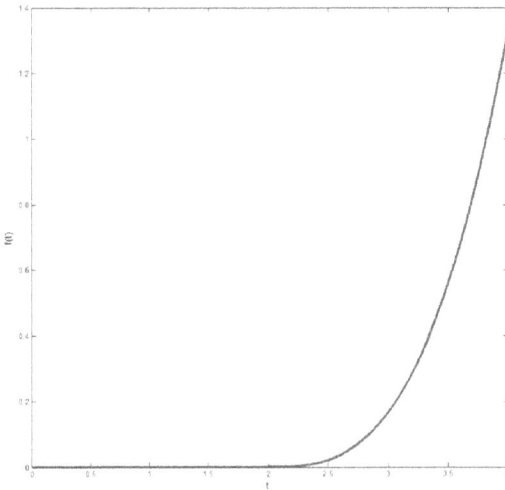

11. $f(t) = \begin{cases} 0 & (0 < t < 1) \\ \dfrac{1}{2}\sinh 2(t-1) & (1 < t < 2) \\ \dfrac{1}{2}\left\{\sinh 2(t-1) - \sinh 2(t-2)\right\} & (t > 2) \end{cases}$

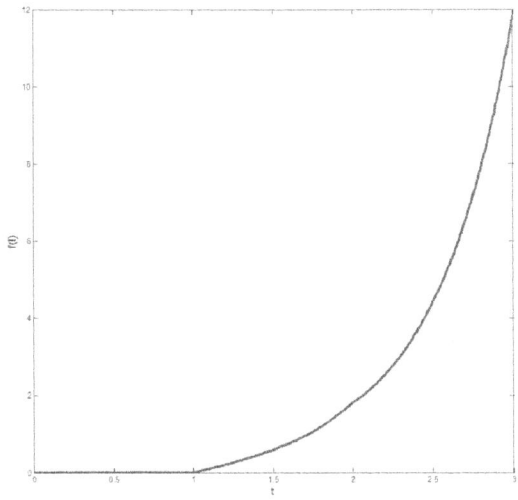

12. $f(t) = \begin{cases} e^{-t}\sin 2t & (0 < t < \pi) \\ \left\{e^{-t} + e^{-(t-\pi)}\right\}\sin 2t & (t > \pi) \end{cases}$

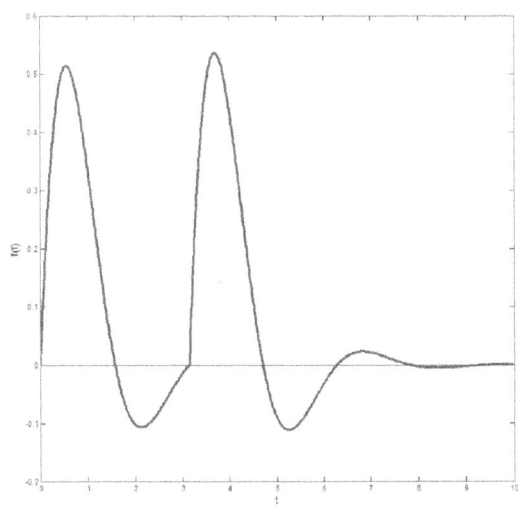

13. $y(t) = \begin{cases} 2t - \sin 2t & (0 < t < 1) \\ 2 - \sin 2t + \sin(2t - 2) & (t > 1) \end{cases}$

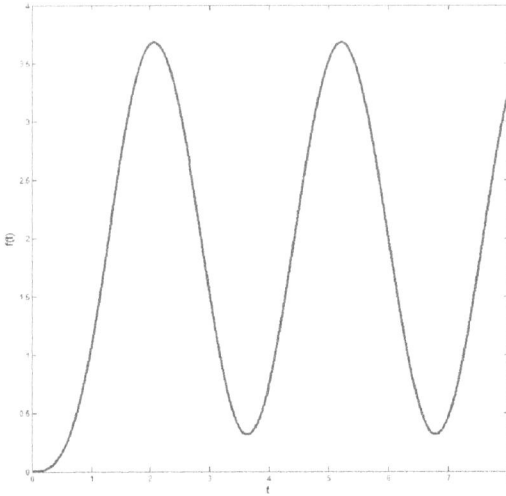

14. $y(t) = \begin{cases} \dfrac{8}{3}(\cos t - \cos 2t) & (0 < t < \pi) \\ -\dfrac{16}{3}\cos 2t & (t > \pi) \end{cases}$

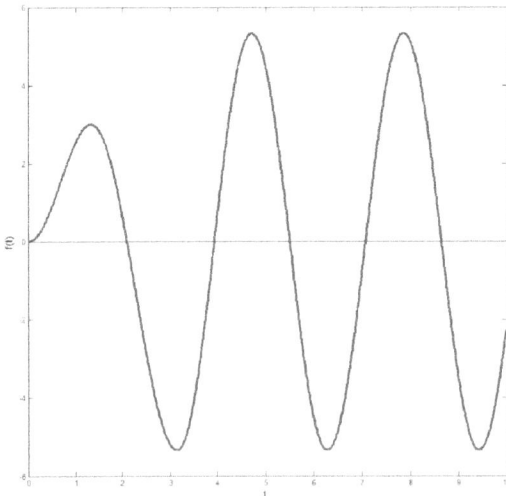

15. $y(t) = \begin{cases} \dfrac{1}{2}(1 - 2e^{-t} + e^{-2t}) & (0 < t < 1) \\[3mm] \dfrac{1}{2}\left\{-2e^{-t} + e^{-2t} + 2e^{-(t-1)} - e^{-2(t-1)}\right\} & (t > 1) \end{cases}$

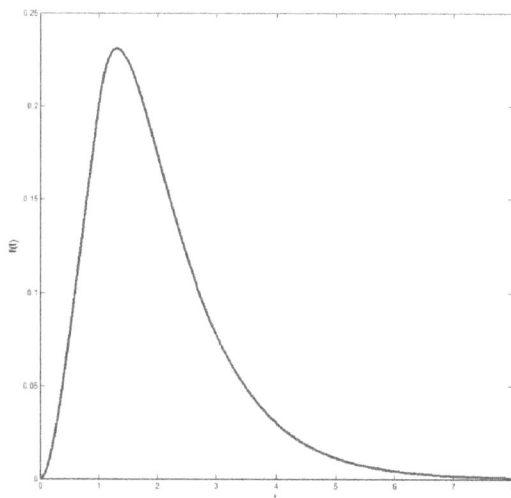

16. $y(t) = \begin{cases} -\dfrac{3}{2} + t + 2e^{-t} - \dfrac{1}{2}e^{-2t} & (0 < t < 1) \\[3mm] 2 + 2e^{-t} - \dfrac{1}{2}e^{-2t} - 4e^{-(t-1)} + \dfrac{3}{2}e^{-2(t-1)} & (t > 1) \end{cases}$

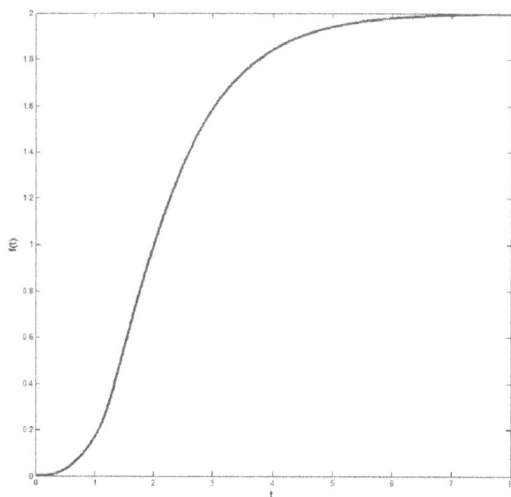

17. $y(t) = \begin{cases} -\dfrac{1}{2} + \dfrac{1}{2}e^{2t} & (0 < t < 2) \\ \dfrac{1}{2}e^{2t} - \dfrac{1}{3}e^{-(t-2)} - \dfrac{1}{6}e^{2(t-2)} & (t > 2) \end{cases}$

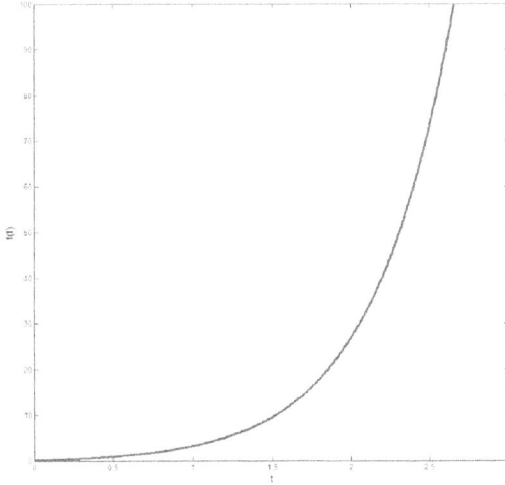

18. $y(t) = \begin{cases} -\dfrac{1}{3}e^{-t} + \dfrac{2}{15}e^{2t} + \dfrac{1}{5}(\cos t - 3\sin t) & (0 < t < 2\pi) \\ -\dfrac{1}{3}e^{-t} + \dfrac{1}{3}e^{-(t-2\pi)} + \dfrac{2}{15}e^{2t} - \dfrac{2}{15}e^{2(t-2\pi)} & (t > 2\pi) \end{cases}$

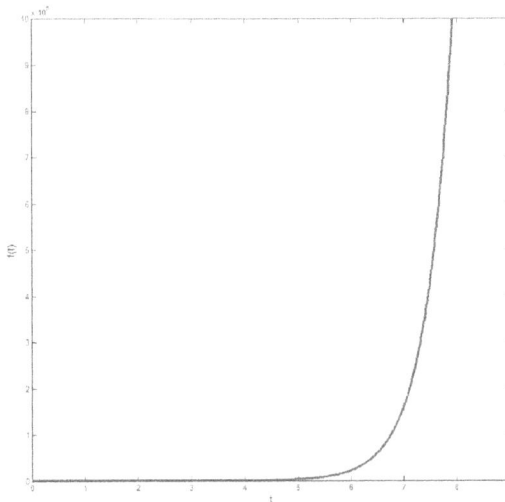

19. $y(t) = \begin{cases} \dfrac{1}{2}t - \dfrac{1}{2}\cos 2(t-1) + \dfrac{1}{4}\sin 2(t-1) & (0 < t < 2) \\[3mm] -\dfrac{1}{2}\cos 2(t-1) + \dfrac{1}{4}\sin 2(t-1) + \cos 2(t-2) + \dfrac{1}{4}\sin 2(t-2) & (t > 2) \end{cases}$

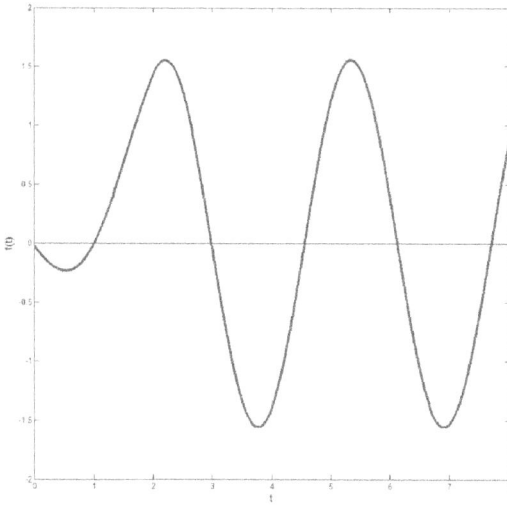

20. $y(t) = \begin{cases} -4\cos t + 2\sin t + e^{-(t-\pi)}(4\cos t + 2\sin t) & (0 < t < 2\pi) \\[3mm] \left\{ e^{-(t-\pi)} - e^{-(t-2\pi)} \right\}(4\cos t + 2\sin t) & (t > 2\pi) \end{cases}$

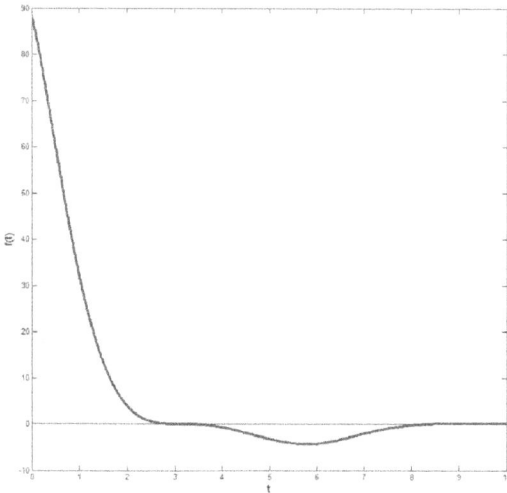

21. $y(t) = \dfrac{1}{3} e^{-t} \sin 3t \qquad (t > 0)$

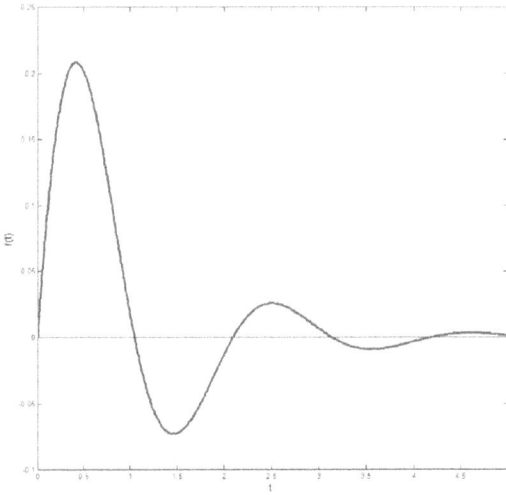

22. $y(t) = \begin{cases} 0 & (0 < t < \pi) \\ \dfrac{1}{2} \sin 2t & (t > \pi) \end{cases}$

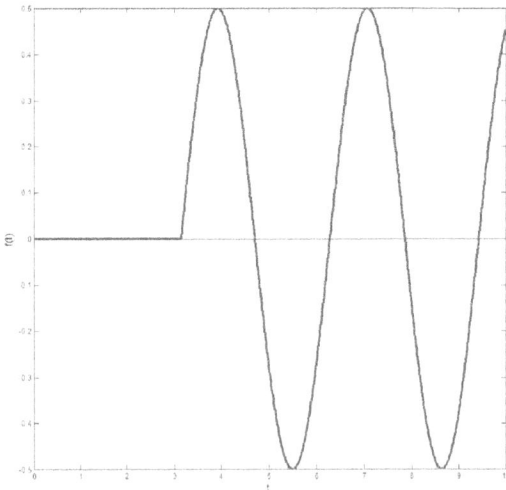

23. $y(t) = \begin{cases} \cos 4t & (0 < t < 2\pi) \\ \cos 4t + \sin 4t & (t > 2\pi) \end{cases}$

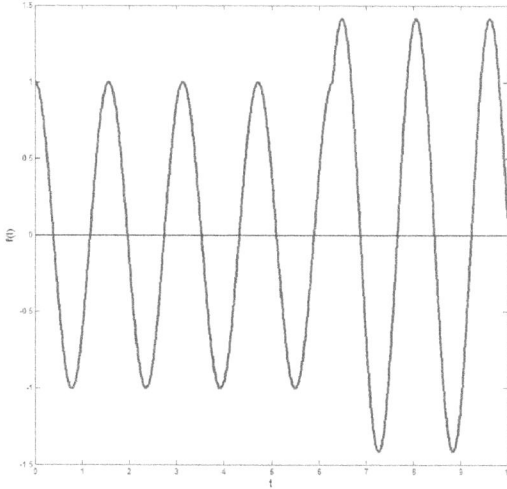

24. $y(t) = \begin{cases} \dfrac{1}{2}e^{-t}\sin 2t & (0 < t < 1) \\ \dfrac{1}{2}e^{-t}\sin 2t + \dfrac{1}{2}e^{-(t-1)}\sin 2(t-1) & (t > 1) \end{cases}$

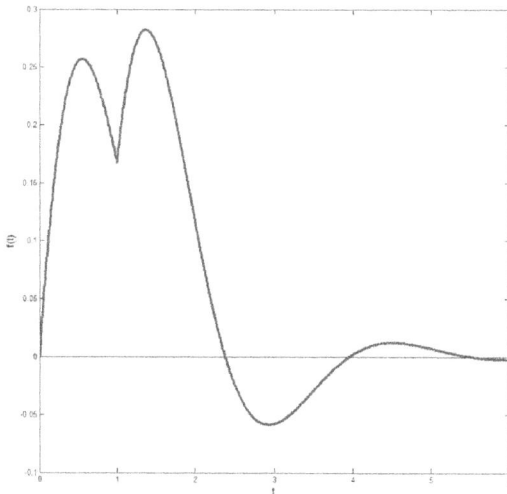

25. $y(t) = \begin{cases} -2\cos t + \sin t + 3e^{-t}(\cos t + \sin t) & (0 < t < \pi) \\ -2\cos t + \sin t + 3e^{-t}(\cos t + \sin t) - 8e^{-(t-\pi)}\sin t & (t > \pi) \end{cases}$

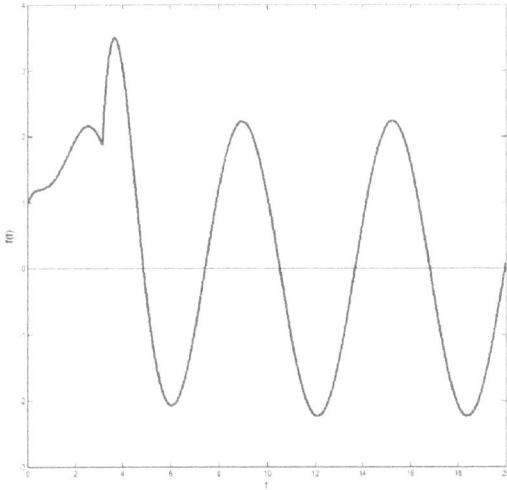

26. $y(t) = \begin{cases} 0 & (0 < t < \pi) \\ e^{-(t-\pi)} - e^{-2(t-\pi)} & (\pi < t < 2\pi) \\ e^{-(t-\pi)} - e^{-2(t-\pi)} + \dfrac{1}{2}\left\{1 - 2e^{-(t-2\pi)} + e^{-2(t-2\pi)}\right\} & (t > 2\pi) \end{cases}$

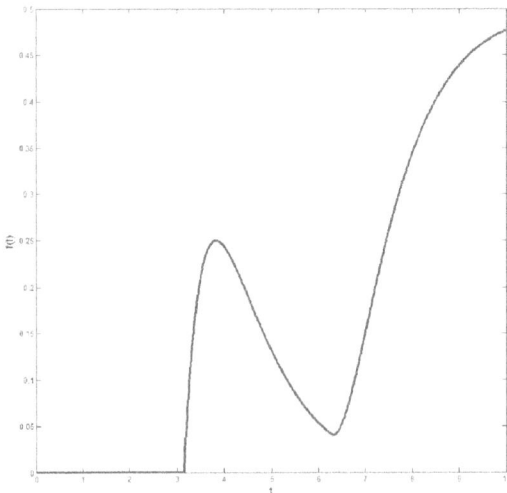

27. $y(t) = \begin{cases} e^{-t} - e^{-2t}(\cos t + \sin t) & (0 < t < 1) \\ e^{-t} - e^{-2t}(\cos t + \sin t) + e^{-2(t-1)}\sin(t-1) & (t > 1) \end{cases}$

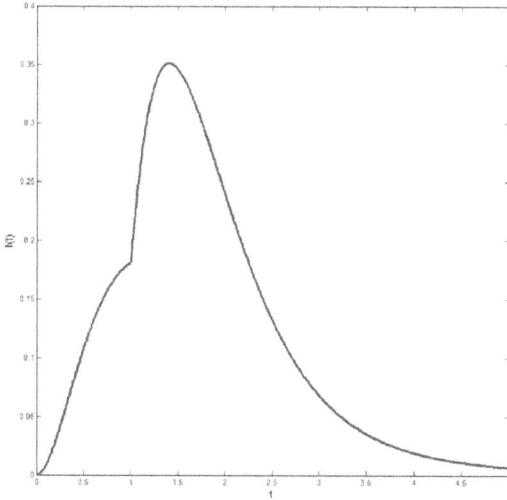

28. $y(t) = \begin{cases} -3 + 2t + 5e^{-t} - 2e^{-2t} & (0 < t < 1) \\ -3 + 2t + 5e^{-t} - 2e^{-2t} - 6\left\{e^{-(t-1)} - e^{-2(t-1)}\right\} & (t > 1) \end{cases}$

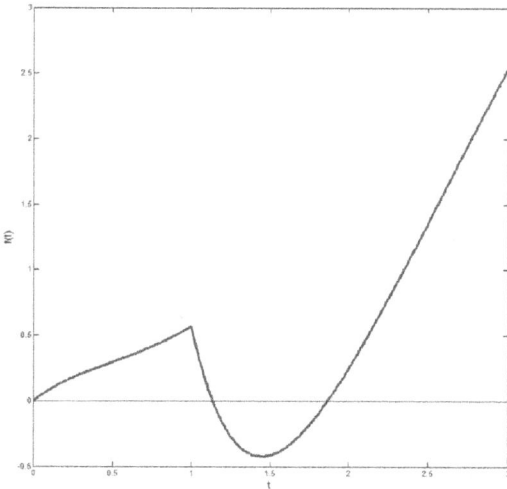

Problem 4.4

1. t^2

2. $\dfrac{t^3}{6}$

3. $\dfrac{\sin \omega t}{\omega}$

4. $\dfrac{t}{\omega} - \dfrac{\sin \omega t}{\omega^2}$

5. $-t - 1 + e^t$

6. $\dfrac{1}{2} t \sin t$

7. $1 - e^{-t}$

8. $\dfrac{e^{at} - e^{bt}}{a - b}$

9. $\dfrac{t}{2\omega} \sin \omega t$

10. $\dfrac{1}{2\omega^3} \left(-\omega t \cos \omega t + \sin \omega t \right)$

11. $\cosh t - 1$

12. $e^{(t-a)} - 1$

13. $y(t) = 2e^{-t}$

14. $y(t) = \cos t$

15. $y(t) = t - \dfrac{t^2}{2}$

16. $y(t) = \dfrac{1}{2} t \cos t + \dfrac{1}{2} \sin t$

17. $y(t) = 1 - \dfrac{2}{\sqrt{3}} e^{-\frac{1}{2}t} \sin \dfrac{\sqrt{3}}{2} t$

18. $y(t) = \dfrac{12}{25} e^t + \dfrac{4}{5} t e^t - \dfrac{12}{25} \cos 2t + \dfrac{9}{25} \sin 2t$

19. $y(t) = \cosh t$

20. $y(t) = \begin{cases} 0 & (0 < t < 1) \\ 2(t-1)^2 & (t > 1) \end{cases}$

21. $f(t) = \dfrac{2 \sinh t}{t}$

22. $f(t) = \dfrac{1 - 2\cos \omega t}{t}$

23. $f(t) = \dfrac{2(1 - e^{-2t} \cos t)}{t}$

24. $f(t) = \dfrac{2(e^t - \cos t)}{t}$

25. $f(t) = \dfrac{4e^{-t} \cos 2t - e^{-3t}}{t}$

26. $f(t) = \dfrac{2e^{-t}\left(2\cos 2t - \cos \sqrt{2}t\right)}{t}$

27. $\dfrac{t}{2\omega} \sin \omega t$

28. $-\dfrac{t}{2\omega^2} \cos \omega t + \dfrac{1}{2\omega^3} \sin \omega t$

29. $\dfrac{t}{2} \cos \omega t + \dfrac{1}{2\omega} \sin \omega t$

30. $t \sinh 3t$

31. $\dfrac{4s}{(s^2 + 4)^2}$

32. $\dfrac{4s}{(s^2 - 4)^2}$

33. $\dfrac{2(s + 1)}{(s^2 + 2s + 2)^2}$

34. $\dfrac{s^2 + 4s - 5}{(s^2 + 4s + 13)^2}$

35. $\dfrac{2}{(s - 3)^3}$

36. $\dfrac{2s(s^2 - 27)}{(s^2 + 9)^3}$

Problem 4.5

1. $x(t) = -\cos t + 50\sin t + e^{-t}\left(\cos 10t - \dfrac{49}{10}\sin 10t\right)$ [mm]

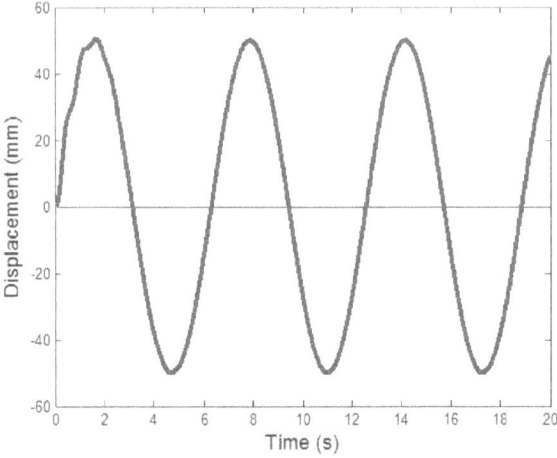

2. $x(t) = \begin{cases} 0 & (0 < t < 1) \\ 1 - e^{-2(t-1)}\left\{\cos 10(t-1) + \dfrac{1}{5}\sin 10(t-1)\right\}[\text{m}] & (t > 1) \end{cases}$

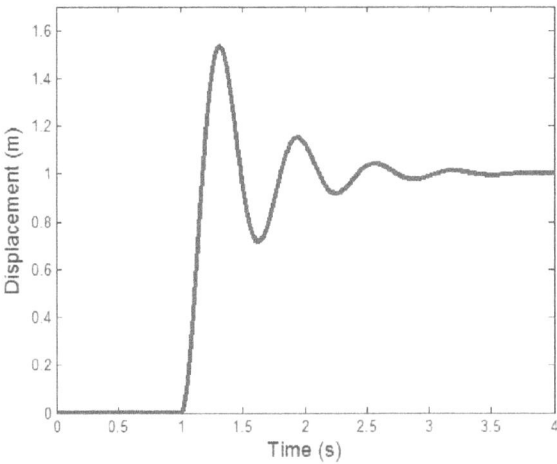

3. $i(t) = \sin 10t \ [A]$

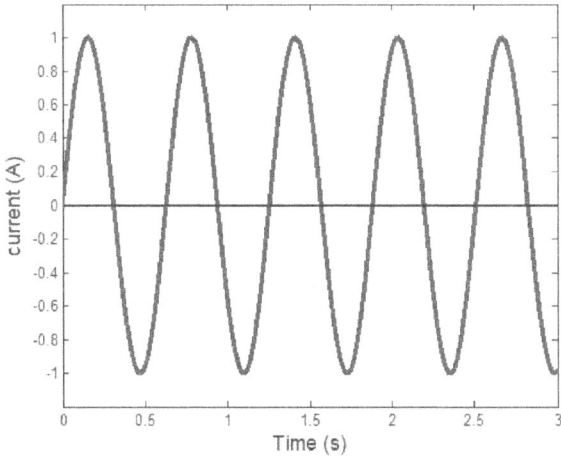

4. $i(t) = 100(10\sin 10t + \sin t)\{u(t - \pi) - u(t - 3\pi)\} \ [A]$

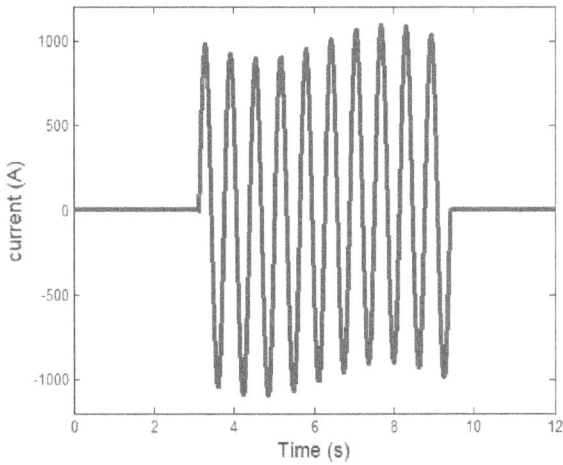

5. $i(t) = e^{-4t}\left(\dfrac{3}{26}\cos 3t - \dfrac{10}{39}\sin 3t\right) - \dfrac{3}{26}\cos 10t + \dfrac{8}{65}\sin 10t \ [\text{A}]$

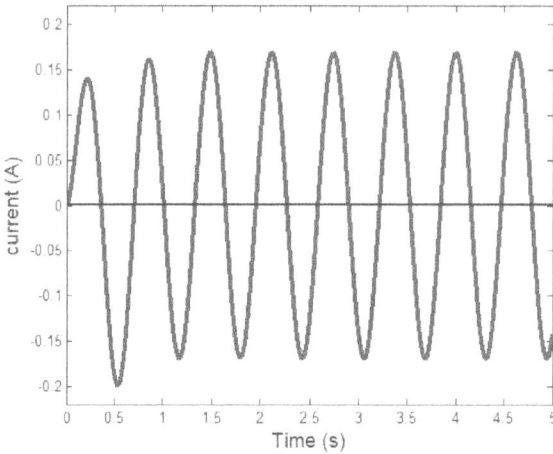

6. $i(t) = 1000e^{-t}\sin t - 1000e^{-(t-2)}\sin(t-2)u(t-2) \ [\text{A}]$

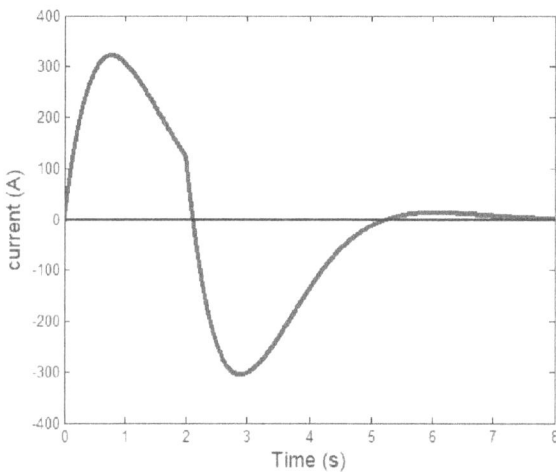

Problem 4.6

1. $F(s) = \dfrac{1}{(s-a)^2}$

2. $F(s) = \dfrac{s^2 - \omega^2}{(s^2 + \omega^2)^2}$

3. $F(s) = \dfrac{\omega}{s^2 - \omega^2}$

4. $F(s) = \dfrac{2s(s^2 - 3)}{(s^2 + 1)^3}$

5. $f(t) = te^{at}$

6. $f(t) = \dfrac{1}{\omega} e^{at} \sin \omega t$

7. $f(t) = \dfrac{\omega t - \sin \omega t}{\omega^3}$

8. $f(t) = \dfrac{t \sin \omega t}{2\omega}$

5 Series solutions of ordinary differential equations

In Chapter 1 through Chapter 4, we have seen that the general solutions to ODEs with constant coefficients, such as x^k, $\cos ax$, $\sin ax$, e^{ax}, etc., are composed in terms of combinations of elementary functions. However, solutions to most higher-order ODEs with variable coefficients cannot be expressed in terms of combinations of elementary functions.

5.1 Power series method

5.1.1 Technique of the power series method

The *power series method* is the standard method for solving linear ODEs with variable coefficients. A power series is an infinite series in powers of $(x - x_0)$ as follows:

$$\sum_{k=0}^{\infty} c_k (x - x_0)^k = c_0 + c_1(x - x_0) + c_2(x - x_0)^2 + \cdots \tag{5.1}$$

where x is a variable and coefficient c_k of the series is constant $(c_0 \neq 0)$. And x_0 is also a constant, called the center of the series. In particular, if $x_0 = 0$, we obtain a power series in powers of x

$$\sum_{k=0}^{\infty} c_k x^k = c_0 + c_1 x + c_2 x^2 + c_3 x^3 + \cdots, \tag{5.2}$$

where we assume that all variables and coefficients are real.

Representative examples include the following familiar power series:

$$\frac{1}{1-x} = \sum_{k=0}^{\infty} x^k = 1 + x + x^2 + x^3 + \cdots \qquad (|x| < 1) \tag{5.3}$$

$$e^x = \sum_{k=0}^{\infty} \frac{x^k}{k!} = 1 + x + \frac{x^2}{2!} + \frac{x^3}{3!} + \cdots \tag{5.4}$$

$$\cos x = \sum_{k=0}^{\infty} \frac{(-1)^k x^{2k}}{(2k)!} = 1 - \frac{x^2}{2!} + \frac{x^4}{4!} - \frac{x^6}{6!} + - \cdots \tag{5.5}$$

$$\sin x = \sum_{k=0}^{\infty} \frac{(-1)^k x^{2k+1}}{(2k+1)!} = x - \frac{x^3}{3!} + \frac{x^5}{5!} - \frac{x^7}{7!} + - \cdots \tag{5.6}$$

DOI: 10.1201/9781003608912-5

And utilizing power series, we can determine the derivatives of each function as follows:

$$\frac{d}{dx}e^x = e^x. \tag{5.7}$$

Check

$$\frac{d}{dx}e^x = \frac{d}{dx}\left(1 + x + \frac{x^2}{2!} + \frac{x^3}{3!} + \cdots\right)$$

$$= 1 + x + \frac{x^2}{2!} + \frac{x^3}{3!} + \cdots = e^x$$

$$\frac{d}{dx}(\cos x) = -\sin x \tag{5.8}$$

Check

$$\frac{d}{dx}(\cos x) = \frac{d}{dx}\left(1 - \frac{x^2}{2!} + \frac{x^4}{4!} - \frac{x^6}{6!} + \cdots\right)$$

$$= -\left(x - \frac{x^3}{3!} + \frac{x^5}{5!} - \frac{x^7}{7!} + \cdots\right) = -\sin x$$

$$\frac{d}{dx}(\sin x) = \cos x \tag{5.9}$$

Check

$$\frac{d}{dx}(\sin x) = \frac{d}{dx}\left(x - \frac{x^3}{3!} + \frac{x^5}{5!} - \frac{x^7}{7!} + \cdots\right)$$

$$= \left(1 - \frac{x^2}{2!} + \frac{x^4}{4!} - \frac{x^6}{6!} + \cdots\right) = \cos x$$

Moreover, two power series can be multiplied term by term as follows:

$$e^x \cdot \cos x = \left(1 + x + \frac{x^2}{2!} + \frac{x^3}{3!} + \cdots\right)\left(1 - \frac{x^2}{2!} + \frac{x^4}{4!} - \frac{x^6}{6!} + - \cdots\right)$$

$$= 1 + x - \frac{1}{3}x^3 - \frac{5}{24}x^4 - \frac{1}{24}x^5 - \cdots.$$

We can apply the power series method for solving linear ODEs. Let's understand the power series method through the following examples.

Example 5.1

Solve the following ODE using the power series method:

$$y' + y = 0.$$

Solution

In the first step, we insert

$$y = \sum_{k=0}^{\infty} c_k x^k,$$

and the series obtained by term-wise differentiation

$$y' = \sum_{k=0}^{\infty} k c_k x^{k-1} = \sum_{k=0}^{\infty} (k+1) c_{k+1} x^k$$

into the given ODE and then we collect like powers of x, finding

$$y' + y = \sum_{k=0}^{\infty} (k+1) c_{k+1} x^k + \sum_{k=0}^{\infty} c_k x^k = \sum_{k=0}^{\infty} \left\{(k+1) c_{k+1} + c_k\right\} x^k = 0.$$

Equating the coefficient of each power of x to zero, we have

$$c_{k+1} = -\frac{1}{k+1} c_k. \qquad (k = 0, 1, 2, \cdots).$$

Then, $$c_1 = -c_0,$$

$$c_2 = -\frac{1}{2}c_1 = \frac{1}{2}c_0 = \frac{1}{2!}c_0,$$

$$c_3 = -\frac{1}{3}c_2 = -\frac{1}{3!}c_0,$$

$$c_4 = -\frac{1}{4}c_3 = \frac{1}{4!}c_0.$$

With these values of the coefficients, the series solution becomes the familiar general solution similar to Eq. (5.4),

$$y = c_0\left(1 - x + \frac{x^2}{2!} - \frac{x^3}{3!} + \frac{x^4}{4!} - +\cdots\right) = c_0 e^{-x}.$$

Answer $y = c_0 e^{-x}$

5.1.2 Theory of the power series method

The n-th partial sum of Eq. (5.1) at the center x_0 is

$$S_n(x) = \sum_{k=0}^{n} c_k(x - x_0)^k = c_0 + c_1(x - x_0) + c_2(x - x_0)^2 + \cdots + c_n(x - x_0)^n \quad (5.10)$$

where $n = 0, 1, 2, \cdots$.

When we omit the terms of $S_n(x)$ from Eq. (5.1), the remaining term is

$$R_n(x) = \sum_{k=n+1}^{\infty} c_k(x - x_0)^k = c_{n+1}(x - x_0)^{n+1} + c_{n+2}(x - x_0)^{n+2} + \cdots. \quad (5.11)$$

Then, a power series (5.1) is expressed as follows:

$$S(x) = S_n(x) + R_n(x). \quad (5.12)$$

If for some $x = x_1$, this sequence converges, that is,

$$S(x_1) = \lim_{n \to \infty} S_n(x_1), \quad (5.13)$$

then the power series (5.1) is called convergent at $x = x_1$.

If Eq. (5.13) is not satisfied, it is said that the power series (5.1) is divergent.

If there exists a range around the center x_0 where the series converges, that range is called the *convergence interval*. The *convergence radius* is defined when the following equation is satisfied (refer to Figure 5.1):

$$|x - x_0| < R. \quad (5.14)$$

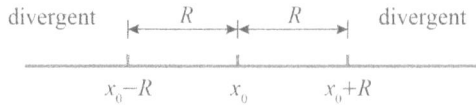

Figure 5.1 Convergence interval of a power series at the center x_0.

Remark   Convergence radius and convergence interval

Power series $\displaystyle\sum_{k=0}^{\infty} c_k(x-x_0)^k = c_0 + c_1(x-x_0) + c_2(x-x_0)^2 + \cdots$ at the center x_0 has the following convergence interval:

$$\text{convergence interval: } |x-x_0| < R, \tag{5.15}$$

where the convergence radius R can be determined from the coefficients c_n of the series as follows:

$$\text{convergence radius: } R = \frac{1}{\displaystyle\lim_{n\to\infty}\left|\frac{c_{n+1}}{c_n}\right|} \tag{5.16a}$$

or

$$R = \frac{1}{\displaystyle\lim_{n\to\infty}\sqrt[n]{|c_n|}}. \tag{5.16b}$$

For example, the following power series at the center $x_0 = 0$

$$\frac{1}{1-x} = \sum_{k=0}^{\infty} x^k = 1 + x + x^2 + x^3 + \cdots$$

has the coefficient

$$c_n = 1.$$

Then we can determine the convergence radius:

$$R = \frac{1}{\displaystyle\lim_{n\to\infty}\left|\frac{c_{n+1}}{c_n}\right|} = 1.$$

Therefore, the convergence interval is $|x| < 1$.

Example 5.2

Determine the convergence radius and the convergence interval of the following power series.

$$\sum_{n=0}^{\infty} \frac{3^n}{n+1} x^n.$$

Solution

Since $c_n = \dfrac{3^n}{n+1}$, the convergence radius is

$$R = \frac{1}{\lim\limits_{n \to \infty} \left| \dfrac{c_{n+1}}{c_n} \right|} = \frac{1}{\lim\limits_{n \to \infty} \left| \dfrac{3^{n+1}/(n+2)}{3^n/(n+1)} \right|} = \frac{1}{3},$$

and the convergence interval is

$$|x| < \frac{1}{3}.$$

$$\textbf{Answer } \frac{1}{3}, \ |x| < \frac{1}{3}$$

Problem 5.1

Determine the convergence radius and the convergence interval of the following power series. [1 ~ 6]

1. $\displaystyle\sum_{n=0}^{\infty} \frac{n(n+1)}{(2n-1)} x^n$

2. $\displaystyle\sum_{n=0}^{\infty} \frac{2^n}{n!} (x-1)^n$

3. $\displaystyle\sum_{k=0}^{\infty} \frac{(-1)^k}{(2k)!} (x+1)^k$

4. $\displaystyle\sum_{k=1}^{\infty} \frac{(-1)^k}{2^{k-1}} (x+3)^k$

5. $\displaystyle\sum_{n=0}^{\infty} n \left(\frac{2}{3} \right)^n x^{2n}$

6. $\displaystyle\sum_{m=0}^{\infty} (2m+1)(x-1)^{2m+1}$

Find a power series solution of the following ODE. [7 ~ 12]

7. $y'' - y = 0$
8. $y'' + y = 0$
9. $y'' - xy = 0$
10. $y'' - xy' + y = 0$
11. $y'' + y' - xy = 0$
12. $y'' - (x^2 + 1)y = 0$

5.2 Frobenius method

The *Frobenius method*, an extension of the power series method, provides a new approach to finding solutions to ODEs that involve fractional functions, logarithmic functions, and other complexities. It is particularly useful for solving equations like the Bessel equation and the hypergeometric equation.

The Frobenius method is defined as follows:

Remark   Frobenius method

Let $p(x)$ and $q(x)$ be any functions that are analytic at the center $x = 0$, as follows.

$$p(x) = \sum_{k=0}^{\infty} p_k x^k$$

and

$$q(x) = \sum_{k=0}^{\infty} q_k x^k.$$

Then the ODE

$$x^2 y'' + xp(x)y' + q(x)y = 0 \qquad (5.17)$$

has at least one solution that can be represented in the form

$$y_1(x) = x^r \sum_{k=0}^{\infty} c_k x^k = \sum_{k=0}^{\infty} c_k x^{k+r} \qquad (5.18)$$

where the exponent r may be any number (and r is chosen so that $c_0 \neq 0$). The series converge at least over some interval $0 < x < R$, and have another solution $y_2(x)$ in a similar form to $y_1(x)$ (with a different exponent r) or including a logarithmic term. Here, $y_1(x)$ and $y_2(x)$ are linearly independent of each other.

At $x_0 = 0$, we expand $p(x)$ and $q(x)$ of Eq. (5.15) in power series,

$$p(x) = p_0 + p_1 x + p_2 x^2 + p_3 x^3 + \cdots$$

$$q(x) = q_0 + q_1 x + q_2 x^2 + q_3 x^3 + \cdots.$$

Then we differentiate Eq. (5.18) term by term, finding

$$y_1(x) = \sum_{k=0}^{\infty} c_k x^{k+r} = x^r \left\{ c_0 + c_1 x + c_2 x^2 + \cdots \right\}$$

$$y_1'(x) = \sum_{k=0}^{\infty} (k+r) c_k x^{k+r-1} = x^{r-1} \left\{ r c_0 + (r+1) c_1 x + \cdots \right\}$$

$$y_1''(x) = \sum_{k=0}^{\infty} (k+r)(k+r-1) c_k x^{k+r-2} = x^{r-2} \left\{ r(r-1) c_0 + (r+1) r c_1 x + \cdots \right\}.$$

By inserting all these series into Eq. (5.17), we obtain

$$x^r \left\{ r(r-1) c_0 + (r+1) r c_1 x + \cdots \right\} + \left\{ p_0 + p_1 x + p_2 x^2 + \cdots \right\} x^r \left\{ r c_0 + (r+1) c_1 x + \cdots \right\}$$
$$+ \left\{ q_0 + q_1 x + q_2 x^2 + \cdots \right\} x^r \left\{ c_0 + c_1 x + c_2 x^2 + \cdots \right\} = 0.$$

Now we equate the sum of the coefficients of each power x^r, x^{r+1}, x^{r+2}, $\cdots$ to zero.

Organizing the coefficients of the smallest power term x^r leads us to obtain the following indicial equation by assumption $c_0 \neq 0$.

$$r(r-1) + p_0 r + q_0 = 0. \tag{5.19}$$

From this quadratic equation (5.19), we can solve for the roots r. The Frobenius method yields a basis of solutions.

Remark Frobenius method

Depending on the roots r of the indicial equation, the following three cases can be classified:

(Case I) Distinct roots r_1, r_2 not differing by an integer 1, 2, 3, $\cdots$.
 Two bases are

$$y_1(x) = \sum_{k=0}^{\infty} c_k x^{k+n} = x^n \left(c_0 + c_1 x + c_2 x^2 + \cdots \right) \tag{5.20a}$$

and

$$y_2(x) = \sum_{k=0}^{\infty} A_k x^{k+r_2} = x^{r_2}\left(A_0 + A_1 x + A_2 x^2 + \cdots\right). \qquad (5.20b)$$

(Case II) Roots r_1, r_2 differing by an integer $1, 2, 3, \cdots$. $(r_1 > r_2)$
   Two bases are

$$y_1(x) = \sum_{k=0}^{\infty} c_k x^{k+r_1} = x^{r_1}\left(c_0 + c_1 x + c_2 x^2 + \cdots\right) \qquad (5.21a)$$

and

$$y_2(x) = K y_1 \ln x + x^{r_2}\left(A_0 + A_1 x + A_2 x^2 + \cdots\right) \qquad (x > 0). \qquad (5.21b)$$

where K may turn out to be zero.

(Case III) Double root $r_1 = r_2 = \dfrac{1}{2}(1 - p_0)$.
   Two bases are

$$y_1(x) = \sum_{k=0}^{\infty} c_k x^{k+r_1} = x^{r_1}\left(c_0 + c_1 x + c_2 x^2 + \cdots\right) \qquad (5.22a)$$

and

$$y_2(x) = y_1(x) \ln x + x^{r_2}\left(A_1 x + A_2 x^2 + \cdots\right). \qquad (x > 0). \qquad (5.22b)$$

Remark

When the difference between the two roots is an integer (including 0), here's how you can find $y_2(x)$.

Method 1: Method of Reduction of Order

When one basis $y_1(x)$ is known in the ODE, another basis $y_2(x)$ can be obtained using the method in Section 2.1.
   In the ODE $y'' + P(x)y' + Q(x)y = 0$, another basis is

$$y_2 = y_1 \int v\, dx. \qquad \left(\text{where } v = \frac{1}{y_1^2} e^{-\int P(x)\, dx}\right) \qquad (2.5)$$

Method 2: Frobenius method

This method is employed when the method of reduction of order cannot be applied.

$$y_2(x) = y_1(x)\ln x + x^{r_2}\left(A_0 + A_1 x + A_2 x^2 + \cdots\right) \qquad (5.22b)$$

After substituting Eq. (5.22b) into the given ODE $x^2 y'' + xp(x)y' + q(x)y = 0$, we eliminate the equation $x^2 y_1'' + xp(x)y_1' + q(x)y_1 = 0$ and then determine coefficients such as A_0, A_1, A_2, $\cdots$ through coefficient comparison.

Example 5.4

Solve the following ODE.

$$2x^2 y'' - xy' + y = 0$$

Solution

For the Euler-Cauchy equation $x^2 y'' + xp(x)y' + q(x)y = 0$, the given equation $2x^2 y'' - xy' + y = 0$ has $p(x) = -\dfrac{1}{2}$ and $q(x) = \dfrac{1}{2}$.

Now let's use the Frobenius method.

We differentiate $y = \displaystyle\sum_{k=0}^{\infty} c_k x^{k+r}$ term by term, finding

$$y' = \sum_{k=0}^{\infty}(k+r)c_k x^{k+r-1},$$

and

$$y'' = \sum_{k=0}^{\infty}(k+r)(k+r-1)c_k x^{k+r-2}.$$

By inserting all these series into the given equation, we obtain

$$2x^2 y'' - xy' + y = 2\sum_{k=0}^{\infty}(k+r)(k+r-1)c_k x^{k+r} - \sum_{k=0}^{\infty}(k+r)c_k x^{k+r} + \sum_{k=0}^{\infty}c_k x^{k+r}$$

$$= \sum_{k=0}^{\infty}\left\{2(k+r)(k+r-1)-(k+r)+1\right\}c_k x^{k+r} = 0, \qquad (c_0 \neq 0)$$

and then obtain the following indicial equation by assumption $c_0 \neq 0$,

$$2r(r-1) - r + 1 = 0, \qquad (k=0)$$

or the roots are

$$r = 1, \ \frac{1}{2}.$$

These roots correspond to (Case I) of the Frobenius method. And we obtain

$$c_k = 0. \qquad (k=1, \ 2, \ \cdots)$$

i. For $r_1 = 1$ and taking $c_0 = 1$, we obtain $y_1(x) = x$ as a first basis.

ii. For $r_2 = \dfrac{1}{2}$ and taking $c_0 = 1$, we obtain $y_2(x) = \sqrt{x}$ as a second basis.

Answer $y(x) = Ax + B\sqrt{x}$

Remark

Using MATLAB®, solutions can be obtained easily (refer to the M_prob 5.3).

Example 5.5

Find a power series solution of the following ODE:

$$\left(x - x^2\right)y'' + xy' - y = 0.$$

Solution

The given ODE $\left(x - x^2\right)y'' + xy' - y = 0$ is an example of a hypergeometric equation.

For the Euler-Cauchy equation $x^2 y'' + xp(x)y' + q(x)y = 0$, the given equation has $p(x) = \dfrac{x}{1-x}$ and $q(x) = -\dfrac{x}{1-x}$.

Now let's use the Frobenius method at $x \neq 1$.

We differentiate $y = \displaystyle\sum_{k=0}^{\infty} c_k x^{k+r}$ term by term, finding

$$y' = \sum_{k=0}^{\infty} (k+r)c_k x^{k+r-1},$$

and

$$y'' = \sum_{k=0}^{\infty}(k+r)(k+r-1)c_k x^{k+r-2}.$$

By inserting all these series into the given equation, we obtain

$$\left(x-x^2\right)y'' + xy' - y$$

$$= \sum_{k=0}^{\infty}(k+r)(k+r-1)c_k x^{k+r-1} - \sum_{k=0}^{\infty}(k+r)(k+r-1)c_k x^{k+r}$$

$$+ \sum_{k=0}^{\infty}(k+r)c_k x^{k+r} - \sum_{k=0}^{\infty}c_k x^{k+r}$$

$$= -r(r-1)c_0 x^{r-1} + \sum_{k=1}^{\infty}(k+r)(k+r-1)c_k x^{k+r-1} - \sum_{k=0}^{\infty}(k+r)(k+r-1)c_k x^{k+r}$$

$$+ \sum_{k=0}^{\infty}(k+r)c_k x^{k+r} - \sum_{k=0}^{\infty}c_k x^{k+r}$$

$$= -r(r-1)c_0 x^{r-1} + \sum_{k=0}^{\infty}(k+r+1)(k+r)c_{k+1} x^{k+r} - \sum_{k=0}^{\infty}(k+r)(k+r-1)c_k x^{k+r}$$

$$+ \sum_{k=0}^{\infty}(k+r)c_k x^{k+r} - \sum_{k=0}^{\infty}c_k x^{k+r}$$

$$= -r(r-1)c_0 x^{r-1} + \sum_{k=0}^{\infty}\left[(k+r+1)(k+r)c_{k+1}\right.$$

$$\left. -\{(k+r)(k+r-1)-(k+r)+1\}c_k\right]x^{k+r}.$$

and then obtain the following indicial equation by assumption $c_0 \neq 0$,

$$r(r-1) = 0, \qquad (k=0)$$

or the roots are

$$r = 1, \ 0.$$

These roots correspond to (Case II) of the Frobenius method.

And we obtain

$$c_{k+1} = \frac{(k+r-1)^2}{(k+r+1)(k+r)}c_k. \qquad (k = 0, 1, 2, \cdots)$$

i. For $r_1 = 1$ and taking $c_0 = 1$, we obtain

$$c_{k+1} = \frac{k^2}{(k+2)(k+1)}c_k. \qquad (k = 0, 1, 2, \cdots)$$

Hence, since $c_1 = c_2 = c_3 = \cdots = 0$, we obtain $y_1(x) = x$ as a first basis.

ii. Now let's use the method of reduction of order to find the second basis.

Since $P(x) = -\dfrac{1}{x-1}$ and $y_1(x) = x$, we obtain

$$v = \frac{1}{y_1^2}e^{-\int P(x)dx}$$

$$= \frac{1}{x^2}e^{-\int -\frac{1}{x-1}dx} = \frac{1}{x^2}e^{\ln|x-1|} = \frac{x-1}{x^2}.$$

Therefore, we obtain $y_2(x)$ as a second basis as follows:

$$y_2 = y_1\int v\,dx$$

$$= x\int \frac{x-1}{x^2}dx = x\left(\ln|x| + \frac{1}{x}\right) = x\ln|x| + 1.$$

Answer $y(x) = Ax + B(x\ln|x| + 1)$

Remark

Using MATLAB®, solutions can be obtained easily (refer to the M_prob 5.4).

Example 5.6

Find a power series solution of the following ODE:

$$xy'' - y = 0.$$

Solution

Multiplying both sides of the given ODE by x yields the follows,

$$x^2 y'' - xy = 0.$$

For the Euler-Cauchy equation $x^2 y'' + xp(x)y' + q(x)y = 0$, the given equation has $p(x) = 0$ and $q(x) = -x$.

Now let's use the Frobenius method.

We differentiate $y = \sum_{k=0}^{\infty} c_k x^{k+r}$ term by term, finding

$$y'' = \sum_{k=0}^{\infty} (k+r)(k+r-1)c_k\, x^{k+r-2}.$$

By inserting all these series into the given equation, we obtain

$$xy'' - y = \sum_{k=0}^{\infty} (k+r)(k+r-1)c_k\, x^{k+r-1} - \sum_{k=0}^{\infty} c_k\, x^{k+r}$$

$$= r(r-1)c_0 x^{r-1} + \sum_{k=0}^{\infty} (k+r+1)(k+r)c_{k+1}\, x^{k+r} - \sum_{k=0}^{\infty} c_k\, x^{k+r}$$

$$= r(r-1)c_0 x^{r-1} + \sum_{k=0}^{\infty} \left\{ (k+r+1)(k+r)c_{k+1} - c_k \right\} x^{k+r} = 0.$$

and then obtain the following indicial equation by assumption $c_0 \neq 0$,

$$r(r-1) = 0, \qquad (k=0)$$

or the roots are

$$r = 1,\ 0.$$

These roots correspond to (Case II) of the Frobenius method.
And we obtain

$$c_{k+1} = \frac{1}{(k+r+1)(k+r)} c_k. \qquad (k = 0,\ 1,\ 2,\ \cdots)$$

i. For $r_1 = 1$ and taking $c_0 = 1$, we obtain

$$c_{k+1} = \frac{1}{(k+2)(k+1)} c_k. \qquad (k = 0,\ 1,\ 2,\ \cdots)$$

Hence,

$$c_1 = \frac{1}{2 \cdot 1} c_0 = \frac{1}{2!}$$

$$c_2 = \frac{1}{3 \cdot 2} c_1 = \frac{1}{3 \cdot 2 \cdot 2} = \frac{1}{3!2!}$$

$$c_3 = \frac{1}{4 \cdot 3} c_2 = \frac{1}{4 \cdot 3 \cdot 3 \cdot 2 \cdot 2} = \frac{1}{4!3!}$$

...

Therefore, we obtain the first basis as follows.

$$y_1(x) = x\left(1 + \frac{x}{2!} + \frac{x^2}{3!2!} + \frac{x^3}{4!3!} + \cdots\right) = x + \frac{x^2}{2!} + \frac{x^3}{3!2!} + \frac{x^4}{4!3!} + \cdots$$

ii. For $r_2 = 0$ and taking $c_0 = 1$, we obtain

$$c_{k+1} = \frac{1}{(k+1)k} c_k. \qquad (k = 0, 1, 2, \cdots)$$

Substituting $k = 0$ into the above equation results in a denominator of 0, making it impossible to determine the value of c_1.

Check

Let's use the method of reduction of order to obtain another basis $y_2(x)$, when one basis $y_1(x)$ is known in the ODE.

Since $P(x) = 0$ and $y_1(x) = x + \frac{x^2}{2!} + \frac{x^3}{3!2!} + \frac{x^4}{4!3!} + \cdots$, we obtain

$$v = \frac{1}{y_1^2} e^{-\int P\,dx} = \frac{1}{y_1^2} e^{-\int 0\,dx} = \frac{1}{y_1^2} = \frac{1}{\left(x + \frac{x^2}{2} + \frac{x^3}{12} + \frac{x^4}{144} + \cdots\right)^2},$$

and $y_2 = y_1 \int v\,dx = y_1 \int \frac{1}{y_1^2}\,dx.$

(Further computation is not possible!)

Now let's use the Frobenius method

iii. For $r_2 = 0$ and $K = 1$, Eq. (5.21b) becomes

$$y_2(x) = y_1 \ln x + \sum_{k=0}^{\infty} A_k x^k.$$

We differentiate the above equation term by term, finding

$$y_2' = y_1' \ln x + y_1' \frac{1}{x} + \sum_{k=0}^{\infty} k A_k x^{k-1},$$

and

$$y_2'' = y_1'' \ln x + 2y_1' \frac{1}{x} - y_1 \frac{1}{x^2} + \sum_{k=0}^{\infty} k(k-1) A_k x^{k-2}.$$

By inserting all these series into the given equation, we obtain

$$xy_2'' - y = (xy_1'' - y_1) \ln x + 2y_1' - y_1 \frac{1}{x} + \sum_{k=0}^{\infty} k(k-1) A_k x^{k-1} - \sum_{k=0}^{\infty} A_k x^k = 0.$$

Since $xy_1'' - y_1 = 0$ and $y_1(x) = x + \dfrac{x^2}{2} + \dfrac{x^3}{12} + \dfrac{x^4}{144} + \cdots$, then we obtain

$$2\left(1 + x + \frac{x^2}{4} + \frac{x^3}{36} + \cdots\right) - \left(1 + \frac{x}{2} + \frac{x^2}{12} + \frac{x^3}{144} + \cdots\right)$$
$$+ \sum_{k=0}^{\infty} \left\{ k(k+1) A_{k+1} - A_k \right\} x^k = 0.$$

i. For $k = 0$, we get

$$1 - A_0 = 0,$$

or

$$A_0 = 1.$$

ii. For $k = 1$, we get

$$2 - \frac{1}{2} + 2A_2 - A_1 = 0.$$

Since $A_1 = 0$, then

$$A_2 = -\frac{3}{4}.$$

iii. For $k = 2$, we get

$$\frac{1}{2} - \frac{1}{12} + 6A_3 - A_2 = 0.$$

Since $A_2 = -\frac{3}{4}$, then

$$A_3 = -\frac{7}{36}.$$

$\cdots$

Therefore, the second basis is

$$y_2(x) = y_1 \ln|x| + 1 - \frac{3}{4}x^2 - \frac{7}{36}x^3 - \cdots.$$

Answer In $y(x) = Ay_1(x) + By_2(x)$,

$$y_1(x) = x + \frac{x^2}{2} + \frac{x^3}{12} + \frac{x^4}{144} + \cdots, \quad y_2(x) = y_1 \ln|x| + 1 - \frac{3}{4}x^2 - \frac{7}{36}x^3 - \cdots$$

Remark

Using MATLAB®, solutions can be obtained easily (refer to M_prob 5.5).

Problem 5.2

Solve the following ODE using the Frobenius method. [1 ~ 12]

[(Case I) Distinct roots r_1, r_2 not differing by an integer 1, 2, 3, $\cdots$]

1. $2xy'' - y' + y = 0$
2. $2xy'' + 5y' - xy = 0$

[(Case II) Roots r_1, r_2 differing by an integer 1, 2, 3, $\cdots$. $(r_1 > r_2)$]

3. $x^2 y'' + xy' - y = 0$
4. $x^2 y'' - 2xy' + 2y = 0$
5. $x(x-1)y'' - xy' + y = 0$
6. $x^2 y'' + 2xy' - x^2 y = 0$
7. $xy'' + y = 0$

[(Case III) Double root $r_1 = r_2 = \frac{1}{2}(1 - p_0)$.]

8. $x^2 y'' - xy' + y = 0$
9. $x^2 y'' - 3xy' + 4y = 0$
10. $x^2 y'' + 3xy' + y = 0$
11. $(x^2 - x)y'' + (3x - 1)y' + y = 0$
12. $xy'' + y' - y = 0$

5.3 Bessel function and Legendre polynomials

The followings are forms of ODEs widely applied in applied mathematics, physics, and other fields.

$$x^2 y'' + xy' + (x^2 - v^2)y = 0 \qquad (v \geq 0) \qquad (5.23)$$

$$(1 - x^2)y'' - 2xy' + n(n+1)y = 0 \qquad (n \geq 0 \text{ integer}) \qquad (5.24)$$

Eq. (5.23) is referred to as *Bessel's equation*, and Eq. (5.24) is called the Legendre's equation. The solutions to these equations are respectively called *Bessel functions* and *Legendre polynomials*.

5.3.1 Bessel function of the first kind $J_v(x)$

In order to obtain the general solution of the Bessel's equation, we apply the power series method.

Bessel equation has a solution of the form

$$y = \sum_{k=0}^{\infty} c_k x^{k+r}. \qquad (5.25)$$

We differentiate $y = \sum_{k=0}^{\infty} c_k x^{k+r}$ term by term, finding

$$y' = \sum_{k=0}^{\infty} (k+r) c_k x^{k+r-1},$$

and

$$y'' = \sum_{k=0}^{\infty} (k+r)(k+r-1) c_k x^{k+r-2}.$$

By inserting all these series into the given Bessel equation, we obtain

$$x^2 y'' + xy' + \left(x^2 - v^2\right)y$$

$$= \sum_{k=0}^{\infty} (k+r)(k+r-1)c_k x^{k+r} + \sum_{k=0}^{\infty} (k+r)c_k x^{k+r} + \sum_{k=0}^{\infty} c_k x^{k+r+2} - v^2 \sum_{k=0}^{\infty} c_k x^{k+r}$$

$$= \left\{ r(r-1) + r - v^2 \right\} c_0 x^r + \left\{ (r+1)r + (r+1) - v^2 \right\} c_1 x^{r+1}$$

$$+ \sum_{k=0}^{\infty} \left[\left\{ (k+r+2)^2 - v^2 \right\} c_{k+2} + c_k \right] x^{k+r+2}$$

$$= \left\{ r^2 - v^2 \right\} c_0 x^r + \left\{ (r+1)^2 - v^2 \right\} c_1 x^{r+1} + \sum_{k=0}^{\infty} \left[\left\{ (k+r+2)^2 - v^2 \right\} c_{k+2} + c_k \right] x^{k+r+2}$$

$$= 0.$$

and then obtain the following indicial equation by assumption $c_0 \neq 0$,

$$r^2 - v^2 = 0, \tag{5.26}$$

or the roots are

$$r = v, \ -v. \qquad \left(v \geq 0\right).$$

And we also obtain

$$\left\{ (r+1)^2 - v^2 \right\} c_1 = 0, \tag{5.27}$$

and

$$\left\{ (k+r+2)^2 - v^2 \right\} c_{k+2} + c_k = 0. \qquad (k = 0, \ 1, \ 2, \ \cdots) \tag{5.28}$$

i. For $r_1 = v$, Eq. (5.27) reduces to $(2v+1)c_1 = 0$
   Hence, $c_1 = 0$, since $v \geq 0$.
   And Eq. (5.28) becomes

$$c_{k+2} = -\frac{1}{(k+2)(k+2v+2)} c_k \qquad (k = 0, \ 1, \ 2, \ \cdots). \tag{5.29}$$

Since $c_1 = 0$ and $v \geq 0$, it follows from Eq. (5.29) that

$$c_1 = c_3 = c_5 = \cdots = 0.$$

And since $c_0 \neq 0$, it follows from Eq. (5.29) that

$$c_2 = -\frac{1}{2(2v+2)}c_0 = -\frac{1}{2^2(v+1)}c_0$$

$$c_4 = -\frac{1}{4(2v+4)}c_2 = \frac{1}{2^4 2!(v+2)(v+1)}c_0$$

$$c_6 = -\frac{1}{6(2v+6)}c_4 = -\frac{1}{2^6 3!(v+3)(v+2)(v+1)}c_0$$

and so on, and in general

$$c_{2k} = \frac{(-1)^k}{2^{2k}k!(v+k)\cdots(v+2)(v+1)}c_0 \qquad (k = 1,\ 2,\ \cdots). \qquad (5.30)$$

On the other hand, let's assume c_0 to be expressed as a value incorporating the gamma function $\Gamma(1+v)$, as follows:

$$c_0 = \frac{1}{2^v \Gamma(v+1)} \qquad\qquad (5.31)$$

where *gamma function* $\Gamma(v+1)$ is defined by

$$\Gamma(v+1) = \int_0^\infty e^{-t}t^v\,dt. \qquad (v > -1) \qquad (5.32)$$

Remark **Properties of gamma functions**

$$\Gamma(v+1) = v\Gamma(v) \qquad\qquad (5.33)$$

$$\Gamma(1) = 1 \qquad\qquad (5.34)$$

$$\Gamma(n+1) = n! \qquad (n\text{: positive integer}) \qquad (5.35)$$

$$\Gamma\left(\frac{1}{2}\right) = \sqrt{\pi} \qquad\qquad (5.36)$$

i. Proof of Eq. (5.33)
   Using integration by parts, Eq. (5.33) is derived as follows:

$$\Gamma(v+1) = \int_0^\infty e^{-t}t^v \, dt = -e^{-t}t^v \Big|_0^\infty - \int_0^\infty \left(-e^{-t}\right)\left(vt^{v-1}\right)dt = v\Gamma(v). \quad (5.33)$$

ii. Proof of Eq. (5.34)
   From Eq. (5.32) with $v = 0$ and then by Eq. (5.33) we obtain

$$\Gamma(1) = \int_0^\infty e^{-t}t^0 \, dt = -e^{-t} \Big|_0^\infty = 1. \quad (5.34)$$

iii. Proof of Eq. (5.35)
   From Eq. (5.33),

$$\Gamma(2) = 1\Gamma(1) = 1,$$

$$\Gamma(3) = 2\Gamma(2) = 2!,$$

$$\Gamma(4) = 3\Gamma(3) = 3!.$$

In general, we obtain

$$\Gamma(n+1) = n\Gamma(n) = \cdots = n!. \quad (5.35)$$

iv. Proof of Eq. (5.36)
   From Eq. (5.32), we obtain

$$\Gamma\left(\frac{1}{2}\right) = \int_0^\infty e^{-t}t^{-1/2} \, dt.$$

Substituting $t = u^2$ into the above equation, then $dt = 2u \, du$ and we get:

$$I = \Gamma\left(\frac{1}{2}\right) = \int_0^\infty e^{-u^2}u^{-1}2u \, du = 2\int_0^\infty e^{-u^2} \, du.$$

Similarly, replacing u by v in the above equation, we get:

$$I = \Gamma\left(\frac{1}{2}\right) = 2\int_0^\infty e^{-v^2} \, dv.$$

Therefore, multiplying the two equations together, we get:

$$I^2 = 4\int_0^\infty e^{-u^2}\,du \int_0^\infty e^{-v^2}\,dv = 4\int_0^\infty \int_0^\infty e^{-\left(u^2+v^2\right)}\,du\,dv.$$

Now we use polar coordinates r, θ by setting $u = r\cos\theta$, $v = r\sin\theta$. Then the element of area is $du\,dv = r\,dr\,d\theta$ and we have to integrate over r from 0 to inf. and over θ from 0 to $\pi/2$, corresponding to the first quadrant:

$$I^2 = 4\int_0^{\pi/2}\int_0^\infty e^{-r^2} r\,dr\,d\theta = 4\int_0^{\pi/2} d\theta \int_0^\infty e^{-r^2} r\,dr = 4\cdot\frac{\pi}{2}\cdot\frac{1}{2} = \pi.$$

Therefore, by taking the square root on both sides, we obtain

$$\Gamma\left(\frac{1}{2}\right) = \sqrt{\pi}. \tag{5.36}$$

Now, let's return to Eq. (5.30) for Bessel functions.

$$c_{2k} = \frac{(-1)^k}{2^{2k}k!(v+k)\cdots(v+2)(v+1)}c_0 \qquad (k = 1,\ 2,\ \cdots) \tag{5.30}$$

i. For $v = n$ (n is 0 or positive integer),
   Substituting $\Gamma(n+1) = n!$ into Eq. (5.31), we obtain

$$c_0 = \frac{1}{2^n n!}.$$

Since $(n+k)\cdots(n+2)(n+1)n \ne (n+k)!$ in Eq. (5.30), Eq. (5.30) simply becomes

$$c_{2k} = \frac{(-1)^k}{2^{2k+n}k!(n+k)!}. \qquad (k = 1,\ 2,\ \cdots) \tag{5.37}$$

By inserting these coefficients into Eq. (5.25), we obtain a particular solution of Bessel's equation that is denoted by $J_n(x)$:

$$J_n(x) = x^n \sum_{k=0}^\infty \frac{(-1)^k}{2^{2k+n}k!(n+k)!}x^{2k}. \qquad (n \ge 0) \tag{5.38}$$

We call $J_n(x)$ as the Bessel function of the first kind of order n.

Remark   Bessel function of the first kind $J_n(x)$

In Bessel's equation $x^2 y'' + xy' + (x^2 - v^2)y = 0,$ $\qquad\qquad$ (5.23)

i. For $v = n$ (n: 0 or a positive integer), we obtain the particular solution of Bessel's equation: $J_n(x)$

$$J_n(x) = x^n \sum_{k=0}^{\infty} \frac{(-1)^k}{2^{2k+n} k!(n+k)!} x^{2k} \qquad (n \geq 0). \qquad (5.38)$$

For $n = 0$, the Bessel function of the first kind of order 0 is

$$J_0(x) = \sum_{k=0}^{\infty} \frac{(-1)^k}{2^{2k}(k!)^2} x^{2k} = 1 - \frac{x^2}{2^2(1!)^2} + \frac{x^4}{2^4(2!)^2} - \frac{x^6}{2^6(3!)^2} + - \cdots .$$
$$(5.39)$$

For $n = 1$, the Bessel function of the first kind of order 1 is

$$J_1(x) = x \sum_{k=0}^{\infty} \frac{(-1)^k}{2^{2k+1} k!(k+1)!} x^{2k} = \frac{x}{2} - \frac{x^3}{2^3 1!2!} + \frac{x^5}{2^5 2!3!} -$$
$$\frac{x^7}{2^7 3!4!} + - \cdots .$$
$$(5.40)$$

Figure 5.2 shows the Bessel function of the first kind, J_0 and J_1.

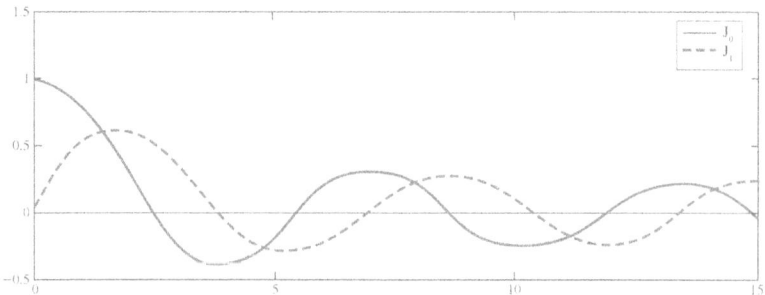

Figure 5.2 Bessel function of the first kind, J_0 and J_1.

ii. For arbitrary positive v $(v > 0)$,
Substituting c_0 into Eq. (5.30), we obtain

$$c_{2k} = \frac{(-1)^k}{2^{2k}\,k!(v+k)\cdots(v+2)(v+1)\,2^v\,\Gamma(v+1)}$$

or

$$c_{2k} = \frac{(-1)^k}{2^{2k+v}\,k!\,\Gamma(v+k+1)} \qquad (k=1,\ 2,\ \cdots). \qquad (5.41)$$

Therefore, by inserting these coefficients into Eq. (5.25), we obtain a particular solution of Bessel's equation that is denoted by $J_v(x)$:

$$J_v(x) = x^v \sum_{k=0}^{\infty} \frac{(-1)^k}{2^{2k+v}\,\Gamma(v+k+1)} x^{2k} \qquad (v > 0). \qquad (5.42)$$

Remark   **Bessel function of the first kind $J_v(x)$ and $J_{-v}(x)$**

In Bessel's equation $x^2 y'' + xy' + (x^2 - v^2)y = 0,$ $\qquad (5.23)$

ii. For arbitrary positive v $(v > 0)$, we obtain the particular solution of Bessel's equation: $J_v(x)$

$$J_v(x) = x^v \sum_{k=0}^{\infty} \frac{(-1)^k}{2^{2k+v}\,\Gamma(v+k+1)} x^{2k} \qquad (v > 0) \qquad (5.42)$$

Also, Bessel's equation has a second linearly independent solution, denoted as $J_{-v}(x)$, in addition to $J_v(x)$:

$$J_{-v}(x) = x^{-v} \sum_{k=0}^{\infty} \frac{(-1)^k}{2^{2k-v}\,\Gamma(-v+k+1)} x^{2k} \qquad (5.43)$$

Therefore, the general solution of the Bessel's equation is

$$y(x) = c_1 J_v(x) + c_2 J_{-v}(x) \qquad (x \neq 0,\ \text{not integer } v). \qquad (5.44)$$

Figure 5.3 shows the Bessel function of the first kind, $J_{1/3}$ and $J_{-1/3}$.

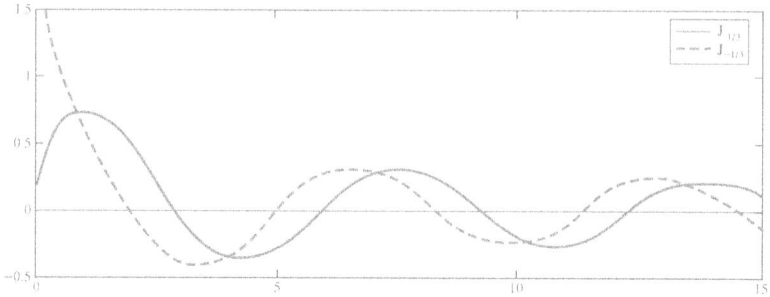

Figure 5.3 Bessel function of the first kind, $J_{1/3}$ and $J_{-1/3}$.

For integer $v = n$, Eq. (5.44) becomes

$$y(x) = c_1 J_n(x) + c_2 J_{-n}(x),$$

where $J_{-n}(x) = (-1)^n J_n(x)$. ($n = 1, 2, \cdots$)

But the above equation cannot be the general solution of the Bessel's equation for integer $v = n$, because, in this case, we have linear dependence.

To obtain the general solution of the Bessel equation, including integer $v = n$ or not, a new Bessel function of the second kind $Y_v(x)$ will be needed, as introduced in the next section.

Example 5.7

Solve the following ODE:

$$x^2 y'' + xy' + \left(x^2 - \frac{1}{4} \right) y = 0.$$

Solution

The general solution of the Bessel's equation $x^2 y'' + xy' + \left(x^2 - v^2 \right) y = 0$ with $v = \frac{1}{2}$ is

$$y(x) = c_1 J_{1/2}(x) + c_2 J_{-1/2}(x),$$

where

$$J_{1/2}(x) = x^{1/2} \sum_{k=0}^{\infty} \frac{(-1)^k}{2^{2k+1/2} \Gamma(k + 3/2)} x^{2k},$$

and

$$J_{-1/2}(x) = x^{-1/2} \sum_{k=0}^{\infty} \frac{(-1)^k}{2^{2k-1/2}\Gamma(k+1/2)} x^{2k}.$$

Answer $y(x) = c_1 J_{1/2}(x) + c_2 J_{-1/2}(x)$

Remark

Using MATLAB®, solutions can be obtained easily (refer to M_prob 5.6).

5.3.2 *Bessel function of the second kind* $Y_0(x)$

To obtain a general solution of Bessel's equation (5.23), we introduce a Bessel function of the second kind $Y_0(x)$ with $v = n = 0$.

$$xy'' + y' + xy = 0 \tag{5.45}$$

Then the indicial equation (5.26) has a double root $r = 0$.

As mentioned earlier, Bessel's equation (5.23) for $n = 0$ has the Bessel function of the first kind of order 0

$$J_0(x) = \sum_{k=0}^{\infty} \frac{(-1)^k}{2^{2k}(k!)^2} x^{2k} = 1 - \frac{x^2}{2^2(1!)^2} + \frac{x^4}{2^4(2!)^2} - \frac{x^6}{2^6(3!)^2} + \cdots, \tag{5.39}$$

and as mentioned earlier in Case III of Frobenius, the desired second solution becomes the following form

$$y_2(x) = J_0(x)\ln x + \sum_{k=1}^{\infty} A_k x^k. \tag{5.46}$$

We substitute $y_2(x)$ ands its derivatives

$$y_2' = J_0'\ln x + \frac{J_0}{x} + \sum_{k=1}^{\infty} kA_k x^{k-1}$$

$$y_2'' = J_0''\ln x + 2\frac{J_0'}{x} - \frac{J_0}{x^2} + \sum_{k=1}^{\infty} k(k-1)A_k x^{k-2}$$

into Eq. (5.45). Then we obtain

$$\left\{ xJ_0'' \ln x + 2J_0' - \frac{J_0}{x} + \sum_{k=1}^{\infty} k(k-1) A_k x^{k-1} \right\}$$

$$+ \left\{ J_0' \ln x + \frac{J_0}{x} + \sum_{k=1}^{\infty} k A_k x^{k-1} \right\} + \left\{ xJ_0 \ln x + \sum_{k=1}^{\infty} A_k x^{k+1} \right\} = 0.$$

Since $xJ_0'' + J_0' + xJ_0 = 0$, the above equation becomes

$$2J_0' + \sum_{k=1}^{\infty} k(k-1) A_k x^{k-1} + \sum_{k=1}^{\infty} k A_k x^{k-1} + \sum_{k=1}^{\infty} A_k x^{k+1} = 0,$$

or

$$2J_0' + \sum_{k=1}^{\infty} k^2 A_k x^{k-1} + \sum_{k=1}^{\infty} A_k x^{k+1} = 0. \qquad (5.47)$$

On the other hand, differentiating $J_0(x) = \sum_{k=0}^{\infty} \dfrac{(-1)^k}{2^{2k}(k!)^2} x^{2k}$ of Eq. (5.38) yields

$$J_0' = \sum_{k=1}^{\infty} \frac{(-1)^k 2k}{2^{2k}(k!)^2} x^{2k-1} = \sum_{k=1}^{\infty} \frac{(-1)^k}{2^{2k-1} k!(k-1)!} x^{2k-1}. \qquad (5.48)$$

Substituting Eq. (5.48) into Eq. (5.47) yields

$$\sum_{k=1}^{\infty} \frac{(-1)^k}{2^{2k-2} k!(k-1)!} x^{2k-1} + \sum_{k=1}^{\infty} k^2 A_k x^{k-1} + \sum_{k=1}^{\infty} A_k x^{k+1} = 0$$

or

$$\sum_{k=1}^{\infty} \frac{(-1)^k}{2^{2k-2} k!(k-1)!} x^{2k-1} + A_1 + 2^2 A_2 x + \sum_{k=1}^{\infty} \left\{ (k+2)^2 A_{k+2} + A_k \right\} x^{k+1} = 0. \qquad (5.49)$$

Firstly, the coefficient of the term x^0 must be zero, hence we obtain

$$A_1 = 0.$$

Similarly, the sum of the coefficients of the term x^{2k} ($k = 1, 2, \cdots$), respectively, must be 0, hence we obtain

$$3^2 A_3 + A_1 = 0, \; 5^2 A_5 + A_3 = 0, \text{ and so on.}$$

Then we obtain

$$A_3 = A_5 = \cdots = 0.$$

Substituting these values into all terms of Eq. (5.49) except for the first term, we get

$$A_1 + 2^2 A_2 x + \sum_{k=1}^{\infty} \left\{ (k+2)^2 A_{k+2} + A_k \right\} x^{k+1}$$

$$= 2^2 A_2 x + \left\{ 4^2 A_4 + A_2 \right\} x^3 + \left\{ 6^2 A_6 + A_4 \right\} x^5 + \left\{ 8^2 A_8 + A_6 \right\} x^7 + \cdots$$

$$= -A_0 x + \left\{ 2^2 A_2 + A_0 \right\} x + \left\{ 4^2 A_4 + A_2 \right\} x^3 + \left\{ 6^2 A_6 + A_4 \right\} x^5$$

$$+ \left\{ 8^2 A_8 + A_6 \right\} x^7 + \cdots$$

$$= -A_0 x + \sum_{k=1}^{\infty} \left\{ (2k)^2 A_{2k} + A_{2k-2} \right\} x^{2k-1}.$$

Substituting the above values back into Eq. (5.49) and simplifying, we get the following:

$$\sum_{k=1}^{\infty} \frac{(-1)^k}{2^{2k-2} k! (k-1)!} x^{2k-1} - A_0 x + \sum_{k=1}^{\infty} \left\{ (2k)^2 A_{2k} + A_{2k-2} \right\} x^{2k-1} = 0$$

or

$$-A_0 x + \sum_{k=1}^{\infty} \left\{ \frac{(-1)^k}{2^{2k-2} k! (k-1)!} + (2k)^2 A_{2k} + A_{2k-2} \right\} x^{2k-1} = 0.$$

Since the sum of the coefficients of the term x^{2k-1} ($k = 1, 2, \cdots$) must be 0, the sum of the coefficient of the term x becomes

$$-A_0 + \left\{ \frac{(-1)}{1!0!} + (2)^2 A_2 + A_0 \right\} = 0,$$

or

$$A_2 = \frac{1}{4}.$$

And the sum of the coefficient of the term x^3 becomes

$$\frac{1}{2^2 2!1!} + 4^2 A_4 + A_2 = 0,$$

or

$$A_4 = -\frac{3}{128}.$$

Similarly, the sum of the coefficient of the term x^5 becomes

$$\frac{-1}{2^4 3!2!} + 6^2 A_6 + A_4 = 0,$$

or

$$A_6 = \frac{11}{13824},$$

$$\cdots$$

In general, we obtain

$$A_{2k} = \frac{(-1)^{k-1}}{2^{2k}(k!)^2}\left(1 + \frac{1}{2} + \frac{1}{3} + \cdots + \frac{1}{k}\right). \qquad (k = 1, 2, \cdots)$$

Then by inserting the coefficients into Eq. (5.46), we obtain

$$y_2(x) = J_0(x)\ln x + \sum_{k=1}^{\infty} \frac{(-1)^{k-1}}{2^{2k}(k!)^2}\left(1 + \frac{1}{2} + \frac{1}{3} + \cdots + \frac{1}{k}\right)x^{2k} \qquad (5.50)$$

$$= J_0(x)\ln x + \frac{1}{4}x^2 - \frac{3}{128}x^4 + \frac{11}{13824}x^6 \mp \cdots.$$

Since $J_0(x)$ and $y_2(x)$ are linearly independent functions, they form a basis of the Bessel's equation (5.23) $xy'' + y' + xy = 0$ for $x > 0$.

If replacing $y_2(x)$ by an independent particular solution of the form $\frac{2}{\pi}\{y_2 + (\gamma - \ln 2)\}$, another basis $Y_0(x)$ is obtained as follows:

$$Y_0(x) = \frac{2}{\pi}J_0(x)\left(\ln\frac{x}{2} + \gamma\right) + \frac{2}{\pi}\sum_{k=1}^{\infty}\frac{(-1)^{k-1}}{2^{2k}(k!)^2}\left(1 + \frac{1}{2} + \frac{1}{3} + \cdots + \frac{1}{k}\right)x^{2k} \qquad (5.51)$$

where $\gamma = 0.5772156649\cdots$ (Euler constant).

Remark General solution of the Bessel's equation with $v = 0$

The general solution of Bessel's equation $xy'' + y' + xy = 0$ is

$$y(x) = c_1 J_0(x) + c_2 Y_0(x). \quad \text{(where } x \neq 0\text{)} \quad (5.52)$$

where $J_0(x) = \sum_{k=0}^{\infty} \frac{(-1)^k}{2^{2k}(k!)^2} x^{2k} = 1 - \frac{x^2}{2^2(1!)^2} + \frac{x^4}{2^4(2!)^2} - \frac{x^6}{2^6(3!)^2} + - \cdots$

$$(5.39)$$

and $Y_0(x) = \frac{2}{\pi} J_0(x) \left(\ln \frac{x}{2} + \gamma \right) + \frac{2}{\pi} \sum_{k=1}^{\infty} \frac{(-1)^{k-1}}{2^{2k}(k!)^2} \left(1 + \frac{1}{2} + \frac{1}{3} + \cdots + \frac{1}{k} \right) x^{2k}$

$$(5.51)$$

Figure 5.4 shows the Bessel function of the first kind J_0 and the second kind Y_0.

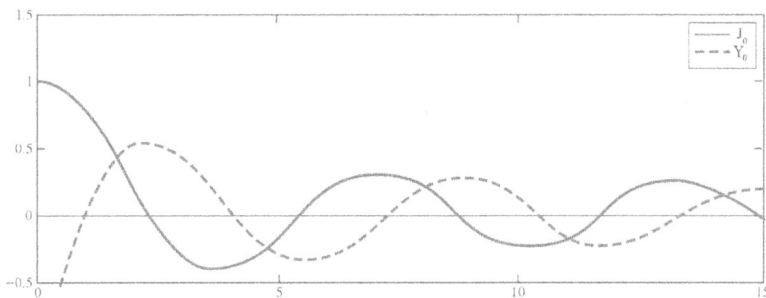

Figure 5.4 Bessel function of the first kind J_0 and the second kind Y_0.

5.3.3 Bessel function of the second kind $Y_n(x)$

To obtain a general solution of Bessel's equation (5.23), we introduce a Bessel function of the second kind $Y_n(x)$ with $v = n = 1, 2, \cdots$.

$$x^2 y'' + xy' + (x^2 - n^2) y = 0 \quad (5.23)$$

We already obtain the particular solution of Bessel's equation $J_n(x)$ as follows:

$$J_n(x) = x^n \sum_{k=0}^{\infty} \frac{(-1)^k}{2^{2k+n} k!(n+k)!} x^{2k}. \quad (n \geq 0) \quad (5.38)$$

And we introduce a standard second solution $Y_n(x)$ defined for all n, which is also independent of $J_n(x)$.

$$Y_n(x) = \frac{2}{\pi} J_n(x) \left(\ln \frac{x}{2} + \gamma \right) + \frac{x^n}{\pi} \sum_{k=1}^{\infty} \frac{(-1)^{k-1}}{2^{2k+n} k!(k+n)!} \left(\sum_{m=1}^{\infty} \frac{1}{m} + \sum_{m=1}^{\infty} \frac{1}{m+n} \right) x^{2k}$$

$$(5.53)$$

Remark   General solution of the Bessel's equation (n: positive integer)

General solution of Bessel's equation $x^2 y'' + xy' + \left(x^2 - n^2 \right) y = 0$ is

$$y(x) = c_1 J_n(x) + c_2 Y_n(x) \qquad (x \neq 0) \qquad (5.54)$$

where $J_n(x) = x^n \sum_{k=0}^{\infty} \frac{(-1)^k}{2^{2k+n} k!(n+k)!} x^{2k}. \qquad (n \geq 0) \qquad (5.38)$

and

$$Y_n(x) = \frac{2}{\pi} J_n(x) \left(\ln \frac{x}{2} + \gamma \right) + \frac{x^n}{\pi} \sum_{k=1}^{\infty} \frac{(-1)^{k-1}}{2^{2k+n} k!(k+n)!} \left(\sum_{m=1}^{\infty} \frac{1}{m} + \sum_{m=1}^{\infty} \frac{1}{m+n} \right) x^{2k}$$

$$(5.53)$$

Figure 5.5 shows the Bessel function of the first kind J_1 and the second kind Y_1.

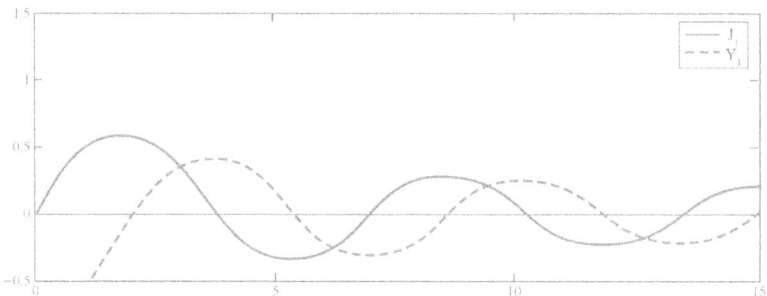

Figure 5.5 Bessel function of the first kind J_1 and the second kind Y_1.

Example 5.8

Solve the following ODE:

$$x^2 y'' + xy' + \left(x^2 - 1\right)y = 0.$$

Solution

The general solution of the Bessel equation $x^2 y'' + xy' + \left(x^2 - v^2\right)y = 0$ is

$$y(x) = c_1 J_1(x) + c_2 Y_1(x)$$

where

$$J_1(x) = x \sum_{k=0}^{\infty} \frac{(-1)^k}{2^{2k+1} k!(1+k)!} x^{2k},$$

and

$$Y_1(x) = \frac{2}{\pi} J_1(x)\left(\ln\frac{x}{2} + \gamma\right) + \frac{x}{\pi}\sum_{k=1}^{\infty} \frac{(-1)^{k-1}}{2^{2k+1} k!(k+1)!}\left(\sum_{m=1}^{\infty}\frac{1}{m} + \sum_{m=1}^{\infty}\frac{1}{m+1}\right)x^{2k}.$$

Answer $y(x) = c_1 J_1(x) + c_2 Y_1(x)$

5.3.4 *Legendre polynomials*

Legendre's equation:

$$\left(1 - x^2\right)y'' - 2xy' + n(n+1)y = 0 \qquad (n \text{ constant}) \qquad (5.24)$$

is one of the important ODEs in physics.

In order to obtain the general solution of the Legendre's equation, we apply the power series method.

Legendre's equation has a solution of the form

$$y = \sum_{k=0}^{\infty} c_k x^k. \qquad (5.55)$$

We differentiate $y = \sum_{k=0}^{\infty} c_k x^k$ term by term, finding

$$y' = \sum_{k=0}^{\infty} k c_k x^{k-1},$$

and

$$y'' = \sum_{k=0}^{\infty} k(k-1)c_k x^{k-2}.$$

Substituting the above equations into the given Eq. (5.24) yields

$$(1-x^2)y'' - 2xy' + n(n+1)y$$

$$= \sum_{k=0}^{\infty} k(k-1)c_k x^{k-2} - \sum_{k=0}^{\infty} k(k-1)c_k x^k - \sum_{k=0}^{\infty} 2kc_k x^k + \sum_{k=0}^{\infty} n(n+1)c_k x^k$$

$$= \sum_{k=0}^{\infty} (k+2)(k+1)c_{k+2} x^k - \sum_{k=0}^{\infty} k(k-1)c_k x^k - \sum_{k=0}^{\infty} 2kc_k x^k$$

$$+ \sum_{k=0}^{\infty} n(n+1)c_k x^k$$

$$= \sum_{k=0}^{\infty} \left[(k+2)(k+1)c_{k+2} - \left\{ k(k+1) - n(n+1) \right\} c_k \right] x^k = 0.$$

Similarly, the sum of the coefficients of the term x^k ($k = 0, 1, 2, \cdots$), respectively, must be 0, hence we obtain

$$c_{k+2} = -\frac{n(n+1)-(k+1)k}{(k+2)(k+1)} c_k = -\frac{(n-k)(n+k+1)}{(k+2)(k+1)} c_k.$$

Firstly, the coefficients of the even-degree terms of x^k are

$$c_2 = -\frac{n(n+1)}{2!} c_0 \qquad (\text{for } k = 0),$$

$$c_4 = -\frac{(n-2)(n+3)}{4\cdot3} c_2 = \frac{(n-2)n(n+1)(n+3)}{4!} c_0 \qquad (\text{for } k = 2),$$

and so on.

And the coefficients of the odd-degree terms of x^k are

$$c_3 = -\frac{(n-1)(n+2)}{3!} \qquad (\text{for } k = 1),$$

$$c_5 = -\frac{(n-3)(n+4)}{5\cdot4} c_3 = -\frac{(n-3)(n-1)(n+2)(n+4)}{5!} \qquad (\text{for } k = 3)$$

and so on.

Therefore, the general solution of the Legendre's equation is as follows.

Remark **General solution of Legendre's equation**

General solution of Legendre's equation $(1-x^2)y'' - 2xy' + n(n+1)y = 0$ is

$$y(x) = c_0 y_1(x) + c_1 y_2(x), \qquad (|x| < 1) \qquad (5.56)$$

where $y_1(x) = 1 - \dfrac{n(n+1)}{2!}x^2 + \dfrac{(n-2)n(n+1)(n+3)}{4!}x^4 - +\cdots$

and $y_2(x) = x - \dfrac{(n-1)(n+2)}{3!}x^3 + \dfrac{(n-3)(n-1)(n+2)(n+4)}{5!}x^5 - +\cdots$

Example 5.9

Solve the following ODE.

$$(1-x^2)y'' - 2xy' = 0 \qquad (|x| < 1)$$

Solution

The given ODE is a special Legendre's equation with $n = 0$.
    Therefore, from Eq. (5.50), the general solution of the given ODE is

$$y(x) = c_0 y_1(x) + c_1 y_2(x),$$

where

$$y_1(x) = 1, \text{ and } y_2(x) = x + \frac{x^3}{3} + \frac{x^5}{5} + \frac{x^7}{7} + \cdots.$$

$$\textbf{Answer } y(x) = c_0 + c_1\left(x + \frac{x^3}{3} + \frac{x^5}{5} + \frac{x^7}{7} + \cdots\right)$$

Problem 5.3

Solve the following Bessel equation. [1 ~ 4]

1. $x^2 y'' + xy' + \left(x^2 - \dfrac{4}{9}\right)y = 0$

2. $4x^2 y'' + 4xy' + (4x^2 - 1)y = 0$

3. $x^2 y'' + xy' + (x^2 - 4)y = 0$

4. $x^2 y'' + xy' + (x^2 - 9)y = 0$

Solve the following Bessel equation using a hint. [5 ~ 8]

5. $4x^2y'' + 4xy' + (x^2 - 1)y = 0$,   (hint: $x = 2z$)

6. $x^2y'' + xy' + (9x^2 - 4)y = 0$,   (hint: $3x = z$)

7. $xy'' + y' + y = 0$,   (hint: $2\sqrt{x} = z$)

8. $(x-1)^2y'' + (x-1)y' + x(x-2)y = 0$,   (hint: $x - 1 = z$)

Solve the following Legendre equation using the power series method. [9 ~ 10]

9. $(1 - x^2)y'' - 2xy' + 2y = 0$, $(|x| < 1)$

10. $(1 - x^2)y'' - 2xy' + 6y = 0$, $(|x| < 1)$

5.4   Utilizing MATLAB®

M-Example 5.1

Calculate the value of the following gamma function:

a. $\Gamma(4)$ b. $\Gamma(0.5)$

Solution   % gamma.m

a.

---

gamma(4)

Answer 6

b.

---

gamma(0.5)

Answer 1.7725

M-Example 5.2

Plot y_1 and y_2 of the following Bessel equation:

a. $x^2y'' + xy' + \left(x^2 - \dfrac{1}{4}\right)y = 0$

b. $xy'' + y' + xy = 0$

c. $x^2y'' + xy' + (x^2 - 4)y = 0$

Solution

a. Since the given equation $x^2y'' + xy' + \left(x^2 - \dfrac{1}{4}\right)y = 0$ is a Bessel's equation with $v = 1/2$, its general solution is

$$y(x) = c_1 J_{1/2}(x) + c_2 J_{-1/2}(x).$$

Answer $y_1 = J_{1/2}(x)$, $y_2 = J_{-1/2}(x)$.

```
close all clear all
x=[ 0: 0.2: 15];
subplot(211)
y1=besselj(1/2, x);                           % besselj.m
plot(x, y1, ' black', 'linewidth', 3); hold on
plot(x, 0*x, 'black')
axis([0 15 −.5 1.5])
legend('J_{1/2}'); xlabel('J_{1/2}(x)', 'fontsize', 12)

subplot(212)
y2=besselj(−1/2, x);                           % besselj.m
plot(x, y2, ':', 'linewidth', 3); hold on
plot(x, 0*x, 'black')
axis([0 15 −.5 1.5])
legend('J_{−1/2}'); xlabel('J_{−1/2}(x)', 'fontsize', 12)
```

Figure 5.6 shows the solutions in M-Example 5.2(a).

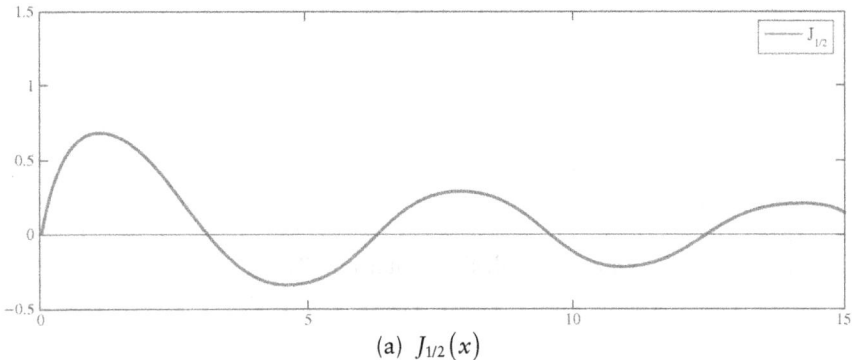

(a) $J_{1/2}(x)$

Figure 5.6 The solutions in M-Example 5.2(a). *(Continued)*

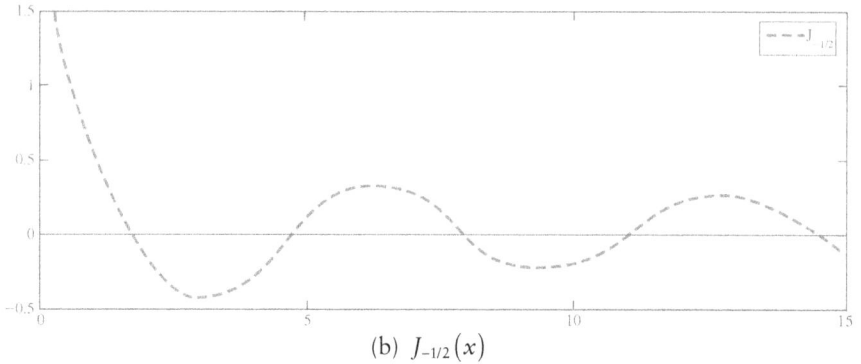

(b) $J_{-1/2}(x)$

Figure 5.6 (Continued)

b. Since the given equation $xy'' + y' + xy = 0$ is a Bessel's equation with $v = 0$, its general solution is

$$y(x) = c_1 J_0(x) + c_2 Y_0(x).$$

Answer $y_1 = J_0(x)$, $y_2 = Y_0(x)$

```
% Bessel function of the first kind
close all; clear all;
x=[0: 0.2: 15];
y1=besselj(0, x);                          % besselj.m
subplot(211)
plot(x, y1, ' black', 'linewidth', 3); hold on
plot(x, 0*x, 'black')
axis([0 15 -.5 1.5])
legend('J_{0}'); xlabel('J_{0}(x)', 'fontsize', 12)

% Bessel function of the second kind
y2=bessely(0, x);                          % bessely.m
subplot(212)
plot(x, y2, ':', 'linewidth', 3); hold on
plot(x, 0*x, 'black')
axis([0 15 -.5 1.5])
legend('Y_{0}'); xlabel('Y_{0}(x)', 'fontsize', 12)
```

Figure 5.7 shows the solutions in M-Example 5.2(b).

(a) $J_0(x)$

(b) $Y_0(x)$.

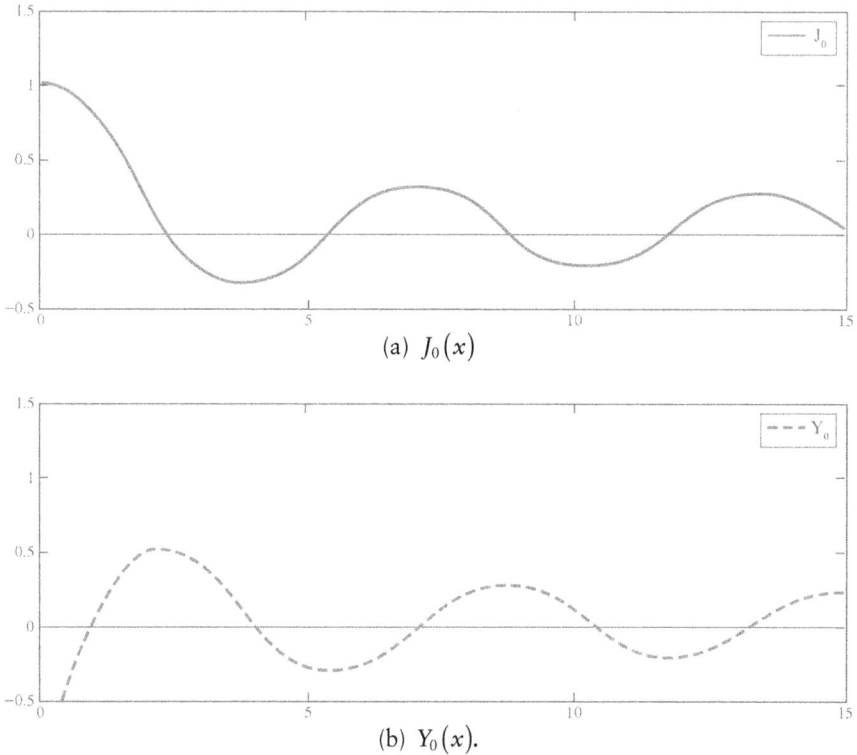

Figure 5.7 The solutions in M-Example 5.2(b).

c. Since the given equation $x^2 y'' + xy' + (x^2 - 4)y = 0$ is a Bessel's equation with $v = 2$, its general solution is

$$y(x) = c_1 J_2(x) + c_2 Y_2(x).$$

Answer $y_1 = J_2(x), y_2 = Y_2(x)$

```
% Bessel function of the first kind
close all; clear all;
x=[0: 0.2: 15];
y1=besselj(2, x);                          % besselj.m
subplot(211)
plot(x, y1, ' black', 'linewidth', 3); hold on
plot(x, 0*x, 'black')
axis([0 15 -.5 1.5])
legend('J_{2}'); xlabel('J_{2}(x)', 'fontsize', 12)
```

```
% Bessel function of the second kind
y2=bessely(2, x);                                    % bessely.m
subplot(212)
plot(x, y2, ':', 'linewidth', 3); hold on
plot(x, 0*x, 'black')
axis([0 15 -.5 1.5])
legend('Y_{2}'); xlabel('Y_{2}(x)', 'fontsize', 12)
```

Figure 5.8 shows the solutions in M-Example 5.2(c).

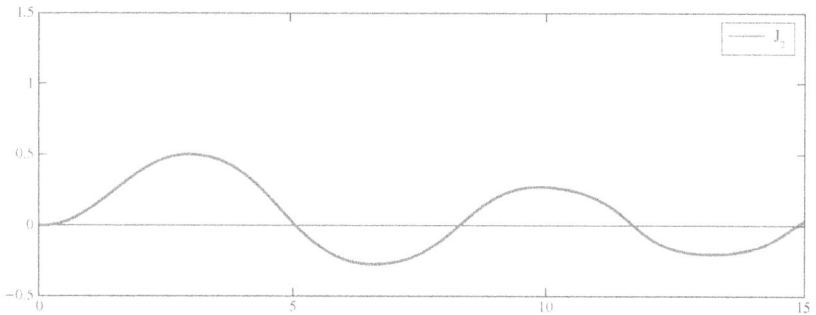

(a) $J_2(x)$

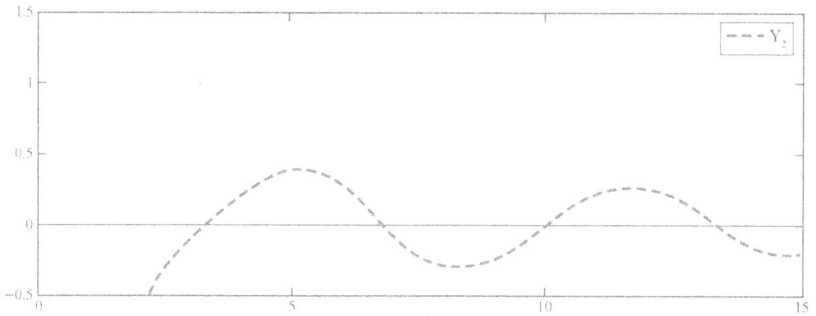

(b) $Y_2(x)$

Figure 5.8 The solutions in M-Example 5.2(c).

M-Example 5.3

Solve the following ODE using MATLAB®:

$$2t^2\ddot{y} - t\dot{y} + y = 0 \qquad (\text{refer to Example 5.4})$$

Solution

If the given ODE is modified into an explicit form with respect to

$$\ddot{y} = \frac{1}{2t}\dot{y} - \frac{1}{2t^2}y,$$

then we can use dsolve.m.

>> dsolve('D2y=1/2/t*Dy−1/2/t/t*y')

---

ans =
    C1*t^(1/2) − 2*C2*t

Answer $y(t) = C_1\sqrt{t} + C_2 t$

M-Example 5.4

Solve the following ODE using MATLAB®:

$$\left(t - t^2\right)\ddot{y} + t\dot{y} - y = 0. \qquad \left(\text{refer to Example 5.5}\right)$$

Solution

If the given ODE is modified into an explicit form with respect to

$$\ddot{y} = -\frac{1}{1-t}\dot{y} + \frac{1}{t-t^2}y,$$

then we can use dsolve.m.

>> dsolve('D2y=−1/(1−t)*Dy+1/(t−t^2)*y')

---

ans =
    C1*t + C2*t*(log(t) + 1/t)

Answer $y(t) = C_1 t + C_2\left(t\ln|t| + 1\right)$

M-Example 5.5

Solve the following ODE using MATLAB®:

$$t\ddot{y} - y = 0. \qquad \left(\text{refer to Example 5.6}\right)$$

Solution

If the given ODE is modified into an explicit form with respect to

$$\ddot{y} = \frac{1}{t} y,$$

then we can use dsolve.m.

>> dsolve('D2y=1/t*y')

---

ans =
    C1*t^(1/2)*besseli(1, 2*t^(1/2)) + C2*t^(1/2)*besselk(1, 2*t^(1/2))
% besseli(NU,Z) is the modified Bessel function of the first kind
% besselk(NU,Z) is the modified Bessel function of the second kind

Answer $y(t) = C_1\sqrt{t}\, J_1\left(2\sqrt{t}\right) + C_2\sqrt{t}\, Y_1\left(2\sqrt{t}\right)$

where $J_1\left(2\sqrt{t}\right)$ and $Y_1\left(2\sqrt{t}\right)$ are Bessel functions represented in Eq. (5.37) and Eq. (5.52), respectively.

M-Example 5.6

Solve the following ODE using MATLAB®:

$$t\ddot{y} + \dot{y} + ty = 0.$$

Solution

If the given ODE is modified into an explicit form with respect to

$$\ddot{y} = -\frac{1}{t}\dot{y} - y,$$

then we can use dsolve.m.

>> dsolve('D2y=−1/t*Dy − y')

---

ans =
    C1*besselj(0, t) + C2*bessely(0, t)
% besselj(NU,Z) is the Bessel function of the first kind
% bessely(NU,Z) is the Bessel function of the second kind

Answer $y(t) = c_1 J_0(t) + c_2 Y_0(t)$

where $J_0(t)$ and $Y_0(t)$ are Bessel functions that are represented in Eq. (5.38a) and Eq. (5.49), respectively.

Answer

Problem 5.1

1. $R = 1, |x| < 1$
2. $R = \infty$, convergence for all x
3. $R = \infty$, convergence for all x
4. $R = 2, -5 < x < -1$
5. $R = \sqrt{\dfrac{3}{2}}, |x| < \sqrt{\dfrac{3}{2}}$
6. $R = 1, 0 < x < 2$
7. $y(x) = Ae^x + Be^{-x}$

or $y = c_0\left(1 + \dfrac{x^2}{2!} + \dfrac{x^4}{4!} + \dfrac{x^6}{6!} + \cdots\right) + c_1\left(x + \dfrac{x^3}{3!} + \dfrac{x^5}{5!} + \dfrac{x^7}{7!} + \cdots\right)$

8. $y(x) = c_0 \cos x + c_1 \sin x$
9. In $y(x) = c_0 y_1(x) + c_1 y_2(x)$,

$y_1(x) = 1 + \dfrac{1}{3\cdot2}x^3 + \dfrac{1}{6\cdot5\cdot3\cdot2}x^6 + \dfrac{1}{9\cdot8\cdot6\cdot5\cdot3\cdot2}x^9 + \cdots,$

$y_2(x) = x + \dfrac{1}{4\cdot3}x^4 + \dfrac{1}{7\cdot6\cdot4\cdot3}x^7 + \dfrac{1}{10\cdot9\cdot7\cdot6\cdot4\cdot3}x^{10} + \cdots.$

10. $y = c_0\left(1 - \dfrac{x^2}{2!} - \dfrac{x^4}{4!} - \dfrac{x^6}{6!} - \dfrac{x^8}{8!} - \cdots\right) + c_1 x$

11. In $y(x) = c_0 y_1(x) + c_1 y_2(x)$,

$y_1(x) = 1 + \dfrac{1}{6}x^3 - \dfrac{1}{24}x^4 + \dfrac{1}{120}x^5 + \dfrac{7}{1800}x^6 + \cdots,$

$y_2(x) = x + \dfrac{1}{6}x^3 + \dfrac{1}{24}x^4 - \dfrac{1}{30}x^5 + \dfrac{11}{900}x^6 + \cdots$

12. In $y(x) = c_0 y_1(x) + c_1 y_2(x)$,

$y_1(x) = 1 + \dfrac{1}{2}x^2 + \dfrac{1}{8}x^4 + \dfrac{1}{48}x^6 + \cdots,$

$y_2(x) = x + \dfrac{1}{6}x^3 + \dfrac{7}{120}x^5 + \dfrac{3}{560}x^7 + \cdots$

Problem 5.2

1. In $y(x) = Ay_1(x) + By_2(x)$,

$y_1(x) = x^{3/2}\left(1 - \dfrac{x}{5} + \dfrac{x^2}{70} - \dfrac{x^3}{1890} + \cdots\right),$

$y_2(x) = 1 + x - \dfrac{x^2}{2} - \dfrac{x^3}{9} - \dfrac{x^4}{20} - \cdots$

2. In $y(x) = Ay_1(x) + By_2(x)$,

$$y_1(x) = 1 + \frac{x^2}{14} + \frac{x^4}{616} + \cdots,$$

$$y_2(x) = x^{-3/2}\left(1 + \frac{x^2}{2} + \frac{x^4}{40} + \frac{x^6}{2160} + \cdots\right)$$

3. $y(x) = Ax + Bx^{-1}$

4. $y(x) = Ax^2 + Bx$

5. $y(x) = Ax + B(x\ln|x| + 1)$

6. In $y(x) = Ay_1(x) + By_2(x)$,

$$y_1(x) = 1 + \frac{x^2}{2^2} + \frac{x^4}{4^2 2^2} + \frac{x^6}{6^2 4^2 2^2} + \cdots,$$

$$y_2(x) = x^{-1}\left(1 + x^2 + \frac{x^4}{3^2} + \frac{x^6}{5^2 3^2} + \cdots\right)$$

7. In $y(x) = Ay_1(x) + By_2(x)$,

$$y_1(x) = x + \frac{x^2}{2} + \frac{x^3}{12} + \frac{x^4}{144} + \cdots,$$

$$y_2(x) = y_1\ln|x| + 1 - \frac{3}{4}x^2 - \frac{7}{36}x^3 - \cdots$$

8. $y(x) = Ax + Bx\ln|x|$

9. $y(x) = Ax^2 + Bx^2\ln|x|$

10. $y(x) = \frac{A}{x} + \frac{B}{x}\ln|x|$

11. $y(x) = \frac{A}{1-x} + \frac{B\ln|x|}{1-x}$

12. $y(x) = Ay_1(x) + By_2(x)$,

$$y_1(x) = 1 + x + \frac{x^2}{(2!)^2} + \frac{x^3}{(3!)^2} + \frac{x^4}{(4!)^2} + \cdots,$$

$$y_2(x) = y_1\ln|x| + 1 - x - \frac{1}{2}x^2 + \frac{1}{27}x^3 - \cdots.$$

Problem 5.3

1. $y(x) = c_1 J_{2/3}(x) + c_2 J_{-2/3}(x)$

2. $y(x) = c_1 J_{1/2}(x) + c_2 J_{-1/2}(x)$

3. $y(x) = c_1 J_2(x) + c_2 Y_2(x)$

4. $y(x) = c_1 J_3(x) + c_2 Y_3(x)$

5. $y = c_1 J_{1/2}\left(\dfrac{x}{2}\right) + c_2 J_{-1/2}\left(\dfrac{x}{2}\right)$

6. $y = c_1 J_2(3x) + c_2 Y_2(3x)$

7. $y = c_1 J_0\left(2\sqrt{x}\right) + c_2 Y_0\left(2\sqrt{x}\right)$

8. $y = c_1 J_1(x-1) + c_2 Y_1(x-1)$

9. $y(x) = c_0\left(1 - x^2 - \dfrac{1}{3}x^4 - \cdots\right) + c_1 x$

10. $y(x) = c_0\left(1 - x^2\right) + c_1\left(x - \dfrac{2x^3}{3} - \dfrac{x^5}{5} - \cdots\right)$

6 Linear algebra
Matrices, Determinants, and Eigenvalue Problems

Linear algebra is a field that systematically organizes a wealth of information using matrices and vectors. With the advancement of computers, linear algebra computations have become easier, making it useful for engineering design, analysis, and experimentation.

6.1 Matrices fundamentals

A vector of n dimensions consisting of a single horizontal row, as in Eq. (6.1), is referred to as a *row vector*.

$$\mathbf{a} = \{a_1, a_2, \cdots, a_n\} \tag{6.1}$$

Furthermore, a vector of m dimensions consisting of a single vertical column, as in Eq. (6.2), is referred to as a *column vector*.

$$\mathbf{b} = \begin{Bmatrix} b_1 \\ b_2 \\ \vdots \\ b_m \end{Bmatrix} \tag{6.2}$$

Generally, an $m \times n$ matrix refers to a *matrix* composed of m row vectors and n column vectors, and it can be represented as follows:

$$\mathbf{A}_{m \times n} = \begin{bmatrix} a_{ij} \end{bmatrix} = \begin{bmatrix} a_{11} & a_{12} & \cdots & a_{1n} \\ a_{21} & a_{22} & \cdots & a_{2n} \\ a_{31} & a_{32} & \cdots & a_{3n} \\ \vdots & \vdots & \ddots & \vdots \\ a_{m1} & a_{m2} & \cdots & a_{mn} \end{bmatrix}. \tag{6.3}$$

Here, a_{ij} denotes the element at the i-th row and j-th column. Here, $i = 1, 2, \ldots, m$ and $j = 1, 2, \ldots, n$. Furthermore, a matrix with $m = n$ is called a *square matrix*. A square matrix where all elements except those on the diagonal are

DOI: 10.1201/9781003608912-6

0 is called a *diagonal matrix,* and a matrix where all elements on the diagonal are 1 is called an *identity matrix I.*

$$\text{Diagonal matrix: } \mathbf{A} = \begin{bmatrix} a_{11} & 0 & \cdots & 0 \\ 0 & a_{22} & \cdots & 0 \\ \vdots & \vdots & \ddots & \vdots \\ 0 & 0 & \cdots & a_{nn} \end{bmatrix}$$

$$\text{Identity matrix: } \mathbf{I} = \begin{bmatrix} 1 & 0 & \cdots & 0 \\ 0 & 1 & \cdots & 0 \\ \vdots & \vdots & \ddots & \vdots \\ 0 & 0 & \cdots & 1 \end{bmatrix}$$

In this chapter, matrices are represented using [ ], while vectors are represented using { } or (). A 2×2 matrix is denoted as shown in Eq. (6.4), and a 2×2 identity matrix is represented as in Eq. (6.5).

$$\mathbf{A}_{2\times2} = \left(a_{ij} \right) = \begin{bmatrix} a_{11} & a_{12} \\ a_{21} & a_{22} \end{bmatrix} \tag{6.4}$$

$$\mathbf{I} = \begin{bmatrix} 1 & 0 \\ 0 & 1 \end{bmatrix} \tag{6.5}$$

6.1.1 Equality of matrices

The *equivalence* or *equality* of two matrices, $\mathbf{A}_{m_1 \times n_1}$ and $\mathbf{B}_{m_2 \times n_2}$, is defined when both matrices have the same dimensions. That is, when the sizes of the matrices are equal, with $m_1 = m_2$ and $n_1 = n_2$, and for all i and j, if the element in the i-th row and j-th column of matrix $\mathbf{A}$ is equal to the element in the i-th row and j-th column of matrix $\mathbf{B}$, then matrices $\mathbf{A}$ and $\mathbf{B}$ are considered equivalent or equal. This can be expressed in the following formula:

$$a_{ij} = b_{ij} \ (i = 1, 2, \cdots, m, \ j = 1, 2, \cdots, n). \tag{6.6}$$

If matrices $\mathbf{A}$ and $\mathbf{B}$ are equivalent or equal, we can express this as $\mathbf{A} = \mathbf{B}$.

6.1.2 Sum and difference of matrices, scalar multiplication

The sum and difference of two matrices are defined only when the matrices have the same dimensions. The sum and difference of matrices are calculated by adding and subtracting corresponding elements, respectively. That is, the sum $A + B$ of matrices A and B is calculated as $a_{ij} + b_{ij}$ for all i and j, and the difference $A - B$ of matrices A and B is calculated as $a_{ij} - b_{ij}$ for all i and j.

And the multiplication cA of matrix $A_{m \times n}$ by scalar c is defined as multiplying each element of matrix A by the scalar c, i.e., it is calculated as $c a_{ij}$ for all i and j.

The operational rules for matrix addition and scalar multiplication are as follows:

$$A + B = B + A, \qquad \text{(commutative law)} \tag{6.7a}$$

$$(A + B) + C = A + (B + C), \qquad \text{(associative law)} \tag{6.7b}$$

$$c(A + B) = cA + cB, \qquad \text{(distributive law)} \tag{6.7c}$$

$$A + (-A) = 0. \tag{6.7d}$$

Here, the matrix 0 denotes a matrix where all elements are 0, for all i and j. This matrix is called the zero matrix or the null matrix.

6.1.3   The product of two matrices

The *product* $C = AB$ of an $m \times n$ matrix A times an $n \times l$ matrix B is defined and is then the $m \times l$ matrix.

$$C = AB \tag{6.8}$$

$$= \begin{bmatrix} a_{11} & a_{12} & \cdots & a_{1n} \\ a_{21} & a_{22} & \cdots & a_{2n} \\ a_{31} & a_{32} & \cdots & a_{3n} \\ \vdots & \vdots & \ddots & \vdots \\ a_{m1} & a_{m2} & \cdots & a_{mn} \end{bmatrix} \begin{bmatrix} b_{11} & b_{12} & \cdots & b_{1l} \\ b_{21} & b_{22} & \cdots & b_{2l} \\ b_{31} & b_{32} & \cdots & b_{3l} \\ \vdots & \vdots & \ddots & \vdots \\ b_{n1} & b_{n2} & \cdots & b_{nl} \end{bmatrix}$$

where $\left[c_{ij} \right] = a_{i1}b_{1j} + a_{i2}b_{2j} + \cdots + a_{in}b_{nj}$.

For example, the product of a 2×3 matrix times a 3×1 matrix is then a 2×1 matrix.

$$C_{2 \times 1} = A_{2 \times 3} B_{3 \times 1}$$

$$= \begin{bmatrix} a_{11} & a_{12} & a_{13} \\ a_{21} & a_{22} & a_{23} \end{bmatrix} \begin{bmatrix} b_{11} \\ b_{21} \\ b_{31} \end{bmatrix} = \begin{bmatrix} a_{11}b_{11} + a_{12}b_{21} + a_{13}b_{31} \\ a_{21}b_{11} + a_{22}b_{21} + a_{23}b_{31} \end{bmatrix}$$

The operational rules for matrix multiplication are as follows:

$$(AB)C = A(BC), \qquad \text{(associative law)} \qquad (6.9a)$$

$$(A + B)C = AC + BC, \qquad \text{(distributive law)} \qquad (6.9b)$$

$$C(A + B) = CA + CB. \qquad \text{(distributive law)} \qquad (6.9c)$$

6.1.4 Transposition matrix

The *transposition matrix* of A, obtained by swapping the rows and columns of a matrix A, is denoted by A^T. The laws governing transpose operations are as follows:

$$(A^T)^T = A, \qquad (6.10a)$$

$$(A + B)^T = A^T + B^T, \qquad (6.10b)$$

$$(cA)^T = cA^T, \qquad (c \text{ arbitrary}) \qquad (6.10c)$$

$$(AB)^T = B^T A^T. \qquad (6.10d)$$

Example 6.1

Solve the problem.

a. When $A = \begin{bmatrix} a_{ij} \end{bmatrix} = \begin{bmatrix} 2 & 3 & -1 \\ 3 & -2 & 1 \end{bmatrix}$ and $B = \begin{bmatrix} b_{ij} \end{bmatrix} = \begin{bmatrix} -1 & 1 & 2 \\ 3 & 1 & -4 \end{bmatrix}$, find

$A + B$, $A - B$, and $3A$, respectively.

b. Find the transposition matrix A^T of

$$A = \begin{bmatrix} a_{ij} \end{bmatrix} = \begin{bmatrix} 2 & 3 & -1 \\ 3 & -2 & 1 \end{bmatrix}.$$

c. When $A = \begin{bmatrix} a_{ij} \end{bmatrix} = \begin{bmatrix} 2 & 3 \\ 3 & -2 \end{bmatrix}$, $B = \begin{bmatrix} b_{ij} \end{bmatrix} = \begin{bmatrix} -1 & 1 \\ 3 & 1 \end{bmatrix}$, find the product of

two matrices AB.

d. When $A = \begin{bmatrix} 2 & 1 \\ 3 & 2 \\ 1 & -1 \end{bmatrix}$, $B = \begin{bmatrix} 1 & -1 & 3 \\ 3 & 2 & 1 \end{bmatrix}$, find the product of two

matrices AB.

Answer a. $\begin{bmatrix} 1 & 4 & 1 \\ 6 & -1 & -3 \end{bmatrix}$, $\begin{bmatrix} 3 & 2 & -3 \\ 0 & -3 & 5 \end{bmatrix}$, $\begin{bmatrix} 6 & 9 & -3 \\ 9 & -6 & 3 \end{bmatrix}$,

b. $\mathbf{A}^T = \begin{bmatrix} 2 & 3 \\ 3 & -2 \\ -1 & 1 \end{bmatrix}$, c. $\mathbf{AB} = \begin{bmatrix} 2 & 3 \\ 3 & -2 \end{bmatrix}\begin{bmatrix} -1 & 1 \\ 3 & 1 \end{bmatrix} = \begin{bmatrix} 7 & 5 \\ -9 & 1 \end{bmatrix}$,

d. $\mathbf{AB} = \begin{bmatrix} 2 & 1 \\ 3 & 2 \\ 1 & -1 \end{bmatrix}\begin{bmatrix} 1 & -1 & 3 \\ 3 & 2 & 1 \end{bmatrix} = \begin{bmatrix} 5 & 0 & 7 \\ 9 & 1 & 11 \\ -2 & -3 & 2 \end{bmatrix}$

Problem 6.1

Solve the problem. [1 ~ 10]

1. When $\mathbf{A} = \begin{bmatrix} 1 & 3 \\ 2 & -1 \end{bmatrix}$ and $\mathbf{B} = \begin{bmatrix} -1 & 1 \\ 3 & 1 \end{bmatrix}$, find $\mathbf{A} + 2\mathbf{B}$.

2. When $\mathbf{A} = \begin{bmatrix} 1 & -3 \\ 2 & 2 \\ -2 & -1 \end{bmatrix}$ and $\mathbf{B} = \begin{bmatrix} 1 & 2 \\ 2 & -1 \\ 1 & 3 \end{bmatrix}$, find $2\mathbf{A} - \mathbf{B}$.

3. Find the transposition matrix $\mathbf{A}^T$ of $\mathbf{A} = \{ 3 \ 1 \ -1 \}$.

4. Find the transposition matrix $\mathbf{A}^T$ of $\mathbf{A} = \begin{bmatrix} 3 & 1 \\ 2 & -1 \\ -2 & 3 \end{bmatrix}$.

5. When $\mathbf{A} = \begin{bmatrix} -2 & 1 \\ 2 & -2 \end{bmatrix}$ and $\mathbf{B} = \begin{bmatrix} -2 & -1 \\ 1 & 3 \end{bmatrix}$, find the product of two matrices $\mathbf{AB}$.

6. When $\mathbf{A} = \begin{bmatrix} -2 & 1 \\ 2 & -2 \end{bmatrix}$ and $\mathbf{B} = \begin{bmatrix} -2 & -1 \\ 1 & 3 \end{bmatrix}$, find the product of two matrices $\mathbf{BA}$.

7. When $\mathbf{A} = \{ 1 \ 1 \ -2 \}$ and $\mathbf{B} = \begin{Bmatrix} 1 \\ 2 \\ -1 \end{Bmatrix}$, find the product of two matrices $\mathbf{AB}$.

8. When $\mathbf{A} = \{ 1 \ 1 \ -2 \}$ and $\mathbf{B} = \begin{Bmatrix} 1 \\ 2 \\ -1 \end{Bmatrix}$, find the product of two matrices $\mathbf{BA}$.

9. When $\mathbf{A} = \begin{bmatrix} 1 & -1 & 3 \\ 3 & 2 & 1 \end{bmatrix}$ and $\mathbf{B} = \begin{bmatrix} 2 & 1 \\ 3 & 2 \\ 1 & -1 \end{bmatrix}$, find the product of two matrices $\mathbf{AB}$.

10. When $\mathbf{A} = \begin{bmatrix} 1 & -1 & 3 \\ 3 & 2 & 1 \end{bmatrix}$ and $\mathbf{B} = \begin{bmatrix} 2 & 1 \\ 3 & 2 \\ 1 & -1 \end{bmatrix}$, find the product of two matrices $\mathbf{BA}$.

6.2 System of linear equations, determinants, and inverse matrix

6.2.1 System of linear equations

A *system of linear equations* with n unknowns $x_1, x_2, \cdots, x_n$ can be represented as follows:

$$
\begin{aligned}
a_{11}x_1 + a_{12}x_2 + \cdots + a_{1n}x_n &= b_1 \\
a_{21}x_1 + a_{22}x_2 + \cdots + a_{2n}x_n &= b_2 \\
&\vdots
\end{aligned}
\qquad ,
\tag{6.11}
$$

$$
a_{m1}x_1 + a_{m2}x_2 + \cdots + a_{mn}x_n = b_m
$$

where a_{ij} $(i = 1, 2, \cdots, m,\ j = 1, 2, \cdots, n)$ and b_i $(i = 1, 2, \cdots, m)$ are the coefficients given in the problem, and $x_1, x_2, \cdots, x_n$ are the unknowns we want to solve for in this system of linear equations. This can be represented in matrix form as follows:

$$
\mathbf{Ax} = \mathbf{B},
\tag{6.12}
$$

where the coefficient matrix $\mathbf{A}$, and the column vectors $\mathbf{x}$ and $\mathbf{B}$ are

$$
\mathbf{A} =
\begin{bmatrix}
a_{11} & a_{12} & \cdots & a_{1n} \\
a_{21} & a_{22} & \cdots & a_{2n} \\
\vdots & \vdots & \ddots & \vdots \\
a_{m1} & a_{m2} & \cdots & a_{mn}
\end{bmatrix},
\quad
\mathbf{x} =
\begin{Bmatrix}
x_1 \\
x_2 \\
\vdots \\
x_n
\end{Bmatrix},
\quad
\mathbf{B} =
\begin{Bmatrix}
b_1 \\
b_2 \\
\vdots \\
b_n
\end{Bmatrix}.
\tag{6.13}
$$

6.2.2 Linear system of equations: Gauss elimination

Let's solve a problem to find the solution of the following *system of linear equations*:

$$
-x_1 + 2x_2 = 0,
\tag{a}
$$

$$
x_1 + x_2 = 3.
\tag{b}
$$

Let's solve it step by step. If we represent the system of equations in matrix form, it would look like this:

$$
\begin{bmatrix}
-1 & 2 \\
1 & 1
\end{bmatrix}
\begin{Bmatrix}
x_1 \\
x_2
\end{Bmatrix}
=
\begin{Bmatrix}
0 \\
3
\end{Bmatrix}.
\tag{c}
$$

If we add Eq. (a) and Eq. (b), we get

$$
3x_2 = 3.
$$

Then substituting it into Eq. (a) or Eq. (b), we get

$$x_1 = 2.$$

Therefore, the answer is

$$x_1 = 2 \quad \text{and} \quad x_2 = 1.$$

Now, let's express this process using *Gauss elimination.*
First, let's represent the system of equations in augmented matrix form:

$$\left[\begin{array}{cc:c} -1 & 2 & 0 \\ 1 & 1 & 3 \end{array} \right]. \tag{d}$$

Step 1: We'll use the first row to eliminate the second row's first element (1). To do this, we'll add the first row and the second row:

$$\left[\begin{array}{cc:c} -1 & 2 & 0 \\ 0 & 3 & 3 \end{array} \right]. \tag{e}$$

Step 2: Next, we'll divide the second row by 3 to make the leading coefficient of the second row 1:

$$\left[\begin{array}{cc:c} -1 & 2 & 0 \\ 0 & 1 & 1 \end{array} \right]. \tag{f}$$

Step 3: Now, we'll use the second row to eliminate the first row's second element (2). We'll multiply the second row by -2 and add it to the first row:

$$\left[\begin{array}{cc:c} -1 & 0 & -2 \\ 0 & 1 & 1 \end{array} \right]. \tag{g}$$

Step 4: Finally, we'll divide each row by its leading coefficient:

$$\left[\begin{array}{cc:c} 1 & 0 & 2 \\ 0 & 1 & 1 \end{array} \right]. \tag{h}$$

Now, we have the solution for the system of equations:

$$x_1 = 2 \quad \text{and} \quad x_2 = 1.$$

Remark

When finding a solution to a system of equations, *using the inverse matrix* described in Section 6.2.5 is simpler than Gauss elimination as described in Section 6.2.2.

Additionally, utilizing MATLAB® in Section 6.4 enables us to solve a system of equations more efficiently. (Refer to M_Example 6.5 for details.)

Example 6.2

Solve the system of equations using Gauss elimination.

$$3x + 2y + z = 7$$
$$x - y + 2z = -1$$
$$2x + y - z = 6$$

Solution

First, let's represent the system of equations in augmented matrix form:

$$\left[\begin{array}{ccc:c} 3 & 2 & 1 & 7 \\ 1 & -1 & 2 & -1 \\ 2 & 1 & -1 & 6 \end{array} \right].$$

Step 1: We'll eliminate the second row's first element (1) and the third row's first element (2).

$$\left[\begin{array}{ccc:c} 3 & 2 & 1 & 7 \\ 0 & 5 & -5 & 10 \\ 0 & -3 & 5 & -8 \end{array} \right].$$

Step 2: Next, we'll divide the second row by 5 to make the second row's second coefficient of 1:

$$\left[\begin{array}{ccc:c} 3 & 2 & 1 & 7 \\ 0 & 1 & -1 & 2 \\ 0 & -3 & 5 & -8 \end{array} \right].$$

Step 3: We'll eliminate the third row's second element (−3):

$$\begin{bmatrix} 3 & 2 & 1 & : & 7 \\ 0 & 1 & -1 & : & 2 \\ 0 & 0 & 2 & : & -2 \end{bmatrix}.$$

Step 4: We'll divide the third row by 2 to make the third row's third coefficient of 1:

$$\begin{bmatrix} 3 & 2 & 1 & : & 7 \\ 0 & 1 & -1 & : & 2 \\ 0 & 0 & 1 & : & -1 \end{bmatrix}.$$

Step 5: We'll eliminate the second row's third element (−1) and the first row's third element (1).

$$\begin{bmatrix} 3 & 3 & 0 & : & 9 \\ 0 & 1 & 0 & : & 1 \\ 0 & 0 & 1 & : & -1 \end{bmatrix}.$$

Step 6: We'll divide the first row by 3 to make the first row's first coefficient of 1:

$$\begin{bmatrix} 1 & 1 & 0 & : & 3 \\ 0 & 1 & 0 & : & 1 \\ 0 & 0 & 1 & : & -1 \end{bmatrix}.$$

Step 7: We'll eliminate the first row's second element (1):

$$\begin{bmatrix} 1 & 0 & 0 & : & 2 \\ 0 & 1 & 0 & : & 1 \\ 0 & 0 & 1 & : & -1 \end{bmatrix}.$$

Now, we have the solution for the system of equations:

$$x = 2, \ y = 1, \ z = -1.$$

Answer $x = 2, \ y = 1, \ z = -1$

6.2.3   *Linearly independent and rank of matrix*

In this section, we will learn concepts such as linearly independence, rank of matrix, and other criteria necessary to determine whether solutions exist or are unique for a system of linear equations.

Given any set of n vectors $\mathbf{a}_{(1)}, \mathbf{a}_{(2)}, \cdots, \mathbf{a}_{(n)}$, a linear combination of these vectors is in the following form:

$$c_1 \mathbf{a}_{(1)} + c_2 \mathbf{a}_{(2)} + \cdots + c_n \mathbf{a}_{(n)},$$

where coefficients $c_1, c_2, \cdots, c_n$ are any scalars.

Now we consider the equation

$$c_1 \mathbf{a}_{(1)} + c_2 \mathbf{a}_{(2)} + \cdots + c_n \mathbf{a}_{(n)} = 0. \tag{6.14}$$

Of course, if all coefficients $c_1, c_2, \cdots, c_n$ are zero, then Eq. (6.9) becomes the zero vector.

If all coefficients $c_1, c_2, \cdots, c_n$ are zero, satisfying Eq. (6.9) with a unique value, we refer to these vectors $\mathbf{a}_{(1)}, \mathbf{a}_{(2)}, \cdots, \mathbf{a}_{(n)}$ as *linearly independent*. On the other hand, if at least one coefficient is nonzero, then those vectors are *linearly dependent*.

The maximum number of linearly independent row vectors in a matrix $\mathbf{A}$ is referred to as the *rank of the matrix* $\mathbf{A}$. This is denoted by $\operatorname{rank}(\mathbf{A})$ in notation.

Example 6.3

Find the rank.

$$\mathbf{A} = \begin{bmatrix} 5 & 0 & 7 \\ 9 & 1 & 11 \\ -2 & -3 & 2 \end{bmatrix}$$

Solution

We'll eliminate the second row's first element (9) and the third row's first element (-2):

$$\begin{bmatrix} 5 & 0 & 7 \\ 0 & -5 & 8 \\ 0 & -25 & 40 \end{bmatrix}.$$

We'll eliminate the third row's second element (-25):

$$\begin{bmatrix} 5 & 0 & 7 \\ 0 & -5 & 8 \\ 0 & 0 & 0 \end{bmatrix}.$$

Then the last matrix, which is in row-echelon form, has two nonzero rows. Hence, $\operatorname{rank}(\mathbf{A}) = 2$.

Answer 2

6.2.4 Determinant

Determinant is a scalar quantity defined for square matrices. First, the determinant for a 2×2 matrix is defined as

$$|\mathbf{A}| = \det(\mathbf{A}) = \begin{vmatrix} a_{11} & a_{12} \\ a_{21} & a_{22} \end{vmatrix} = a_{11}a_{22} - a_{12}a_{21}. \tag{6.15}$$

Furthermore, the determinant for a 3×3 matrix is defined as

$$|\mathbf{A}| = \det(\mathbf{A}) = \begin{vmatrix} a_{11} & a_{12} & a_{13} \\ a_{21} & a_{22} & a_{23} \\ a_{31} & a_{32} & a_{33} \end{vmatrix} \tag{6.16}$$

$$= a_{11}C_{11} + a_{12}C_{12} + a_{13}C_{13}$$

$$= a_{11} \begin{vmatrix} a_{22} & a_{23} \\ a_{32} & a_{33} \end{vmatrix} - a_{12} \begin{vmatrix} a_{21} & a_{23} \\ a_{31} & a_{33} \end{vmatrix} + a_{13} \begin{vmatrix} a_{21} & a_{22} \\ a_{31} & a_{32} \end{vmatrix}.$$

Expanding upon that, for an $n \times n$ matrix, we can also get the determinant as follows.

Remark Determinant $|\mathbf{A}|$

$$|\mathbf{A}| = \det(\mathbf{A}) = \sum_{k=1}^{n} a_{ik}C_{ik} \tag{6.17a}$$

or

$$|\mathbf{A}| = \det(\mathbf{A}) = \sum_{k=1}^{n} a_{kj}C_{kj} \tag{6.17b}$$

where the *cofactors* of $\mathbf{A}$, C_{ik} and C_{kj} are defined as $C_{ik} = (-1)^{i+k} A_{ik}$ and $C_{kj} = (-1)^{k+j} A_{kj}$, respectively.

Also, A_{ik} and A_{kj} are *minors*, corresponding to the determinant of the submatrix obtained by excluding the row and column of the elements a_{ik} and a_{kj} from the matrix $\mathbf{A}$.

Example 6.4

Get the determinant $|\mathbf{A}|$.

a. $\mathbf{A} = \begin{bmatrix} 2 & 1 \\ 3 & -4 \end{bmatrix}$

b. $\mathbf{A} = \begin{bmatrix} 2 & 1 & 2 \\ 1 & 3 & 2 \\ 3 & -2 & 0 \end{bmatrix}$.

Solution

a.

$$|\mathbf{A}| = \begin{vmatrix} 2 & 1 \\ 3 & -4 \end{vmatrix} = 2 \cdot (-4) - 1 \cdot 3 = -11$$

Answer −11

b.

$$|\mathbf{A}| = \begin{vmatrix} 2 & 1 & 2 \\ 1 & 3 & 2 \\ 3 & -2 & 0 \end{vmatrix} = 2 \cdot \begin{vmatrix} 3 & 2 \\ -2 & 0 \end{vmatrix} - \begin{vmatrix} 1 & 2 \\ 3 & 0 \end{vmatrix} + 2 \cdot \begin{vmatrix} 1 & 3 \\ 3 & -2 \end{vmatrix} = -8$$

Answer −8

6.2.5 *Inverse matrix*

The *inverse matrix* is defined for square matrices, similar to the determinant. The inverse matrix of an $n \times n$ matrix $\mathbf{A}$ is denoted as $\mathbf{A}^{-1}$, and it is defined as

$$\mathbf{A}^{-1}\mathbf{A} = \mathbf{A}\,\mathbf{A}^{-1} = \mathbf{I}. \tag{6.18}$$

However, not all matrices have an inverse. First, the inverse of a 2×2 matrix $\mathbf{A} = \begin{bmatrix} a & b \\ c & d \end{bmatrix}$ is calculated as

$$\mathbf{A}^{-1} = \frac{1}{\det(\mathbf{A})}\begin{bmatrix} d & -b \\ -c & a \end{bmatrix}. \tag{6.19}$$

The inverse matrix of an $n \times n$ matrix $\mathbf{A}$ is defined as

$$\mathbf{A}^{-1} = \frac{1}{\det(\mathbf{A})}[C_{ij}]^T, \tag{6.20}$$

where C_{ij} represents the *cofactor* of matrix $\mathbf{A}$, defined as $C_{ij} = (-1)^{i+j} A_{ij}$. Additionally, A_{ij} is the minor of the matrix $\mathbf{A}$.

Example 6.5

Determine the inverse matrix $\mathbf{A}^{-1}$.

a. $\mathbf{A} = \begin{bmatrix} 2 & 2 \\ 3 & 4 \end{bmatrix}$

b. $\mathbf{A} = \begin{bmatrix} 2 & 1 & 2 \\ 1 & 3 & 2 \\ 3 & -2 & 0 \end{bmatrix}$

Solution

a.

$$\mathbf{A}^{-1} = \frac{1}{2}\begin{bmatrix} 4 & -2 \\ -3 & 2 \end{bmatrix} = \begin{bmatrix} 2 & -1 \\ -1.5 & 1 \end{bmatrix}$$

$$\text{Answer} \quad \begin{bmatrix} 2 & -1 \\ -1.5 & 1 \end{bmatrix}$$

b.

$$\mathbf{A}^{-1} = \frac{1}{\det(\mathbf{A})}\begin{bmatrix} \begin{vmatrix} 3 & 2 \\ -2 & 0 \end{vmatrix} & -\begin{vmatrix} 1 & 2 \\ -2 & 0 \end{vmatrix} & \begin{vmatrix} 1 & 2 \\ 3 & 2 \end{vmatrix} \\ -\begin{vmatrix} 1 & 2 \\ 3 & 0 \end{vmatrix} & \begin{vmatrix} 2 & 2 \\ 3 & 0 \end{vmatrix} & -\begin{vmatrix} 2 & 2 \\ 1 & 2 \end{vmatrix} \\ \begin{vmatrix} 1 & 3 \\ 3 & -2 \end{vmatrix} & -\begin{vmatrix} 2 & 1 \\ 3 & -2 \end{vmatrix} & \begin{vmatrix} 2 & 1 \\ 1 & 3 \end{vmatrix} \end{bmatrix} = \frac{1}{-8}\begin{bmatrix} 4 & -4 & -4 \\ 6 & -6 & -2 \\ -11 & 7 & 5 \end{bmatrix}$$

$$\text{Answer} \quad \begin{bmatrix} -0.5 & 0.5 & 0.5 \\ -0.75 & 0.75 & 0.25 \\ 1.375 & -0.875 & -0.625 \end{bmatrix}$$

6.2.6   Solution of a system of linear equations: Using inverse matrix

---

Remark   Solution of a system of linear equations

If the given *system of linear equations* is represented as $\mathbf{Ax} = \mathbf{B}$, the inverse of matrix $\mathbf{A}$ exists, and $\mathbf{A}$ is a square matrix, then the solution to the system of equations can be obtained as follows:

$$\mathbf{x} = \mathbf{A}^{-1}\mathbf{B} \qquad (6.21)$$

where $\mathbf{A}^{-1}$ is the inverse matrix of $\mathbf{A}$.

For example, let's find the solution to the system of linear equations represented as follows:

$$\begin{bmatrix} -1 & 2 \\ 1 & 1 \end{bmatrix} \begin{Bmatrix} x_1 \\ x_2 \end{Bmatrix} = \begin{Bmatrix} 0 \\ 3 \end{Bmatrix}.$$

To solve for x_1 and x_2, we can use matrix algebra. We'll denote the coefficient matrix as $\mathbf{A}$, the variable matrix as $\mathbf{x}$, and the constant matrix as $\mathbf{B}$.

$$\mathbf{Ax} = \mathbf{B}.$$

where $\mathbf{A} = \begin{bmatrix} -1 & 2 \\ 1 & 1 \end{bmatrix}$, $\mathbf{x} = \begin{Bmatrix} x_1 \\ x_2 \end{Bmatrix}$, and $\mathbf{B} = \begin{Bmatrix} 0 \\ 3 \end{Bmatrix}$.

We can solve for $\mathbf{x}$ by multiplying both sides by the inverse of $\mathbf{A}$:

$$\mathbf{x} = \mathbf{A}^{-1}\mathbf{B} \qquad (6.22)$$

Then, we find the inverse of $\mathbf{A}$ and multiply it by $\mathbf{B}$ to get the solution for $\mathbf{x}$ as follows:

$$\begin{Bmatrix} x_1 \\ x_2 \end{Bmatrix} = \begin{bmatrix} -1 & 2 \\ 1 & 1 \end{bmatrix}^{-1} \begin{Bmatrix} 0 \\ 3 \end{Bmatrix}$$

$$= \frac{1}{-3}\begin{bmatrix} 1 & -2 \\ -1 & -1 \end{bmatrix} \begin{Bmatrix} 0 \\ 3 \end{Bmatrix} = \frac{1}{-3}\begin{Bmatrix} -6 \\ -3 \end{Bmatrix} = \begin{Bmatrix} 2 \\ 1 \end{Bmatrix}.$$

So, the solution to the system of equations is

$$x_1 = 2, \; x_2 = 1.$$

Example 6.6

Solve a system of the following equations.

$$3x + 2y + z = 7$$
$$x - y + 2z = -1$$
$$2x + y - z = 6$$

Solution

From $\begin{bmatrix} 3 & 2 & 1 \\ 1 & -1 & 2 \\ 2 & 1 & -1 \end{bmatrix} \begin{Bmatrix} x \\ y \\ z \end{Bmatrix} = \begin{Bmatrix} 7 \\ -1 \\ 6 \end{Bmatrix}$, we obtain

$$\begin{Bmatrix} x \\ y \\ z \end{Bmatrix} = \begin{bmatrix} 3 & 2 & 1 \\ 1 & -1 & 2 \\ 2 & 1 & -1 \end{bmatrix}^{-1} \begin{Bmatrix} 7 \\ -1 \\ 6 \end{Bmatrix} = \frac{1}{10} \begin{bmatrix} -1 & 2 & 5 \\ 5 & -5 & -5 \\ 3 & 1 & -5 \end{bmatrix} \begin{Bmatrix} 7 \\ -1 \\ 6 \end{Bmatrix} = \frac{1}{10} \begin{Bmatrix} 20 \\ 10 \\ -10 \end{Bmatrix} = \begin{Bmatrix} 2 \\ 1 \\ -1 \end{Bmatrix}.$$

Answer $x = 2$, $y = 1$, $z = -1$

Problem 6.2

Solve the system by Gauss elimination. [1 ~ 6]

1. $3x_1 - 5x_2 = 3$
   $x_1 + 2x_2 = 1$

2. $x + 2y = 3$
   $-2x + y = -1$

3. $x + z = 4$
   $2x + y - 2z = 3$
   $x - 3y + z = 7$

4. $3x + 2y + z = 3$
   $x - 2y - 3z = -3$
   $x + z = 3$

5. $2x + y + 2z = 3$
   $x - y = -4$
   $2y - z = 5$

6. $-2x + y + z = -1$

   $x - 2y - 3z = 0$

   $4x + y + 2z = 4$

Find the rank. [7 ~ 10]

7. $\begin{bmatrix} 2 & -1 & 3 \\ -4 & 2 & 1 \end{bmatrix}$

8. $\begin{bmatrix} 0 & 0 & 1 \\ 1 & 1 & 0 \\ -2 & 0 & 2 \end{bmatrix}$

9. $\begin{bmatrix} 0 & 0 & 3 \\ 1 & 0 & 1 \\ -2 & 0 & -2 \end{bmatrix}$

10. $\begin{bmatrix} 0 & 2 & 1 \\ 1 & -1 & 4 \\ 2 & 3 & -1 \end{bmatrix}$

Are the following sets of vectors linearly independent? [11 ~ 16]

11. $\{1 \quad -2\}, \{-1 \quad 3\}, \{0 \quad 2\}$

12. $\{2 \quad 0 \quad 1\}, \{1 \quad 2 \quad 0\}, \{0 \quad 2 \quad 1\}$

13. $\{2 \quad -2 \quad 1\}, \{-1 \quad 3 \quad -1\}, \{1 \quad 1 \quad 0\}$

14. $\{2 \quad 1 \quad -2 \quad 1\}, \{1 \quad -1 \quad 3 \quad 2\}, \{0 \quad -1 \quad 2 \quad 0\}$

15. $\{1 \quad 2 \quad 3 \quad 4\}, \{-2 \quad 1 \quad 2 \quad -3\}, \{3 \quad -2 \quad 1 \quad 0\}$

16. $\{1 \quad 2 \quad 3 \quad 4\}, \{0 \quad 1 \quad 2 \quad 3\}, \{2 \quad 3 \quad 4 \quad 5\}, \{3 \quad 4 \quad 5 \quad 6\}$

Determine the inverse matrix. [17 ~ 22]

17. $\begin{bmatrix} 1 & 2 \\ 3 & 4 \end{bmatrix}$

18. $\begin{bmatrix} 3 & -2 \\ 1 & 4 \end{bmatrix}$

19. $\begin{bmatrix} 1 & 0 & -2 \\ 0 & 2 & 1 \\ 1 & -1 & 2 \end{bmatrix}$

20. $\begin{bmatrix} 1 & 1 & -2 \\ 1 & 2 & 2 \\ 1 & -2 & 4 \end{bmatrix}$

21. $\begin{bmatrix} \cos\theta & \sin\theta \\ -\sin\theta & \cos\theta \end{bmatrix}$

22. $\begin{bmatrix} \cosh 2\theta & \sinh 2\theta \\ \sinh 2\theta & \cosh 2\theta \end{bmatrix}$

Solve the system by inverse matrix. [23 ~ 28]

23. $3x_1 - 5x_2 = 3$
$x_1 + 2x_2 = 1$

24. $x + 2y = 3$
$-2x + y = -1$

25. $x + z = 4$
$2x + y - 2z = 3$
$x - 3y + z = 7$

26. $3x + 2y + z = 2$
$x - 2y - z = 2$
$x + z = 4$

27. $2y - z = 5$
$2x + y + 2z = 3$
$x - y = -4$

28. $-2x + y + z = -2$
$x - 2y - 3z = -5$
$4x + y + 2z = 13$

6.3 Eigenvalue problem (EVP)

6.3.1 Eigenvalues and eigenvectors for the basic form I in linear algebra

Let's review the following basic form I in linear algebra:

$$\mathbf{AX} = \lambda\mathbf{X}, \tag{6.23}$$

where **A** is a square matrix. When considering the state $\mathbf{X} = \{X_1 \ X_2 \ \cdots \ \}^T$, Eq. (6.23) becomes

$$(\mathbf{A} - \lambda \mathbf{I})\mathbf{X} = \mathbf{O}. \tag{6.24}$$

In order to have a *nontrivial solution* $\mathbf{X} \neq 0$, the determinant of the coefficient matrix $\mathbf{A} - \lambda \mathbf{I}$ must be zero. From Eq. (6.24), we can obtain the characteristic equation

$$\det(\mathbf{A} - \lambda \mathbf{I}) = 0. \tag{6.25}$$

The process of finding the *eigenvalues* and the *corresponding eigenvectors* of matrix **A** is similar to what we will learn in Section 7.1.3.

The set of all eigenvalues of a matrix **A** is called the *spectrum* of matrix **A**, and among them, the eigenvalue with the largest magnitude is referred to as the *spectrum radius* of matrix **A**.

Example 6.7

Find the eigenvalues and their corresponding eigenvectors for the basic form I of linear algebra.

$$\mathbf{A} = \begin{bmatrix} 2 & 3 \\ 1 & 4 \end{bmatrix}$$

Solution

The eigenvalues are the solution of the characteristic equation

$$\det(\mathbf{A} - \lambda \mathbf{I}) = \det\left(\begin{bmatrix} 2 & 3 \\ 1 & 4 \end{bmatrix} - \lambda \begin{bmatrix} 1 & 0 \\ 0 & 1 \end{bmatrix} \right) = \begin{vmatrix} 2 - \lambda & 3 \\ 1 & 4 - \lambda \end{vmatrix} = 0,$$

or

$$(2 - \lambda)(4 - \lambda) - 3 = 0.$$

This gives the eigenvalues

$$\lambda = 1, \ 5.$$

i. For the first eigenvalue $\lambda_1 = 1$, we obtain

$$\begin{bmatrix} 1 & 3 \\ 1 & 3 \end{bmatrix} \begin{Bmatrix} X_1 \\ X_2 \end{Bmatrix} = \begin{Bmatrix} 0 \\ 0 \end{Bmatrix},$$

or

$$X_1 + 3X_2 = 0.$$

To obtain a simple value, let's set $X_1 = 1$ then $X_2 = -1/3$. Therefore, its corresponding eigenvector becomes

$$\mathbf{v}_1 = \begin{Bmatrix} 1 \\ -1/3 \end{Bmatrix}.$$

ii. For the second eigenvalue $\lambda_2 = 5$, we obtain

$$\begin{bmatrix} -3 & 3 \\ 1 & -1 \end{bmatrix} \begin{Bmatrix} X_1 \\ X_2 \end{Bmatrix} = \begin{Bmatrix} 0 \\ 0 \end{Bmatrix}$$

or

$$X_1 - X_2 = 0.$$

To obtain a simple value, let's set $X_1 = 1$ then $X_2 = 1$. Therefore, its corresponding eigenvector becomes

$$\mathbf{v}_2 = \begin{Bmatrix} 1 \\ 1 \end{Bmatrix}.$$

Answer For $\lambda_1 = 1$, $\mathbf{v}_1 = \begin{Bmatrix} 1 \\ -1/3 \end{Bmatrix}$, and for $\lambda_2 = 5$, $\mathbf{v}_2 = \begin{Bmatrix} 1 \\ 1 \end{Bmatrix}$

6.3.2   *Eigenvalues and eigenvectors for the basic form II in linear algebra*

Now let's review the following basic form II in linear algebra:

$$\mathbf{AX} = \lambda \mathbf{BX} \tag{6.26}$$

where $\mathbf{A}$ and $\mathbf{B}$ are square matrixes. When considering the state $\mathbf{X} = \{X_1 \ X_2 \ \cdots \ \}^T$, Eq. (6.26) becomes

$$(\mathbf{A} - \lambda \mathbf{B})\mathbf{X} = \mathbf{O}. \tag{6.27}$$

In order to have a *nontrivial solution* $\mathbf{X} \neq 0$, the determinant of the coefficient matrix $\mathbf{A} - \lambda\mathbf{B}$ must be zero. From Eq. (6.27), we can obtain the characteristic equation

$$\det(\mathbf{A} - \lambda\mathbf{B}) = 0. \tag{6.28}$$

Using Eq. (6.27) and Eq. (6.28), we can find the eigenvalues $\lambda_1, \lambda_2, \cdots$ and the corresponding eigenvectors $\mathbf{v}_1, \mathbf{v}_2, \cdots$.

Example 6.8

Find the eigenvalues and their corresponding eigenvectors for the basic form II of linear algebra.

$$\mathbf{A} = \begin{bmatrix} 8 & -2 \\ -5 & 8 \end{bmatrix}, \quad \mathbf{B} = \begin{bmatrix} 1 & 0 \\ 0 & 2 \end{bmatrix}$$

Solution

The eigenvalues are the solution of the characteristic equation

$$\det(\mathbf{A} - \lambda\mathbf{B}) = \det\left(\begin{bmatrix} 8 & -2 \\ -5 & 8 \end{bmatrix} - \lambda \begin{bmatrix} 1 & 0 \\ 0 & 2 \end{bmatrix} \right) = \begin{vmatrix} 8-\lambda & -2 \\ -5 & 8-2\lambda \end{vmatrix} = 0,$$

or

$$(8-\lambda)(8-2\lambda) - 10 = 0.$$

This gives the eigenvalues

$$\lambda_1 = 3 \quad \text{and} \quad \lambda_2 = 9.$$

i. For the first eigenvalue $\lambda_1 = 3$, we obtain

$$\begin{bmatrix} 8-3 & -2 \\ -5 & 8-2\cdot3 \end{bmatrix} \begin{Bmatrix} X_1 \\ X_2 \end{Bmatrix} = \begin{Bmatrix} 0 \\ 0 \end{Bmatrix},$$

or

$$5X_1 - 2X_2 = 0.$$

To obtain a simple value, let's set $X_1 = 1$ then $X_2 = 2.5$.
Therefore, its corresponding eigenvector becomes

$$\mathbf{v}_1 = \begin{Bmatrix} 1 \\ 2.5 \end{Bmatrix}.$$

ii. For the second eigenvalue $\lambda_2 = 9$, we obtain

$$\begin{bmatrix} 8-9 & -2 \\ -5 & 8-2 \cdot 9 \end{bmatrix} \begin{Bmatrix} X_1 \\ X_2 \end{Bmatrix} = \begin{Bmatrix} 0 \\ 0 \end{Bmatrix},$$

or

$$-X_1 - 2X_2 = 0.$$

To obtain a simple value, let's set $X_1 = 1$, then $X_2 = -0.5$.
Therefore, its corresponding eigenvector becomes

$$\mathbf{v}_2 = \begin{Bmatrix} 1 \\ -0.5 \end{Bmatrix}.$$

Answer For $\lambda_1 = 3$, $\mathbf{v}_1 = \begin{Bmatrix} 1 \\ 2.5 \end{Bmatrix}$, and for $\lambda_2 = 9$, $\mathbf{v}_2 = \begin{Bmatrix} 1 \\ -0.5 \end{Bmatrix}$

6.3.3 Normalization

Since multiplying a nonzero constant c to the eigenvector $\mathbf{v}$ still satisfies Eq. (6.24) or Eq. (6.27), $c\mathbf{v}$ remains an eigenvector. Therefore, instead of representing the first eigenvector obtained in Example 6.7 as $\mathbf{v}_1 = \{1 \quad 2.5\}^T$, it's permissible to represent it as $\mathbf{v}_1 = \dfrac{1}{\sqrt{29}}\{2 \quad 5\}^T$, and similarly, the second eigenvector $\mathbf{v}_2 = \{1 \quad -0.5\}^T$ can be represented as $\mathbf{v}_2 = \{-2 \quad 1\}^T$.

The process of scaling each eigenvector by a nonzero constant to make the norm of the eigenvector 1, or alternatively setting the diagonal elements of the eigenvector matrix to 1, is called *normalization*. Eigenvectors that undergo this normalization process are referred to as *normalized eigenvectors*.

The most commonly used method for normalizing eigenvectors is to scale each eigenvector to have a norm of 1, and these are referred to as unit eigenvectors.

Example 6.9

Solve the problem.

a. Normalize the eigenvector $v = \begin{bmatrix} 2 & 3 \\ 1 & -4 \end{bmatrix}$ so that its diagonal element becomes 1.

b. Normalize the eigenvector $v = \begin{bmatrix} 1 & 1 \\ 0.6 & -2 \end{bmatrix}$ so that each eigenvector has a norm of 1.

Solution

a. In $v = \begin{bmatrix} 2 & 3 \\ 1 & -4 \end{bmatrix}$, the first vector is $\begin{Bmatrix} 2 \\ 1 \end{Bmatrix}$ and the second vector is $\begin{Bmatrix} 3 \\ -4 \end{Bmatrix}$.

Dividing each element of the first vector by the first diagonal element 2 yields

$$\begin{Bmatrix} 1 \\ 0.5 \end{Bmatrix}.$$

Similarly, dividing each element of the second vector by the second diagonal element −4 yields

$$\begin{Bmatrix} -0.75 \\ 1 \end{Bmatrix}.$$

Therefore, we obtain the normalized eigenvectors

$$v = \begin{bmatrix} 1 & -0.75 \\ 0.5 & 1 \end{bmatrix}.$$

$$\textbf{Answer } v = \begin{bmatrix} 1 & -0.75 \\ 0.5 & 1 \end{bmatrix}$$

b. In $v = \begin{bmatrix} 1 & 1 \\ 0.6 & -2 \end{bmatrix}$, the first vector is $\begin{Bmatrix} 1 \\ 0.6 \end{Bmatrix}$ and the second vector is $\begin{Bmatrix} 1 \\ -2 \end{Bmatrix}$.

Dividing each element of the first vector by its norm $\sqrt{1^2 + 0.6^2}$ yields vector

$$\begin{Bmatrix} 0.8575 \\ 0.5145 \end{Bmatrix}.$$

Similarly, dividing each element of the second vector by its norm $\sqrt{1^2 + (-2)^2}$ yields vector

$$\begin{Bmatrix} 0.4472 \\ -0.8944 \end{Bmatrix}.$$

Therefore we obtain the normalized eigenvectors

$$\mathbf{v} = \begin{bmatrix} 0.8575 & 0.4472 \\ 0.5145 & -0.8944 \end{bmatrix}.$$

$$\mathbf{Answer \ v} = \begin{bmatrix} 0.8575 & 0.4472 \\ 0.5145 & -0.8944 \end{bmatrix}$$

Problem 6.3

Find the eigenvalues and their corresponding eigenvectors for the basic form I of linear algebra. And normalize the eigenvector so that its diagonal element becomes 1. [1 ~ 6]

1. $A = \begin{bmatrix} 2 & -1 \\ -1 & 2 \end{bmatrix}$

2. $A = \begin{bmatrix} 1 & 2 \\ -1 & 4 \end{bmatrix}$

3. $A = \begin{bmatrix} 3 & 1 \\ 2 & 4 \end{bmatrix}$

4. $A = \begin{bmatrix} \cos\theta & \sin\theta \\ -\sin\theta & \cos\theta \end{bmatrix}$

5. $A = \begin{bmatrix} 1 & 1 & 0 \\ 0 & 1 & 2 \\ 0 & 2 & 1 \end{bmatrix}$

6. $A = \begin{bmatrix} 1 & 0 & 2 \\ 0 & 1 & 2 \\ -1 & 2 & 0 \end{bmatrix}$

Find the eigenvalues and their corresponding eigenvectors for the basic form II of linear algebra. [7 ~ 10]

7. $A = \begin{bmatrix} 1 & 2 \\ 3 & 1 \end{bmatrix}, B = \begin{bmatrix} 1 & 0 \\ 0 & 2 \end{bmatrix}$

8. $A = \begin{bmatrix} 3 & -1 \\ -2 & 3 \end{bmatrix}, B = \begin{bmatrix} 2 & 0 \\ 0 & 1 \end{bmatrix}$

9. $A = \begin{bmatrix} 4 & -2 \\ -2 & 3 \end{bmatrix}, B = \begin{bmatrix} 1 & 0 \\ 0 & 2 \end{bmatrix}$

10. $A = \begin{bmatrix} 3 & -1 \\ -2 & 4 \end{bmatrix}, B = \begin{bmatrix} 2 & 0 \\ 0 & 3 \end{bmatrix}$

6.4 Applications of linear algebra

Linear algebra serves as a fundamental discipline not only in engineering but also in all fields of study, including physics, chemistry, economics, and more.

6.4.1 *Statics*

Example 6.10

The point O, connected by two ropes OA and OB, has a mass m hanging from it, as shown in Figure 6.1. Calculate the tensions T_1 and T_2 in the ropes. Here, g represents the acceleration due to gravity.

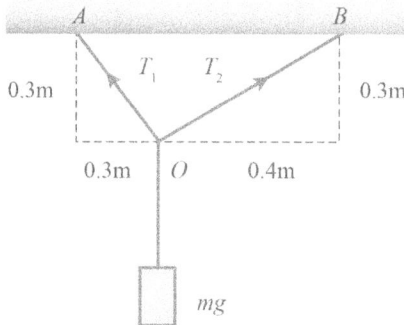

Figure 6.1 Force equilibrium.

Solution

The horizontal equilibrium equation is

$$-T_1 \frac{1}{\sqrt{2}} + T_2 \frac{4}{5} = 0.$$

The vertical equilibrium equation is

$$-mg + T_1 \frac{1}{\sqrt{2}} + T_2 \frac{3}{5} = 0.$$

Then we obtain

$$\begin{bmatrix} -0.707 & 0.8 \\ 0.707 & 0.6 \end{bmatrix} \begin{Bmatrix} T_1 \\ T_2 \end{Bmatrix} = \begin{Bmatrix} 0 \\ mg \end{Bmatrix}.$$

Therefore, the answer is

$$\begin{Bmatrix} T_1 \\ T_2 \end{Bmatrix} = \begin{bmatrix} -0.707 & 0.8 \\ 0.707 & 0.6 \end{bmatrix}^{-1} \begin{Bmatrix} 0 \\ mg \end{Bmatrix} = \begin{Bmatrix} 0.8082 \\ 0.7143 \end{Bmatrix} mg.$$

Answer $T_1 = 0.8082\,mg,\quad T_2 = 0.7143\,mg$

6.4.2   Vibration

Example 6.11

The equation of motion for the undamped-free vibration system is $\mathbf{M\ddot{x}} + \mathbf{Kx} = \mathbf{O}$, with the mass matrix $\mathbf{M} = \begin{bmatrix} 1 & 0 \\ 0 & 2 \end{bmatrix}$ and the stiffness matrix $\mathbf{K} = \begin{bmatrix} 200 & -100 \\ -100 & 300 \end{bmatrix}$, as shown in Figure 6.2.

Find the natural frequencies and the corresponding eigenvectors.

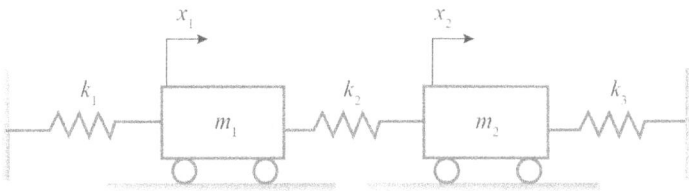

Figure 6.2 Two degree-of-freedom vibration system.

Solution

When complex notation $\mathbf{x} = \mathbf{X}e^{i(\omega t - \varphi)}$ is applied to the given equation of motion, it can be written as follows:

$$\left(\mathbf{K} - \omega^2 \mathbf{M}\right)\mathbf{X} = \mathbf{O},$$

or

$$\mathbf{KX} = \omega^2 \mathbf{MX}.$$

This equation resembles to the basic form II of linear algebra. Therefore, matrices $\mathbf{A}$ and $\mathbf{B}$ can be replaced by matrices $\mathbf{K}$ and $\mathbf{M}$ respectively, and each eigenvalue λ can indeed be replaced by the square of each natural frequency ω. Then we obtain the characteristic equation

$$\det\left(\mathbf{K} - \omega^2\mathbf{M}\right) = \left(200 - \omega^2\right)\left(300 - 2\omega^2\right) - 100^2 = 0.$$

The square of each natural frequency is

$$\omega_1^2 = 100 \quad \text{and} \quad \omega_2^2 = 250.$$

i. For $\omega_1^2 = 100$ (or $\omega_1 = 10$ rad/s), we obtain

$$\begin{bmatrix} 200 - 100 & -100 \\ -100 & 300 - 2 \cdot 100 \end{bmatrix} \begin{Bmatrix} X_1 \\ X_2 \end{Bmatrix} = \begin{Bmatrix} 0 \\ 0 \end{Bmatrix},$$

or

$$100X_1 - 100X_2 = 0.$$

To obtain a simple value, let's set $X_1 = 1$ then $X_2 = 1$. Therefore, its corresponding eigenvector becomes

$$\mathbf{v}_1 = \begin{Bmatrix} 1 \\ 1 \end{Bmatrix}.$$

This eigenmode means that when X_1 moves by 1 unit, X_2 moves in the same direction (in-phase) by 1 unit.

ii. For $\omega_2^2 = 250$ (or $\omega_2 = 15.81$ rad/s), we obtain

$$\begin{bmatrix} 200 - 250 & -100 \\ -100 & 300 - 2 \cdot 250 \end{bmatrix} \begin{Bmatrix} X_1 \\ X_2 \end{Bmatrix} = \begin{Bmatrix} 0 \\ 0 \end{Bmatrix},$$

or

$$-50X_1 - 100X_2 = 0.$$

To obtain a simple value, let's set $X_1 = 1$ then $X_2 = -0.5$. Therefore, its corresponding eigenvector becomes

$$\mathbf{v}_2 = \begin{Bmatrix} 1 \\ -0.5 \end{Bmatrix}.$$

This eigenmode means that when X_1 moves by 1 unit, X_2 moves in the opposite direction (out-of-phase) by 0.5 units.

Answer For $\omega_1 = 10$ rad/s, $\mathbf{v}_1 = \begin{Bmatrix} 1 \\ 1 \end{Bmatrix}$, and for $\omega_2 = 15.81$ rad/s, $\mathbf{v}_2 = \begin{Bmatrix} 1 \\ -0.5 \end{Bmatrix}$

6.4.3   *Electrical circuit*

***

Example 6.12

The circuit consists of 5 resistors ($R_1 = 10$ Ω, $R_2 = 50$ Ω, $R_3 = 100$ Ω, $R_4 = 150$ Ω, $R_5 = 200$ kΩ) and 2 applied voltages ($E_1 = 100$ V, $E_2 = 200$ V), as shown in Figure 6.3.

Assuming the current directions as shown in the diagram, calculate the currents in each branch.

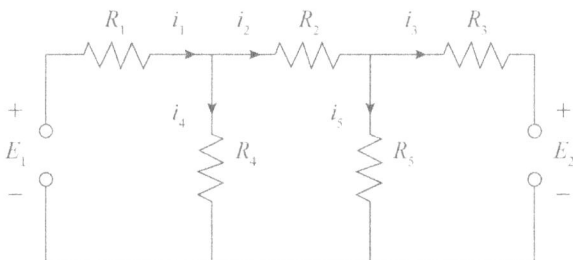

Figure 6.3 Electrical circuit.

***

Solution

Applying *Kirchhoff's voltage law* to each closed loop yields

$$-E_1 + R_1 i_1 + R_4 i_4 = 0,$$
$$-R_4 i_4 + R_2 i_2 + R_5 i_5 = 0,$$
$$-R_5 i_5 + R_3 i_3 + E_2 = 0.$$

Furthermore, applying *Kirchhoff's current law* to each node in the circuit yields

$$i_1 = i_2 + i_4,$$
$$i_2 = i_3 + i_5.$$

Using these two equations, expressing i_4 and i_5 in terms of i_1, i_2, and i_3 by substituting them into the initial three equations yields the following expressions for i_1, i_2, and i_3:

$$(R_1 + R_4)i_1 - R_4 i_2 = E_1,$$
$$-R_4 i_1 + (R_2 + R_4 + R_5)i_2 - R_5 i_3 = 0,$$
$$R_5 i_2 - (R_3 + R_5)i_3 = E_2,$$

or

$$\begin{bmatrix} R_1 + R_4 & -R_4 & 0 \\ -R_4 & R_2 + R_4 + R_5 & -R_5 \\ 0 & R_5 & -(R_3 + R_5) \end{bmatrix} \begin{Bmatrix} i_1 \\ i_2 \\ i_3 \end{Bmatrix} = \begin{Bmatrix} E_1 \\ 0 \\ E_2 \end{Bmatrix}.$$

Then we obtain

$$\begin{bmatrix} 160 & -150 & 0 \\ -150 & 200150 & -200000 \\ 0 & 200000 & -200100 \end{bmatrix} \begin{Bmatrix} i_1 \\ i_2 \\ i_3 \end{Bmatrix} = \begin{Bmatrix} 100 \\ 0 \\ 200 \end{Bmatrix}.$$

Therefore, the answer is

$$\begin{Bmatrix} i_1 \\ i_2 \\ i_3 \end{Bmatrix} = \begin{bmatrix} 160 & -150 & 0 \\ -150 & 200150 & -200000 \\ 0 & 200000 & -200100 \end{bmatrix}^{-1} \begin{Bmatrix} 100 \\ 0 \\ 200 \end{Bmatrix} = \begin{Bmatrix} 0.4617 \\ -0.5227 \\ -0.5235 \end{Bmatrix}.$$

Answer $i_1 = 0.4617$ A, $i_2 = -0.5227$ A, $i_3 = -0.5235$ A,

$i_4 = 0.9844$ A, $i_5 = 8 \times 10^{-4}$ A

Problem 6.4

1. **Vibration** The equation of motion for the undamped-free vibration system is $M\ddot{x} + Kx = O$, with the mass matrix $M = \begin{bmatrix} 1 & 0 \\ 0 & 1 \end{bmatrix}$ and the stiffness matrix $K = \begin{bmatrix} 3 & -1 \\ -2 & 4 \end{bmatrix}$. Find the natural frequencies and the corresponding eigenvectors.

2. **Vibration** In the undamped-free vibration system with $m_1 = m_2 = 1$ kg and $k_1 = k_2 = k_3 = 4$ N/m, as shown in Figure 6.4, find the natural frequencies and the corresponding eigenvectors.

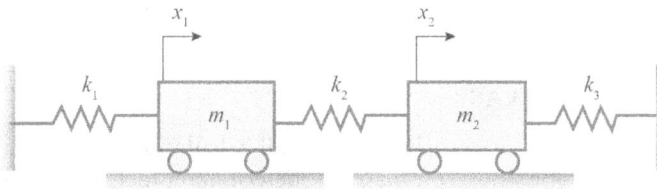

Figure 6.4 Two degree-of-freedom vibration system.

3. **Electric circuit**   The circuit consists of 4 resistors ($R_1 = 30\ \Omega$, $R_2 = 50\ \Omega$, and $R_3 = 100\ \Omega$) and one applied voltage ($E = 100\ V$). Assuming the current directions as shown in Figure 6.5, calculate the currents in each branch.

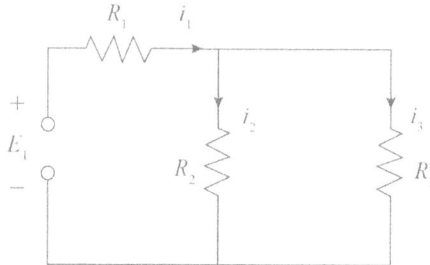

Figure 6.5 Electrical circuit.

4. **Electric circuit**   The circuit consists of 5 resistors ($R_1 = 10\ \Omega$, $R_2 = 50\ \Omega$, $R_3 = 30\ \Omega$, $R_4 = 40\ \Omega$, and $R_5 = 1\ k\Omega$) and one applied voltage ($E = 100\ V$). Assuming the current directions as shown in Figure 6.6, calculate the currents in each branch.

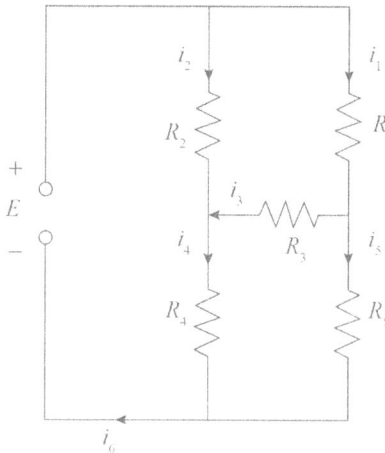

Figure 6.6 Electrical circuit.

6.5   Utilizing MATLAB®

With MATLAB®, we can easily perform all calculations related to matrices, such as various operations on matrices, determinant (det.m), inverse (inv.m), norm of the vector (norm.m), solutions to systems of equations, eigenvalues and eigenvectors (eig.m), and more.

M-Example 6.1

Solve the problem.

a. When $A = (a_{ij}) = \begin{bmatrix} 2 & 3 & -1 \\ 3 & -2 & 1 \end{bmatrix}$ and $B = (b_{ij}) = \begin{bmatrix} -1 & 1 & 2 \\ 3 & 1 & -4 \end{bmatrix}$, find

$2A + B$.

b. Find the transposition matrix A^T of

$$A = (a_{ij}) = \begin{bmatrix} 2 & 3 & -1 \\ 3 & -2 & 1 \end{bmatrix}.$$

c. When $A = (a_{ij}) = \begin{bmatrix} 2 & 3 \\ 3 & -2 \end{bmatrix}$ and $B = (b_{ij}) = \begin{bmatrix} -1 & 1 \\ 3 & 1 \end{bmatrix}$, find the product

of two matrices AB.

d. When $A = \begin{bmatrix} 2 & 1 \\ 3 & 2 \\ 1 & -1 \end{bmatrix}$ and $B = \begin{bmatrix} 1 & -1 & 3 \\ 3 & 2 & 1 \end{bmatrix}$, find the product of two

matrices AB.

e. Find the norm of the vector.

$$A = \begin{bmatrix} 2 & 1 & -3 \end{bmatrix}$$

Solution

a.

```
a=[ 2 3 –1; 3 –2 1]; b=[ –1 1 2; 3 1 –4];
2*a+b
```

Answer $\begin{bmatrix} 3 & 7 & 0 \\ 9 & -3 & -2 \end{bmatrix}$

b.

```
a=[ 2 3 –1; 3 –2 1];
a'
```

Answer $\begin{bmatrix} 2 & 3 \\ 3 & -2 \\ -1 & 1 \end{bmatrix}$

c.

```
a=[ 2 3; 3 –2]; b=[–1 1; 3 1];
a*b
```

Answer $\begin{bmatrix} 7 & 5 \\ -9 & 1 \end{bmatrix}$

d.

```
a=[ 2 1; 3 2; 1 –1]; b=[ 1 –1 3; 3 2 1];
a*b
```

Answer $\begin{bmatrix} 5 & 0 & 7 \\ 9 & 1 & 11 \\ -2 & -3 & 2 \end{bmatrix}$

e.

    % norm.m

```
a=[ 2 1 –3 ];
norm(a)
```

Answer 3.7417

M-Example 6.2

Find the rank.

$$A = \begin{bmatrix} 5 & 0 & 7 \\ 9 & 1 & 11 \\ -2 & -3 & 2 \end{bmatrix}$$

Solution

% rank.m

```
a=[ 5 0 7; 9 1 11; –2 –3 2];
rank(a)
```

Answer 2

M-Example 6.3

Get the determinant.

a. $A = \begin{bmatrix} 2 & 1 \\ 3 & -4 \end{bmatrix}$

b. $B = \begin{bmatrix} 2 & 1 & 2 \\ 1 & 3 & 2 \\ 3 & -2 & 0 \end{bmatrix}$

Solution

a. % det.m

```
a=[ 2 1; 3 −4]
det(a)
```

Answer −11

b. % det.m

```
b=[ 2 1 2; 1 3 2; 3 −2 0]
det(b)
```

Answer −8

M-Example 6.4

Determine the inverse matrix A^{-1}.

a. $A = \begin{bmatrix} 2 & 2 \\ 3 & 4 \end{bmatrix}$

b. $B = \begin{bmatrix} 2 & 1 & 2 \\ 1 & 3 & 2 \\ 3 & -2 & 0 \end{bmatrix}$

Solution

a. % inv.m

```
a=[ 2 2; 3 4]
inv(a)
```

Answer $\begin{bmatrix} 2 & -1 \\ -1.5 & 1 \end{bmatrix}$

b. % inv.m

```
b=[ 2 1 2; 1 3 2; 3 -2 0]
inv(b)
```

Answer $\begin{bmatrix} -0.5 & 0.5 & 0.5 \\ -0.75 & 0.75 & 0.25 \\ 1.375 & -0.875 & -0.625 \end{bmatrix}$

M-Example 6.5

Solve a system of equations.

$$3x + 2y + z = 7$$
$$x - y + 2z = -1$$
$$2x + y - z = 6$$

Solution

Representing the given system of equations in matrix form gives

$$\begin{bmatrix} 3 & 2 & 1 \\ 1 & -1 & 2 \\ 2 & 1 & -1 \end{bmatrix} \begin{Bmatrix} x \\ y \\ z \end{Bmatrix} = \begin{Bmatrix} 7 \\ -1 \\ 6 \end{Bmatrix}.$$

```
a=[ 3 2 1; 1 -1 2; 2 1 -1]; b=[7; -1; 6];
x=inv(a)*b
```

Answer $x = 2, y = 1, z = -1$

M-Example 6.6

Solve a system of equations.

$$ax + by = 1$$
$$cx + dy = 2$$

Solution

When dealing with equations involving variables, we can use symbolic (syms) designation in MATLAB® to solve a system of equations.

Representing the given system of equations in matrix form gives

$$\begin{bmatrix} a & b \\ c & d \end{bmatrix} \begin{Bmatrix} x \\ y \end{Bmatrix} = \begin{Bmatrix} 1 \\ 2 \end{Bmatrix}.$$

```
syms a b c d
a=[ a b; c d ]; b=[1; 2];
x=inv(A)*B
```

(result in MATLAB®)
x =
 d/(a*d – b*c) – (2*b)/(a*d – b*c)
 (2*a)/(a*d – b*c) – c/(a*d – b*c)

Answer $x = \dfrac{d - 2b}{ad - bc}, y = \dfrac{2a - c}{ad - bc}$

M-Example 6.7

Find the inner product of two vectors.

a. $(2, \ -2, \ 1), (-1, \ 3, \ -1)$

b. $\begin{Bmatrix} 2 \\ -1 \\ 3 \end{Bmatrix}, \begin{Bmatrix} 0 \\ 2 \\ 1 \end{Bmatrix}$

Solution

a.

```
a=[ 2 -2 1]; b=[-1 3 -1];
x=a*b'
```

ans =
   -9

Answer -9

b.

```
a=[ 2 -1 3]'; b=[0 2 1]';
x=a'*b
```

ans =
   1

Answer 1

M-Example 6.8

Find the outer product of two vectors.

a. $(1, \quad 2, \quad 0), (-1, \quad 3, \quad 0)$

b. $a = 2i - 2j + k, \, b = i + 3j + 2k$

Solution

a.

```
a=[ 1 2 0]; b=[-1 3 0];
x=cross(a, b)                    %cross.m
```

x =
   0   0   5

Answer 5k

b.

```
a=[ 2 -2 1]; b=[1 3 2];
x=cross(a, b)
```

x=
   -7   -3   8

Answer $-7i - 3j + 8k$

M-Example 6.9

Find the eigenvalues and their corresponding eigenvectors for the basic form I of linear algebra.

$$A = \begin{bmatrix} 2 & 3 \\ 1 & 4 \end{bmatrix}$$

Solution

% eig.m

```
a=[2 3; 1 4];
[v, d]=eig(A)
```

(result in MATLAB®)

```
v =  -0.9487    -0.7071
      0.3161    -0.7071
d =   1          0
      0          5
```

Answer For $\lambda_1 = 1$, $v_1 = \left\{ \begin{array}{c} -0.9487 \\ 0.3161 \end{array} \right\}$, and for $\lambda_2 = 5$, $v_2 = \left\{ \begin{array}{c} -0.7071 \\ -0.7071 \end{array} \right\}$

M-Example 6.10

Find the eigenvalues and their corresponding eigenvectors for the basic form II of linear algebra.

$$A = \begin{bmatrix} 8 & -2 \\ -5 & 8 \end{bmatrix}, B = \begin{bmatrix} 1 & 0 \\ 0 & 2 \end{bmatrix}$$

Solution

% eig.m

```
a=[8 -2; -5 8]; b=[ 1 0; 0 2];
[v, d]=eig(a, b)
```

(result in MATLAB®)

v =	1	0.4
	−0.5	1
d =	9	0
	0	3

Answer For $\lambda_1 = 9$, $\mathbf{v}_1 = \left\{ \begin{array}{c} 1 \\ -0.5 \end{array} \right\}$, and for $\lambda_2 = 3$, $\mathbf{v}_2 = \left\{ \begin{array}{c} 0.4 \\ 1 \end{array} \right\}$

Answer

Problem 6.1

1. $\begin{bmatrix} -1 & 5 \\ 8 & 1 \end{bmatrix}$

2. $\begin{bmatrix} 1 & -8 \\ 2 & 5 \\ -5 & -5 \end{bmatrix}$

3. $\begin{Bmatrix} 3 \\ 1 \\ -1 \end{Bmatrix}$

4. $\begin{bmatrix} 3 & 2 & -2 \\ 1 & -1 & 3 \end{bmatrix}$

5. $\begin{bmatrix} 5 & 5 \\ -6 & -8 \end{bmatrix}$

6. $\begin{bmatrix} 2 & 0 \\ 4 & -5 \end{bmatrix}$

7. 5

8. $\begin{bmatrix} 1 & 1 & -2 \\ 2 & 2 & -4 \\ -1 & -1 & 2 \end{bmatrix}$

9. $\begin{bmatrix} 2 & -4 \\ 13 & 6 \end{bmatrix}$

10. $\begin{bmatrix} 5 & 0 & 7 \\ 9 & 1 & 11 \\ -2 & -3 & 2 \end{bmatrix}$

Problem 6.2

1. $x_1 = 1, x_2 = 0$
2. $x = 1, y = 1$
3. $x = 3, y = -1, z = 1$
4. $x = 1, y = -1, z = 2$
5. $x = -1, y = 3, z = 1$
6. $x = 1, y = 2, z = -1$
7. 2
8. 3

9. 2

10. 3

11. Linearly dependent

12. Linearly independent

13. Linearly dependent

14. Linearly independent

15. Linearly independent

16. Linearly dependent

17. $\begin{bmatrix} -2 & 1 \\ 1.5 & -0.5 \end{bmatrix}$

18. $\dfrac{1}{14}\begin{bmatrix} 4 & 2 \\ -1 & 3 \end{bmatrix}$

19. $\dfrac{1}{9}\begin{bmatrix} 5 & 2 & 4 \\ 1 & 4 & -1 \\ -2 & 1 & 2 \end{bmatrix}$

20. $\dfrac{1}{18}\begin{bmatrix} 12 & 0 & 6 \\ -2 & 6 & -4 \\ -4 & 3 & 1 \end{bmatrix}$

21. $\begin{bmatrix} \cos\theta & -\sin\theta \\ \sin\theta & \cos\theta \end{bmatrix}$

22. $\begin{bmatrix} \cosh 2\theta & -\sinh 2\theta \\ -\sinh 2\theta & \cosh 2\theta \end{bmatrix}$

23. $x_1 = 1,\ x_2 = 0$

24. $x = 1,\ y = 1$

25. $x = 3,\ y = -1,\ z = 1$

26. $x = 1,\ y = -2,\ z = 3$

27. $x = -1,\ y = 3,\ z = 1$

28. $x = 2,\ y = -1,\ z = 3$

Problem 6.3

1. For $\lambda_1 = 1$, $v_1 = \begin{Bmatrix} 1 \\ 1 \end{Bmatrix}$, and for $\lambda_2 = 3$, $v_2 = \begin{Bmatrix} -1 \\ 1 \end{Bmatrix}$.

2. For $\lambda_1 = 2$, $v_1 = \begin{Bmatrix} 1 \\ 0.5 \end{Bmatrix}$, and for $\lambda_2 = 3$, $v_2 = \begin{Bmatrix} 1 \\ 1 \end{Bmatrix}$.

3. For $\lambda_1 = 2$, $v_1 = \begin{Bmatrix} 1 \\ -1 \end{Bmatrix}$, and for $\lambda_2 = 5$, $v_2 = \begin{Bmatrix} 0.5 \\ 1 \end{Bmatrix}$

4. $\lambda_1 = \cos\theta - i\sin\theta$, $v_1 = \begin{Bmatrix} 1 \\ -i \end{Bmatrix}$, and for $\lambda_2 = \cos\theta + i\sin\theta$, $v_2 = \begin{Bmatrix} i \\ 1 \end{Bmatrix}$.

5. For $\lambda_1 = 1$, $v_1 = \begin{Bmatrix} 1 \\ 0 \\ 0 \end{Bmatrix}$, for $\lambda_2 = 3$, $v_2 = \begin{Bmatrix} 0.5 \\ 1 \\ 1 \end{Bmatrix}$, and for $\lambda_3 = -1$, $v_3 = \begin{Bmatrix} 0.5 \\ -1 \\ 1 \end{Bmatrix}$.

6. For $\lambda_1 = 1$, $v_1 = \begin{Bmatrix} 1 \\ 0.5 \\ 0 \end{Bmatrix}$, for $\lambda_2 = 2$, $v_2 = \begin{Bmatrix} 1 \\ 1 \\ 0.5 \end{Bmatrix}$, and for $\lambda_3 = -1$, $v_3 = \begin{Bmatrix} -1 \\ -1 \\ 1 \end{Bmatrix}$.

7. For $\lambda_1 = -1$, $v_1 = \begin{Bmatrix} 1 \\ -1 \end{Bmatrix}$, and for $\lambda_2 = 2.5$, $v_2 = \begin{Bmatrix} 1 \\ 0.75 \end{Bmatrix}$

8. For $\lambda_1 = 1$, $v_1 = \begin{Bmatrix} 1 \\ 1 \end{Bmatrix}$, and for $\lambda_2 = 3.5$, $v_2 = \begin{Bmatrix} 1 \\ -4 \end{Bmatrix}$.

9. For $\lambda_1 = 0.8625$, $v_1 = \begin{Bmatrix} 1 \\ 1.5688 \end{Bmatrix}$, and for $\lambda_2 = 4.6375$, $v_2 = \begin{Bmatrix} -3.1375 \\ 1 \end{Bmatrix}$.

10. For $\lambda_1 = 0.8333$, $v_1 = \begin{Bmatrix} 1 \\ 1.3333 \end{Bmatrix}$, and for $\lambda_2 = 2$, $v_2 = \begin{Bmatrix} -1 \\ 1 \end{Bmatrix}$.

Problem 6.4

1. For $\omega_1 = 1.414$ rad/s, $v_1 = \begin{Bmatrix} 1 \\ 1 \end{Bmatrix}$, and for $\omega_2 = 2.236$ rad/s, $v_2 = \begin{Bmatrix} 1 \\ -2 \end{Bmatrix}$

2. For $\omega_1 = 2$ rad/s, $v_1 = \begin{Bmatrix} 1 \\ 1 \end{Bmatrix}$, and for $\omega_2 = \sqrt{12}$ rad/s, $v_2 = \begin{Bmatrix} 1 \\ -1 \end{Bmatrix}$

3. $i_1 = 1.5789$ A, $i_2 = 1.0526$ A

4. $i_1 = 0.9687$ A, $i_2 = 0.7207$ A, $i_3 = 0.8783$ A, $i_4 = 1.5990$ A, $i_5 = 0.0904$ A, $i_6 = 1.6894$ A

7 Linear system of ordinary differential equations

In engineering, when multiple systems are coupled, the behavior of one system affects the behavior of neighboring systems. In this case, a system of ODEs for each system can be constructed by sequentially listing the ODEs for each system. Representing such a system of ODEs in matrix form resembles a first-order ODE, allowing for easy solution using linear algebra concepts such as determinant, inverse matrix, eigenvalue problem, and others.

7.1 System of two first-order ODEs

7.1.1 Second-order ODE

In Section 2.5, we learned the method of converting a second-order ODE into a system of two first-order ODEs. Summarizing it again, it can be stated as follows:

$$x'' + a(t)x' + b(t)x = r(t). \tag{7.1}$$

Since $x'' = -b(t)x - a(t)x' + r(t)$ is obtained from Eq. (7.1), let's denote it as the state $\mathbf{x} = \begin{Bmatrix} x \\ x' \end{Bmatrix}$. Then, we obtain a system of ODEs as follows:

$$\begin{Bmatrix} x' \\ x'' \end{Bmatrix} = \begin{bmatrix} 0 & 1 \\ -b(t) & -a(t) \end{bmatrix} \begin{Bmatrix} x \\ x' \end{Bmatrix} + \begin{Bmatrix} 0 \\ r(t) \end{Bmatrix}. \tag{7.2}$$

Rewriting this equation in matrix form yields the following form of a system of first-order ODEs:

$$\mathbf{x}' = \mathbf{A}\mathbf{x} + \mathbf{R}, \tag{7.3}$$

where a matrix $\mathbf{A} = \begin{bmatrix} 0 & 1 \\ -b(t) & -a(t) \end{bmatrix}$ and a vector $\mathbf{R} = \begin{Bmatrix} 0 \\ r(t) \end{Bmatrix}$.

This system is called homogeneous if $\mathbf{R} = 0$, so that it is

$$\mathbf{x}' = \mathbf{A}\mathbf{x}.$$

If $\mathbf{R} \neq 0$, then Eq. (7.3) is called nonhomogeneous.

DOI: 10.1201/9781003608912-7

Remark Converting a second-order ODE into a system of two first-order ODEs

Let's use the state $\mathbf{x} = \left\{ \begin{array}{c} x \\ x' \end{array} \right\}$ from the ODE $x'' + a(t)x' + b(t)x = r(t)$.

Then we obtain

$$\mathbf{x}' = \mathbf{A}\mathbf{x} + \mathbf{R} \tag{7.3}$$

where the matrix $\mathbf{A} = \begin{bmatrix} 0 & 1 \\ -b(t) & -a(t) \end{bmatrix}$ and the vector $\mathbf{R} = \left\{ \begin{array}{c} 0 \\ r(t) \end{array} \right\}$.

Example 7.1

Convert the following ODE into a system of first-order ODEs:

$$x'' + x' - 2x = e^t.$$

Solution

From the given ODE $x'' + x' - 2x = e^t$, we obtain

$$x'' = 2x - x' + e^t.$$

When we let the state $\mathbf{x} = \left\{ \begin{array}{c} x \\ x' \end{array} \right\}$, then

$$\mathbf{x}' = \mathbf{A}\mathbf{x} + \mathbf{R}$$

where the matrix $\mathbf{A} = \begin{bmatrix} 0 & 1 \\ 2 & -1 \end{bmatrix}$ and the vector $\mathbf{R} = \left\{ \begin{array}{c} 0 \\ e^t \end{array} \right\}$.

$$\text{Answer } \mathbf{x}' = \mathbf{A}\mathbf{x} + \mathbf{R}, \ \mathbf{x} = \left\{ \begin{array}{c} x \\ x' \end{array} \right\}, \mathbf{A} = \begin{bmatrix} 0 & 1 \\ 2 & -1 \end{bmatrix}, \mathbf{R} = \left\{ \begin{array}{c} 0 \\ e^t \end{array} \right\}.$$

7.1.2 *Higher-order ODE*

In Section 3.3, we learned the method of converting a higher-order ODE into a system of several first-order ODEs. Summarizing it again, it can be stated as follows:

$$x''' + a(t)x'' + b(t)x' + c(t)x = r(t). \tag{7.4}$$

Since we get $x''' = -c(t)x - b(t)x' - a(t)x'' + r(t)$ from the given third-order ODE and let's use $\mathbf{x} = \begin{Bmatrix} x \\ x' \\ x'' \end{Bmatrix}$, we obtain

$$\begin{Bmatrix} x' \\ x'' \\ x''' \end{Bmatrix} = \begin{bmatrix} 0 & 1 & 0 \\ 0 & 0 & 1 \\ -c(t) & -b(t) & -a(t) \end{bmatrix} \begin{Bmatrix} x \\ x' \\ x'' \end{Bmatrix} + \begin{Bmatrix} 0 \\ 0 \\ r(t) \end{Bmatrix}. \tag{7.5}$$

Therefore, we obtain

$$\mathbf{x'} = \mathbf{Ax} + \mathbf{R}, \tag{7.6}$$

where the matrix $\mathbf{A} = \begin{bmatrix} 0 & 1 & 0 \\ 0 & 0 & 1 \\ -c(t) & -b(t) & -a(t) \end{bmatrix}$, and the vector $\mathbf{R} = \begin{Bmatrix} 0 \\ 0 \\ r(t) \end{Bmatrix}$.

This system is called homogeneous if $\mathbf{R} = 0$, and it is called nonhomogeneous if $\mathbf{R} \neq 0$.

Remark   Converting a third-order ODE into a system of three first-order ODEs

Let's use the state $\mathbf{x} = \begin{Bmatrix} x \\ x' \\ x'' \end{Bmatrix}$ from the given third-order ODE

$$x''' + a(t)x'' + b(t)x' + c(t)x = r(t).$$

Then we obtain a system of three first-order ODEs

$$\mathbf{x'} = \mathbf{Ax} + \mathbf{R}, \tag{7.6}$$

where the matrix $\mathbf{A} = \begin{bmatrix} 0 & 1 & 0 \\ 0 & 0 & 1 \\ -c(t) & -b(t) & -a(t) \end{bmatrix}$ and the vector $\mathbf{R} = \begin{Bmatrix} 0 \\ 0 \\ r(t) \end{Bmatrix}$.

Example 7.2

Convert a third-order ODE into a system of three first-order ODEs.

$$x''' - x'' + 3x' + 2x = \cos t$$

Solution

From the given ODE $x''' - x'' + 3x' + 2x = \cos t$, we obtain

$$x''' = -2x - 3x' + x'' + \cos t.$$

Now we set the state $\mathbf{x} = \left\{ \begin{array}{c} x \\ x' \\ x'' \end{array} \right\}$, then we obtain

$$\mathbf{x}' = \mathbf{A}\mathbf{x} + \mathbf{R},$$

where the matrix $\mathbf{A} = \begin{bmatrix} 0 & 1 & 0 \\ 0 & 0 & 1 \\ -2 & -3 & 1 \end{bmatrix}$ and the vector $\mathbf{R} = \begin{bmatrix} 0 \\ 0 \\ \cos t \end{bmatrix}$.

$$\text{Answer } \mathbf{x}' = \mathbf{A}\mathbf{x} + \mathbf{R}, \; \mathbf{x} = \left\{ \begin{array}{c} x \\ x' \\ x'' \end{array} \right\}, \mathbf{A} = \begin{bmatrix} 0 & 1 & 0 \\ 0 & 0 & 1 \\ -2 & -3 & 1 \end{bmatrix}, \mathbf{R} = \left\{ \begin{array}{c} 0 \\ 0 \\ \cos t \end{array} \right\}$$

Similarly, let's apply the same method to the higher-order ODE with $n \geq 4$.

Remark   Converting a Fourth-order ODE into a System of Four First-order ODEs

Let's use the state $\mathbf{x} = \left\{ \begin{array}{c} x \\ x' \\ x'' \\ x''' \end{array} \right\}$ from $x^{(4)} + a(t)x''' + b(t)x'' + c(t)x' + d(t) = r(t)$. Then we obtain a system of four first-order ODEs

$$\mathbf{x}' = \mathbf{A}\mathbf{x} + \mathbf{R} \tag{7.7}$$

where the matrix $\mathbf{A} = \begin{bmatrix} 0 & 1 & 0 & 0 \\ 0 & 0 & 1 & 0 \\ 0 & 0 & 0 & 1 \\ -d(t) & -c(t) & -b(t) & -a(t) \end{bmatrix}$ and the vector

$\mathbf{R} = \begin{Bmatrix} 0 \\ 0 \\ 0 \\ r(t) \end{Bmatrix}.$

7.1.3   *Homogeneous solution of the ODE*

Let's find a *homogeneous solution* of the second-order ODE in the homogeneous matrix equation with $\mathbf{R} = 0$ in Eq. (7.3).

$$\mathbf{x}' = \mathbf{A}\mathbf{x}, \tag{7.8}$$

where A is a given nonzero square matrix of dimension 2×2.

Since it is a second-order ODE, we can set the state $\mathbf{x} = \begin{Bmatrix} x \\ x' \end{Bmatrix} = \begin{Bmatrix} X \\ \lambda X \end{Bmatrix} e^{\lambda t}$,

where λ is constant.

Then, we get

$$\mathbf{x}' = \lambda \mathbf{x}. \tag{7.9}$$

From Eq. (7.8), a system of two first-order ODEs is transformed into the following linear algebraic equation:

$$\mathbf{A}\mathbf{x} = \lambda \mathbf{x}. \tag{7.10}$$

Therefore, we can write

$$(\mathbf{A} - \lambda \mathbf{I})\mathbf{x} = 0. \tag{7.11}$$

This equation involves n linear algebraic equations. In order to have a nontrivial solution $\mathbf{x} \neq 0$, the determinant of the coefficient matrix $\mathbf{A} - \lambda \mathbf{I}$ must be zero. Then we obtain the following *characteristic equation*:

$$\det(\mathbf{A} - \lambda \mathbf{I}) = 0. \tag{7.12}$$

To solve Eq. (7.8) means to determine the *eigenvalues* λ of the matrix $\mathbf{A}$ and its *corresponding eigenvectors* $\mathbf{v}$ for each eigenvalue.

Therefore, each linear algebraic equation has its own solution as follows:

$$\mathbf{x}_1 = \mathbf{v}_1 e^{\lambda_1 t}, \quad \mathbf{x}_2 = \mathbf{v}_2 e^{\lambda_2 t}. \tag{7.13}$$

Therefore, the homogeneous solution of the matrix equation (7.9) is represented as a linear combination of these solutions. This is referred to as the superposition principle.

$$\mathbf{x}_h = c_1 \mathbf{x}_1 + c_2 \mathbf{x}_2 = c_1 \mathbf{v}_1 e^{\lambda_1 t} + c_2 \mathbf{v}_2 e^{\lambda_2 t}, \quad \left(c_1, c_2 \text{ arbitrary} \right) \tag{7.14}$$

where $\mathbf{x}_1$ and $\mathbf{x}_2$ are basis solutions.

Remark   Homogeneous solution of the ODE with different real eigenvalues

From a system of homogeneous first-order ODEs $\mathbf{x}' = \mathbf{A}\mathbf{x}$ with the state $\mathbf{x} = \left\{ \begin{array}{c} x \\ x' \\ \vdots \end{array} \right\}$, we obtain the following characteristic equation:

$$\det\left(\mathbf{A} - \lambda \mathbf{I}\right) = 0. \tag{7.15}$$

This quadratic equation in λ is called the characteristic equation of the matrix $\mathbf{A}$, whose solutions are the eigenvalues $\lambda_1, \lambda_2, \cdots$ of $\mathbf{A}$.

And we can obtain the eigenvectors $\mathbf{v}_1, \mathbf{v}_2, \cdots$ of $\mathbf{A}$ corresponding to $\lambda_1, \lambda_2, \cdots$, respectively.

Therefore, we obtain the following homogeneous solution of the ODE:

$$\mathbf{x}_h = c_1 \mathbf{x}_1 + c_2 \mathbf{x}_2 + \cdots = c_1 \mathbf{v}_1 e^{\lambda_1 t} + c_2 \mathbf{v}_2 e^{\lambda_2 t} + \cdots \left(c_1, c_2, \cdots \text{ arbitrary}\right),$$
$$\tag{7.16}$$

where $\mathbf{x}_1, \mathbf{x}_2, \cdots$ are basis solutions.

Example 7.3

Convert the following second-order ODE into a system of first-order ODEs and then find the homogeneous solution through eigenvalue analysis.

$$x'' - 2x' - 3x = 0$$

Solution

From the given ODE $x'' - 2x' - 3x = 0$, we get

$$x'' = 3x + 2x'.$$

Now we set $\mathbf{x} = \begin{Bmatrix} x \\ x' \end{Bmatrix}$, then we obtain a system of first-order ODEs

$$\mathbf{x}' = \mathbf{Ax},$$

where the matrix $\mathbf{A} = \begin{bmatrix} 0 & 1 \\ 3 & 2 \end{bmatrix}$.

Then, we can write

$$(\mathbf{A} - \lambda \mathbf{I})\mathbf{x} = 0.$$

The eigenvalues are the solution of the characteristic equation

$$\det(\mathbf{A} - \lambda \mathbf{I}) = \begin{vmatrix} -\lambda & 1 \\ 3 & 2 - \lambda \end{vmatrix} = 0.$$

This gives the eigenvalues

$$\lambda_1 = -1 \quad \text{and} \quad \lambda_2 = 3.$$

i. For the first eigenvalue $\lambda_1 = -1$, we obtain

$$(\mathbf{A} - \lambda_1 \mathbf{I})\mathbf{x} = \begin{bmatrix} -\lambda_1 & 1 \\ 3 & 2 - \lambda_1 \end{bmatrix} \begin{Bmatrix} x \\ x' \end{Bmatrix} = \begin{bmatrix} 1 & 1 \\ 3 & 3 \end{bmatrix} \begin{Bmatrix} x \\ x' \end{Bmatrix} = 0,$$

or

$$x + x' = 0.$$

To obtain a simple value, let's set $x = 1$, then $x' = -1$.
Therefore, its corresponding eigenvector becomes

$$\mathbf{v}_1 = \begin{Bmatrix} 1 \\ -1 \end{Bmatrix}.$$

And the first basis solution is

$$\mathbf{x}_1 = \mathbf{v}_1 e^{\lambda_1 t} = \begin{Bmatrix} 1 \\ -1 \end{Bmatrix} e^{-t}.$$

ii. For the second eigenvalue $\lambda_2 = 3$, we obtain

$$(A - \lambda_2 I)x = \begin{bmatrix} -\lambda_2 & 1 \\ 3 & 2-\lambda_2 \end{bmatrix} \begin{Bmatrix} x \\ x' \end{Bmatrix} = \begin{bmatrix} -3 & 1 \\ 3 & -1 \end{bmatrix} \begin{Bmatrix} x \\ x' \end{Bmatrix} = 0,$$

or

$$-3x + x' = 0.$$

To obtain a simple value, let's set $x = 1$, then $x' = 3$. Therefore, its corresponding eigenvector becomes

$$v_2 = \begin{Bmatrix} 1 \\ 3 \end{Bmatrix}.$$

And the second basis solution is

$$x_2 = v_2 e^{\lambda_2 t} = \begin{Bmatrix} 1 \\ 3 \end{Bmatrix} e^{3t}.$$

Therefore, we obtain the homogeneous solution of the ODE

$$x_h = c_1 x_1 + c_2 x_2$$

or

$$x = \begin{Bmatrix} x \\ x' \end{Bmatrix} = c_1 \begin{Bmatrix} 1 \\ -1 \end{Bmatrix} e^{-t} + c_2 \begin{Bmatrix} 1 \\ 3 \end{Bmatrix} e^{3t},$$

where c_1 and c_2 are constants that are determined using initial conditions.

Answer $x(t) = c_1 e^{-t} + c_2 e^{3t}$

Check

The general solution of the ODE where the first row of x is

$$x(t) = c_1 e^{-t} + c_2 e^{3t},$$

it can be verified by differentiating it that

$$x'(t) = -c_1 e^{-t} + 3c_2 e^{3t}$$

becomes the second row of x.

When the eigenvalues obtained from the characteristic equation (7.12) are double, the following process can be used to obtain new eigenvectors.

When the first eigenvalue λ_1 is a double real root and its corresponding eigenvector is $\mathbf{v}_1$, we assume the second solution

$$\mathbf{x}_2 = (\mathbf{v}_1 t + \mathbf{v}_2)e^{\lambda_1 t}. \tag{7.17}$$

When Eq. (7.17) is substituted into the matrix equation $\mathbf{x}' = \mathbf{Ax}$, then

$$(\lambda_1 \mathbf{v}_1)te^{\lambda_1 t} + (\mathbf{v}_1 + \lambda_1 \mathbf{v}_2)e^{\lambda_1 t} = (A\mathbf{v}_1)te^{\lambda_1 t} + (A\mathbf{v}_2)e^{\lambda_1 t}. \tag{7.18}$$

When comparing coefficients in Eq. (7.18), it separates into the following two equations:

$$(\mathbf{A} - \lambda_1 \mathbf{I})\mathbf{v}_1 = 0 \tag{7.19}$$

$$(\mathbf{A} - \lambda_1 \mathbf{I})\mathbf{v}_2 = \mathbf{v}_1 \tag{7.20}$$

Eq. (7.19) represents the expressions for previously obtained eigenvalue λ_1 and its corresponding eigenvector $\mathbf{v}_1$, thus, from Eq. (7.20), the second eigenvector $\mathbf{v}_2$ can be determined.

Example 7.4

(double real root) Convert the following second-order ODE into a system of first-order ODEs and then find the homogeneous solution through eigenvalue analysis.

$$x'' + 6x' + 9x = 0$$

Solution

From the given ODE $x'' + 6x' + 9x = 0$, we get

$$x'' = -9x - 6x'.$$

Now we set $\mathbf{x} = \begin{Bmatrix} x \\ x' \end{Bmatrix}$, then we obtain a system of first-order ODEs

$$\mathbf{x}' = \mathbf{Ax},$$

where the matrix $\mathbf{A} = \begin{bmatrix} 0 & 1 \\ -9 & -6 \end{bmatrix}$.

Then, we can write

$$(A - \lambda I)x = 0.$$

The eigenvalues are the solution of the characteristic equation $\det(A - \lambda I) = 0$

$$\det(A - \lambda I) = \begin{vmatrix} -\lambda & 1 \\ -9 & -6-\lambda \end{vmatrix} = 0.$$

This gives the eigenvalues

$$\lambda_{1,2} = -3. \text{ (double real root)}$$

i. For the first eigenvalue $\lambda_1 = -3$, we obtain

$$(A - \lambda_1 I)x = \begin{bmatrix} -\lambda_1 & 1 \\ -9 & -6-\lambda_1 \end{bmatrix} \begin{Bmatrix} x \\ x' \end{Bmatrix} = \begin{bmatrix} 3 & 1 \\ -9 & -3 \end{bmatrix} \begin{Bmatrix} x \\ x' \end{Bmatrix} = 0$$

or

$$3x + x' = 0.$$

To obtain a simple value, let's set $x = 1$, then $x' = -3$.
Therefore, its corresponding eigenvector becomes

$$v_1 = \begin{Bmatrix} 1 \\ -3 \end{Bmatrix}.$$

And the first basis solution is

$$x_1 = v_1 e^{\lambda_1 t} = \begin{Bmatrix} 1 \\ -3 \end{Bmatrix} e^{-3t}.$$

ii. Let's set $x_2 = (v_1 t + v_2)e^{\lambda_1 t} = \begin{Bmatrix} 1 \\ -3 \end{Bmatrix} t e^{-3t} + \begin{Bmatrix} a \\ b \end{Bmatrix} e^{-3t}.$

From Eq. (6.26), we get

$$(A - \lambda_1 I)v_2 = \begin{bmatrix} -\lambda_1 & 1 \\ -9 & -6-\lambda_1 \end{bmatrix} \begin{Bmatrix} a \\ b \end{Bmatrix} = \begin{bmatrix} 3 & 1 \\ -9 & -3 \end{bmatrix} \begin{Bmatrix} a \\ b \end{Bmatrix} = \begin{Bmatrix} 1 \\ -3 \end{Bmatrix},$$

or

$$3a + b = 1.$$

To obtain a simple value, let's set $a = 0$, then $b = 1$.

And the second basis solution is

$$\mathbf{v}_2 = \left\{ \begin{matrix} 0 \\ 1 \end{matrix} \right\}.$$

And the second basis solution is

$$\mathbf{x}_2 = (\mathbf{v}_1 t + \mathbf{v}_2)e^{\lambda_1 t} = \left\{ \begin{matrix} 1 \\ -3 \end{matrix} \right\} t e^{-3t} + \left\{ \begin{matrix} 0 \\ 1 \end{matrix} \right\} e^{-3t}.$$

Therefore, we obtain the homogeneous solution of the ODE

$$\mathbf{x}_h = c_1 \mathbf{x}_1 + c_2 \mathbf{x}_2,$$

or

$$\mathbf{x} = \left\{ \begin{matrix} x \\ x' \end{matrix} \right\} = c_1 \left\{ \begin{matrix} 1 \\ -3 \end{matrix} \right\} e^{-3t} + c_2 \left\{ \begin{matrix} t \\ -3t+1 \end{matrix} \right\} e^{-3t},$$

where c_1 and c_2 are constants that are determined using initial conditions.

Answer $x(t) = c_1 e^{-3t} + c_2 t e^{-3t}$

Check

The general solution of the ODE where the first row of **x** is

$$x(t) = c_1 e^{-3t} + c_2 t e^{-3t},$$

it can be verified by differentiating it that

$$x'(t) = -3c_1 e^{-3t} + c_2 (-3t+1) e^{-3t}$$

becomes the second row of **x**.

7.1.4　*General solution of nonhomogeneous ODEs*

Let's find solutions to the ODE in the nonhomogeneous matrix equation with $\mathbf{R} \neq 0$

$$\mathbf{x}' = \mathbf{Ax} + \mathbf{R} \tag{7.21}$$

where $\mathbf{A}$ is a given nonzero square matrix of dimension $n \times n$ and $\mathbf{R}$ is a given nonzero vector of dimension $n \times 1$.

The general solution of a system of nonhomogeneous ODEs is expressed as the sum of the *homogeneous solution* $\mathbf{x}_h$ of the corresponding system and the *particular solution* $\mathbf{x}_p$ of the system of nonhomogeneous ODEs.

$$\mathbf{x} = \mathbf{x}_h + \mathbf{x}_p \tag{7.22}$$

We can express the general solution by substituting the homogeneous solution Eq. (7.16) into Eq. (7.21) as follows:

$$\mathbf{x} = (c_1\mathbf{x}_1 + c_2\mathbf{x}_2 + \cdots) + \mathbf{x}_p \quad (c_1, c_2, \cdots \text{ arbitrary}) \tag{7.23}$$

where $c_1, c_2, \cdots$ are constants that are determined using initial conditions, and $\mathbf{x}_p$ can be obtained using either the *method of undetermined coefficients* or the *method of variation of parameters*, as learned in Section 2.4.

Remark General solution of the nonhomogeneous ODE

The general solution of a system of nonhomogeneous ODEs is expressed as the sum of the homogeneous solution $\mathbf{x}_h$ of the corresponding system and the particular solution $\mathbf{x}_p$ of the system of nonhomogeneous ODEs.

$$\mathbf{x} = \mathbf{x}_h + \mathbf{x}_p$$

$$= (c_1\mathbf{x}_1 + c_2\mathbf{x}_2 + \cdots) + \mathbf{x}_p \quad (c_1, c_2, \cdots \text{ arbitrary}) \tag{7.23}$$

where $\mathbf{x}_1, \mathbf{x}_2, \cdots$ are basis solutions.

Example 7.5

Using the system of two first-order ODEs, find the general solution to the following second-order ODE:

$$x'' - 2x' - 3x = 10\cos t.$$

Solution

From the given ODE $x'' - 2x' - 3x = 10\cos t$, we get

$$x'' = 3x + 2x' + 10\cos t.$$

Now we set $\mathbf{x} = \begin{Bmatrix} x \\ x' \end{Bmatrix}$, then we obtain a system of first-order ODEs

$$\mathbf{x}' = \mathbf{Ax} + \mathbf{R},$$

where $\mathbf{A} = \begin{bmatrix} 0 & 1 \\ 3 & 2 \end{bmatrix}$ and $\mathbf{R} = \begin{Bmatrix} 0 \\ 10\cos t \end{Bmatrix}$.

From Example 7.3, we get the homogeneous solution

$$x_h = c_1 e^{-t} + c_2 e^{3t}$$

Now we choose

$$x_p = a\cos t + b\sin t.$$

By differentiating it, we obtain

$$x_p' = b\cos t - a\sin t,$$
$$x_p'' = -a\cos t - b\sin t.$$

We substitute these expressions into the given ODE. This yields

$$x_p'' - 2x_p' - 3x_p = (-4a - 2b)\cos t + (2a - 4b)\sin t = 10\cos t.$$

Comparing the coefficients of cos and sin gives

$$-4a - 2b = 10, \quad 2a - 4b = 0,$$

or

$$a = -2, \quad b = -1.$$

This gives the particular solution

$$x_p = -2\cos t - \sin t.$$

Therefore, the given ODE has the general solution

$$x = x_h + x_p = c_1 e^{-t} + c_2 e^{3t} - 2\cos t - \sin t.$$

where c_1 and c_2 are constants that are determined using initial conditions.

Answer $x = x_h + x_p = c_1 e^{-t} + c_2 e^{3t} - 2\cos t - \sin t$

Check

The particular solution of the ODE x is

$$x_p = -2\cos t - \sin t.$$

By differentiating it, we obtain

$$x_p' = -\cos t + 2\sin t,$$
$$x_p'' = 2\cos t + \sin t.$$

If we substitute them into the given equation, we get

LHS: $x_p'' - 2x_p' - 3x_p$

$$= (2\cos t + \sin t) - 2(-\cos t + 2\sin t) - 3(-2\cos t - \sin t)$$

$$= 10\cos t.$$

RHS: $10\cos t$

Therefore, LHS = RHS.

Example 7.6

Using the system of two first-order ODEs, find the general solution to the following second-order ODE:

$$x'' - x = e^t.$$

Solution

From the given ODE $x'' - x = e^t$, we get

$$x'' = x + e^t.$$

Now we set $x = \begin{Bmatrix} x \\ x' \end{Bmatrix}$, then we obtain a system of first-order ODEs

$$x' = Ax + R,$$

where $A = \begin{bmatrix} 0 & 1 \\ 1 & 0 \end{bmatrix}$ and $R = \begin{Bmatrix} 0 \\ e^t \end{Bmatrix}$.

From the homogeneous equation $\mathbf{x}' = \mathbf{A}\mathbf{x}$, we obtain the characteristic equation

$$\det(\mathbf{A} - \lambda\mathbf{I}) = 0.$$

This gives the eigenvalues

$$\lambda^2 - 1 = 0,$$

or

$$\lambda_1 = -1, \quad \lambda_2 = 1.$$

i. For the first eigenvalue $\lambda_1 = -1$, we obtain

$$(\mathbf{A} - \lambda\mathbf{I})\mathbf{x} = \begin{bmatrix} -\lambda & 1 \\ 1 & -\lambda \end{bmatrix} \begin{Bmatrix} x \\ x' \end{Bmatrix} = \begin{bmatrix} 1 & 1 \\ 1 & 1 \end{bmatrix} \begin{Bmatrix} x \\ x' \end{Bmatrix} = 0$$

or

$$x + x' = 0.$$

To obtain a simple value, let's set $x = 1$, then $x' = -1$. Therefore, its corresponding eigenvector becomes

$$\mathbf{v}_1 = \begin{Bmatrix} 1 \\ -1 \end{Bmatrix}.$$

And the first basis solution is

$$\mathbf{x}_1 = \mathbf{v}_1 e^{\lambda_1 t} = \begin{Bmatrix} 1 \\ -1 \end{Bmatrix} e^{-t}.$$

ii. For the second eigenvalue $\lambda_2 = 1$, we obtain

$$(\mathbf{A} - \lambda\mathbf{I})\mathbf{x} = \begin{bmatrix} -\lambda & 1 \\ 1 & -\lambda \end{bmatrix} \begin{Bmatrix} x \\ x' \end{Bmatrix} = \begin{bmatrix} -1 & 1 \\ 1 & -1 \end{bmatrix} \begin{Bmatrix} x \\ x' \end{Bmatrix} = 0,$$

or

$$-x + x' = 0.$$

To obtain a simple value, let's set $x = 1$, then $x' = 1$. Therefore, its corresponding eigenvector becomes

$$\mathbf{v}_2 = \begin{Bmatrix} 1 \\ 1 \end{Bmatrix}.$$

And the second basis solution is

$$\mathbf{x}_2 = \mathbf{v}_2 e^{\lambda_2 t} = \begin{Bmatrix} 1 \\ 1 \end{Bmatrix} e^t.$$

Therefore, we obtain the homogeneous solution of the ODE

$$\mathbf{x}_h = c_1 \mathbf{x}_1 + c_2 \mathbf{x}_2$$

or

$$\mathbf{x}_h = \begin{Bmatrix} x \\ x' \end{Bmatrix} = c_1 \begin{Bmatrix} 1 \\ -1 \end{Bmatrix} e^{-t} + c_2 \begin{Bmatrix} 1 \\ 1 \end{Bmatrix} e^t,$$

where c_1 and c_2 are constants that are determined using initial conditions.

$$\text{Answer } \mathbf{x}_h = \begin{Bmatrix} x \\ x' \end{Bmatrix} = c_1 \begin{Bmatrix} 1 \\ -1 \end{Bmatrix} e^{-t} + c_2 \begin{Bmatrix} 1 \\ 1 \end{Bmatrix} e^t$$

Check

The homogeneous solution of the ODE where the first row of **x** is

$$x(t) = c_1 e^{-t} + c_2 e^t,$$

it can be verified by differentiating it that

$$x'(t) = -c_1 e^{-t} + c_2 e^t$$

becomes the second row of **x**.

Now, using the *method of variation of parameters* (*Wronskian method*), let's find the particular solution x_p instead of using the method of undetermined coefficients.

Since Wronskian $W = \begin{vmatrix} e^{-t} & e^t \\ -e^{-t} & e^t \end{vmatrix} = 2$ and $r(t) = e^t$, we obtain the particular solution as follows:

$$x_p(t) = -x_1 \int \frac{x_2 r}{W} dt + x_2 \int \frac{x_1 r}{W} dt$$

$$= -e^{-t} \int \frac{e^t \cdot e^t}{2} dt + e^t \int \frac{e^{-t} \cdot e^t}{2} dt$$

$$= -e^{-t} \cdot \frac{e^{2t}}{4} + e^t \cdot \frac{t}{2} = \frac{e^t}{4}(2t - 1).$$

Meanwhile, since $-\dfrac{1}{4}e^t$ is involved in $x_h(t)$, we can obtain

$$x_p(t) = \frac{1}{2}te^t.$$

Therefore, we obtain the general solution of the ODE

$$x = x_h + x_p = c_1e^{-t} + c_2e^t + \frac{1}{2}te^t,$$

where c_1 and c_2 are constants that are determined using initial conditions.

Answer $x = c_1e^{-t} + c_2e^t + \dfrac{1}{2}te^t$

Check

The particular solution of the ODE **x** is

$$x_p(t) = \frac{1}{2}te^t.$$

By differentiating it, we obtain

$$x_p' = \frac{1}{2}(1+t)e^t,$$

$$x_p'' = \frac{1}{2}(2+t)e^t.$$

If we substitute them into the given equation, we get

LHS: $x_p'' - x_p = \dfrac{1}{2}(2+t)e^t - \dfrac{1}{2}te^t = e^t.$

RHS: e^t

Therefore, LHS = RHS.

Problem 7.1

Convert the following ODE into a system of first-order ODEs. [1 ~ 4]

1. $x'' - x' - 2x = e^{-2t}$
2. $x'' - 2x' + x = \sin 2t$
3. $x''' - 2x'' + 3x' - 2x = e^{-t}\cos 2t$
4. $x^{(4)} + 2x''' + 3x' - x = \sin t$

Convert the following ODE into a system of first-order ODEs, and find the homogeneous solution. [5 ~ 10]

5. $x'' + 3x' + 2x = 0$

6. $x'' - 4x' + 3x = 0$

7. $x'' - 2x' + x = 0$

8. $x'' + 4x' + 4x = 0$

9. $x''' + 2x'' - x' - 2x = 0$

10. $x''' + x'' - 4x' - 4x = 0$

Convert the following ODE into a system of first-order ODEs, and find the general solution. [11 ~ 14]

11. $x'' - 5x' + 4x = 4e^{2t}$

12. $x'' - 2x' + x = 2\cos t$

13. $x'' - 3x' + 2x = e^t$

14. $x'' + 2x' + x = e^{-t}$

7.2 Linear system of homogeneous ODEs

Consider the following *linear system* of first-order ODEs involving linear variables $y_1, y_2, \cdots, y_n$.

$$\frac{dy_1}{dt} = a_{11}(t)y_1 + a_{12}(t)y_2 + \cdots + a_{1n}(t)y_n + r_1(t)$$

$$\frac{dy_2}{dt} = a_{21}(t)y_1 + a_{22}(t)y_2 + \cdots + a_{2n}(t)y_n + r_2(t) \qquad (7.24)$$

$$\vdots$$

$$\frac{dy_n}{dt} = a_{n1}(t)y_1 + a_{n2}(t)y_2 + \cdots + a_{nn}(t)y_n + r_n(t)$$

Expressing these equations in matrix form yields

$$\mathbf{y}' = \mathbf{A}\mathbf{y} + \mathbf{R}, \qquad (7.25)$$

where the state $\mathbf{y} = \begin{Bmatrix} y_1 \\ y_2 \\ \vdots \\ y_n \end{Bmatrix}$, the matrix $\mathbf{A} = \begin{bmatrix} a_{11} & a_{12} & \cdots & a_{1n} \\ a_{21} & a_{22} & \cdots & a_{2n} \\ \vdots & \vdots & \ddots & \vdots \\ a_{n1} & a_{n2} & \cdots & a_{nn} \end{bmatrix}$, and the

vector $\mathbf{R} = \begin{Bmatrix} r_1(t) \\ r_2(t) \\ \vdots \\ r_n(t) \end{Bmatrix}$.

If $\mathbf{R} = 0$, this system is called homogeneous, so that it is

$$\mathbf{y}' = \mathbf{A}\mathbf{y}. \tag{7.26}$$

If $\mathbf{R} \neq 0$, then Eq. (7.25) is called nonhomogeneous.

Now since a single ODE $y' = \lambda y$ has the solution $y = Ce^{\lambda t}$ (C arbitrary), we can set

$$y = xe^{\lambda t}. \tag{7.27}$$

Substituting Eq. (7.27) into Eq. (7.26) yields

$$\mathbf{A}\mathbf{x} = \lambda\mathbf{x}. \tag{7.28}$$

From Eq. (7.28), we can write

$$(\mathbf{A} - \lambda\mathbf{I})\mathbf{x} = 0. \tag{7.29}$$

Similar to Section 7.1, from Eq. (7.29), we obtain the following *characteristic equation*:

$$\det(\mathbf{A} - \lambda\mathbf{I}) = 0. \tag{7.30}$$

Depending on the types of eigenvalues obtained from the characteristic equation (such as different real roots, double roots, conjugate complex roots, etc.), different methods are employed to find the general solution.

7.2.1   *General solution for distinct real eigenvalues*

If the characteristic equation (7.30) has distinct real eigenvalues $\lambda_1, \lambda_2, \cdots, \lambda_n$, we can find corresponding eigenvectors $\mathbf{v}_1, \mathbf{v}_2, \cdots, \mathbf{v}_n$ for each eigenvalue.

Then we obtain the following basis solutions.

$$\mathbf{x}_1 = \mathbf{v}_1 e^{\lambda_1 t}, \quad \mathbf{x}_2 = \mathbf{v}_2 e^{\lambda_2 t}, \quad \cdots, \quad \mathbf{x}_n = \mathbf{v}_n e^{\lambda_n t} \tag{7.31}$$

Therefore, the homogeneous solution of the matrix equation (7.26) is represented as a *linear combination* of these solutions. This is referred to as the superposition principle.

$$\mathbf{x} = c_1 \mathbf{x}_1 + c_2 \mathbf{x}_2 + \cdots + c_n \mathbf{x}_n \tag{7.32}$$

Remark   Linearly independent solution

The necessary and sufficient condition for the solution vectors $\mathbf{x}_1, \mathbf{x}_2, \cdots, \mathbf{x}_n$ to be *linearly independent* is that they satisfy the following condition for all t:

$$W(\mathbf{x}_1, \mathbf{x}_2, \cdots, \mathbf{x}_n) = \begin{vmatrix} x_{11} & x_{12} & \cdots & x_{1n} \\ x_{21} & x_{22} & \cdots & x_{2n} \\ \vdots & \vdots & \ddots & \vdots \\ x_{n1} & x_{n2} & \cdots & x_{nn} \end{vmatrix} \neq 0 \qquad (7.33)$$

where $\mathbf{x}_1 = \begin{Bmatrix} x_{11} \\ x_{21} \\ \vdots \\ x_{n1} \end{Bmatrix}, \mathbf{x}_2 = \begin{Bmatrix} x_{12} \\ x_{22} \\ \vdots \\ x_{n2} \end{Bmatrix}, \cdots, \text{and } \mathbf{x}_n = \begin{Bmatrix} x_{1n} \\ x_{2n} \\ \vdots \\ x_{nn} \end{Bmatrix}.$

Example 7.7

Solve the linear system of ODEs.

$$\frac{dx_1}{dt} = -x_1 + 2x_2$$

$$\frac{dx_2}{dt} = 3x_1 - 2x_2$$

Solution

Now we set $\mathbf{x} = \begin{Bmatrix} x_1 \\ x_2 \end{Bmatrix}$, then we obtain a system of first-order ODEs

$$\begin{Bmatrix} x_1' \\ x_2' \end{Bmatrix} = \begin{bmatrix} -1 & 2 \\ 3 & -2 \end{bmatrix} \begin{Bmatrix} x_1 \\ x_2 \end{Bmatrix}.$$

Since $\mathbf{A} - \lambda\mathbf{I} = \begin{bmatrix} -1-\lambda & 2 \\ 3 & -2-\lambda \end{bmatrix}$ in the characteristic equation $\det(\mathbf{A} - \lambda\mathbf{I}) = 0$, we get

$$\lambda^2 + 3\lambda - 4 = 0,$$

or

$$\lambda_1 = 1, \quad \lambda_2 = -4.$$

i. From the first eigenvalue $\lambda_1 = 1$, we obtain

$$(\mathbf{A} - \lambda_1\mathbf{I})\mathbf{x} = \begin{bmatrix} -1-\lambda_1 & 2 \\ 3 & -2-\lambda_1 \end{bmatrix} \begin{Bmatrix} x_1 \\ x_2 \end{Bmatrix} = \begin{bmatrix} -2 & 2 \\ 3 & -3 \end{bmatrix} \begin{Bmatrix} x_1 \\ x_2 \end{Bmatrix} = 0$$

or

$$-x + x' = 0.$$

To obtain a simple value, let's set $x = 1$, then $x' = 1$.
Therefore, its corresponding eigenvector becomes

$$\mathbf{v}_1 = \begin{Bmatrix} 1 \\ 1 \end{Bmatrix}.$$

And the first basis solution is

$$\mathbf{x}_1 = \begin{Bmatrix} 1 \\ 1 \end{Bmatrix} e^t.$$

ii. From the first eigenvalue $\lambda_2 = -4$, we obtain

$$(\mathbf{A} - \lambda_2\mathbf{I})\mathbf{x} = \begin{bmatrix} -1-\lambda_2 & 2 \\ 3 & -2-\lambda_2 \end{bmatrix} \begin{Bmatrix} x_1 \\ x_2 \end{Bmatrix} = \begin{bmatrix} 3 & 2 \\ 3 & 2 \end{bmatrix} \begin{Bmatrix} x_1 \\ x_2 \end{Bmatrix} = 0,$$

or

$$3x + 2x' = 0.$$

To obtain a simple value, let's set $x = 2$, then $x' = -3$.
Therefore, its corresponding eigenvector becomes

$$\mathbf{v}_2 = \begin{Bmatrix} 2 \\ -3 \end{Bmatrix}.$$

And the second basis solution is

$$\mathbf{x}_2 = \begin{Bmatrix} 2 \\ -3 \end{Bmatrix} e^{-4t}.$$

Therefore, we obtain the general solution of the ODE

$$\mathbf{x} = \begin{Bmatrix} x_1 \\ x_2 \end{Bmatrix} = c_1 \begin{Bmatrix} 1 \\ 1 \end{Bmatrix} e^t + c_2 \begin{Bmatrix} 2 \\ -3 \end{Bmatrix} e^{-4t},$$

where c_1 and c_2 are constants that are determined using initial conditions.

Answer $x_1(t) = c_1 e^t + 2c_2 e^{-4t}$, $x_2(t) = c_1 e^t - 3c_2 e^{-4t}$

Check

If we substitute $x_1(t)$ and $x_2(t)$ into $\begin{Bmatrix} x_1' \\ x_2' \end{Bmatrix} = \begin{bmatrix} -1 & 2 \\ 3 & -2 \end{bmatrix} \begin{Bmatrix} x_1 \\ x_2 \end{Bmatrix}$, it becomes

LHS: $\begin{Bmatrix} x_1' \\ x_2' \end{Bmatrix} = \begin{bmatrix} c_1 & -8c_2 \\ c_1 & 12c_2 \end{bmatrix} \begin{Bmatrix} e^t \\ e^{-4t} \end{Bmatrix}$,

RHS: $\begin{bmatrix} -1 & 2 \\ 3 & -2 \end{bmatrix} \begin{Bmatrix} x_1 \\ x_2 \end{Bmatrix} = \begin{bmatrix} -1 & 2 \\ 3 & -2 \end{bmatrix} \begin{bmatrix} c_1 & 2c_2 \\ c_1 & -3c_2 \end{bmatrix} \begin{Bmatrix} e^t \\ e^{-4t} \end{Bmatrix}$

$$= \begin{bmatrix} c_1 & -8c_2 \\ c_1 & 12c_2 \end{bmatrix} \begin{Bmatrix} e^t \\ e^{-4t} \end{Bmatrix}.$$

Therefore, LHS = RHS.

Check   Linearly independence

To confirm the independence of the solutions, let's apply the solutions obtained in Example 7.2. Since

$$\text{Wronskian } W(\mathbf{x}_1, \mathbf{x}_2) = \begin{vmatrix} x_{11} & x_{12} \\ x_{21} & x_{22} \end{vmatrix} = \begin{vmatrix} e^t & 2e^{-4t} \\ e^t & -3e^{-4t} \end{vmatrix} \neq 0,$$

we can see that they are linearly independent.

7.2.2 General solution for double real eigenvalues

The method for finding solutions for double real eigenvalues is as previously explained in Section 7.1. Summarizing it again, we have the following.

Remark   General solution for double real eigenvalues

In a system of homogeneous first-order ODEs x' = Ax, we obtain the characteristic equation

$$\det(\mathbf{A} - \lambda \mathbf{I}) = 0. \tag{7.34}$$

When the system has double real eigenvalue λ_1 and the corresponding eigenvector $\mathbf{v}_1$, the first solution is $\mathbf{x}_1 = \mathbf{v}_1 e^{\lambda_1 t}$ and the second solution is

$$\mathbf{x}_2 = (\mathbf{v}_1 t + \mathbf{v}_2) e^{\lambda_1 t}, \tag{7.35}$$

where the second eigenvector $\mathbf{v}_2$ satisfy

$$(\mathbf{A} - \lambda_1 \mathbf{I}) \mathbf{v}_2 = \mathbf{v}_1. \tag{7.36}$$

Example 7.8

Solve a system of homogeneous first-order ODEs.

$$\frac{dx_1}{dt} = 3x_1 - 18x_2$$

$$\frac{dx_2}{dt} = 2x_1 - 9x_2$$

Solution

Now we set $\mathbf{x} = \begin{Bmatrix} x_1 \\ x_2 \end{Bmatrix}$, then we obtain a system of first-order ODEs

$$\begin{Bmatrix} x_1' \\ x_2' \end{Bmatrix} = \begin{bmatrix} 3 & -18 \\ 2 & -9 \end{bmatrix} \begin{Bmatrix} x_1 \\ x_2 \end{Bmatrix}.$$

Since $\mathbf{A} - \lambda\mathbf{I} = \begin{bmatrix} 3-\lambda & -18 \\ 2 & -9-\lambda \end{bmatrix}$ in the characteristic equation $\det(\mathbf{A} - \lambda\mathbf{I}) = 0$, we get

$$\lambda^2 + 6\lambda + 9 = 0,$$

or

$$\lambda_1 = \lambda_2 = -3 \quad (\text{double root})$$

i. From the first eigenvalue $\lambda_1 = -3$, we obtain

$$(\mathbf{A} - \lambda_1\mathbf{I})\mathbf{x} = \begin{bmatrix} 3-\lambda_1 & -18 \\ 2 & -9-\lambda_1 \end{bmatrix}\begin{Bmatrix} x_1 \\ x_2 \end{Bmatrix} = \begin{bmatrix} 6 & -18 \\ 2 & -6 \end{bmatrix}\begin{Bmatrix} x_1 \\ x_2 \end{Bmatrix} = 0,$$

or

$$x_1 - 3x_2 = 0.$$

To obtain a simple value, let's set $x_1 = 3$, then $x_2 = 1$.
Therefore, its corresponding eigenvector becomes

$$\mathbf{v}_1 = \begin{Bmatrix} 3 \\ 1 \end{Bmatrix}.$$

And the first basis solution is

$$\mathbf{x}_1 = \begin{Bmatrix} 3 \\ 1 \end{Bmatrix} e^{-3t}.$$

ii. The second eigenvalue is same as the first eigenvalue $\lambda_1 = \lambda_2 = -3$.
Then, from Eq. (6.26) $(\mathbf{A} - \lambda_1\mathbf{I})\mathbf{v}_2 = \mathbf{v}_1$, we obtain

$$(\mathbf{A} - \lambda_1\mathbf{I})\mathbf{x} = \begin{bmatrix} 6 & -18 \\ 2 & -6 \end{bmatrix}\begin{Bmatrix} x_1 \\ x_2 \end{Bmatrix} = \begin{Bmatrix} 3 \\ 1 \end{Bmatrix}.$$

We can choose any pairs of x_1 and x_2 that satisfy $2x_1 - 6x_2 = 1$, but to obtain a simple value, let's set $x_2 = 0$, then $x_1 = 0.5$.
Therefore, its corresponding eigenvector becomes

$$\mathbf{v}_2 = \begin{Bmatrix} 0.5 \\ 0 \end{Bmatrix}.$$

From Eq. (6.26), we get

$$\mathbf{x}_2 = (\mathbf{v}_1 t + \mathbf{v}_2)e^{\lambda_1 t} = \begin{Bmatrix} 3t + 0.5 \\ t \end{Bmatrix} e^{-3t}.$$

Therefore, we obtain the general solution

$$
\mathbf{x} = \begin{Bmatrix} x_1 \\ x_2 \end{Bmatrix} = c_1 \begin{Bmatrix} 3 \\ 1 \end{Bmatrix} e^{-3t} + c_2 \begin{Bmatrix} 3t + 0.5 \\ t \end{Bmatrix} e^{-3t},
$$

where c_1 and c_2 are constants that are determined using initial conditions.

Answer $x_1(t) = 3c_1 e^{-3t} + c_2(3t + 0.5)e^{-3t}$, $x_2(t) = (c_1 + c_2 t)e^{-3t}$

Check

If we substitute $x_1(t)$ and $x_2(t)$ into $\begin{Bmatrix} x_1' \\ x_2' \end{Bmatrix} = \begin{bmatrix} 3 & -18 \\ 2 & -9 \end{bmatrix} \begin{Bmatrix} x_1 \\ x_2 \end{Bmatrix}$, it becomes

$$
\text{LHS: } \begin{Bmatrix} x_1' \\ x_2' \end{Bmatrix} = \begin{bmatrix} -9c_1 + 1.5c_2 & -9c_2 \\ -3c_1 + c_2 & -3c_2 \end{bmatrix} \begin{Bmatrix} e^{-3t} \\ te^{-3t} \end{Bmatrix}
$$

$$
\text{RHS: } \begin{bmatrix} 3 & -18 \\ 2 & -9 \end{bmatrix} \begin{Bmatrix} x_1 \\ x_2 \end{Bmatrix} = \begin{bmatrix} 3 & -18 \\ 2 & -9 \end{bmatrix} \begin{bmatrix} 3c_1 + 0.5c_2 & 3c_2 \\ c_1 & c_2 \end{bmatrix} \begin{Bmatrix} e^{-3t} \\ te^{-3t} \end{Bmatrix}
$$

$$
= \begin{bmatrix} -9c_1 + 1.5c_2 & -9c_2 \\ -3c_1 + c_2 & -3c_2 \end{bmatrix} \begin{Bmatrix} e^{-3t} \\ te^{-3t} \end{Bmatrix}
$$

Therefore, LHS = RHS.

Example 7.9

Solve a system of homogeneous first-order ODEs.

$$
\frac{dx_1}{dt} = -x_1 + x_2 + x_3
$$

$$
\frac{dx_2}{dt} = -x_2 + x_3
$$

$$
\frac{dx_3}{dt} = 2x_1 + x_2 - x_3
$$

Solution

Now we set $\mathbf{x} = \begin{Bmatrix} x_1 \\ x_2 \\ x_3 \end{Bmatrix}$, then we obtain a system of first-order ODEs

$$\begin{Bmatrix} x_1' \\ x_2' \\ x_3' \end{Bmatrix} = \begin{bmatrix} -1 & 1 & 1 \\ 0 & -1 & 1 \\ 2 & 1 & -1 \end{bmatrix} \begin{Bmatrix} x_1 \\ x_2 \\ x_3 \end{Bmatrix}.$$

Since $\mathbf{A} - \lambda \mathbf{I} = \begin{bmatrix} -1-\lambda & 1 & 1 \\ 0 & -1-\lambda & 1 \\ 2 & 1 & -1-\lambda \end{bmatrix}$ in the characteristic equation

$\det(\mathbf{A} - \lambda \mathbf{I}) = 0$, we get

$$\lambda^3 + 3\lambda^2 - 4 = 0$$

or

$$\lambda_1 = 1, \quad \lambda_2 = \lambda_3 = -2 \quad \text{(double root)}$$

i. From the first eigenvalue $\lambda_1 = 1$, we obtain

$$(\mathbf{A} - \lambda_1 \mathbf{I})\mathbf{x} = \begin{bmatrix} -1-\lambda_1 & 1 & 1 \\ 0 & -1-\lambda_1 & 1 \\ 2 & 1 & -1-\lambda_1 \end{bmatrix} \begin{Bmatrix} x_1 \\ x_2 \\ x_3 \end{Bmatrix} = \begin{bmatrix} -2 & 1 & 1 \\ 0 & -2 & 1 \\ 2 & 1 & -2 \end{bmatrix} \begin{Bmatrix} x_1 \\ x_2 \\ x_3 \end{Bmatrix} = 0,$$

or

$$-2x_1 + x_2 + x_3 = 0, \quad -2x_2 + x_3 = 0, \quad 2x_1 + x_2 - 2x_3 = 0.$$

To obtain a simple value, let's set $x_1 = 3$, then $x_2 = 2$ and $x_3 = 4$. Therefore, its corresponding eigenvector becomes

$$\mathbf{v}_1 = \begin{Bmatrix} 3 \\ 2 \\ 4 \end{Bmatrix}.$$

And the first basis solution is

$$\mathbf{x}_1 = \begin{Bmatrix} 3 \\ 2 \\ 4 \end{Bmatrix} e^t.$$

ii. From the second eigenvalue $\lambda_2 = -2$ (double root), we obtain

$$\left(\mathbf{A} - \lambda_2\mathbf{I}\right)\mathbf{x} = \begin{bmatrix} -1-\lambda_2 & 1 & 1 \\ 0 & -1-\lambda_2 & 1 \\ 2 & 1 & -1-\lambda_2 \end{bmatrix} \begin{Bmatrix} x_1 \\ x_2 \\ x_3 \end{Bmatrix} = \begin{bmatrix} 1 & 1 & 1 \\ 0 & 1 & 1 \\ 2 & 1 & 1 \end{bmatrix} \begin{Bmatrix} x_1 \\ x_2 \\ x_3 \end{Bmatrix} = 0,$$

or

$$x_1 + x_2 + x_3 = 0, \quad x_2 + x_3 = 0, \quad 2x_1 + x_2 + x_3 = 0.$$

To obtain a simple value, let's set $x_1 = 0$ and $x_2 = 1$, then $x_3 = -1$.
Therefore, its corresponding eigenvector becomes

$$\mathbf{v}_2 = \begin{Bmatrix} 0 \\ 1 \\ -1 \end{Bmatrix}.$$

And the second basis solution is

$$\mathbf{x}_2 = \begin{Bmatrix} 0 \\ 1 \\ -1 \end{Bmatrix} e^{-2t}.$$

iii. The third eigenvalue is also same as the second eigenvalue $\lambda_2 = \lambda_3 = -3$.
Then, from Eq. (6.26) $\left(\mathbf{A} - \lambda_2\mathbf{I}\right)\mathbf{v}_3 = \mathbf{v}_2$, we obtain

$$\left(\mathbf{A} - \lambda_2\mathbf{I}\right)\mathbf{x} = \begin{bmatrix} 1 & 1 & 1 \\ 0 & 1 & 1 \\ 2 & 1 & 1 \end{bmatrix} \begin{Bmatrix} x_1 \\ x_2 \\ x_3 \end{Bmatrix} = \begin{Bmatrix} 0 \\ 1 \\ -1 \end{Bmatrix},$$

or

$$x_1 + x_2 + x_3 = 0, \quad x_2 + x_3 = 1, \quad 2x_1 + x_2 + x_3 = -1.$$

To obtain a simple value, let's set $x_1 = -1$.
And we can choose any pairs of x_2 and x_3 that satisfy $x_2 + x_3 = 1$, but to obtain a simple value, let's set $x_3 = 0$. Then $x_2 = 1$.
Therefore, its corresponding eigenvector becomes

$$\mathbf{v}_3 = \begin{Bmatrix} -1 \\ 1 \\ 0 \end{Bmatrix}.$$

From Eq. (7.26), we get

$$\mathbf{x}_3 = \mathbf{v}_2 t e^{\lambda_2 t} + \mathbf{v}_3 e^{\lambda_2 t} = \left\{ \begin{array}{c} 0 \\ 1 \\ -1 \end{array} \right\} t e^{-2t} + \left\{ \begin{array}{c} -1 \\ 1 \\ 0 \end{array} \right\} e^{-2t}.$$

Therefore, we obtain the general solution

$$\mathbf{x} = \left\{ \begin{array}{c} x_1 \\ x_2 \\ x_3 \end{array} \right\} = c_1 \left\{ \begin{array}{c} 3 \\ 2 \\ 4 \end{array} \right\} e^t + c_2 \left\{ \begin{array}{c} 0 \\ 1 \\ -1 \end{array} \right\} e^{-2t} + c_3 \left\{ \begin{array}{c} -1 \\ t+1 \\ -t \end{array} \right\} e^{-2t},$$

where c_1, c_2 and c_3 are constants that are determined using initial conditions.

$$\text{Answer } x_1(t) = 3 c_1 e^t - c_3 e^{-2t},$$
$$x_2(t) = 2 c_1 e^t + (c_2 + c_3 + c_3 t) e^{-2t},$$
$$x_3(t) = 4 c_1 e^t - (c_2 + c_3 t) e^{-2t}$$

Check

If we substitute $x_1(t), x_2(t)$, and $x_3(t)$ into $\left\{ \begin{array}{c} x_1' \\ x_2' \\ x_3' \end{array} \right\} = \left[\begin{array}{ccc} -1 & 1 & 1 \\ 0 & -1 & 1 \\ 2 & 1 & -1 \end{array} \right] \left\{ \begin{array}{c} x_1 \\ x_2 \\ x_3 \end{array} \right\}$,

it becomes

$$\text{LHS: } \left\{ \begin{array}{c} x_1' \\ x_2' \\ x_3' \end{array} \right\} = \left[\begin{array}{ccc} 3c_1 & 2c_3 & 0 \\ 2c_1 & -2c_2 - c_3 & -2c_3 \\ 4c_1 & 2c_2 - c_3 & 2c_3 \end{array} \right] \left\{ \begin{array}{c} e^t \\ e^{-2t} \\ t e^{-2t} \end{array} \right\}$$

$$\text{RHS: } \left[\begin{array}{ccc} -1 & 1 & 1 \\ 0 & -1 & 1 \\ 2 & 1 & -1 \end{array} \right] \left[\begin{array}{ccc} 3c_1 & -c_3 & 0 \\ 2c_1 & c_2 + c_3 & c_3 \\ 4c_1 & -c_2 & -c_3 \end{array} \right] \left\{ \begin{array}{c} e^t \\ e^{-2t} \\ t e^{-2t} \end{array} \right\}$$

$$= \left[\begin{array}{ccc} 3c_1 & 2c_3 & 0 \\ 2c_1 & -2c_2 - c_3 & -2c_3 \\ 4c_1 & 2c_2 - c_3 & 2c_3 \end{array} \right] \left\{ \begin{array}{c} e^t \\ e^{-2t} \\ t e^{-2t} \end{array} \right\}$$

Therefore, LHS = RHS.

7.2.3 (*optional) General solution for complex conjugate eigenvalues

The characteristic equation (7.32) has the first eigenvalue $\lambda_1 = \alpha + \beta i$ and the corresponding eigenvector v_1.

Additionally, the second eigenvalue of the characteristic equation is the complex conjugate of λ_1, that is $\lambda_2 = \widetilde{\lambda}_1 = \alpha - \beta i$, and its corresponding eigenvector is the complex conjugate eigenvector of v_1, that is $v_2 = \tilde{v}_1$.

Therefore, the general solution of the system of ODEs with complex conjugate eigenvalues is summarized as follows:

Remark   General solution for complex conjugate eigenvalues

If a system of ODEs $x' = Ax$ has complex conjugate eigenvalues λ_1, $\widetilde{\lambda}_1$ and their corresponding eigenvectors v_1, $\tilde{v}_1$, then the general solution of the system is

$$x = c_1 v_1 e^{\lambda_1 t} + c_2 \tilde{v}_1 e^{\widetilde{\lambda}_1 t} \qquad (c_1, c_2 \text{ arbitrary}) \qquad (7.37)$$

When the characteristic equation has a complex eigenvector $\lambda_1 = \alpha + \beta i$, we obtain the solutions of a system of ODEs as follows:

$$v_1 e^{\lambda_1 t} = v_1 e^{\alpha t} e^{i\beta t} = v_1 e^{\alpha t} \left(\cos \beta t + i \sin \beta t \right),$$

and

$$\tilde{v}_1 e^{\widetilde{\lambda}_1 t} = \tilde{v}_1 e^{\alpha t} e^{-i\beta t} = \tilde{v}_1 e^{\alpha t} \left(\cos \beta t - i \sin \beta t \right).$$

The real and imaginary parts of the above two solutions also serve as solutions of the system.

So, when you add the two solutions, you get one real solution, and when you subtract them, you get another real solution.

$$x_1 = \frac{1}{2}\left(v_1 e^{\lambda_1 t} + \tilde{v}_1 e^{\widetilde{\lambda}_1 t} \right) = \frac{1}{2}\left(v_1 + \tilde{v}_1 \right) e^{\alpha t} \cos \beta t + \frac{i}{2}\left(v_1 - \tilde{v}_1 \right) e^{\alpha t} \sin \beta t$$

$$(7.38a)$$

$$x_2 = \frac{1}{2}\left(-v_1 e^{\lambda_1 t} + \tilde{v}_1 e^{\widetilde{\lambda}_1 t} \right) = -\frac{i}{2}\left(v_1 - \tilde{v}_1 \right) e^{\alpha t} \cos \beta t + \frac{1}{2}\left(v_1 + \tilde{v}_1 \right) e^{\alpha t} \sin \beta t$$

$$(7.38b)$$

When the eigenvector v_1 is complex, $\frac{1}{2}(v_1 + \tilde{v}_1)$ and $-\frac{i}{2}(v_1 - \tilde{v}_1)$ become real. Thus, we define new real eigenvectors b_1 and b_2 as follows:

$$b_1 = \frac{1}{2}(v_1 + \tilde{v}_1) = \text{Re}(v_1) \tag{7.39a}$$

$$b_2 = -\frac{i}{2}(v_1 - \tilde{v}_1) = \text{Im}(v_1) \tag{7.39b}$$

where b_1 is the real part, and b_2 is the imaginary part of the complex eigenvector v_1.

Therefore, the basis solutions of the system, Eq. (7.38a) and Eq. (7.38b), are organized as follows.

$$x_1 = e^{\alpha t}(b_1 \cos \beta t - b_2 \sin \beta t) \tag{7.40a}$$

$$x_2 = e^{\alpha t}(b_2 \cos \beta t + b_1 \sin \beta t) \tag{7.40b}$$

Remark   Real basis solutions for complex conjugate eigenvalues

If a system of ODEs $x' = Ax$ has a complex eigenvalue $\lambda_1 = \alpha + \beta i$ and its corresponding eigenvector v_1, we obtain two real basis solutions

$$x_1 = e^{\alpha t}(b_1 \cos \beta t - b_2 \sin \beta t), \tag{7.40a}$$

$$x_2 = e^{\alpha t}(b_2 \cos \beta t + b_1 \sin \beta t), \tag{7.40b}$$

where $b_1 = \text{Re}(v_1)$ and $b_2 = \text{Im}(v_1)$.

Example 7.10

Solve a system of homogeneous first-order ODEs.

$$\frac{dx_1}{dt} = 2x_1 + 4x_2$$

$$\frac{dx_2}{dt} = -2x_1 - 2x_2$$

Solution

Now we set $\mathbf{x} = \begin{Bmatrix} x_1 \\ x_2 \end{Bmatrix}$, then we obtain a system of first-order ODEs

$$\begin{Bmatrix} x_1' \\ x_2' \end{Bmatrix} = \begin{bmatrix} 2 & 4 \\ -2 & -2 \end{bmatrix} \begin{Bmatrix} x_1 \\ x_2 \end{Bmatrix}.$$

Since $\mathbf{A} - \lambda\mathbf{I} = \begin{bmatrix} 2-\lambda & 4 \\ -2 & -2-\lambda \end{bmatrix}$ in the characteristic equation $\det(\mathbf{A} - \lambda\mathbf{I}) = 0$, we get

$$\lambda^2 + 4 = 0,$$

or

$$\lambda_1 = 2i \quad (\alpha = 0, \beta = 2).$$

For the first eigenvalue $\lambda_1 = 2i$, we obtain

$$(\mathbf{A} - \lambda_1\mathbf{I})\mathbf{x} = \begin{bmatrix} 2-\lambda_1 & 4 \\ -2 & -2-\lambda_1 \end{bmatrix} \begin{Bmatrix} x_1 \\ x_2 \end{Bmatrix} = \begin{bmatrix} 2-2i & 4 \\ -2 & -2-2i \end{bmatrix} \begin{Bmatrix} x_1 \\ x_2 \end{Bmatrix} = 0.$$

To obtain a simple value, let's set $x_1 = 3$, then $x_2 = 2$ and $x_3 = 4$. Therefore, its corresponding eigenvector becomes

$$\mathbf{v}_1 = \begin{Bmatrix} 1+i \\ -1 \end{Bmatrix}.$$

When we substitute $b_1 = \mathrm{Re}(\mathbf{v}_1) = \begin{Bmatrix} 1 \\ -1 \end{Bmatrix}$ and $b_2 = \mathrm{Im}(\mathbf{v}_1) = \begin{Bmatrix} 1 \\ 0 \end{Bmatrix}$ into $x_1 = e^{\alpha t}(b_1 \cos \beta t - b_2 \sin \beta t)$ and $x_2 = e^{\alpha t}(b_2 \cos \beta t + b_1 \sin \beta t)$, we get

$$x_1 = \begin{Bmatrix} 1 \\ -1 \end{Bmatrix} \cos 2t - \begin{Bmatrix} 1 \\ 0 \end{Bmatrix} \sin 2t,$$

$$x_2 = \begin{Bmatrix} 1 \\ 0 \end{Bmatrix} \cos 2t + \begin{Bmatrix} 1 \\ -1 \end{Bmatrix} \sin 2t.$$

Therefore, we obtain the general solution

$$\mathbf{x} = c_1 \left[\begin{Bmatrix} 1 \\ -1 \end{Bmatrix} \cos 2t - \begin{Bmatrix} 1 \\ 0 \end{Bmatrix} \sin 2t \right] + c_2 \left[\begin{Bmatrix} 1 \\ 0 \end{Bmatrix} \cos 2t + \begin{Bmatrix} 1 \\ -1 \end{Bmatrix} \sin 2t \right],$$

where c_1 and c_2 are constants that are determined using initial conditions.

Answer $x_1(t) = (c_1 + c_2)\cos 2t + (-c_1 + c_2)\sin 2t,$

$$x_2(t) = -c_1 \cos 2t - c_2 \sin 2t$$

Check

If we substitute $x_1(t)$ and $x_2(t)$ into $\begin{Bmatrix} x_1' \\ x_2' \end{Bmatrix} = \begin{bmatrix} 2 & 4 \\ -2 & -2 \end{bmatrix} \begin{Bmatrix} x_1 \\ x_2 \end{Bmatrix}$, it becomes

LHS: $\begin{Bmatrix} x_1' \\ x_2' \end{Bmatrix} = \begin{bmatrix} 2(-c_1+c_2) & -2(c_1+c_2) \\ -2c_2 & 2c_1 \end{bmatrix} \begin{Bmatrix} \cos 2t \\ \sin 2t \end{Bmatrix}$

RHS: $\begin{bmatrix} 2 & 4 \\ -2 & -2 \end{bmatrix} \begin{Bmatrix} x_1 \\ x_2 \end{Bmatrix} = \begin{bmatrix} c_1+c_2 & -c_1+c_2 \\ -c_1 & -c_2 \end{bmatrix} \begin{Bmatrix} \cos 2t \\ \sin 2t \end{Bmatrix}$

$= \begin{bmatrix} 2(-c_1+c_2) & -2(c_1+c_2) \\ -2c_2 & 2c_1 \end{bmatrix} \begin{Bmatrix} \cos 2t \\ \sin 2t \end{Bmatrix}$

Therefore, LHS = RHS.

7.2.4   (*optional) Phase plane

From a system of ODEs $\mathbf{x'} = \mathbf{Ax}$, we obtain its solutions, $x_1(t)$ and $x_2(t)$.

The plane with the horizontal axis labeled as $x_1(t)$ and the vertical axis as $x_2(t)$ is called the *phase plane*, parameterized by t.

i. For the two eigenvalues with opposite signs, $(\lambda_2 < 0 < \lambda_1)$
   In Example 6.7, we obtain

$$x_1(t) = c_1 e^t + 2c_2 e^{-4t},$$

$$x_2(t) = c_1 e^t - 3c_2 e^{-4t}.$$

Assuming $c_1 = 1$ and $c_2 = 1$ gives

$$x_1(t) = e^t + 2e^{-4t},$$

$$x_2(t) = e^t - 3e^{-4t}.$$

If you were to graph them, they would result in Figure 7.1(a) and Figure 7.1(b), respectively.

Figure 7.2 shows the phase plane for $x_1(t)$ and $x_2(t)$, where the two asymptotic lines correspond to $y = x$ determined by the first eigenvector $\mathbf{v}_1 = \{1 \quad 1\}^T$ and $y = -\dfrac{3}{2}x$ determined by the second eigenvector $\mathbf{v}_2 = \{2 \quad -3\}^T$.

When $c_1 > 0$ and $c_2 > 0$, as t increases, $x_1(t)$ and $x_2(t)$ both tend to infinity (the first quadrant of Figure 7.2(b)), so the direction of the arrows points away from the origin, moving farther away from it. And when $c_1 < 0$ and $c_2 < 0$, as t increases, $x_1(t) \to -\infty$ and $x_2(t) \to -\infty$ (the third quadrant of Figure 7.2(b)), so the direction of the arrows points away from the origin, moving farther away from it.

The phase plane in Figure 7.2 indeed illustrates the typical behavior when the eigenvalues have opposite signs $(\lambda_2 < 0 < \lambda_1)$, and an unstable critical point is referred to as a *saddle point*.

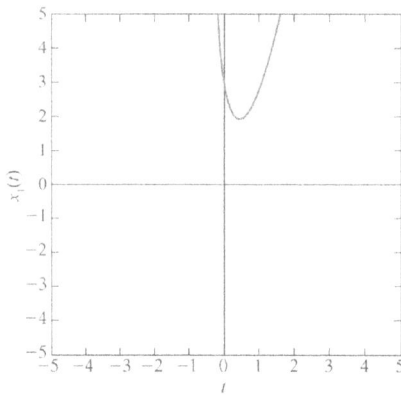

(a) $x_1(t) = e^t + 2e^{-4t}$

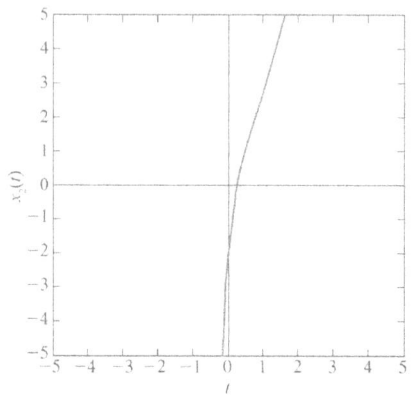

(b) $x_2(t) = e^t - 3e^{-4t}$.

Figure 7.1 Graphs of $x_1(t)$ and $x_2(t)$.

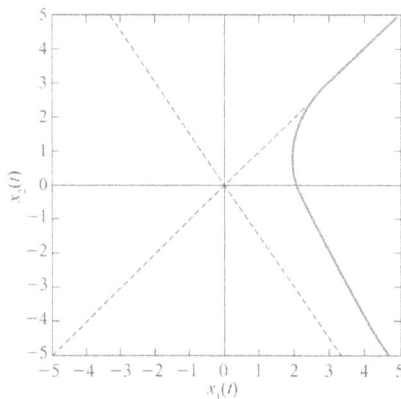

(a) $x_1(t) = e^t + 2e^{-4t}$ and

$x_2(t) = e^t - 3e^{-4t}$

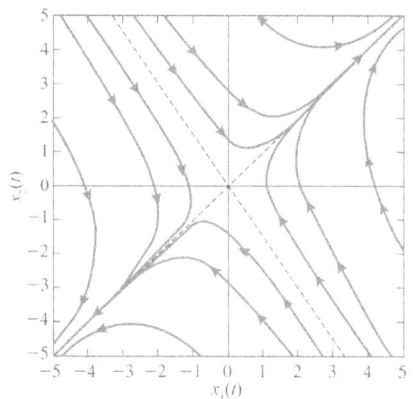

(b) $x_1(t) = c_1 e^t + 2c_2 e^{-4t}$ and

$x_2(t) = c_1 e^t - 3c_2 e^{-4t}$.

Figure 7.2 Phase plane with a saddle point.

ii. For the two negative eigenvalues, $(\lambda_1, \lambda_2 < 0)$
In Example 7.8, we obtain

$$x_1(t) = (3c_1 + 3c_2 t)e^{-3t} + 0.5c_2 e^{-3t},$$

$$x_2(t) = (c_1 + c_2 t)e^{-3t}.$$

Assuming $c_1 = 1$ and $c_2 = 1$ gives

$$x_1(t) = (3.5 + 3t)e^{-3t},$$

$$x_2(t) = (1 + t)e^{-3t}.$$

If you were to graph them, they would result in Figure 7.3(a) and Figure 7.3(b), respectively.

Figure 7.4 shows the phase plane for $x_1(t)$ and $x_2(t)$. This is a typical phase plane graph representing a negative double eigenvalue, where the two asymptotic lines correspond to $y = x/3$ determined by the first eigenvector $v_1 = \{3 \quad 1\}^T$ and $y = 0$ determined by the second eigenvector $v_2 = \{0.5 \quad 0\}^T$.

When $c_1 > 0$ and $c_2 > 0$, as t increases, $x_1(t) \to +0$, $x_2(t) \to +0$ (the first quadrant of Figure 7.4(b)), so the direction of the arrows points toward the origin. And when $c_1 < 0$ and $c_2 < 0$, as t increases, $x_1(t) \to -0$, $x_2(t) \to -0$ (the third quadrant of Figure 7.4(b)), so the direction of the arrows points toward the origin.

The phase plane in Figure 7.4 indeed illustrates the typical behavior when the eigenvalues are negative $(\lambda_1, \lambda_2 < 0)$, and a stable critical point is referred to as a *stable node*.

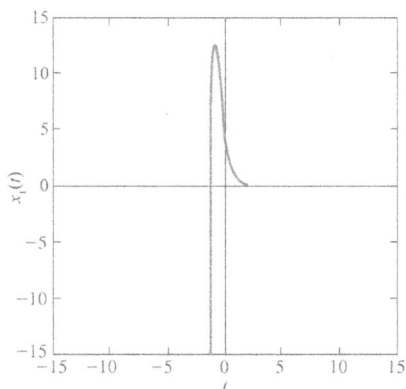

(a) $x_1(t) = (3.5 + 3t)e^{-3t}$

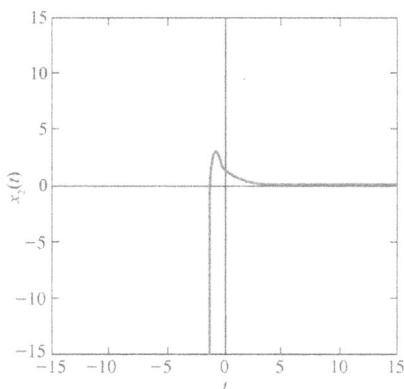

(b) $x_2(t) = (1 + t)e^{-3t}$.

Figure 7.3 Graphs of $x_1(t)$ and $x_2(t)$.

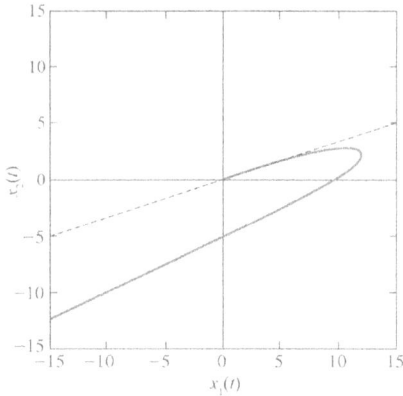

(a) $x_1(t) = (3.5 + 3t)e^{-3t}$ and
$x_2(t) = (1+t)e^{-3t}$

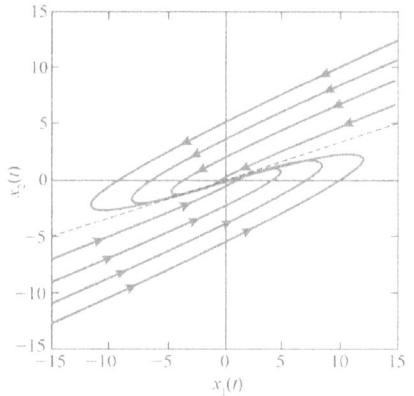

(b) $x_1(t) = (3c_1 + 3c_2t)e^{-3t} + 0.5c_2e^{-3t}$ and
$x_2(t) = (c_1 + c_2t)e^{-3t}$

Figure 7.4 Phase plane with a stable node.

iii. For the two positive eigenvalues, $(\lambda_1, \lambda_2 > 0)$

The phase plane in Figure 7.5 indeed illustrates the typical behavior when the eigenvalues are positive $(\lambda_1, \lambda_2 > 0)$, and a stable critical point is referred to as a stable node.

Its shape resembles Figure 7.4 but with arrows pointing in the opposite direction. Such an unstable critical point is referred to as an *unstable node*.

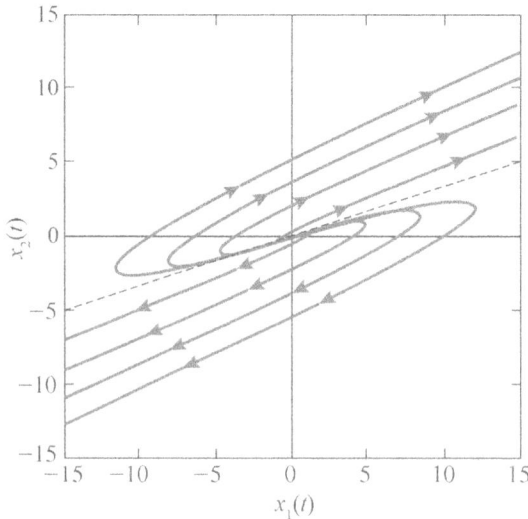

Figure 7.5 Phase plane with an unstable node.

iv. For the complex conjugate eigenvalue
In Example 7.10, we obtain

$$x_1(t) = (c_1 + c_2)\cos 2t + (-c_1 + c_2)\sin 2t,$$
$$x_2(t) = -c_1 \cos 2t - c_2 \sin 2t.$$

Assuming $c_1 = 1$ and $c_2 = 1$ gives

$$x_1(t) = 2\cos 2t,$$
$$x_2(t) = -\cos 2t - \sin 2t.$$

Figure 7.6 shows the phase plane for $x_1(t)$ and $x_2(t)$, where the arrows rotate clockwise.

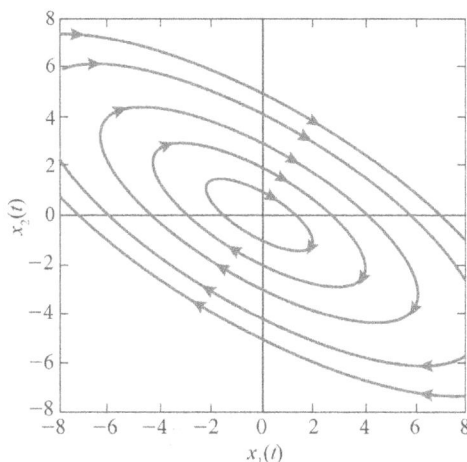

Figure 7.6 Phase plane for a complex conjugate eigenvalue (ellipse type).

Problem 7.2

Solve a linear system of ODEs (for distinct real eigenvalues). [1 ~ 8]

1. $\dfrac{dx_1}{dt} = x_1 + 2x_2 \quad \dfrac{dx_2}{dt} = 3x_1 + 2x_2$

2. $\dfrac{dx_1}{dt} = 2x_1 + 3x_2 \quad \dfrac{dx_2}{dt} = 4x_1 + 3x_2$

3. $\dfrac{dx_1}{dt} = -3x_1 + x_2 \quad \dfrac{dx_2}{dt} = 2x_1 - 2x_2$

4. $\dfrac{dx_1}{dt} = -6x_1 + 2x_2 \quad \dfrac{dx_2}{dt} = -3x_1 + x_2$

5. $\dfrac{dx_1}{dt} = -2x_1 - 3x_2 \quad \dfrac{dx_2}{dt} = -4x_1 + 2x_2$

6. $\dfrac{dx_1}{dt} = 2x_1 + x_2$ $\dfrac{dx_2}{dt} = x_1 + 2x_2$

7. $\dfrac{dx_1}{dt} = -x_1 + 2x_2,$ $\dfrac{dx_2}{dt} = x_1 + 3x_2 + x_3,$ $\dfrac{dx_3}{dt} = 3x_2 - x_3$

8. $\dfrac{dx_1}{dt} = x_1 + 2x_2 + x_3,$ $\dfrac{dx_2}{dt} = x_2,$ $\dfrac{dx_3}{dt} = x_1 + x_3$

Solve a linear system of ODEs (for double eigenvalues). [9 ~ 12]

9. $\dfrac{dx_1}{dt} = -x_1 + x_2,$ $\dfrac{dx_2}{dt} = -x_1 - 3x_2$

10. $\dfrac{dx_1}{dt} = 3x_1 - x_2,$ $\dfrac{dx_2}{dt} = x_1 + x_2$

11. $\dfrac{dx_1}{dt} = 2x_1 - x_2,$ $\dfrac{dx_2}{dt} = 4x_1 - 2x_2$

12. $\dfrac{dx_1}{dt} = -6x_1 - 5x_2,$ $\dfrac{dx_2}{dt} = 5x_1 + 4x_2$

Solve a linear system of ODEs (for complex conjugate eigenvalues). [13 ~ 16]

13. $\dfrac{dx_1}{dt} = 6x_1 + x_2,$ $\dfrac{dx_2}{dt} = -5x_1 + 2x_2$

14. $\dfrac{dx_1}{dt} = x_1 + 5x_2,$ $\dfrac{dx_2}{dt} = -x_1 - x_2$

15. $\dfrac{dx_1}{dt} = 3x_1 + 2x_2,$ $\dfrac{dx_2}{dt} = -x_1 + x_2$

16. $\dfrac{dx_1}{dt} = 6x_1 + 2x_2,$ $\dfrac{dx_2}{dt} = -5x_1 + 4x_2$

7.3   Linear system of nonhomogeneous ODEs

General form of a *linear system of nonhomogeneous ODEs* is

$$\mathbf{x}' = \mathbf{A}\mathbf{x} + \mathbf{R} \tag{7.41}$$

where the state $\mathbf{x} = \begin{Bmatrix} x_1 \\ x_2 \\ \vdots \\ x_n \end{Bmatrix},$ matrix $\mathbf{A} = \begin{bmatrix} a_{11} & a_{12} & \cdots & a_{1n} \\ a_{21} & a_{22} & \cdots & a_{2n} \\ \vdots & \vdots & \ddots & \vdots \\ a_{n1} & a_{n2} & \cdots & a_{nn} \end{bmatrix},$ and vector

$\mathbf{R} = \begin{Bmatrix} r_1(t) \\ r_2(t) \\ \vdots \\ r_n(t) \end{Bmatrix}.$

The general solution of Eq. (7.41) consists of the homogeneous solution $\mathbf{x}_h$ and the *particular solution* $\mathbf{x}_p$.

We have learned two methods for finding particular solutions: the *method of undetermined coefficients* and the *method of variation of parameters*, as discussed in Section 2.4. However, for the sake of convenience, let's focus solely on the method of undetermined coefficients in this section.

The method of undetermined coefficients assumes that when the components of the vector **R** are constants or exponential, sine, or cosine functions, the components of the particular solution x_p have similar forms, as shown in Table 7.1.

Table 7.1 Method of undetermined coefficients

Vector R	Particular solution x_p
a (constant vector)	$\mathbf{x}_p = \mathbf{u}$ (constant vector)
$\mathbf{a}t + \mathbf{b}$	$\mathbf{x}_p = \mathbf{u}t + \mathbf{v}$
$\mathbf{a}\cos qt$	$\mathbf{x}_p = \mathbf{u}\cos qt + \mathbf{v}\sin qt$
$\mathbf{b}\sin qt$	
$\mathbf{a}e^{qt}$	$\mathbf{x}_p = \mathbf{u}e^{qt}$

(where **a**, **b**, **u**, and **v** are constant vectors, and q is a real constant)

Example 7.11

Solve the following system of ODEs.

$$\frac{dx_1}{dt} = -2x_1 + x_2 + 8e^t, \quad \frac{dx_2}{dt} = x_1 - 2x_2$$

Solution

Now we set $\mathbf{x} = \begin{Bmatrix} x_1 \\ x_2 \end{Bmatrix}$, then we obtain a system of first-order ODEs

$$\begin{Bmatrix} x_1' \\ x_2' \end{Bmatrix} = \begin{bmatrix} -2 & 1 \\ 1 & -2 \end{bmatrix}\begin{Bmatrix} x_1 \\ x_2 \end{Bmatrix} + \begin{Bmatrix} 8 \\ 0 \end{Bmatrix}e^t.$$

First, let's find the homogeneous solution.

Since $\mathbf{A} - \lambda\mathbf{I} = \begin{bmatrix} -2-\lambda & 1 \\ 1 & -2-\lambda \end{bmatrix}$ in the characteristic equation $\det(\mathbf{A} - \lambda\mathbf{I}) = 0$, we get

$$\lambda^2 + 4\lambda + 3 = 0,$$

or

$$\lambda_1 = -1, \quad \lambda_2 = -3.$$

i. For the first eigenvalue $\lambda_1 = -1$, we obtain

$$(A - \lambda_1 I)x = \begin{bmatrix} -2 - \lambda_1 & 1 \\ 1 & -2 - \lambda_1 \end{bmatrix} \begin{Bmatrix} x_1 \\ x_2 \end{Bmatrix} = \begin{bmatrix} -1 & 1 \\ 1 & -1 \end{bmatrix} \begin{Bmatrix} x_1 \\ x_2 \end{Bmatrix} = 0,$$

or

$$-x_1 + x_2 = 0.$$

To obtain a simple value, let's set $x_1 = 1$, then $x_2 = 1$. Therefore, its corresponding eigenvector becomes

$$v_1 = \begin{Bmatrix} 1 \\ 1 \end{Bmatrix}.$$

And the first basis solution is

$$x_1 = \begin{Bmatrix} 1 \\ 1 \end{Bmatrix} e^{-t}.$$

ii. For the second eigenvalue $\lambda_2 = -3$, we obtain

$$(A - \lambda_2 I)x = \begin{bmatrix} -2 - \lambda_2 & 1 \\ 1 & -2 - \lambda_2 \end{bmatrix} \begin{Bmatrix} x_1 \\ x_2 \end{Bmatrix} = \begin{bmatrix} 1 & 1 \\ 1 & 1 \end{bmatrix} \begin{Bmatrix} x_1 \\ x_2 \end{Bmatrix} = 0$$

or

$$x_1 + x_2 = 0.$$

To obtain a simple value, let's set $x_1 = 1$, then $x_2 = -1$. Therefore, its corresponding eigenvector becomes

$$v_2 = \begin{Bmatrix} 1 \\ -1 \end{Bmatrix}.$$

And the second basis solution is

$$x_2 = \begin{Bmatrix} 1 \\ -1 \end{Bmatrix} e^{-3t}.$$

Therefore, we obtain the homogeneous solution

$$x_h = c_1 \begin{Bmatrix} 1 \\ 1 \end{Bmatrix} e^{-t} + c_2 \begin{Bmatrix} 1 \\ -1 \end{Bmatrix} e^{-3t}.$$

Now, let's find the particular solution.

Since the vector $\mathbf{R}$ has the exponential function e^t, we can assume

$$\mathbf{x}_p = \begin{Bmatrix} a \\ b \end{Bmatrix} e^t.$$

Substituting this into the system equation $\mathbf{x}' = \mathbf{Ax} + \mathbf{R}$, we obtain

LHS: $\mathbf{x}_p' = \begin{Bmatrix} a \\ b \end{Bmatrix} e^t$

RHS: $\mathbf{Ax}_p + \mathbf{R} = \begin{bmatrix} -2 & 1 \\ 1 & -2 \end{bmatrix} \begin{Bmatrix} a \\ b \end{Bmatrix} e^t + \begin{Bmatrix} 8 \\ 0 \end{Bmatrix} e^t$

Comparing the coefficients, we get

$$a = -2a + b + 8, \quad b = a - 2b,$$

or

$$a = 3, \quad b = 1.$$

Therefore, we obtain the particular solution

$$\mathbf{x}_p = \begin{Bmatrix} 3 \\ 1 \end{Bmatrix} e^t.$$

Hence, the answer is

$$\mathbf{x} = \mathbf{x}_h + \mathbf{x}_p = c_1 \begin{Bmatrix} 1 \\ 1 \end{Bmatrix} e^{-t} + c_2 \begin{Bmatrix} 1 \\ -1 \end{Bmatrix} e^{-3t} + \begin{Bmatrix} 3 \\ 1 \end{Bmatrix} e^t.$$

Answer $x_1 = c_1 e^{-t} + c_2 e^{-3t} + 3e^t$, $x_2 = c_1 e^{-t} - c_2 e^{-3t} + e^t$

Check

If we substitute $x_1(t)$ and $x_2(t)$ into $\dfrac{dx_1}{dt} = -2x_1 + x_2 + 8e^t$, it becomes

LHS: $-c_1 e^{-t} - 3c_2 e^{-3t} + 3e^t$

RHS: $-2\left(c_1 e^{-t} + c_2 e^{-3t} + 3e^t\right) + \left(c_1 e^{-t} - c_2 e^{-3t} + e^t\right) + 8e^t$

$= -c_1 e^{-t} - 3c_2 e^{-3t} + 3e^t$

Therefore, LHS = RHS.

If we substitute $x_1(t)$ and $x_2(t)$ into $\dfrac{dx_2}{dt} = x_1 - 2x_2$, it becomes

LHS: $-c_1 e^{-t} + 3c_2 e^{-3t} + e^t$

RHS: $\left(c_1 e^{-t} + c_2 e^{-3t} + 3e^t\right) - 2\left(c_1 e^{-t} - c_2 e^{-3t} + e^t\right) = -c_1 e^{-t} + 3c_2 e^{-3t} + e^t$

Therefore, LHS = RHS.

Example 7.12

Solve a system of nonhomogeneous ODEs.

$$\frac{dx_1}{dt} = -x_1 + 2x_2 + 3e^t, \quad \frac{dx_2}{dt} = 3x_1 - 2x_2 - 2e^t$$

Solution

Now we set $\mathbf{x} = \begin{Bmatrix} x_1 \\ x_2 \end{Bmatrix}$, then we obtain a system of first-order ODEs

$$\begin{Bmatrix} x_1' \\ x_2' \end{Bmatrix} = \begin{bmatrix} -1 & 2 \\ 3 & -2 \end{bmatrix} \begin{Bmatrix} x_1 \\ x_2 \end{Bmatrix} + \begin{Bmatrix} 3 \\ -2 \end{Bmatrix} e^t.$$

First, let's find the homogeneous solution.

Since $\quad \mathbf{A} - \lambda\,\mathbf{I} = \begin{bmatrix} -1-\lambda & 2 \\ 3 & -2-\lambda \end{bmatrix} \quad$ in   the   characteristic   equation $\det(\mathbf{A} - \lambda\,\mathbf{I}) = 0$, we get

$$\lambda^2 + 3\lambda - 4 = 0,$$

or

$$\lambda_1 = 1, \quad \lambda_2 = -4.$$

i. From the first eigenvalue $\lambda_1 = 1$, we obtain

$$(\mathbf{A} - \lambda_1 \mathbf{I})\mathbf{x} = \begin{bmatrix} -1-\lambda_1 & 2 \\ 3 & -2-\lambda_1 \end{bmatrix} \begin{Bmatrix} x_1 \\ x_2 \end{Bmatrix} = \begin{bmatrix} -2 & 2 \\ 3 & -3 \end{bmatrix} \begin{Bmatrix} x_1 \\ x_2 \end{Bmatrix} = 0,$$

or

$$-x_1 + x_2 = 0.$$

To obtain a simple value, let's set $x_1 = 1$, then $x_2 = 1$.
Therefore, its corresponding eigenvector becomes

$$\mathbf{v}_1 = \begin{Bmatrix} 1 \\ 1 \end{Bmatrix}.$$

And the first basis solution is

$$\mathbf{x}_1 = \begin{Bmatrix} 1 \\ 1 \end{Bmatrix} e^t.$$

ii. From the second eigenvalue $\lambda_2 = -4$, we obtain

$$(\mathbf{A} - \lambda_2 \mathbf{I})\mathbf{x} = \begin{bmatrix} -1 - \lambda_2 & 2 \\ 3 & -2 - \lambda_2 \end{bmatrix} \begin{Bmatrix} x_1 \\ x_2 \end{Bmatrix} = \begin{bmatrix} 3 & 2 \\ 3 & 3 \end{bmatrix} \begin{Bmatrix} x_1 \\ x_2 \end{Bmatrix} = 0,$$

or

$$3x_1 + 2x_2 = 0.$$

To obtain a simple value, let's set $x_1 = 2$, then $x_2 = -3$.
Therefore, its corresponding eigenvector becomes

$$\mathbf{v}_2 = \begin{Bmatrix} 2 \\ -3 \end{Bmatrix}.$$

And the second basis solution is

$$\mathbf{x}_2 = \begin{Bmatrix} 2 \\ -3 \end{Bmatrix} e^{-4t}.$$

Therefore, we obtain the homogeneous solution

$$\mathbf{x}_h = c_1 \begin{Bmatrix} 1 \\ 1 \end{Bmatrix} e^t + c_2 \begin{Bmatrix} 2 \\ -3 \end{Bmatrix} e^{-4t}.$$

Now, let's find the particular solution.
Since the vector $\mathbf{R}$ contains the function e^t, it duplicates a term of the homogeneous solution.

Thus, we assume the particular solution to be $\mathbf{x}_p = \begin{Bmatrix} at+b \\ ct+d \end{Bmatrix} e^t$. Substituting it into $\mathbf{x}' = \mathbf{Ax} + \mathbf{R}$ gives,

$$\text{LHS: } \mathbf{x}_p' = \begin{Bmatrix} at+a+b \\ ct+c+d \end{Bmatrix} e^t$$

$$\text{RHS: } \mathbf{Ax}_p + \mathbf{R} = \begin{bmatrix} -1 & 2 \\ 3 & -2 \end{bmatrix} \begin{Bmatrix} at+b \\ ct+d \end{Bmatrix} e^t + \begin{Bmatrix} 3 \\ -2 \end{Bmatrix} e^t$$

Comparing the coefficients gives

$$a = -a + 2c,$$
$$c = 3a - 2c,$$
$$a + b = -b + 2d + 3,$$
$$c + d = 3b - 2d - 2,$$

or

$$a = 1, \quad b = 2, \quad c = 1, \quad d = 1.$$

Therefore, we obtain the particular solution

$$\mathbf{x}_p = \begin{Bmatrix} t+2 \\ t+1 \end{Bmatrix} e^t.$$

Hence, the answer is

$$\mathbf{x} = \mathbf{x}_h + \mathbf{x}_p = c_1 \begin{Bmatrix} 1 \\ 1 \end{Bmatrix} e^t + c_2 \begin{Bmatrix} 2 \\ -3 \end{Bmatrix} e^{-4t} + \begin{Bmatrix} t+2 \\ t+1 \end{Bmatrix} e^t.$$

Answer $x_1 = c_1 e^t + 2c_2 e^{-4t} + (t+2)e^t, \quad x_2 = c_1 e^t - 3c_2 e^{-4t} + (t+1)e^t$

Check

If we substitute $x_1 = c_1 e^t + 2c_2 e^{-4t} + (t+2)e^t$

and $x_2 = c_1 e^t - 3c_2 e^{-4t} + (t+1)e^t$ into $\dfrac{dx_1}{dt} = -x_1 + 2x_2 + 3e^t$,

it becomes

LHS: $c_1 e^t - 8c_2 e^{-3t} + (t+3)e^t$

RHS: $-(c_1 e^t + 2c_2 e^{-4t} + (t+2)e^t) + 2(c_1 e^t - 3c_2 e^{-4t} + (t+1)e^t) + 3e^t$

Therefore, LHS = RHS.

If we substitute $x_1 = c_1 e^t + 2c_2 e^{-4t} + (t+2)e^t$

and $x_2 = c_1 e^t - 3c_2 e^{-4t} + (t+1)e^t$ into $\dfrac{dx_2}{dt} = 3x_1 - 2x_2 - 2e^t$,

it becomes

LHS: $c_1 e^t + 12c_2 e^{-4t} + (t+2)e^t$

RHS: $3\left(c_1 e^t + 2c_2 e^{-4t} + (t+2)e^t\right) - 2\left(c_1 e^t - 3c_2 e^{-4t} + (t+1)e^t\right) - 2e^t$

Therefore, LHS = RHS.

Problem 7.3

Solve a system of ODEs. [1 ~ 6]

1. $\dfrac{dx_1}{dt} = x_2 - 2e^{2t}, \quad \dfrac{dx_2}{dt} = x_1 + e^{2t}$

2. $\dfrac{dx_1}{dt} = x_1 + x_2 - 5e^{-t}, \quad \dfrac{dx_2}{dt} = 3x_1 - x_2 - 6e^{-t}$

3. $\dfrac{dx_1}{dt} = 2x_1 - 4x_2, \quad \dfrac{dx_2}{dt} = x_1 - 3x_2 + t$

4. $\dfrac{dx_1}{dt} = 4x_1 + 3x_2 - t - 2, \quad \dfrac{dx_2}{dt} = -2x_1 - x_2 - t + 1$

5. $\dfrac{dx_1}{dt} = 2x_2 - 5\sin t, \quad \dfrac{dx_2}{dt} = 2x_1$

6. $\dfrac{dx_1}{dt} = x_1 - x_2 - 5\cos t, \quad \dfrac{dx_2}{dt} = 3x_1 - x_2 + 5\sin t$

7.4   Applications of a system of ODEs

The system of ODEs is applied in various fields such as solution mixing in liquids, electrical circuits, and so on.

7.4.1   Mixing in liquids

Example 7.13

Each of the tanks T_1 and T_2 contains initially 1000 l of water. 30 kg of fertilizer are dissolved in the tank T_1, whereas the water is pure in tank T_2, as shown in Figure 7.7.

Assuming the liquid circulates at the rate of 20 l/min, the quantity of fertilizer in T_1 and the quantity of fertilizer in T_2 vary with time t. How much time is needed for the quantity of fertilizer in T_1 to reach 25 kg?

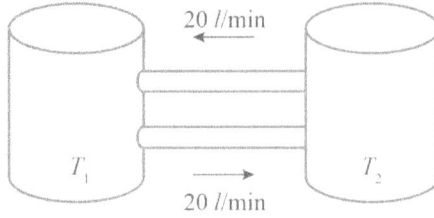

Figure 7.7 Mixing in liquids.

Solution

Initial conditions are $x_1(0) = 30$, $x_2(0) = 0$.
The quantity of fertilizer in each tank satisfies

$$x_1' = -\frac{20}{1000}x_1 + \frac{20}{1000}x_2,$$

$$x_2' = \frac{20}{1000}x_1 - \frac{20}{1000}x_2.$$

When we set $\mathbf{x} = \begin{Bmatrix} x_1 \\ x_2 \end{Bmatrix}$, we obtain

$$\begin{Bmatrix} x_1' \\ x_2' \end{Bmatrix} = \begin{bmatrix} -0.02 & 0.02 \\ 0.02 & -0.02 \end{bmatrix} \begin{Bmatrix} x_1 \\ x_2 \end{Bmatrix}.$$

Since $\mathbf{A} - \lambda\mathbf{I} = \begin{bmatrix} -0.02 - \lambda & 0.02 \\ 0.02 & -0.02 - \lambda \end{bmatrix}$, we get

$$\lambda^2 + 0.04\lambda = 0,$$

or

$$\lambda_1 = 0, \quad \lambda_2 = -0.04.$$

i. For the first eigenvalue $\lambda_1 = 0$, we obtain

$$(\mathbf{A} - \lambda_1\mathbf{I})\mathbf{x} = \begin{bmatrix} -0.02 - \lambda_1 & 0.02 \\ 0.02 & -0.02 - \lambda_1 \end{bmatrix} \begin{Bmatrix} x_1 \\ x_2 \end{Bmatrix} = \begin{bmatrix} -0.02 & 0.02 \\ 0.02 & -0.02 \end{bmatrix} \begin{Bmatrix} x_1 \\ x_2 \end{Bmatrix} = 0.$$

Then the corresponding eigenvector is

$$\mathbf{v}_1 = \begin{Bmatrix} 1 \\ 1 \end{Bmatrix}.$$

Therefore, the first basis solution is

$$\mathbf{x}_1 = \begin{Bmatrix} 1 \\ 1 \end{Bmatrix}.$$

ii. For the second eigenvalue $\lambda_2 = -0.04$, we obtain

$$(\mathbf{A} - \lambda_2 \mathbf{I})\mathbf{x} = \begin{bmatrix} -0.02 - \lambda_2 & 0.02 \\ 0.02 & -0.02 - \lambda_2 \end{bmatrix} \begin{Bmatrix} x_1 \\ x_2 \end{Bmatrix} = \begin{bmatrix} 0.02 & 0.02 \\ 0.02 & 0.02 \end{bmatrix} \begin{Bmatrix} x_1 \\ x_2 \end{Bmatrix} = 0.$$

Then the corresponding eigenvector is

$$\mathbf{v}_2 = \begin{Bmatrix} 1 \\ -1 \end{Bmatrix}.$$

Therefore, the second basis solution is

$$\mathbf{x}_2 = \begin{Bmatrix} 1 \\ -1 \end{Bmatrix} e^{-0.04t}.$$

Hence, the answer is

$$\mathbf{x} = \begin{Bmatrix} x_1 \\ x_2 \end{Bmatrix} = c_1 \begin{Bmatrix} 1 \\ 1 \end{Bmatrix} + c_2 \begin{Bmatrix} 1 \\ -1 \end{Bmatrix} e^{-0.04t},$$

or

$$x_1(t) = c_1 + c_2 e^{-0.04t},$$

$$x_2(t) = c_1 - c_2 e^{-0.04t}.$$

Coefficients c_1 and c_2 are obtained from the initial conditions $x_1(0) = 30$, $x_2(0) = 0$ as follows:

$$x_1(0) = 30 = c_1 + c_2,$$

$$x_2(0) = 0 = c_1 - c_2.$$

Then $c_1 = c_2 = 15$, and we get

$$x_1(t) = 15 + 15e^{-0.04t},$$

$$x_2(t) = 15 - 15e^{-0.04t}.$$

After time t is passes, the remaining quantity of fertilizer is 25 kg. Then

$$25 = 15 + 15e^{-0.04t}.$$

The answer is

$$t = \frac{\ln(3/2)}{0.04} = 10.14 \text{ min.}$$

Answer 10.14 min

Check

Substituting $x_1(t) = c_1 + c_2 e^{-0.04t}$ and $x_2(t) = c_1 - c_2 e^{-0.04t}$ into

$$\begin{Bmatrix} x_1' \\ x_2' \end{Bmatrix} = \begin{bmatrix} -0.02 & 0.02 \\ 0.02 & -0.02 \end{bmatrix} \begin{Bmatrix} x_1 \\ x_2 \end{Bmatrix} \text{ gives}$$

LHS: $\begin{Bmatrix} x_1' \\ x_2' \end{Bmatrix} = \begin{Bmatrix} -0.04 c_2 e^{-0.04t} \\ 0.04 c_2 e^{-0.04t} \end{Bmatrix}$

RHS: $\begin{bmatrix} -0.02 & 0.02 \\ 0.02 & -0.02 \end{bmatrix} \begin{Bmatrix} c_1 + c_2 e^{-0.04t} \\ c_1 - c_2 e^{-0.04t} \end{Bmatrix} = \begin{Bmatrix} -0.04 c_2 e^{-0.04t} \\ 0.04 c_2 e^{-0.04t} \end{Bmatrix}$

Therefore, LHS = RHS.

Figure 7.8 shows the solution in Example 7.13.

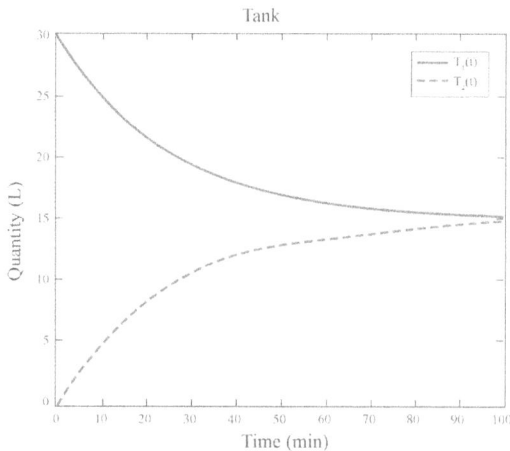

Figure 7.8 The solution in Example 7.13.

7.4.2 Electric circuit

Example 7.14

Find the currents $i_1(t)$ and $i_2(t)$ in the circuit, as shown in Figure 7.9. Assume all currents and charges to be zero at $t = 0$, the instant when the switch is closed.

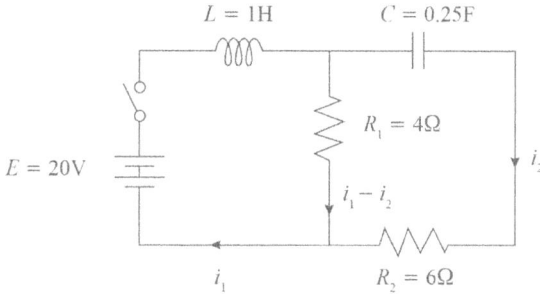

Figure 7.9 Electric circuit.

Solution

From *Kirchhoff's voltage law*, we obtain

$$E = Li_1' + R_1(i_1 - i_2),$$ ①

$$0 = \frac{1}{C}\int i_2\, dt + R_2 i_2 - R_1(i_1 - i_2).$$ ②

Substituting the initial condition into Eq. ① gives

$$20 = i_1' + 4(i_1 - i_2),$$

or

$$i_1' = -4i_1 + 4i_2 + 20.$$ ③

Differentiating Eq. ② gives

$$0 = \frac{1}{C}i_2 + R_2 i_2' - R_1(i_1' - i_2').$$ ④

Substituting the initial condition into Eq. ④ gives

$$0 = 4i_2 + 6i_2' - 4(i_1' - i_2').$$ ⑤

Then substituting Eq. ③ into Eq. ⑤ gives

$$i_2' = -1.6i_1 + 1.2i_2 + 8.$$ ⑥

When we set $\mathbf{x} = \left\{ \begin{array}{c} i_1 \\ i_2 \end{array} \right\}$ in Eq. ③ and Eq. ⑥, we obtain

$$\left\{ \begin{array}{c} i_1' \\ i_2' \end{array} \right\} = \left[\begin{array}{cc} -4 & 4 \\ -1.6 & 1.2 \end{array} \right] \left\{ \begin{array}{c} i_1 \\ i_2 \end{array} \right\} + \left\{ \begin{array}{c} 20 \\ 8 \end{array} \right\}.$$

Since $\mathbf{A} - \lambda \mathbf{I} = \left[\begin{array}{cc} -4 - \lambda & 4 \\ -1.6 & 1.2 - \lambda \end{array} \right]$, we obtain

$$\lambda^2 + 2.8\lambda + 1.6 = (\lambda + 0.8)(\lambda + 2) = 0,$$

or

$$\lambda_1 = -0.8, \quad \lambda_2 = -2.$$

i. For the first eigenvalue $\lambda_1 = -0.8$, we obtain

$$(\mathbf{A} - \lambda_1 \mathbf{I})\mathbf{x} = \left[\begin{array}{cc} -4 - \lambda_1 & 4 \\ -1.6 & 1.2 - \lambda_1 \end{array} \right] \left\{ \begin{array}{c} i_1 \\ i_2 \end{array} \right\} = \left[\begin{array}{cc} -3.2 & 4 \\ -1.6 & 2 \end{array} \right] \left\{ \begin{array}{c} i_1 \\ i_2 \end{array} \right\} = 0.$$

Then the corresponding eigenvector is

$$\mathbf{v}_1 = \left\{ \begin{array}{c} 1 \\ 0.8 \end{array} \right\}.$$

Therefore, we obtain the first basis solution

$$\mathbf{x}_1 = \left\{ \begin{array}{c} 1 \\ 0.8 \end{array} \right\} e^{-0.8t}.$$

ii. For the first eigenvalue $\lambda_2 = -2$, we obatain

$$(\mathbf{A} - \lambda_2 \mathbf{I})\mathbf{x} = \left[\begin{array}{cc} -4 - \lambda_2 & 4 \\ -1.6 & 1.2 - \lambda_2 \end{array} \right] \left\{ \begin{array}{c} i_1 \\ i_2 \end{array} \right\} = \left[\begin{array}{cc} -2 & 4 \\ -1.6 & 3.2 \end{array} \right] \left\{ \begin{array}{c} i_1 \\ i_2 \end{array} \right\} = 0.$$

Then the corresponding eigenvector is

$$\mathbf{v}_2 = \left\{ \begin{array}{c} 2 \\ 1 \end{array} \right\}.$$

Therefore, we obtain the second basis solution

$$\mathbf{x}_2 = \left\{ \begin{array}{c} 2 \\ 1 \end{array} \right\} e^{-2t}.$$

Hence, the homogeneous solution is

$$\mathbf{x}_h = c_1 \begin{Bmatrix} 1 \\ 0.8 \end{Bmatrix} e^{-0.8t} + c_2 \begin{Bmatrix} 2 \\ 1 \end{Bmatrix} e^{-2t}.$$

Since the vector $\mathbf{R}$ is constant, we assume the particular solution

$$\mathbf{x}_p = \begin{Bmatrix} a \\ b \end{Bmatrix}.$$

Substituting this into the system $\mathbf{x}' = \mathbf{Ax} + \mathbf{R}$, we obtain

$$\text{LHS: } \mathbf{x}_p' = \begin{Bmatrix} 0 \\ 0 \end{Bmatrix}$$

$$\text{RHS: } \mathbf{Ax}_p + \mathbf{R} = \begin{bmatrix} -4 & 4 \\ -1.6 & 1.2 \end{bmatrix} \begin{Bmatrix} a \\ b \end{Bmatrix} + \begin{Bmatrix} 20 \\ 8 \end{Bmatrix}$$

Then we get

$$0 = -4a + 4b + 20,$$
$$0 = -1.6a + 1.2b + 8,$$

or

$$a = 5, \quad b = 0.$$

Hence, the particular solution is

$$\mathbf{x}_p = \begin{Bmatrix} 5 \\ 0 \end{Bmatrix}.$$

Therefore, the general solution is

$$\mathbf{x} = \mathbf{x}_h + \mathbf{x}_p = c_1 \begin{Bmatrix} 1 \\ 0.8 \end{Bmatrix} e^{-0.8t} + c_2 \begin{Bmatrix} 2 \\ 1 \end{Bmatrix} e^{-2t} + \begin{Bmatrix} 5 \\ 0 \end{Bmatrix},$$

or

$$i_1(t) = c_1 e^{-0.8t} + 2c_2 e^{-2t} + 5,$$
$$i_2(t) = 0.8 c_1 e^{-0.8t} + c_2 e^{-2t}.$$

Coefficients c_1 and c_2 are obtained from the initial conditions $i_1(0) = 0$, $i_2(0) = 0$.

$$0 = c_1 + 2c_2 + 5,$$
$$0 = 0.8 c_1 + c_2.$$

Then we get

$$c_1 = \frac{25}{3}, \quad c_2 = -\frac{20}{3}.$$

Therefore, the answer is

$$i_1(t) = \frac{25}{3}e^{-0.8t} - \frac{40}{3}e^{-2t} + 5,$$

$$i_2(t) = \frac{20}{3}e^{-0.8t} - \frac{20}{3}e^{-2t}.$$

Answer $i_1(t) = \frac{25}{3}e^{-0.8t} - \frac{40}{3}e^{-2t} + 5, i_2(t) = \frac{20}{3}e^{-0.8t} - \frac{20}{3}e^{-2t}$

Figure 7.10 shows the solution in Example 7.14.

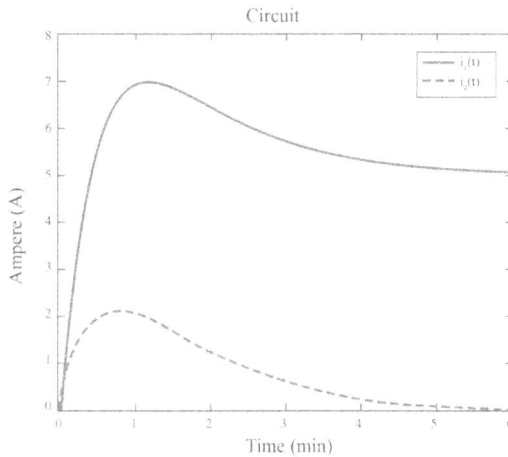

Figure 7.10 The solution in Example 7.14.

Problem 7.4

1. **Mixing in liquids**   Each of the tanks T_1 and T_2 contains initially 100 *l* of water. 2 kg of fertilizer are dissolved in the tank T_1, whereas the water is pure in tank T_2, as shown in Figure 7.11.

Assuming the liquid circulates at the rate of 5 *l*/min, find the quantity $x_1(t)$ of fertilizer in T_1 and the quantity $x_2(t)$ of fertilizer in T_2.

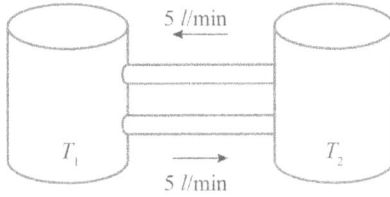

Figure 7.11 Mixing in liquids.

2. **Mixing in liquids**   Each of the tanks T_1 and T_2 contains initially 100 l of water. 1 kg of fertilizer are dissolved in the tank T_1, whereas 1 kg of fertilizer are dissolved in the tank T_2, as shown in Figure 7.12.

   Assuming the liquid circulates at the rate of 2 l/min, the tank T_1 receives pure water at a rate of 3 l/min, while a solution is discharged from the tank T_2 at a rate of 3 l/min.

   Find the quantity $x_1(t)$ of fertilizer in T_1 and the quantity $x_2(t)$ of fertilizer in T_2.

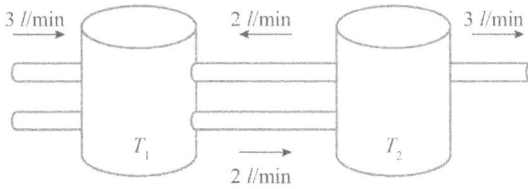

Figure 7.12 Mixing in liquids.

3. **Electric circuit**   Find the currents $i_1(t)$ and $i_2(t)$ in the circuit, as shown in Figure 7.13. Assume all currents and charges to be zero at $t = 0$, the instant when the switch is closed.

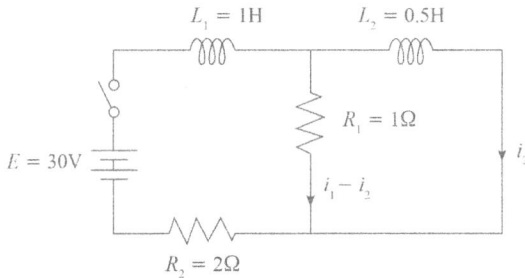

Figure 7.13 Electric circuit.

4. **Electric circuit** Find the currents $i_1(t)$ and $i_2(t)$ in the circuit, as shown in Figure 7.14. Assume all currents and charges to be zero at $t = 0$, the instant when the switch is closed.

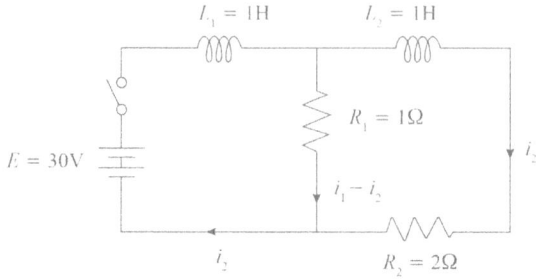

Figure 7.14 Electric circuit.

5. **Electric circuit** Find the currents $i_1(t)$ and $i_2(t)$ in the circuit, as shown in Figure 7.15. Assume all currents and charges to be zero at $t = 0$, the instant when the switch is closed.

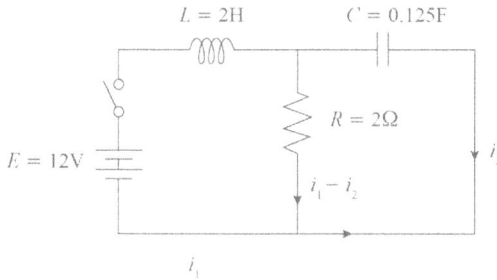

Figure 7.15 Electric circuit.

6. **Electric circuit** Find the currents $i_1(t)$ and $i_2(t)$ in the circuit, as shown in Figure 7.16. Initial charge is $Q(0) = 0$ C and initial current is $i(0) = \dfrac{dQ(0)}{dt} = 0$ A.

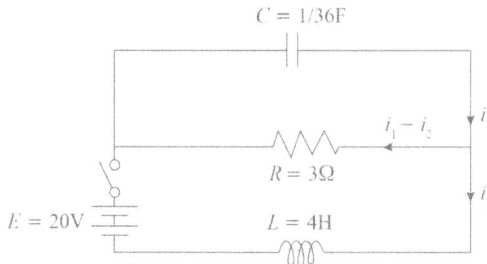

Figure 7.16 Electric circuit.

7.5 Utilizing MATLAB®

M-Example 7.1

Solve a system of ODEs using MATLAB®.

$$\frac{dx_1}{dt} = -x_1 + 2x_2$$

$$\frac{dx_2}{dt} = 3x_1 - 2x_2$$

Solution

>> dsolve('Dx1=-x1+2*x2', 'Dx2=3*x1-2*x2') % without initial conditions

---

$[x1, x2] =$
x1=C1*exp(t)+2*C2*exp(-4*t), x2=C1*exp(t)-3*C2*exp(-4*t)

Answer $x_1(t) = c_1 e^t + 2c_2 e^{-4t}$,
$x_2(t) = c_1 e^t - 3c_2 e^{-4t}$

M-Example 7.2

Solve a system of ODEs using MATLAB®.

$$\frac{dx}{dt} = -x + 2y, \quad x(0) = 4$$

$$\frac{dy}{dt} = 3x - 2y, \quad y(0) = -1$$

Solution

>> dsolve('Dx=-x+2*y', 'Dy=3*x-2*y', 'x(0)=4', 'y(0)=-1')

$[x1, x2] =$
x1=2*exp(t)+2*exp(-4*t), x2=2*exp(t)-3*exp(-4*t)

Answer $x_1(t) = 2e^t + 2e^{-4t}$,
$x_2(t) = 2e^t - 3e^{-4t}$

M-Example 7.3

Solve a system of ODEs using MATLAB®.

$$\frac{dx_1}{dt} = 2x_1 + 4x_2$$

$$\frac{dx_2}{dt} = -2x_1 - 2x_2$$

Solution

```
>> dsolve('Dx1=2*x1+4*x2', 'Dx2=-2*x1-2*x2')
```

$[x1, x2] =$
x1=(C1+C2)*cos(2*t)+(-C1+C2)*sin(2*t), x2= -C1*cos(2*t)-C2*sin(2*t)

Answer $x_1(t) = (c_1 + c_2)\cos 2t + (-c_1 + c_2)\sin 2t,$

$$x_2(t) = -c_1 \cos 2t - c_2 \sin 2t$$

M-Example 7.4

Solve a system of ODEs using MATLAB®.

$$\frac{dx}{dt} = -x + 2y + 3e^t, \quad x(0) = 1$$

$$\frac{dy}{dt} = 3x - 2y - 2e^t, \quad y(0) = 5$$

Solution

```
>> dsolve('Dx=-x+2*y+3*exp(t)', 'Dy=3*x-2*y-2*exp(t)', 'x(0)=1',
'y(0)=5')
```

$[x1, x2] =$
x1=exp(t)-2*exp(-4*t)+(t+2)*exp(t), x2=exp(t)+3*exp(-4*t)+(t+1)*exp(t)

Answer $x_1 = e^t - 2e^{-4t} + (t+2)e^t,$

$$x_2 = e^t + 3e^{-4t} + (t+1)e^t$$

Answer

Problem 7.1

1. $x' = Ax + R$, where $x = \left\{ \begin{array}{c} x \\ x' \end{array} \right\}$, $A = \left[\begin{array}{cc} 0 & 1 \\ 2 & 1 \end{array} \right]$, $R = \left\{ \begin{array}{c} 0 \\ e^{-2t} \end{array} \right\}$

2. $x' = Ax + R$, where $x = \left\{ \begin{array}{c} x \\ x' \end{array} \right\}$, $A = \left[\begin{array}{cc} 0 & 1 \\ -1 & 2 \end{array} \right]$, $R = \left\{ \begin{array}{c} 0 \\ \sin 2t \end{array} \right\}$

3. $x' = Ax + R$, where $x = \left\{ \begin{array}{c} x \\ x' \\ x'' \end{array} \right\}$, $A = \left[\begin{array}{ccc} 0 & 1 & 0 \\ 0 & 0 & 1 \\ 2 & -3 & 2 \end{array} \right]$, $R = \left\{ \begin{array}{c} 0 \\ 0 \\ e^{-t} \cos 2t \end{array} \right\}$

4. $x' = Ax + R$, where $x = \left\{ \begin{array}{c} x \\ x' \\ x'' \\ x''' \end{array} \right\}$, $A = \left[\begin{array}{cccc} 0 & 1 & 0 & 0 \\ 0 & 0 & 1 & 0 \\ 0 & 0 & 0 & 1 \\ 1 & -3 & 0 & -2 \end{array} \right]$, $R = \left\{ \begin{array}{c} 0 \\ 0 \\ 0 \\ \sin t \end{array} \right\}$

5. $x(t) = c_1 e^{-t} + c_2 e^{-2t}$

6. $x(t) = c_1 e^{t} + c_2 e^{3t}$

7. $x(t) = (c_1 + c_2 t) e^{t}$

8. $x(t) = (c_1 + c_2 t) e^{-2t}$

9. $x(t) = c_1 e^{t} + c_2 e^{-t} + c_3 e^{-2t}$

10. $x(t) = c_1 e^{2t} + c_2 e^{-t} + c_3 e^{-2t}$

11. $x(t) = c_1 e^{t} + c_2 e^{4t} - e^{2t}$

12. $x(t) = (c_1 + c_2 t) e^{t} - \sin t$

13. $x(t) = c_1 e^{t} + c_2 e^{2t} - t e^{t}$

14. $x(t) = (c_1 + c_2 t) e^{-t} + \dfrac{1}{2} t^2 e^{-t}$

Problem 7.2

1. $x_1(t) = c_1 e^{-t} + 2 c_2 e^{4t}$,
   $x_2(t) = -c_1 e^{-t} + 3 c_2 e^{4t}$

2. $x_1(t) = c_1 e^{-t} + 3 c_2 e^{6t}$,
   $x_2(t) = -c_1 e^{-t} + 4 c_2 e^{6t}$

3. $x_1(t) = c_1 e^{-t} + c_2 e^{-4t}$,
   $x_2(t) = 2 c_1 e^{-t} - c_2 e^{-4t}$

4. $x_1(t) = c_1 + 2 c_2 e^{-5t}$,
   $x_2(t) = 3 c_1 + c_2 e^{-5t}$

5. $x_1(t) = 3c_1e^{-4t} + c_2e^{4t}$,

   $x_2(t) = 2c_1e^{-4t} - 2c_2e^{4t}$

6. $x_1(t) = c_1e^t + c_2e^{3t}$,

   $x_2(t) = -c_1e^t + c_2e^{3t}$

7. $x_1(t) = c_1e^{-t} + 2c_2e^{-2t} + 2c_3e^{4t}$,

   $x_2(t) = -c_2e^{-2t} + 5c_3e^{4t}$,

   $x_3(t) = -c_1e^{-t} + 3c_2e^{-2t} + 3c_3e^{4t}$

8. $x_1(t) = c_1 + c_3e^{2t}$,

   $x_2(t) = c_2e^t$,

   $x_3(t) = -c_1 - 2c_2e^t + c_3e^{2t}$

9. $x_1(t) = (c_1 + c_2t)e^{-2t}$,

   $x_2(t) = (-c_1 + c_2 - c_2t)e^{-2t}$

10. $x_1(t) = (c_1 + c_2 + c_2t)e^{2t}$,

    $x_2(t) = (c_1 + c_2t)e^{2t}$

11. $x_1(t) = c_1 + c_2t$,

    $x_2(t) = 2c_1 - c_2 + 2c_2t$

12. $x_1(t) = (c_1 - 0.2c_2 + c_2t)e^{-t}$,

    $x_2(t) = -(c_1 + c_2t)e^{-t}$

13. $x_1(t) = e^{4t}(c_1\cos t + c_2\sin t)$,

    $x_2(t) = e^{4t}\{(-2c_1 + c_2)\cos t - (c_1 + 2c_2)\sin t\}$

14. $x_1(t) = (c_1 + 2c_2)\cos 2t + (-2c_1 + c_2)\sin 2t$,

    $x_2(t) = -c_1\cos 2t - c_2\sin 2t$

15. $x_1(t) = e^{2t}\{(c_1 + c_2)\cos t + (-c_1 + c_2)\sin t\}$,

    $x_2(t) = e^{2t}\{-c_1\cos t - c_2\sin t\}$

16. $x_1(t) = e^{5t}\{(c_1 + 3c_2)\cos 3t + (-3c_1 + c_2)\sin 3t\}$,

    $x_2(t) = e^{5t}\{-5c_1\cos 3t - 5c_2\sin 3t\}$

Problem 7.3

1. $x_1 = c_1e^{-t} + c_2e^t - e^{2t}$,

   $x_2 = -c_1e^{-t} + c_2e^t$

2. $x_1 = c_1 e^{-2t} + c_2 e^{2t} + 2e^{-t}$,

$x_2 = -3c_1 e^{-2t} + c_2 e^{2t} + e^{-t}$

3. $x_1 = c_1 e^{-2t} + 4c_2 e^t + 2t + 1$,

$x_2 = c_1 e^{-2t} + c_2 e^t + t$

4. $x_1 = c_1 e^t + 3c_2 e^{2t} - 2t - 3$,

$x_2 = -c_1 e^t - 2c_2 e^{2t} + 3t + 4$

5. $x_1 = c_1 e^{2t} + c_2 e^{-2t} + \cos t$,

$x_2 = c_1 e^{2t} - c_2 e^{-2t} + 2\sin t$

6. $x_1 = c_1 e^{2t} + c_2 e^{-2t} + \cos t - 2\sin t$,

$x_2 = c_1 e^{2t} - 3c_2 e^{-2t} + 2\cos t + \sin t$

Problem 7.4

1. $x_1(t) = 1 + e^{-0.1t}$,

$x_2(t) = 1 - e^{-0.1t}$

2. $x_1(t) = 1.6e^{-0.01t} - 0.6e^{-0.06t}$,

$x_2(t) = 0.8e^{-0.01t} + 1.2e^{-0.06t}$

3. $i_1(t) = -10e^{-t} - 5e^{-4t} + 15$,

$i_2(t) = -20e^{-t} + 5e^{-4t} + 15$

4. $i_1(t) = -24e^{-t} - e^{-6t} + 25$,

$i_2(t) = -12e^{-t} + 2e^{-6t} + 10$

5. $i_1(t) = -6(t+1)e^{-2t} + 6$,

$i_2(t) = 6te^{-2t}$

6. $i_1(t) = 6te^{-6t}$,

$i_2(t) = (-2 - 6t)e^{-6t} + 2$

8 Vector differential calculus
Grad, Div, and Curl

Vector differential calculus deals with vector functions and their rates of change. This field finds extensive application in physics and engineering, including dynamics, fluid mechanics, robotics, electromagnetism, quantum physics, and so on. The primary concepts of vector differential calculus involve finding derivatives of vector functions, such as displacement, velocity, acceleration, angular displacement, angular velocity, angular acceleration, force, and torque. Additionally, divergence and curl are important concepts that characterize the properties of vector functions. Divergence indicates how much a vector field is spreading out, while curl indicates the rotational behavior of the vector field.

8.1 Vectors operations

A *scalar* represents magnitude only and does not indicate direction. It is primarily used to represent physical quantities that are concerned only with magnitude, such as time, mass, length, density, frequency, resistance, voltage, temperature, etc. The magnitude of a scalar, whether it is large or small, is referred to as scalar quantity.

On the other hand, a *vector* represents both magnitude and direction. It is used to represent quantities such as force, displacement, velocity, acceleration, moment of force, angular displacement, angular velocity, angular acceleration, etc. A vector quantity considers both magnitude and direction when representing it.

In vectors, if two vectors have the same direction and magnitude, they are considered identical. Similarly, if vectors have the same direction or undergo parallel displacement, they remain the same. Additionally, if the sign of a vector is reversed, it becomes a vector of equal magnitude but in the opposite direction.

Vectors are denoted in English either by boldface letters or by adding arrows above (or below) the letters. The magnitude (or length) of a vector can be indicated by enclosing the vector notation with absolute value symbols or by using the same letter in italic font. Another name for length is norm.

Boldface notation: $\mathbf{A}, \mathbf{a}, \mathbf{v}$
Arrow notation: $\vec{A}, \vec{a}, \overrightarrow{OA}, \overrightarrow{AB}$
Magnitude notation: $|\mathbf{A}|, |\mathbf{a}|, |\mathbf{v}|, |\vec{A}|, |\vec{a}|, |\overrightarrow{OA}|, |\overrightarrow{AB}|$
Italic notation: a, v, l

As seen in Figure 8.1, the vector $\overrightarrow{OA}$ represents the vector from the origin (reference point) O to point $\mathbf{A}$. Additionally, the vector $\overrightarrow{AB}$ represents the

DOI: 10.1201/9781003608912-8

vector from point A to point B, and when expressed with respect to the origin O, it is as follows:

$$\overrightarrow{OA} + \overrightarrow{AB} = \overrightarrow{OB} \tag{8.1a}$$

$$\overrightarrow{AB} = \overrightarrow{OB} - \overrightarrow{OA}. \tag{8.1b}$$

In this chapter, vectors are represented in boldface, and scalars are represented in italic font. For example, **A** denotes a vector, and A denotes a scalar.

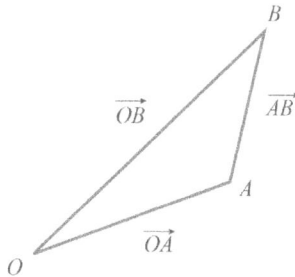

Figure 8.1 Vector.

8.1.1 *Unit vector*

A unit vector is a vector with a magnitude of 1, representing a specific direction. There are unit vectors indicating motion direction and rotation direction.

1. Unit vectors indicating motion direction
   In the Cartesian coordinate system shown in Figure 8.2, the unit vector indicating motion direction is defined as follows:

   i: the unit vector in the x-direction
   j: the unit vector in the y-direction
   k: the unit vector in the z-direction

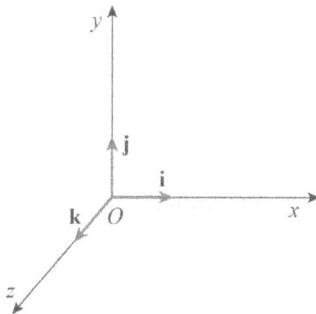

Figure 8.2 Unit vectors indicating motion direction.

2. Unit vectors indicating rotation direction

In the Cartesian coordinate system shown in Figure 8.3, the unit vector indicating rotation direction is defined according to the right-hand rule as follows:

    i: the unit vector for rotation about the x-axis
    j: the unit vector for rotation about the y-axis
    k: the unit vector for rotation about the z-axis

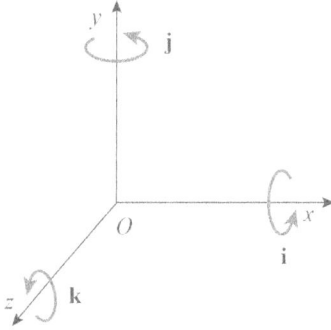

Figure 8.3 Unit vectors indicating rotation direction.

Remark   Unit vectors

In the Cartesian coordinate system, the unit vectors **i**, **j**, and **k** remain constant in magnitude and direction over time. They are always consistent and do not change unless the coordinate system itself changes. These vectors can represent both translation and rotation.

8.1.2 *Vector representation using unit vectors*

Any arbitrary vector in space can be expressed as a linear combination of a scalar representing its magnitude and unit vectors representing each direction.

$$\mathbf{r} = x\mathbf{i} + y\mathbf{j} + z\mathbf{k},$$

where scalars x, y, and z are directional lengths of the vector **r**, respectively.

8.1.3 *Magnitude of vector*

The *magnitude* (length or *norm*) *of vector* **A** is denoted as $|\mathbf{A}|$ or $|\overline{OA}|$, and is a scalar value.

When the coordinate components of a vector are given by $\mathbf{A} = (A_x, A_y, A_z)$, the *length of vector* $\mathbf{A}$ can be expressed as follows, utilizing the Pythagorean theorem due to the perpendicularity of the coordinate axes, as shown as Figure 8.4:

$$|\mathbf{A}| = A = \sqrt{A_x^2 + A_y^2 + A_z^2}. \tag{8.2}$$

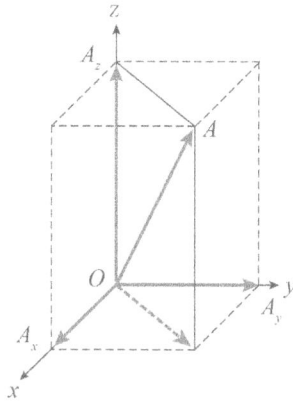

Figure 8.4 Magnitude of vector $\mathbf{A}$.

8.1.4 *Vectors operation*

When adding or subtracting vectors, both algebraic and graphical methods can be used. Algebraically, the method involves dividing each vector into components along mutually perpendicular directions, computing the components in the same direction separately, and then combining them again.

As seen in Figure 8.5, when the directions of vectors are different, graphical methods such as the triangle method or the parallelogram method can be used for vector addition.

The *triangle method* involves representing vectors graphically, connecting the endpoint of one vector to the starting point of the second vector to create a third vector. Then, considering this as one side of a triangle, the magnitude and direction of the third vector can be determined.

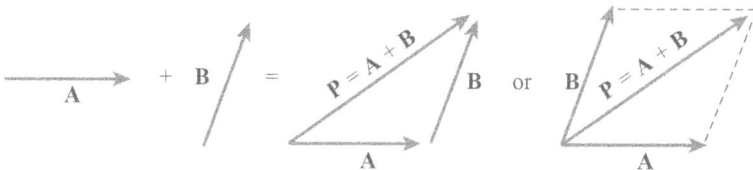

Figure 8.5 Adding vectors (the triangle method or the parallelogram method).

The *parallelogram method* considers two vectors as the adjacent sides of a parallelogram. Drawing the diagonal of the parallelogram allows the determination of the magnitude and direction of the resultant vector.

$$\mathbf{A} + \mathbf{B} = \mathbf{P} \quad (\text{Adding}) \tag{8.3}$$

When performing subtraction, as shown in Figure 8.6, the vector to be subtracted is added with a vector of equal magnitude but in the opposite direction. The negative of a vector $\mathbf{B}$, denoted as $-\mathbf{B}$, represents a vector with the same magnitude but in the opposite direction. Thus, vector subtraction can be computed by adding the negative of the vector being subtracted.

$$\mathbf{A} - \mathbf{B} = \mathbf{A} + (-\mathbf{B}) = \mathbf{Q} \quad (\text{Subtracting}) \tag{8.4}$$

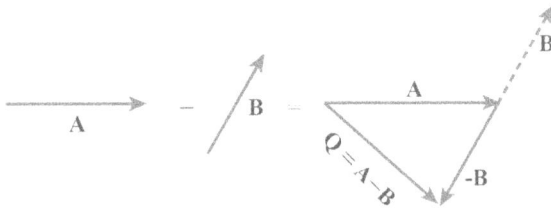

Figure 8.6 Subtracting vectors.

Algebraically, when adding (or subtracting) two vectors, only the components in the same direction are computed using scalar operations.

For example, when the components of two vectors are denoted as $\mathbf{A} = A_x\mathbf{i} + A_y\mathbf{j} + A_z\mathbf{k}$ and $\mathbf{B} = B_x\mathbf{i} + B_y\mathbf{j} + B_z\mathbf{k}$, vector addition or subtraction operations involve computing only the components in the same direction using scalar operations, as follows:

$$\mathbf{A} + \mathbf{B} = (A_x + B_x)\mathbf{i} + (A_y + B_y)\mathbf{j} + (A_z + B_z)\mathbf{k}$$
$$\mathbf{A} - \mathbf{B} = (A_x - B_x)\mathbf{i} + (A_y - B_y)\mathbf{j} + (A_z - B_z)\mathbf{k}.$$

The laws of vector operations are as follows:

$$\mathbf{A} + \mathbf{B} = \mathbf{B} + \mathbf{A} \quad (\text{commutative law}) \tag{8.5a}$$

$$\mathbf{A} + (\mathbf{B} + \mathbf{C}) = (\mathbf{A} + \mathbf{B}) + \mathbf{C} \quad (\text{associative law}) \tag{8.5b}$$

$$\mathbf{A} + 0 = 0 + \mathbf{A} = \mathbf{A} \tag{8.5c}$$

$$\mathbf{A} + (-\mathbf{A}) = (-\mathbf{A}) + \mathbf{A} = 0 \tag{8.5d}$$

where the vector 0 is referred to as the zero vector or null vector.

The product of a vector **A** and a scalar c is denoted as c**A**, where the direction is the same as that of vector **A**, and the length is $c|$**A**$|$. The result of the multiplication of a scalar c and a vector **A** is as follows, where the scalar is multiplied by each directional component of the vector.

$$c\mathbf{A} = cA_x\mathbf{i} + cA_y\mathbf{j} + cA_z\mathbf{k}$$

And in the multiplication of a vector and a scalar, the following operational rules apply.

$$c(\mathbf{A} + \mathbf{B}) = c\mathbf{A} + c\mathbf{B} \qquad (\text{distributive law}) \qquad (8.6)$$

8.1.5 Inner product (dot product)

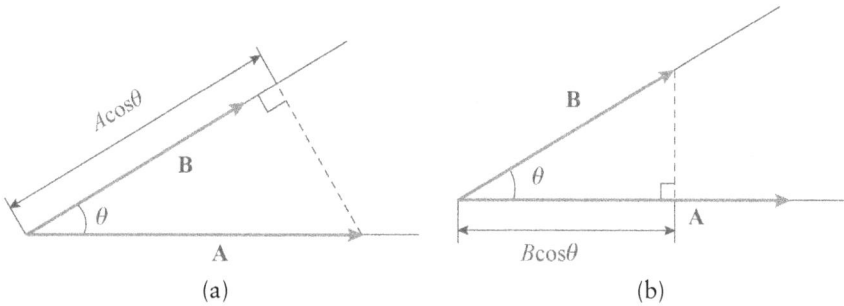

Figure 8.7 Inner product: (a) Project vector **A** onto vector **B** and (b) Project vector **B** onto vector **A**.

The inner product (or dot product) of vectors **A** and **B** is denoted by **A** · **B**, and the result of this operation is a scalar value. The inner product is calculated by projecting one vector onto the direction of the other vector and multiplying the magnitude of the projected vector by the magnitude of the other vector, as shown in Figure 8.7. In other words, when calculating the inner product of two vectors, only the components of the two vectors that are in the same direction are considered.

If the angle between the two vectors is θ, the inner product can be expressed as:

$$\mathbf{A} \cdot \mathbf{B} = AB\cos\theta \qquad (8.7a)$$

$$\mathbf{A} \cdot \mathbf{B} = (A\cos\theta)B \qquad (8.7b)$$

$$\mathbf{A} \cdot \mathbf{B} = A(B\cos\theta) \qquad (8.7c)$$

The order of multiplication can be changed without affecting the result (commutative property). The inner product varies based on the angle between the vectors: it becomes larger as the angle decreases (approaching parallel), and it is zero when the vectors are perpendicular to each other.

To examine the result of the inner product between unit vectors, let's substitute Eq. (8.7) for the inner product operation. The multiplication of unit vectors in the same direction is computed as follows:

$$\mathbf{i} \cdot \mathbf{i} = 1 \cdot 1 \cos 0° = 1.$$

Since unit vectors in different directions are orthogonal, the multiplication of unit vectors in different directions is computed as follows:

$$\mathbf{i} \cdot \mathbf{j} = 1 \cdot 1 \cos 90° = 0.$$

As seen above, the result of the inner product between unit vectors is scalar. The inner product of two unit vectors in perpendicular directions becomes 0, while the inner product of two identical unit vectors becomes 1. Therefore, the inner product results of unit vectors $\mathbf{i}$, $\mathbf{j}$, and $\mathbf{k}$ that are orthogonal to each other are as follows:

$$\mathbf{i} \cdot \mathbf{i} = 1, \qquad \mathbf{j} \cdot \mathbf{j} = 1, \qquad \mathbf{k} \cdot \mathbf{k} = 1 \qquad (8.8a)$$

$$\mathbf{i} \cdot \mathbf{j} = \mathbf{j} \cdot \mathbf{i} = 0, \quad \mathbf{j} \cdot \mathbf{k} = \mathbf{k} \cdot \mathbf{j} = 0, \quad \mathbf{i} \cdot \mathbf{k} = \mathbf{k} \cdot \mathbf{i} = 0. \qquad (8.8b)$$

When computing the inner product of two vectors, the expressions representing the directional components of the two vectors are expanded, and scalar components are operated upon scalar components, while vector components are operated upon vector components. When operating on unit vector components, the rules of the inner product for unit vectors can be applied.

For example, when performing the inner product of two vectors, the inner product can be expressed as follows:

$$\mathbf{A} \cdot \mathbf{B} = \left(A_x \mathbf{i} + A_y \mathbf{j} + A_z \mathbf{k} \right) \cdot \left(B_x \mathbf{i} + B_y \mathbf{j} + B_z \mathbf{k} \right).$$

Expanding the elements of the two vectors yields

$$\begin{aligned} \mathbf{A} \cdot \mathbf{B} = {}& A_x B_x \, \mathbf{i} \cdot \mathbf{i} + A_x B_y \, \mathbf{i} \cdot \mathbf{j} + A_x B_z \, \mathbf{i} \cdot \mathbf{k} \\ & + A_y B_x \, \mathbf{j} \cdot \mathbf{i} + A_y B_y \, \mathbf{j} \cdot \mathbf{j} + A_y B_z \, \mathbf{j} \cdot \mathbf{k} \\ & + A_z B_x \, \mathbf{k} \cdot \mathbf{i} + A_z B_y \, \mathbf{k} \cdot \mathbf{j} + A_z B_z \, \mathbf{k} \cdot \mathbf{k}. \end{aligned}$$

Applying the inner product operation for unit vectors yields

$$\begin{aligned} \mathbf{A} \cdot \mathbf{B} = {}& A_x B_x \cdot 1 + A_x B_y \cdot 0 + A_x B_z \cdot 0 \\ & + A_y B_x \cdot 0 + A_y B_y \cdot 1 + A_y B_z \cdot 0 \\ & + A_z B_x \cdot 0 + A_z B_y \cdot 0 + A_z B_z \cdot 1. \end{aligned}$$

Therefore, we obtain

$$\mathbf{A} \cdot \mathbf{B} = A_x B_x + A_y B_y + A_z B_z. \tag{8.9}$$

As can be inferred from the above results, the components perpendicular to each other are zero, so only the inner product components of vectors in the same direction are valid.

Remark   Inner product

When $\mathbf{A} = A_x\mathbf{i} + A_y\mathbf{j} + A_z\mathbf{k}$ and $\mathbf{B} = B_x\mathbf{i} + B_y\mathbf{j} + B_z\mathbf{k}$, the inner product of them yields

$$\mathbf{A} \cdot \mathbf{B} = \left(A_x, \ A_y, \ A_z\right) \cdot \left(B_x, \ B_y, \ B_z\right)$$
$$= A_x B_x + A_y B_y + A_z B_z. \tag{8.9}$$

If the inner product is performed between identical vectors, only the components in the same direction contribute, so the result is as follows:

$$\mathbf{A} \cdot \mathbf{A} = A_x^2 + A_y^2 + A_z^2. \tag{8.10}$$

The operational rules that apply when performing the inner product on vectors are as follows:

$$\mathbf{A} \cdot \mathbf{B} = \mathbf{B} \cdot \mathbf{A} \qquad \left(\text{commutative law}\right) \tag{8.11a}$$

$$\mathbf{A} \cdot \left(\mathbf{B} + \mathbf{C}\right) = \mathbf{A} \cdot \mathbf{B} + \mathbf{A} \cdot \mathbf{C} \qquad \left(\text{distributive law}\right) \tag{8.11b}$$

Example 8.1

Solve the problem.

a. When $\mathbf{A} = 2\mathbf{i} - 3\mathbf{j}$ and $\mathbf{B} = \mathbf{i} + 2\mathbf{j}$, find the inner product of them.
b. Find the angle between the two vectors, $\mathbf{A} = 3\mathbf{i} - 4\mathbf{j}$ and $\mathbf{B} = \mathbf{i} - 2\mathbf{j}$.

Solution

a. $\mathbf{A} \cdot \mathbf{B} = \left(2\mathbf{i} - 3\mathbf{j}\right) \cdot \left(\mathbf{i} + 2\mathbf{j}\right) = 2 \cdot 1 + (-3) \cdot 2 = -4$

Answer –4

b. From $\mathbf{A} \cdot \mathbf{B} = |\mathbf{A}||\mathbf{B}| \cos\theta$, we obtain

$$(3\mathbf{i} - 4\mathbf{j}) \cdot (\mathbf{i} - 2\mathbf{j}) = \sqrt{3^2 + (-4)^2} \cdot \sqrt{1^2 + (-2)^2} \cos\theta$$
$$3 \cdot 1 + (-4) \cdot (-2) = 5 \cdot \sqrt{5} \cos\theta$$

Answer $\theta = 10.3°$

8.1.6 Application of the inner product

The work done by a constant force is defined as the inner product of force [N] and displacement [m]. Since work is the result of an inner product, it is a scalar quantity.

$$W = \mathbf{F} \cdot \mathbf{d} = |\mathbf{F}||\mathbf{d}| \cos\theta \qquad (8.12)$$

Thus, the magnitude of the force $|\mathbf{F}|$, the magnitude of the displacement $|\mathbf{d}|$, and the cosine of the angle between the two vectors $\mathbf{F}$ and $\mathbf{d}$ are multiplied to calculate it. If the angle between the vectors is less than 90 degrees ($\theta < 90°$), positive work ($W > 0$) is done, while if the angle is 90 degrees (i.e., $\theta = 90°$, the vectors are orthogonal), the work is 0. If the angle is greater than 90 degrees ($\theta > 90°$), negative work ($W < 0$) is done.

Example 8.2

If an object placed on a frictionless surface is pulled by a force $\mathbf{F}$ at an angle θ with the horizontal plane and is moved along the ground by a certain distances as shown in Figure 8.8, find the work done by the force.

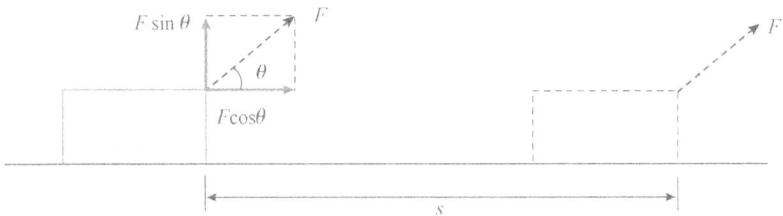

Figure 8.8 Work.

Solution

$$\mathbf{F} \cdot \mathbf{s} = F \cos\theta \cdot s = Fs \cos\theta$$

Answer $Fs \cos\theta$ [J]

8.1.7   *Outer product (cross product)*

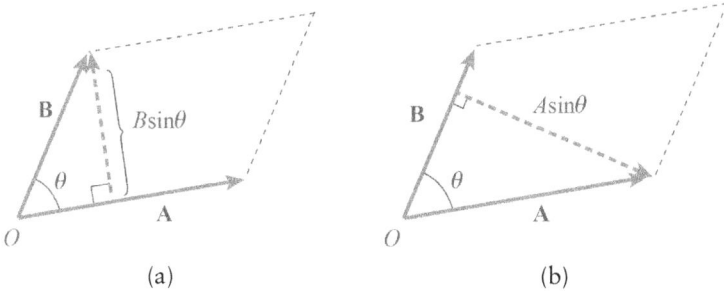

Figure 8.9 Magnitude of the outer product: (a) the component of a vector **B** perpendicular to vector **A** and (b) the component of a vector **A** perpendicular to vector **B**.

The outer product (or cross product) of vectors **A** and **B** is denoted by **A** × **B**, and the result of this operation is a vector. If the angle between the two vectors is θ, the magnitude of the outer product of the two vectors is given by:

$$|A \times B| = AB \sin \theta \qquad (8.13)$$

where θ represents the smaller of the angles formed by the two vectors, ranging between 0 and 180 degrees.

In Figure 8.9(a), $B\sin\theta$ represents the component of vector **B** perpendicular to vector **A**. Moreover, the magnitude of **A** × **B** is the product of the magnitude of vector **A** and this perpendicular component. Additionally, in Figure 8.9(b), the magnitude of **A** × **B** is the product of the magnitude of vector **A**'s perpendicular component, $A\sin\theta$, and the magnitude of vector **B**.

Therefore, the magnitude of **A** × **B** is equal to the area of the parallelogram formed by vectors **A** and **B**.

The direction of the outer product **C** = **A** × **B** of two vectors **A** and **B** is determined by the right-hand rule. Specifically, in the plane formed by vectors **A** and **B**, rotate vector **A** to align with vector **B**, and then curl the fingers of your right hand in the direction of this rotation. Next, with your thumb pointing along the axis perpendicular to this plane, the direction in which your thumb points corresponds to the direction of the cross product **C** of the two vectors.

In other words, the direction of the resulting vector from the outer product of two vectors is perpendicular to the plane formed by the two vectors, as shown in Figure 8.10.

Two unit vectors in different directions form a right angle, so the magnitude of the outer product of two unit vectors in different directions is calculated as follows:

$$|i \times j| = 1 \cdot 1 \sin 90° = 1.$$

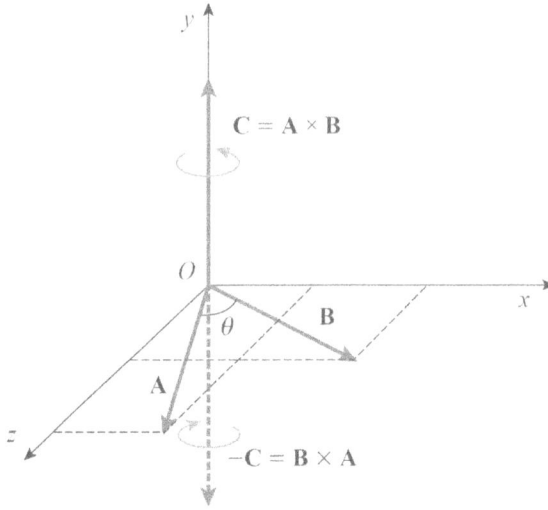

Figure 8.10 Direction of the outer product.

The magnitude of the outer product of unit vectors in the same direction is calculated as follows:

$$|i \times i| = 1 \cdot 1 \sin 0° = 0.$$

As observed earlier, when taking the outer product of unit vectors, the outer product of two unit vectors in the same direction results in 0, while the outer product of two unit vectors in different directions has a magnitude of 1, and the direction is determined by the right-hand rule. The operation of unit vectors that are orthogonal, i, j, and k, yields the following results:

$$i \times j = k, \quad j \times k = i, \quad k \times i = j,$$
$$j \times i = -k, \quad k \times j = -i, \quad i \times k = -j, \quad\quad (8.14)$$
$$i \times i = 0, \quad j \times j = 0, \quad k \times k = 0.$$

In the above equations, the two vectors undergoing the outer product operation represent unit vectors indicating the motion direction, and the resulting vector represents a unit vector indicating the rotation direction.

Figure 8.11 serves as a reference for determining the sign of the outer product. When two vectors are crossed in the direction indicated in the figure, the result is positive; when crossed in the opposite direction, the result is negative.

This principle is applicable not only in determining the moment of force through the outer product of distance and force vectors but also in calculating velocity through the outer product of angular velocity and distance vectors.

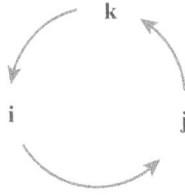

Figure 8.11 Direction of the outer product (counterclockwise).

The cross product of two vectors $\mathbf{A} = A_x\mathbf{i} + A_y\mathbf{j} + A_z\mathbf{k}$ and $\mathbf{B} = B_x\mathbf{i} + B_y\mathbf{j} + B_z\mathbf{k}$ can be expressed as follows:

$$\mathbf{A} \times \mathbf{B} = \left(A_x\mathbf{i} + A_y\mathbf{j} + A_z\mathbf{k}\right) \times \left(B_x\mathbf{i} + B_y\mathbf{j} + B_z\mathbf{k}\right).$$

Expanding the elements of the two vectors, we have:

$$\begin{aligned} \mathbf{A} \times \mathbf{B} = {} & A_xB_x\,\mathbf{i} \times \mathbf{i} + A_xB_y\,\mathbf{i} \times \mathbf{j} + A_xB_z\,\mathbf{i} \times \mathbf{k} \\ & + A_yB_x\,\mathbf{j} \times \mathbf{i} + A_yB_y\,\mathbf{j} \times \mathbf{j} + A_yB_z\,\mathbf{j} \times \mathbf{k} \\ & + A_zB_x\,\mathbf{k} \times \mathbf{i} + A_zB_y\,\mathbf{k} \times \mathbf{j} + A_zB_z\,\mathbf{k} \times \mathbf{k}. \end{aligned}$$

Applying the outer product operation law for unit vectors, Eq. (8.14), we get:

$$\begin{aligned} \mathbf{A} \times \mathbf{B} = {} & A_xB_x \cdot 0 + A_xB_y\,\mathbf{k} + A_xB_z\left(-\mathbf{j}\right) \\ & + A_yB_x\left(-\mathbf{k}\right) + A_yB_y \cdot 0 + A_yB_z\,\mathbf{i} \\ & + A_zB_x\,\mathbf{j} + A_zB_y\left(-\mathbf{i}\right) + A_zB_z \cdot 0. \end{aligned}$$

Therefore, the outer product of two vectors is

$$\mathbf{A} \times \mathbf{B} = \left(A_yB_z - A_zB_y\right)\mathbf{i} + \left(A_zB_x - A_xB_z\right)\mathbf{j} + \left(A_xB_y - A_yB_x\right)\mathbf{k}. \qquad (8.15)$$

The above result can also be represented by the following matrix expression:

Remark   Outer product

The outer product of two vectors $\mathbf{A}$ and $\mathbf{B}$ is calculated as follows:

$$\mathbf{A} \times \mathbf{B} = \begin{vmatrix} \mathbf{i} & \mathbf{j} & \mathbf{k} \\ A_x & A_y & A_z \\ B_x & B_y & B_z \end{vmatrix} \qquad (8.16)$$

$$= \mathbf{i} \begin{vmatrix} A_y & A_z \\ B_y & B_z \end{vmatrix} - \mathbf{j} \begin{vmatrix} A_x & A_z \\ B_x & B_z \end{vmatrix} + \mathbf{k} \begin{vmatrix} A_x & A_y \\ B_x & B_y \end{vmatrix}.$$

The same principle applies not only to the outer product of unit vectors but also to the product of vectors.

Multiplying identical vectors results in the multiplication of vectors in the same direction, yielding 0. Changing the order of operations yields different results (the commutative property does not hold).

In other words, reversing the order of multiplication results in vectors of equal magnitude but in the opposite direction. Additionally, the distributive property holds, and scalars are independent of the order of multiplication.

$$\mathbf{A} \times \mathbf{B} = -\mathbf{B} \times \mathbf{A} \tag{8.17a}$$

$$\mathbf{A} \times (\mathbf{B} + \mathbf{C}) = \mathbf{A} \times \mathbf{B} + \mathbf{A} \times \mathbf{C} \qquad \text{(distributive law)} \tag{8.17b}$$

$$m(\mathbf{A} \times \mathbf{B}) = (m\mathbf{A}) \times \mathbf{B} = \mathbf{A} \times (m\mathbf{B}) = (\mathbf{A} \times \mathbf{B})m \qquad (m \text{ arbitrary}) \tag{8.17c}$$

Example 8.3

Given two vectors $\mathbf{A} = 3\mathbf{i} - 4\mathbf{j}$, $\mathbf{B} = \mathbf{i} + 2\mathbf{j}$, find their outer product, and then calculate the area of the parallelogram formed by these vectors.

Solution

$$\mathbf{A} \times \mathbf{B} = (3\mathbf{i} - 4\mathbf{j}) \times (\mathbf{i} + 2\mathbf{j})$$

$$= 3 \cdot 1\, \mathbf{i} \times \mathbf{i} + 3 \cdot 2\, \mathbf{i} \times \mathbf{j} + (-4) \cdot 1\, \mathbf{j} \times \mathbf{i} + (-4) \cdot 2\, \mathbf{j} \times \mathbf{j} = 10\mathbf{k}$$

or

$$\mathbf{A} \times \mathbf{B} = \begin{vmatrix} \mathbf{i} & \mathbf{j} & \mathbf{k} \\ 3 & -4 & 0 \\ 1 & 2 & 0 \end{vmatrix} = \mathbf{k} \begin{vmatrix} 3 & -4 \\ 1 & 2 \end{vmatrix} = 10\mathbf{k}$$

The area of the parallelogram is equal to the magnitude of the outer product of the two vectors, which is $|10\mathbf{k}| = 10$.

Answer $10\mathbf{k}$, 10

Example 8.4

Find the area of the parallelogram formed by two vectors $\mathbf{a} = \mathbf{i} - 2\mathbf{j} + 3\mathbf{k}$ and $\mathbf{b} = 2\mathbf{i} + \mathbf{j} - \mathbf{k}$ in space, and determine a unit vector perpendicular to this parallelogram.

Solution

Since the outer product

$$a \times b = \begin{vmatrix} i & j & k \\ 1 & -2 & 3 \\ 2 & 1 & -1 \end{vmatrix}$$

$$= i \begin{vmatrix} -2 & 3 \\ 1 & -1 \end{vmatrix} - j \begin{vmatrix} 1 & 3 \\ 2 & -1 \end{vmatrix} + k \begin{vmatrix} 1 & -2 \\ 2 & 1 \end{vmatrix} = -i + 7j + 5k,$$

then the magnitude is

$$\sqrt{1^2 + 7^2 + 5^2} = \sqrt{75} = 5\sqrt{3}.$$

Therefore, the area of the parallelogram formed by two vectors is

$$5\sqrt{3}.$$

And a unit vector perpendicular to this parallelogram is

$$\pm \frac{1}{5\sqrt{3}}(-i + 7j + 5k).$$

$$\text{Answer } 5\sqrt{3}, \pm \frac{\sqrt{3}}{15}(-i + 7j + 5k)$$

8.1.8   *Application of the outer product*

Remark   Moment of force

In mechanics problems, the moment of force M_O, or torque [Nm], is defined as the outer product of displacement d [m] and force F [N].

$$M_O = d_O \times F \qquad\qquad (8.18)$$

where d_O represents the displacement from the reference point O, while M_O denotes the moment of force about the reference point O.

Example 8.5

Find the moment of force about point O, when a force of 100N is applied to the structure shown in Figure 8.12.

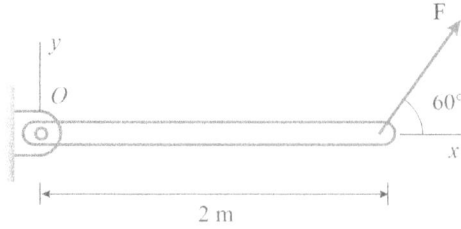

Figure 8.12 Moment of force.

Solution

From the moment of force $\mathbf{M}_O = \mathbf{d}_O \times \mathbf{F}$, we get

$$\mathbf{M}_O = 2\mathbf{i} \times 100\left(\cos 60°\mathbf{i} + \sin 60°\mathbf{j}\right) = 173.2\mathbf{k}\,\text{Nm}.$$

Answer 173.2 Nm ↻

Remark   Velocity of circular motion

The velocity of an object undergoing circular motion is expressed in terms of an outer product as follows:

$$\mathbf{v} = \boldsymbol{\omega} \times \mathbf{r} \tag{8.19}$$

where $\boldsymbol{\omega}$ is angular velocity [rad/s] and $\mathbf{r}$ [m] is the displacement from the center of rotation to the object.

Example 8.6

Find the velocity of point A, located 2 meters to the right from the center O, on a rod rotating clockwise with an angular velocity of 3 rad/s, as shown in Figure 8.13.

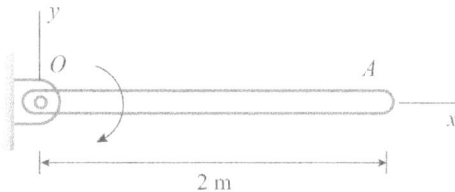

Figure 8.13 Velocity.

Solution

The velocity of the point A is

$$\mathbf{v} = \omega \times \mathbf{r} = (-3\mathbf{k}) \times (2\mathbf{i}) = -6\mathbf{j}.$$

Answer 6 m/s ↓

8.1.9 *Scalar triple product*

> **Remark   Scalar triple product**
>
> The scalar triple product of three vectors **A**, **B**, and **C** is denoted as
> (**A B C**) and is defined as follows. Its result is a scalar value.
>
> $$(\mathbf{A}\ \mathbf{B}\ \mathbf{C}) = \mathbf{A} \cdot (\mathbf{B} \times \mathbf{C}) \tag{8.20}$$

In Eq. (8.19), the magnitude of $\mathbf{B} \times \mathbf{C}$ represents the area of the parallelogram formed by vectors **B** and **C**, while the direction of $\mathbf{B} \times \mathbf{C}$ is perpendicular to the plane formed by **B** and **C**.

The inner product of vector **A** and $\mathbf{B} \times \mathbf{C}$ results in the projection of vector **A** onto the direction of $\mathbf{B} \times \mathbf{C}$, which is $|\mathbf{A}|\cos\theta$ (corresponding to the height h in Figure 8.14).

Therefore, the magnitude of the scalar triple product, $|(\mathbf{A}\ \mathbf{B}\ \mathbf{C})|$, represents the volume of the parallelepiped formed by three vectors **A**, **B**, and **C**.

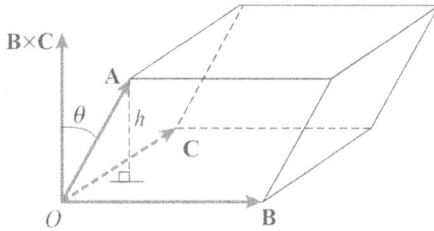

Figure 8.14 The geometric meaning of scalar triple product.

> **Example 8.7**
>
> Find the volume of a tetrahedron formed by three vectors.
>
> $$\mathbf{A} = (2, 1, 0), \quad \mathbf{B} = (1, 0, 3), \quad \mathbf{C} = (0, 2, 2)$$

Solution

Since the scalar triple product $(A\ B\ C) = A \cdot (B \times C)$, first we get

$$B \times C = \begin{vmatrix} i & j & k \\ 1 & 0 & 3 \\ 0 & 2 & 2 \end{vmatrix} = -6i - 2j + 2k.$$

Then

$$A \cdot (B \times C) = (2,\ 1,\ 0) \cdot (-6,\ -2,\ 2) = -14.$$

The volume of the parallelepiped formed by three vectors **A**, **B**, and **C** is 14. The volume of a tetrahedron is one-sixth the volume of a parallelepiped, so the volume of a tetrahedron formed by three vectors **A**, **B**, **C** is $\dfrac{7}{3}$.

$$\text{Answer } \dfrac{7}{3}$$

Problem 8.1

Find the inner product of the following two vectors and calculate the angle between them. [1 ~ 6]

1. $A = i + 2j$, $B = 2i - 2j$
2. $A = i - 3j$, $B = 3i + j$
3. $A = i - 2j + 3k$, $B = 3i - 2j - k$
4. $A = i + 3j + 2k$, $B = 4i - 3k$
5. $A = (3,\ -2)$, $B = (2,\ 3)$
6. $A = (2,\ 1,\ -2)$, $B = (0,\ 1,\ 2)$

Find the area of the parallelogram formed by the following two vectors. [7 ~ 12]

7. $A = i + 2j$, $B = 3i - 2j$
8. $A = i - 3j$, $B = 3i + j$
9. $A = i - 2j + 3k$, $B = 3i - j - k$
10. $A = -i + 3j + 2k$, $B = 2i - 3k$
11. $A = (-2,\ 3,\ 1)$, $B = (2,\ 1,\ 1)$
12. $A = (3,\ 1,\ -2)$, $B = (0,\ 1,\ 2)$

Find the volume of the parallelepiped formed by the following three vectors. [13 ~ 16]

13. $A = (2,\ 0,\ 0)$, $B = (0,\ 3,\ 0)$, $C = (0,\ 0,\ 4)$
14. $A = (1,\ 3,\ 0)$, $B = (0,\ -1,\ 2)$, $C = (2,\ 0,\ 3)$
15. $A = i - 2j + 3k$, $B = 3i - j - k$, $C = 2i + j + 2k$
16. $A = -i + 2j + k$, $B = 2i - 2j + 3k$, $C = i + 2k$

Find the volume of a tetrahedron formed by three vectors. [17 ~ 18]

17. $A = (3, 0, 0)$, $B = (0, 1, 0)$, $C = (0, 0, 2)$
18. $A = i + 2j + 3k$, $B = 2i + k$, $C = i + 2j$

Solve the problem. [19 ~ 24]

19. Find the area of triangle ABC with vertices $A(1, 2, 0)$, $B(5, -1, 0)$, and $C(3, 4, 0)$.
20. Find the area of triangle ABC with vertices $A(2, 3, 1)$, $B(3, -1, 2)$, and $C(1, 2, 0)$.
21. Find the area of the parallelogram $ABCD$ with vertices $A(1, 2, 0)$, $B(5, -1, 0)$, $C(7, 1, 0)$, and $D(3, 4, 0)$.
22. Find the area of the parallelogram $ABCD$ with vertices $A(2, 2, 1)$, $B(3, 0, 2)$, $C(2, 1, 0)$, and $D(1, 3, -1)$.
23. Find the volume of the tetrahedron $ABCD$ with vertices $A(2, 0, 0)$, $B(2, 3, 0)$, $C(0, 3, 0)$, and $D(0, 0, 4)$.
24. Find the volume of the tetrahedron $ABCD$ with vertices $A(2, 2, 1)$, $B(3, 0, 2)$, $C(2, 3, 8)$, and $D(1, 3, -1)$.

8.2 Derivatives of vector functions, direction of vector functions

8.2.1 Vector function

Let's first talk about vector function **r** and scalar function f.

Remark   Vector function

When P is any point in a three-dimensional domain, *vector function* **r** can be represented as:

$$\mathbf{r} = \mathbf{r}(P) = \{x(P),\ y(P),\ z(P)\} = x(P)\mathbf{i} + y(P)\mathbf{j} + z(P)\mathbf{k}, \quad (8.21)$$

where $x(P)$, $y(P)$, and $z(P)$ are scalar functions. Vector functions often describe curves, paths, or trajectories in space.

Similarly, a scalar function f is defined as

$$f = f(P). \quad (8.22)$$

A vector function $\mathbf{r} = \mathbf{r}(P)$ and a scalar function $f = f(P)$ are functions that vary depending on the point P. The region represented by a vector function is called a *vector field*, and the region represented by a scalar function is called a *scalar field*.

In a vector field, there are various types of fields such as the tangent vector field, the normal vector field, the velocity field, etc. In a scalar field, there are fields such as the temperature field of a surface, the pressure field in the water, the energy field, etc.

8.2.2 Chain rule

If the scalar function $f = f(x, y)$ is continuous in the xy-plane domain and has continuous derivatives, and if x and y are represented respectively as functions of variables u and v, $x = x(u, v)$ and $y = y(u, v)$, then the partial derivative $\dfrac{\partial f}{\partial u}$ and $\dfrac{\partial f}{\partial v}$ of the scalar function f, can be expressed by the following equations, respectively:

$$\frac{\partial f}{\partial u} = \frac{\partial f}{\partial x}\frac{\partial x}{\partial u} + \frac{\partial f}{\partial y}\frac{\partial y}{\partial u} \tag{8.23a}$$

$$\frac{\partial f}{\partial v} = \frac{\partial f}{\partial x}\frac{\partial x}{\partial v} + \frac{\partial f}{\partial y}\frac{\partial y}{\partial v}. \tag{8.23b}$$

These equations utilize the *chain rule* to compute the partial derivatives of f with respect to u and v in terms of the partial derivatives of f with respect to x and y, and the partial derivatives of x and y with respect to u and v, respectively.

Similarly, if the scalar function $f = f(x, y, z)$ is continuous in the xyz-space domain and has continuous derivatives, and if x, y, and z are represented respectively as functions of variables u and v, $x = x(u, v)$, $y = y(u, v)$, and $z = z(u, v)$, then the partial derivative $\dfrac{\partial f}{\partial u}$ and $\dfrac{\partial f}{\partial v}$ of the scalar function f, can be expressed by the following equations, respectively.

$$\frac{\partial f}{\partial u} = \frac{\partial f}{\partial x}\frac{\partial x}{\partial u} + \frac{\partial f}{\partial y}\frac{\partial y}{\partial u} + \frac{\partial f}{\partial z}\frac{\partial z}{\partial u} \tag{8.24a}$$

$$\frac{\partial f}{\partial v} = \frac{\partial f}{\partial x}\frac{\partial x}{\partial v} + \frac{\partial f}{\partial y}\frac{\partial y}{\partial v} + \frac{\partial f}{\partial z}\frac{\partial z}{\partial v} \tag{8.24b}$$

Example 8.8

When the scalar function $f = f(x, y) = x^2 + 2y^2$ is expressed as $x = r\cos\theta$ and $y = r\sin\theta$, calculate $\dfrac{\partial f}{\partial r}$ and $\dfrac{\partial f}{\partial \theta}$, respectively.

Solution

Since $\dfrac{\partial f}{\partial r} = \dfrac{\partial f}{\partial x}\dfrac{\partial x}{\partial r} + \dfrac{\partial f}{\partial y}\dfrac{\partial y}{\partial r}$, then we get

$$\frac{\partial f}{\partial r} = 2x \cdot \cos\theta + 4y \cdot \sin\theta$$

$$= 2r\cos\theta \cdot \cos\theta + 4r\sin\theta \cdot \sin\theta = 2r\left(1 + \sin^2\theta\right),$$

and since $\dfrac{\partial f}{\partial\theta} = \dfrac{\partial f}{\partial x}\dfrac{\partial x}{\partial\theta} + \dfrac{\partial f}{\partial y}\dfrac{\partial y}{\partial\theta}$, then we get

$$\frac{\partial f}{\partial\theta} = 2x \cdot \left(-r\sin\theta\right) + 4y \cdot r\cos\theta$$

$$= 2r\cos\theta \cdot \left(-r\sin\theta\right) + 4r\sin\theta \cdot r\cos\theta = 2r^2\sin\theta\cos\theta = r^2\sin 2\theta.$$

Answer $2r\left(1 + \sin^2\theta\right),\ r^2\sin 2\theta$

8.2.3   Derivative of a vector function

Similar to the derivative of a scalar function, the derivative of a vector function is defined as follows:

Remark   Derivative of a vector function

The derivative $\mathbf{r}'(t)$ of the vector function $\mathbf{r}(t)$ with respect to time t is defined as follows:

$$\mathbf{r}'(t) = \frac{d\mathbf{r}}{dt} = \lim_{\Delta t \to 0} \frac{\mathbf{r}(t + \Delta t) - \mathbf{r}(t)}{\Delta t} \tag{8.25}$$

The rules for the derivatives of two vector functions $\mathbf{u}$ and $\mathbf{v}$ are as follows:

$$\left(\mathbf{u} + \mathbf{v}\right)' = \mathbf{u}' + \mathbf{v}' \tag{8.26a}$$

$$\left(\mathbf{u} \cdot \mathbf{v}\right)' = \mathbf{u}' \cdot \mathbf{v} + \mathbf{u} \cdot \mathbf{v}' \tag{8.26b}$$

$$\left(\mathbf{u} \times \mathbf{v}\right)' = \mathbf{u}' \times \mathbf{v} + \mathbf{u} \times \mathbf{v}'. \tag{8.26c}$$

Example 8.9

When $\mathbf{v} = \mathbf{v}(x, y) = \left(x^2 - y^2\right)\mathbf{i} + \left(x^2 + y^2\right)\mathbf{j} + xy\mathbf{k}$, calculate $\dfrac{\partial v}{\partial x}$ and $\dfrac{\partial v}{\partial y}$, respectively.

Solution

We can get

$$\frac{\partial \mathbf{v}}{\partial x} = 2xi + 2xj + yk,$$

and

$$\frac{\partial \mathbf{v}}{\partial y} = -2yi + 2yj + xk.$$

Answer $2xi + 2xj + yk, -2yi + 2yj + xk$

8.2.4　The vector equation of a curve with parametric representations

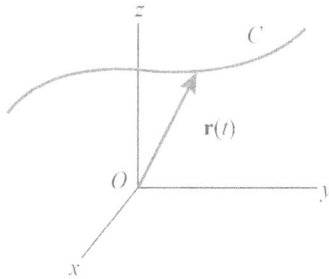

Figure 8.15 Parametric representation of a curve.

As seen in Figure 8.15, the position of path C in space can be represented as follows:

$$\mathbf{r}(t) = \{x(t), y(t), z(t)\} = x(t)\mathbf{i} + y(t)\mathbf{j} + z(t)\mathbf{k} \qquad (8.27)$$

where t is the parameter, and the position at $t = t_0$ is represented as $\mathbf{r}(t_0) = \{x(t_0), y(t_0), z(t_0)\}$.

This method of representation using parameters is called *parametric representation*.

If the vector $\mathbf{r}$ is parallel to the vector $\mathbf{a}$ (i.e., $\mathbf{r} \parallel \mathbf{a}$), the following equation holds true:

$$\mathbf{r} = t\mathbf{a} \qquad (8.28a)$$

where t is an arbitrary parameter.

Moreover, as seen in Figure 8.16, if the line vector $\mathbf{r}$ passing through a specific position r_0 is parallel to the vector $\mathbf{a}$, the following equation holds true:

$$\mathbf{r}(t) - r_0 = t\mathbf{a} \qquad (8.28b)$$

where the vector $\mathbf{a}$ is called the direction vector, and $\dfrac{\mathbf{a}}{|\mathbf{a}|}$ is called the direction cosine.

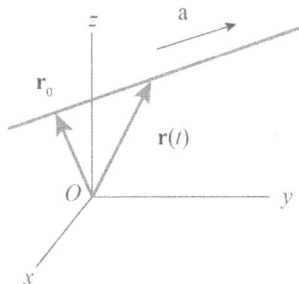

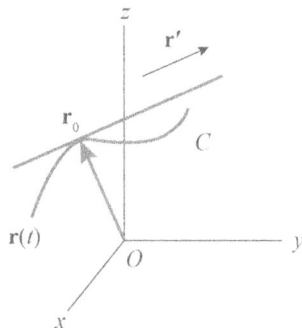

Figure 8.16 Straight line equation.   *Figure 8.17* Tangent equation.

As seen in Figure 8.17, let's express the tangent equation $q(t)$ at a point r_0 on the curve C when the position vector $r(t)$ represents the curve. Since the vector a in Eq. (8.28) is replaced by the tangent vector $r'(t)$, the vector equation of the tangent is represented by the following equation:

$$q(t) = r_0 + tr'. \tag{8.29}$$

Example 8.10

Find the equation of the tangent line at the point $\left(\dfrac{3\sqrt{2}}{2}, \sqrt{2} \right)$ on the ellipse $\dfrac{x^2}{9} + \dfrac{y^2}{4} = 1$.

Solution

Since the equation of ellipse

$$r = (3\cos\theta, \ 2\sin\theta),$$

the point $\left(\dfrac{3\sqrt{2}}{2}, \sqrt{2} \right)$ is at $\theta = \dfrac{\pi}{4}$, and its derivative is

$$r' = (-3\sin\theta, \ 2\cos\theta).$$

Then we get

$$r'\big|_{\theta=\pi/4} = (-3\sin\theta, \ 2\cos\theta)\big|_{\theta=\pi/4} = \left(-\dfrac{3\sqrt{2}}{2}, \ \sqrt{2} \right).$$

Therefore, the equation of the tangent line is expressed as

$$q(t) = \left(\frac{3\sqrt{2}}{2},\ \sqrt{2} \right) + t\left(-\frac{3\sqrt{2}}{2},\ \sqrt{2} \right) = \left(\frac{3\sqrt{2}}{2}(1-t),\ \sqrt{2}(1+t) \right).$$

$$\textbf{Answer } q(t) = \left(\frac{3\sqrt{2}}{2}(1-t),\ \sqrt{2}(1+t) \right)$$

Check

The answer $q(t) = \left(\dfrac{3\sqrt{2}}{2}(1-t),\ \sqrt{2}(1+t) \right)$ gives

$$x = \frac{3\sqrt{2}}{2}(1-t),\ y = \sqrt{2}(1+t).$$

Eliminating the parameter t, we get

$$\frac{x}{3\sqrt{2}} + \frac{y}{2\sqrt{2}} = 1.$$

Another Solution

Since the equation of the tangent line at the point $(x_0,\ y_0)$ on the ellipse $\dfrac{x^2}{a^2} + \dfrac{y^2}{b^2} = 1$ is $\dfrac{x_0 x}{a^2} + \dfrac{y_0 y}{b^2} = 1$, substituting the given values, $x_0 = \dfrac{3\sqrt{2}}{2}$ and $y_0 = \sqrt{2}$, on the equation of the tangent line yields

$$\frac{x}{3\sqrt{2}} + \frac{y}{2\sqrt{2}} = 1.$$

Problem 8.2

Solve the problem. [1 ~ 4]

1. When the scalar function $f = f(x, y) = 2xy$ is expressed as $x = r\cos\theta$ and $y = r\sin\theta$, calculate $\dfrac{\partial f}{\partial r}$ and $\dfrac{\partial f}{\partial \theta}$, respectively.

2. When the scalar function $f = f(x, y, z) = x^2 + y^2 + z^2$ is expressed as $x = r\cos\theta$, $y = r\sin\theta$, and $z = r^2$, calculate $\dfrac{\partial f}{\partial r}$ and $\dfrac{\partial f}{\partial \theta}$, respectively.

3. Find the derivative of the vector function $\mathbf{r} = \mathbf{r}(t) = (2\cos t)\mathbf{i} + (2\sin t)\mathbf{j} + (3t)\mathbf{k}$.
4. Find the partial derivatives of the vector function $\mathbf{v} = \mathbf{v}(x,\ y) = (e^{-x}\cos y)\mathbf{i} + (e^{-x}\sin y)\mathbf{j} + (3xy)\mathbf{k}$.

Express the following in parametric representation. [5 ~ 10]

5. A circle with center at (2, 3) and passing through the origin.
6. A line passing through the point (1, 2, 3) and parallel to the vector (2, –1, 1).
7. A line passing through the points (1, 2, 3) and (0, 1, –1).
8. The line described by the equations $y = 2x + 1$ and $z = 2x$.

9. The shape defined by $x^2 + \dfrac{y^2}{2} = 1$ and $z = 2x$

10. The shape defined by $x^2 - \dfrac{y^2}{2} = 1$ and $z = 2$

Solve the problem. [11 ~ 14]

11. Find the equation of the tangent line at the point $(1,\ \sqrt{3})$ on the circle $x^2 + y^2 = 4$.
12. Find the equation of the tangent line at the point $(\sqrt{2},\ -\dfrac{\sqrt{2}}{2})$ on the ellipse $\dfrac{x^2}{4} + y^2 = 1$.
13. Find the equation of the tangent line at the point $(2,\ \sqrt{3})$ on the hyperbola $x^2 - y^2 = 1$.
14. Find the equation of the tangent line at the point $(2,\ -2\sqrt{3})$ on the hyperbola $x^2 - \dfrac{y^2}{4} = 1$.

8.3 Gradient of a scalar field, directional derivative

8.3.1 Gradient of a scalar field

Remark Gradient of a scalar field

The *gradient* is a prominent function that derives a vector field from a scalar field. We denote the gradient of a differentiable scalar function $f = f(x, y, z)$ by grad f or ∇f. Then the gradient of $f(x, y, z)$ is defined as the vector function

$$\text{grad } f = \nabla f = \left\{ \frac{\partial f}{\partial x},\ \frac{\partial f}{\partial y},\ \frac{\partial f}{\partial z} \right\} = \frac{\partial f}{\partial x}\mathbf{i} + \frac{\partial f}{\partial y}\mathbf{j} + \frac{\partial f}{\partial z}\mathbf{k}. \qquad (8.30)$$

Here, ∇f, read as "**nabla** f", denotes the gradient in space $f = f(x, y, z)$, representing the rate of change (slope) in each direction. The newly defined ∇ is a differential operator defined as

$$\nabla = \frac{\partial}{\partial x}\mathbf{i} + \frac{\partial}{\partial y}\mathbf{j} + \frac{\partial}{\partial z}\mathbf{k}. \tag{8.31}$$

By the way, ∇^2 is defined as the Laplace operator, which is expressed as

$$\nabla^2 = \Delta = \frac{\partial^2}{\partial x^2} + \frac{\partial^2}{\partial y^2} + \frac{\partial^2}{\partial z^2} \tag{8.32}$$

where ∇^2 is read as "nabla squared" or "delta, Δ".

For example, let's find the gradient of a scalar function $f(x, y) = y - 2x - 5 = 0$. Since $\frac{\partial f}{\partial x} = -2$ and $\frac{\partial f}{\partial y} = 1$, we get

$$\text{grad } f = \frac{\partial f}{\partial x}\mathbf{i} + \frac{\partial f}{\partial y}\mathbf{j} = -2\mathbf{i} + \mathbf{j},$$

where this implies that the slope of the line $y = 2x + 5$ is 2 and the *normal slope* is $-1/2$.

Example 8.11

Find the normal slope at the point $(-4, 3)$ on the curve represented by $f = x^2 + y^2 - 25 = 0$.

Solution

At the point $(-4, 3)$ on the given scalar function $f = x^2 + y^2 - 25 = 0$, we get

$$\text{grad} f = \frac{\partial f}{\partial x}\mathbf{i} + \frac{\partial f}{\partial y}\mathbf{j}$$

$$= 2x\mathbf{i} + 2y\mathbf{j}|_{(-4, 3)} = -8\mathbf{i} + 6\mathbf{j}.$$

Therefore, the normal slope is

$$-\frac{3}{4}.$$

Answer $-\dfrac{3}{4}$

Another Solution

Since the equation of the tangent line at the point (x_0, y_0) on the circle $x^2 + y^2 = r^2$ is $x_0 x + y_0 y = r^2$, substituting the given values, $x_0 = -4$ and $y_0 = 3$, on the on the equation of the tangent line yields

$$-4x + 3y = 25,$$

where this implies that its slope is $\dfrac{4}{3}$ and its normal slope is $-\dfrac{3}{4}$.

8.3.2 Surface normal vector

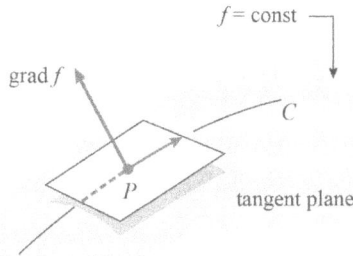

Figure 8.18 Gradient as surface normal vector.

Let S be a surface represented by the scalar function $f = f(x, y, z) = c$ as shown in Figure 8.18, where $f = f(x, y, z)$ is differentiable. Now let C be a curve on S through a point P of S. Then, a curve C in space is represented as $r(t) = \{x(t), y(t), z(t)\}$. For C on the surface S, the components of $r(t)$ must satisfy

$$f\left[x(t), y(t), z(t)\right] = c. \quad (c \text{ constant}) \tag{8.33}$$

Then a tangent vector of the curve C is $r'(t) = \{x'(t), y'(t), z'(t)\}$. And the tangent vectors of all curves on S passing through the point P will form a plane, which is called the *tangent plane* of S at P. The normal of this plane is called the *surface normal* to S at P, and a vector in the direction of the surface normal is called a *surface normal vector* of S at P.

Remark   Surface normal vector

Let f be a differentiable scalar function in space, and let $f(x, y, z) = c$ (c constant) represent a surface S. Then if the grad f at a point P of S is not the zero vector, it is a *surface normal vector* of S at P.

When differentiating Eq. (8.33) with respect to the parameter t, we obtain the following equation:

$$\frac{\partial f}{\partial x}x' + \frac{\partial f}{\partial y}y' + \frac{\partial f}{\partial z}z' = (\text{grad}f) \cdot \mathbf{r}' = 0. \tag{8.34}$$

Since the dot product of grad f and $\mathbf{r}'$ is 0, grad f and $\mathbf{r}'$ are orthogonal to each other. Therefore, grad f becomes a *normal vector* perpendicular to all vectors $\mathbf{r}'$ on the tangent plane.

Remark   **Unit normal vector**

The grad f becomes a *normal vector* perpendicular to all vectors $\mathbf{r}'$ on the tangent plane. Thus, the unit normal vector $\mathbf{n}$ of the curve is given as follows:

$$\mathbf{n} = \frac{\text{grad}f}{|\text{grad}f|} = \frac{\nabla f}{|\nabla f|} \tag{8.35}$$

where $f(x, y, z) = c$ (c constant) represents the equation of the surface.

Example 8.12   Find the unit normal vector $\mathbf{n}$ to the plane represented by

$$3x + 2y + z = 6.$$

Solution

From the given equation $f = 3x + 2y + z - 6 = 0$, we obtain

$$\text{grad } f = \nabla f = 3\mathbf{i} + 2\mathbf{j} + \mathbf{k}.$$

Then the unit normal vector is

$$\mathbf{n} = \frac{\nabla f}{|\nabla f|} = \frac{3\mathbf{i} + 2\mathbf{j} + \mathbf{k}}{\sqrt{3^2 + 2^2 + 1}} = \frac{1}{\sqrt{14}}(3\mathbf{i} + 2\mathbf{j} + \mathbf{k}).$$

$$\text{Answer } \frac{\sqrt{14}}{14}(3\mathbf{i} + 2\mathbf{j} + \mathbf{k})$$

Example 8.13

Find the unit normal vector $\mathbf{n}$ to the surface represented by $z^2 = 2(x^2 + y^2)$ at the point $P(1, -1, 2)$.

Solution

From the given equation $z^2 = 2(x^2 + y^2)$, we obatin

$$f = f(x, y, z) = 2(x^2 + y^2) - z^2 = 0.$$

Then a normal vector is

$$\operatorname{grad} f = \nabla f = \frac{\partial f}{\partial x}\mathbf{i} + \frac{\partial f}{\partial y}\mathbf{j} + \frac{\partial f}{\partial z}\mathbf{k}$$

$$= 4x\mathbf{i} + 4y\mathbf{j} - 2z\mathbf{k}.$$

And a normal vector at the point P is

$$\operatorname{grad} f(P) = 4\mathbf{i} - 4\mathbf{j} - 4\mathbf{k},$$

Therefore, we obtain the unit normal vector

$$\mathbf{n} = \frac{\operatorname{grad} f(P)}{|\operatorname{grad} f(P)|} = \frac{1}{\sqrt{3}}(\mathbf{i} - \mathbf{j} - \mathbf{k}).$$

$$\textbf{Answer } \mathbf{n} = \frac{\sqrt{3}}{3}(\mathbf{i} - \mathbf{j} - \mathbf{k})$$

8.3.3 *Directional derivative*

Remark   Directional derivative

The directional derivative $D_a f$ of a scalar function $f = f(x, y, z)$ at a point $P(x_0, y_0, z_0)$ in the direction of a vector $\mathbf{a}$ is defined by (see Figure 8.19)

$$D_a f = \frac{df}{ds} = \lim_{s \to 0} \frac{f(Q) - f(P)}{s}, \tag{8.36}$$

where a point Q is a variable point on the straight line L in the direction of $\mathbf{a}$, and $|s|$ is the distance between P and Q.

Also, $s > 0$ if Q lies in the direction of $\mathbf{a}$, $s < 0$ if Q lies in the opposite direction of $\mathbf{a}$, and $s = 0$ if $Q = P$.

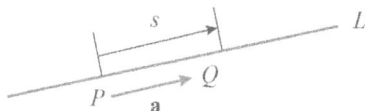

Figure 8.19 Directional derivative.

If **a** is a unit vector in Cartesian coordinates, the line L is given by

$$\mathbf{r}(s) = x(s)\mathbf{i} + y(s)\mathbf{j} + z(s)\mathbf{k} = \mathbf{p}_0 + s\mathbf{a}, \quad (|\mathbf{a}| = 1) \tag{8.37}$$

where $\mathbf{p}_0$ is the position vector of $P(x_0, y_0, z_0)$. Applying the chain rule in Eq. (8.36), we obtain the directional derivative as

$$D_a f = \frac{df}{ds} = \frac{\partial f}{\partial x} x' + \frac{\partial f}{\partial y} y' + \frac{\partial f}{\partial z} z' = \left(\frac{\partial f}{\partial x} \mathbf{i} + \frac{\partial f}{\partial y} \mathbf{j} + \frac{\partial f}{\partial z} \mathbf{k} \right) \cdot (x'\mathbf{i} + y'\mathbf{j} + z'\mathbf{k}),$$

or

$$D_a f = \mathbf{a} \cdot \operatorname{grad} f. \tag{8.38}$$

But, if **a** is not a unit vector, we obtain the directional derivative as

$$D_a f = \frac{\mathbf{a}}{|\mathbf{a}|} \cdot \operatorname{grad} f. \tag{8.39}$$

Therefore, the directional derivative $D_a f$ of a function $f = f(x, y, z)$ is calculated as the inner product of the unit vector $\left(\dfrac{\mathbf{a}}{|\mathbf{a}|} \right)$ and the normal vector (grad f) at a point $P(x_0, y_0, z_0)$ in space.

Example 8.14

Find the directional derivative of a function $f = x^2 + 2y^2 + z^2$ at a point $P(1, -2, 2)$ in the direction of a vector $\mathbf{a} = (2, 1, -3)$.

Solution

A unit vector of **a** is

$$\frac{\mathbf{a}}{|\mathbf{a}|} = \frac{1}{\sqrt{14}} (2\mathbf{i} + \mathbf{j} - 3\mathbf{k}).$$

And we obtain the normal vector as

$$\operatorname{grad} f = \nabla f = \frac{\partial f}{\partial x} \mathbf{i} + \frac{\partial f}{\partial y} \mathbf{j} + \frac{\partial f}{\partial z} \mathbf{k} = 2x\mathbf{i} + 4y\mathbf{j} + 2z\mathbf{k}.$$

Substituting the given point $P(1, -2, 2)$ into the normal vector gives

$$\operatorname{grad} f(P) = 2\mathbf{i} - 8\mathbf{j} + 4\mathbf{k}.$$

Therefore, we obtain the directional derivative as

$$D_a f = \frac{\mathbf{a}}{|\mathbf{a}|} \cdot \operatorname{grad} f$$

$$= \frac{1}{\sqrt{14}}(2\mathbf{i} + \mathbf{j} - 3\mathbf{k}) \cdot (2\mathbf{i} - 8\mathbf{j} + 4\mathbf{k}) = -\frac{16}{\sqrt{14}} = -4.276.$$

Answer −4.276

Problem 8.3

Find $\operatorname{grad} f = \nabla f$ of a scalar field f. [1 ~ 4]

1. $f = f(x, y) = x^2 - xy + y^2$
2. $f = f(x, y) = x^2 + 2y^2$
3. $f = f(x, y, z) = \dfrac{x^2 z}{y}$
4. $f = f(x, y, z) = (x+1)^2 + (y-2)^2 + z^2$

Find the directional derivative of a function f at a point P in the direction of a vector **a**. [5 ~ 10]

5. $f = 2x^2 + y^2$, $P(1, 2)$, $\mathbf{a} = \{2, -1\}$
6. $f = 2x^2 - y$, $P(-1, 2)$, $\mathbf{a} = \{1, 2\}$
7. $f = x^2 + y^2 + z^2$, $P(1, -1, 2)$, $\mathbf{a} = \{1, 2, 3\}$
8. $f = x^2 - y^2 + yz$, $P(2, 0, -1)$, $\mathbf{a} = \{2, -2, 1\}$
9. $f = \ln(x^2 + y^2)$, $P(3, 1)$, $\mathbf{a} = \{2, -1\}$
10. $f = xyz$, $P(-1, 2, 3)$, $\mathbf{a} = \{0, 2, -1\}$

Find the unit normal vector to the surface represented by the equation at the point P. [11 ~ 18]

11. $P(1, 2)$, $y = x^2 + 1$
12. $P(e, e)$, $y = x \ln x$
13. $P(3, 2)$, $x - y = 1$
14. $P(4, 2)$, $(x-1)^2 + (y+2)^2 = 25$
15. $P(1, 2, 5)$, $z = x^2 + y^2$
16. $P(1, 1, -1)$, $3 = x^2 + y^2 + z^2$
17. $P(1, -1, 2)$, $4 = x^2 + y^2 + z$
18. $P(-1, 2, 3)$, $z = x^3 + y^3 - 4$

8.4 Divergence and curl of a vector field

8.4.1 Divergence of a vector field

In Chapter 8.3, we introduced the concept of the normal vector to derive a vector field from a scalar field. In this chapter, we will learn about introducing divergence to a vector field to derive a scalar field from a vector field. Additionally, we will introduce curl to a vector field to derive another vector field from a vector field.

Divergence and curl of a vector field are essential concepts primarily used in physics and fluid dynamics.

Remark   Divergence of a vector field

If $v = v(x, y, z)$ is a differentiable vector function in Cartesian coordinates and its corresponding components of v are v_1, v_2, and v_3, the *divergence* of v is defined as

$$\text{div } v = \frac{\partial v_1}{\partial x} + \frac{\partial v_2}{\partial y} + \frac{\partial v_3}{\partial z}, \tag{8.40}$$

or

$$\text{div } v = \nabla \cdot v = \left\{ \frac{\partial}{\partial x}, \frac{\partial}{\partial y}, \frac{\partial}{\partial z} \right\} \cdot \{v_1, v_2, v_3\}$$

$$= \left(\frac{\partial}{\partial x} i + \frac{\partial}{\partial y} j + \frac{\partial}{\partial z} k \right) \cdot (v_1 i + v_2 j + v_3 k)$$

$$= \frac{\partial v_1}{\partial x} + \frac{\partial v_2}{\partial y} + \frac{\partial v_3}{\partial z},$$

where ∇f means the vector grad f defined in Eq. (8.30).

And the divergence of the normal gradient leads to the following Laplace equation:

$$\text{div } (\text{grad} f) = \nabla \cdot (\text{grad} f) = \nabla \cdot (\nabla f) = \nabla^2 f. \tag{8.41}$$

To understand the physical meaning of divergence, let's consider a fluid dynamics problem similar to what we see in Figure 8.20. (*optional).

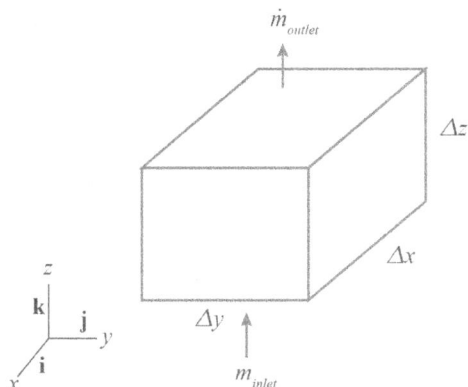

Figure 8.20 Physical meaning of the divergence.

Let's calculate the mass flow per unit time ($\Delta \dot{m} = \rho \Delta \dot{V} = \rho v \Delta A$ [kg/s]) entering and leaving the control volume ΔV of the rectangular prism shown in Figure 8.20, where the density of fluid is ρ, the control volume is $\Delta V = \Delta x \Delta y \Delta z$, each cross-sectional area is composed of $\Delta x \Delta y$, $\Delta y \Delta z$, and $\Delta z \Delta x$.

The infinitesimal mass (Δm, [kg]) flowing into the bottom surface and out of the top surface (diverging) during the time Δt with a velocity vector $\mathbf{v} = \{v_1, v_2, v_3\} = v_1 \mathbf{i} + v_2 \mathbf{j} + v_3 \mathbf{k}$ are given by the following:

$$\Delta m_{inlet} = (\rho v_3)_z \, \Delta x \Delta y \Delta t$$

$$\Delta m_{outlet} = (\rho v_3)_{z+\Delta z} \, \Delta x \Delta y \Delta t.$$

So, the infinitesimal mass diverging from the top and bottom surfaces of the control volume is:

$$\Delta m = \Delta m_{outlet} - \Delta m_{inlet} = \{(\rho v_3)_{z+\Delta z} - (\rho v_3)_z\} \Delta x \Delta y \Delta t$$

$$= \frac{(\rho v_3)_{z+\Delta z} - (\rho v_3)_z}{\Delta z} \Delta V \Delta t$$

$$= \frac{\Delta(\rho v_3)}{\Delta z} \Delta V \Delta t.$$

Similarly, summing up the infinitesimal mass diverging from all side surfaces yields the following:

$$\left\{ \frac{\Delta(\rho v_1)}{\Delta x} + \frac{\Delta(\rho v_2)}{\Delta y} + \frac{\Delta(\rho v_3)}{\Delta z} \right\} \Delta V \Delta t.$$

As Δx, Δy, and Δz approach zero, the expression simplifies to:

$$\left\{\frac{\partial(\rho v_1)}{\partial x}+\frac{\partial(\rho v_2)}{\partial y}+\frac{\partial(\rho v_3)}{\partial z}\right\}\Delta V\,\Delta t = \operatorname{div}(\rho v)\Delta V\Delta t. \qquad (8.42)$$

On the other hand, the mass loss within the control volume ΔV occurs due to changes in density. Then calculating this value yields the following:

$$-\frac{\partial \rho}{\partial t}\Delta V\Delta t. \qquad (8.43)$$

Therefore, from Eq. (8.42) and Eq. (8.43), the following continuity equation is derived:

$$\frac{\partial \rho}{\partial t}+\operatorname{div}(\rho v)=0. \qquad (8.44)$$

This indicates that the divergence of a vector field represents the physical quantity (in this section, it refers to mass) obtained by subtracting inflow from outflow.

Remark   Continuity equation

$$\frac{\partial \rho}{\partial t}+\operatorname{div}(\rho v)=0 \qquad (8.43)$$

The divergence of a vector field represents the physical quantity of net flow, obtained by subtracting inflow from outflow.

Example 8.15

Find the divergence of the vector $v = v(x,\ y,\ z)=\left(x^2,\ -3y,\ 2z^2\right)$ at the point $P(1,\ -1,\ 2)$.

Solution

First, we calculate the div v

$$\operatorname{div} v = \nabla \cdot v = \left(\frac{\partial}{\partial x}i+\frac{\partial}{\partial y}j+\frac{\partial}{\partial z}k\right)\cdot(v_1 i+v_2 j+v_3 k)$$

$$= \frac{\partial v_1}{\partial x}+\frac{\partial v_2}{\partial y}+\frac{\partial v_3}{\partial z}=2x-3+4z.$$

Therefore, we obtain the divergence of the vector at $P(1, -1, 2)$ as follows:

$$\text{div } \mathbf{v} = 2x - 3 + 4z\big|_{x=1, y=-1, z=2}$$

$$= 2 \cdot 1 - 3 + 4 \cdot 2 = 7.$$

Answer 7

8.4.2 Curl of a vector field

Remark **Curl of a vector field**

If $\mathbf{v} = \mathbf{v}(x, y, z)$ is a differentiable vector function in Cartesian coordinates and its corresponding components of $\mathbf{v}$ are v_1, v_2, and v_3, the *curl* of $\mathbf{v}$ is defined as

$$\text{curl } \mathbf{v} = \nabla \times \mathbf{v} = \left(\frac{\partial}{\partial x}\mathbf{i} + \frac{\partial}{\partial y}\mathbf{j} + \frac{\partial}{\partial z}\mathbf{k} \right) \times (v_1\mathbf{i} + v_2\mathbf{j} + v_3\mathbf{k}) \qquad (8.44)$$

$$= \begin{vmatrix} \mathbf{i} & \mathbf{j} & \mathbf{k} \\ \dfrac{\partial}{\partial x} & \dfrac{\partial}{\partial y} & \dfrac{\partial}{\partial z} \\ v_1 & v_2 & v_3 \end{vmatrix} = \left(\frac{\partial v_3}{\partial y} - \frac{\partial v_2}{\partial z} \right)\mathbf{i} + \left(\frac{\partial v_1}{\partial z} - \frac{\partial v_3}{\partial x} \right)\mathbf{j} + \left(\frac{\partial v_2}{\partial x} - \frac{\partial v_1}{\partial y} \right)\mathbf{k}$$

Instead of curl $\mathbf{v}$, rot $\mathbf{v}$ is also used. If curl $\mathbf{v} = 0$, it means the absence of rotation (i.e., irrotation).

The operation for the curl $\mathbf{v}$ of a vector field is as follows:

$$\text{curl}(\mathbf{u} + \mathbf{v}) = \text{curl}(\mathbf{u}) + \text{curl}(\mathbf{v}) \qquad (8.45a)$$

$$\text{curl}(\text{grad} f) = \nabla \times (\nabla f) = 0 \qquad (8.45b)$$

$$\text{div}(\text{curl } \mathbf{v}) = 0. \qquad (8.45c)$$

Example 8.16

Find the curl $\mathbf{v}$ of $\mathbf{v} = xy\mathbf{i} + yz\mathbf{j} - z\mathbf{k}$ with right-handed x, y, z.

Solution

Since curl $\mathbf{v} = \nabla \times \mathbf{v}$, we get

$$\text{curl } \mathbf{v} = \nabla \times \mathbf{v} = \begin{vmatrix} \mathbf{i} & \mathbf{j} & \mathbf{k} \\ \dfrac{\partial}{\partial x} & \dfrac{\partial}{\partial y} & \dfrac{\partial}{\partial z} \\ xy & yz & -z \end{vmatrix} = (0-y)\mathbf{i} - (0-0)\mathbf{j} + (0-x)\mathbf{k}.$$

Answer $-y\mathbf{i} - x\mathbf{k}$

Problem 8.4

Find the divergence of the vector $\mathbf{v}$ at the point P. [1 ~ 4]

1. $\mathbf{v} = \mathbf{v}(x, y) = (2x, -3y^2), P(-1, 1)$

2. $\mathbf{v} = \mathbf{v}(x, y) = \left(\dfrac{x}{\sqrt{x^2 + y^2}}, \dfrac{y}{\sqrt{x^2 + y^2}} \right), P(1, -2)$

3. $\mathbf{v} = \mathbf{v}(x, y, z) = (x+y, xy, 2z), P(1, 3, -2)$

4. $\mathbf{v} = \mathbf{v}(x, y, z) = (x^2 y, xy^2, z^2), P(1, -2, 1)$

Find the curl $\mathbf{v}$ with right-handed x, y, z. [5 ~ 8]

5. $\mathbf{v} = x^2 \mathbf{i} + z^2 \mathbf{j} - y^2 \mathbf{k}$
6. $\mathbf{v} = z \ln y \mathbf{i} + z \ln x \mathbf{j} - z^2 \mathbf{k}$
7. $\mathbf{v} = e^x \cos y \mathbf{i} + e^x \sin y \mathbf{j} + z^2 \mathbf{k}$
8. $\mathbf{v} = e^{-xy} \sin z \mathbf{i} + e^{-xy} \cos z \mathbf{j} + e^{-xy} \mathbf{k}$

8.5   Utilizing MATLAB®

Using MATLAB®, you can easily compute the magnitude of a vector (norm.m), the inner product of two vectors (dot.m), and the outer product of two vectors (cross.m).

M_Example 8.1

Solve the problem.

a. Find the magnitude of vector $\mathbf{A} = [2 \ 3 \ -1]$.
b. Find the inner product of $\mathbf{A} = [2 \ 3 \ -1]$ and $\mathbf{B} = [-1 \ 1 \ 2]$.
c. Find the outer product of $\mathbf{A} = [2 \ 3 \ -1]$, $\mathbf{B} = [-1 \ 1 \ 2]$.

Solution

a. % norm.m

```
a= [ 2 3 –1];
norm(a)
```

Answer 3.7417

b. % dot.m

```
a= [ 2 3 –1]; b= [ –1 1 2];
dot(a,b)
```

Answer –1

c. % cross.m

```
a= [ 2 3 –1]; b= [ –1 1 2];
cross(a,b)
```

Answer $7i - 3j + 5k$

Answer

Problem 8.1

1. $\mathbf{A} \cdot \mathbf{B} = -2, \theta = 108.4°$
2. $\mathbf{A} \cdot \mathbf{B} = 0, \theta = 90°$
3. $\mathbf{A} \cdot \mathbf{B} = 4, \theta = 73.4°$
4. $\mathbf{A} \cdot \mathbf{B} = -2, \theta = 96.1°$
5. $\mathbf{A} \cdot \mathbf{B} = 0, \theta = 90°$
6. $\mathbf{A} \cdot \mathbf{B} = -3, \theta = 116.6°$
7. 8
8. 10
9. $5\sqrt{6}$
10. $\sqrt{118}$
11. $2\sqrt{21}$
12. $\sqrt{61}$
13. 24
14. 9
15. 30
16. 4
17. 1
18. 2
19. 7
20. $\dfrac{5\sqrt{2}}{2}$
21. 14
22. $\sqrt{11}$
23. 4
24. 1

Problem 8.2

1. $2r\sin 2\theta,\ 2r^2 \cos 2\theta$
2. $2r + 4r^3,\ 0$
3. $\mathbf{r}' = (-2\sin t)\mathbf{i} + (2\cos t)\mathbf{j} + 3\mathbf{k}$
4. $\dfrac{\partial \mathbf{v}}{\partial x} = -e^{-x} \cos y\,\mathbf{i} - e^{-x} \sin y\,\mathbf{j} + 3y\mathbf{k},$

    $\dfrac{\partial \mathbf{v}}{\partial y} = -e^{-x} \sin y\,\mathbf{i} + e^{-x} \cos y\,\mathbf{j} + 3x\mathbf{k}$
5. $\mathbf{r}(t) = (2 + \sqrt{13}\cos\theta,\ 3 + \sqrt{13}\sin\theta,\ 0)$
6. $\mathbf{r}(t) = (1 + 2t,\ 2 - t,\ 3 + t)$
7. $\mathbf{r}(t) = (1 + t,\ 2 + t,\ 3 + 4t)$
8. $\mathbf{r}(t) = (t,\ 2t + 1,\ 2t)$
9. $\mathbf{r}(t) = \left(\cos t,\ \sqrt{2}\sin t,\ 2\cos t\right)$
10. $\mathbf{r}(t) = \left(\cosh t,\ \sqrt{2}\sinh t,\ 2\right)$

11. $q(t) = \left(1 - \sqrt{3}t, \ \sqrt{3} + t\right)$ or $x + \sqrt{3}y = 4$

12. $q(t) = \left(\sqrt{2}\left(1 + t\right), \ \dfrac{\sqrt{2}}{2}\left(-1 + t\right)\right)$ or $\dfrac{x}{2\sqrt{2}} - \dfrac{y}{\sqrt{2}} = 1$

13. $q(t) = \left(2 + \sqrt{3}t, \ \sqrt{3} + 2t\right)$ or $2x - \sqrt{3}y = 1$

14. $q(t) = \left(2 - \sqrt{3}t, \ -2\sqrt{3} + 4t\right)$ or $2x + \dfrac{\sqrt{3}y}{2} = 1$

Problem 8.3

1. $(2x - y)\mathbf{i} + (-x + 2y)\mathbf{j}$

2. $2x\mathbf{i} + 4y\mathbf{j}$

3. $\dfrac{2xz}{y}\mathbf{i} - \dfrac{x^2 z}{y^2}\mathbf{j} + \dfrac{x^2}{y}\mathbf{k}$

4. $2(x + 1)\mathbf{i} + 2(y - 2)\mathbf{j} + 2z\mathbf{k}$

5. $\dfrac{4}{\sqrt{5}}$

6. $-\dfrac{6}{\sqrt{5}}$

7. $\dfrac{10}{\sqrt{14}}$

8. $\dfrac{10}{3}$

9. $\dfrac{1}{\sqrt{5}}$

10. $-\dfrac{4}{\sqrt{5}}$

11. $\mathbf{n} = \dfrac{1}{\sqrt{5}}(2\mathbf{i} - \mathbf{j})$

12. $\mathbf{n} = \dfrac{1}{\sqrt{5}}(2\mathbf{i} - \mathbf{j})$

13. $\mathbf{n} = \dfrac{1}{\sqrt{2}}(\mathbf{i} - \mathbf{j})$

14. $\mathbf{n} = \dfrac{1}{5}(3\mathbf{i} + 4\mathbf{j})$

15. $\mathbf{n} = \dfrac{1}{\sqrt{21}}(2\mathbf{i} + 4\mathbf{j} - \mathbf{k})$

16. $\mathbf{n} = \dfrac{1}{\sqrt{3}}(\mathbf{i} + \mathbf{j} - \mathbf{k})$

17. $\mathbf{n} = \dfrac{1}{3}(2\mathbf{i} - 2\mathbf{j} + \mathbf{k})$

18. $\mathbf{n} = \dfrac{1}{\sqrt{154}}(3\mathbf{i} + 12\mathbf{j} - \mathbf{k})$

Problem 8.4

1. -4

2. $\dfrac{1}{\sqrt{5}}$

3. 4

4. -6

5. $-2(y+z)\mathbf{i}$

6. $-\ln x\,\mathbf{i}+\ln y\,\mathbf{j}+\left(\dfrac{z}{x}-\dfrac{z}{y}\right)\mathbf{k}$

7. $(2e^x \sin y)\mathbf{k}$

8. $e^{-xy}\left\{(-x+\sin z)\mathbf{i}+(\cos z+y)\mathbf{j}+(-y\cos z+x\sin z)\mathbf{k}\right\}$

9 Vector integral and integral theorems

Vector integral refers to integrating a vector field, used to describe the properties of a vector field over paths, surfaces, volumes, etc., in space. It's a crucial concept in engineering and physics, particularly in applications such as solid mechanics, fluid dynamics, electromagnetism, and thermodynamics. For instance, in solid mechanics, it's used to integrate stress and strain vector fields to determine areas and volumes. In fluid dynamics, it's used to integrate velocity vector fields and pressure distributions to calculate fluid flow rates, forces, and moments. In electromagnetism, it's used to integrate electric and magnetic field vector fields to determine charges, currents, and magnetic moments.

9.1 Line integrals

9.1.1 Definition of line integrals

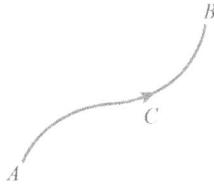

Figure 9.1 Path of integration.

As seen in Figure 9.1, it refers to integrating along a curve C in space coordinates. The curve C is called the *path of integration*. The path of integration goes from its initial point A to its terminal point B. When the points A and B coincide, the curve C becomes a *closed path*.

Remark Line integral

A *line integral* of a vector function $\mathbf{F}(\mathbf{r})$ over a curve $C: \mathbf{r}(t)$ is defined by

$$\int_C \mathbf{F}(\mathbf{r}) \cdot d\mathbf{r} = \int_a^b \mathbf{F}(\mathbf{r}(t)) \cdot \mathbf{r}'(t)\,dt, \qquad (9.1)$$

where $\mathbf{F}(\mathbf{r}) = \{F_x, F_y, F_z\} = F_x\mathbf{i} + F_y\mathbf{j} + F_z\mathbf{k}$ and $d\mathbf{r} = \{dx, dy, dz\} = dx\mathbf{i} + dy\mathbf{j} + dz\mathbf{k}$.

Writing Eq. (9.1) in terms of components $dx = x'dt$, $dy = y'dt$, and $dz = z'dt$, we get

$$\int_C \mathbf{F}(\mathbf{r}) \cdot d\mathbf{r} = \int_C \left(F_x dx + F_y dy + F_z dz \right) = \int_a^b \left(F_x x' + F_y y' + F_z z' \right) dt. \quad (9.2)$$

DOI: 10.1201/9781003608912-9

Remark   Work

If the vector function $\mathbf{F}(\mathbf{r})$ represents a physical quantity of a force vector, then the line integral with respect to it represents a new physical quantity known as *work*.

$$U_{1\to2} = \int_{\mathbf{r}_1}^{\mathbf{r}_2} \mathbf{F}(\mathbf{r})\cdot d\mathbf{r} \qquad (9.3)$$

Example 9.1

Find the value of the line integral $\int_C \mathbf{F}(\mathbf{r})\cdot d\mathbf{r}$, when the vector function $\mathbf{F}(\mathbf{r})=\left[y^2,\ xy\right]$ and the following path moves from $O(0, 0)$ to $A(1, 2)$, as shown in Figure 9.2:

a. path C_1 (line $y = 2x$)
b. path C_2 (curve $y = 2x^2$)

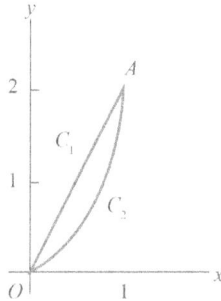

Figure 9.2 The path C_1 and the path C_2.

Solution

a. When the path C_1: $y = 2x$, then $\mathbf{r}(t)=[t,\ 2t]$ and $\mathbf{r}'(t)=[1,\ 2]$.
    Since $\mathbf{F}(\mathbf{r})=\left[y^2,\ xy\right]=\left[4t^2,\ 2t^2\right]$, $(0\le t\le1)$ then we get

$$\int_{C_1} \mathbf{F}(\mathbf{r})\cdot d\mathbf{r} = \int_0^1 \mathbf{F}(\mathbf{r}(t))\cdot \mathbf{r}'(t)\,dt$$

$$= \int_0^1\left[4t^2,\ 2t^2\right]\cdot[1,\ 2]\,dt = \int_0^1 8t^2\,dt = \frac{8}{3}.$$

Answer $\dfrac{8}{3}$

b. When the path C_2: $y = 2x^2$, then $r(t) = [t, 2t^2]$ and $r'(t) = [1, 4t]$.

Since $F(r) = [y^2, xy] = [4t^4, 2t^3]$, $(0 \le t \le 1)$ then we get

$$\int_{C_2} F(r) \cdot dr = \int_0^1 F(r(t)) \cdot r'(t) dt$$

$$= \int_0^1 [4t^4, 2t^3] \cdot [1, 4t] dt = \int_0^1 (4t^4 + 8t^4) dt = \frac{12}{5}.$$

Answer $\dfrac{12}{5}$

Example 9.2

Find the value of the line integral $\int_C F(r) \cdot dr$ when the vector function $F(r) = yz\mathbf{i} + zx\mathbf{j} + z\mathbf{k}$ and the path C is the helix, as shown in Figure 9.3.

$$r(t) = (\cos t, \sin t, 2t) = \cos t \, \mathbf{i} + \sin t \, \mathbf{j} + 2t \, \mathbf{k} \quad (0 < t < \pi).$$

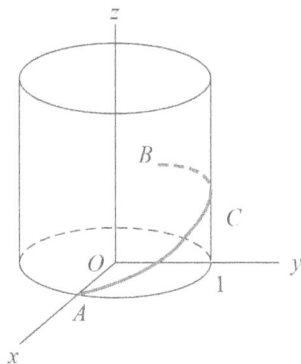

Figure 9.3 The path C.

Solution

From $x(t) = \cos t$, $y(t) = \sin t$, $z(t) = 2t$, we get

$$dx = -\sin t \, dt, \quad dy = \cos t \, dt, \quad dz = 2dt.$$

From the vector function $F(r) = yz\mathbf{i} + zx\mathbf{j} + z\mathbf{k}$, we get

$$F_x = yz = 2t \sin t, \quad F_y = zx = 2t \cos t, \quad F_z = z = 2t.$$

Therefore, we obtain the line integral

$$\int_C \mathbf{F}(\mathbf{r}) \cdot d\mathbf{r} = \int_C \left(F_x dx + F_y dy + F_z dz \right)$$

$$= \int_0^\pi \left\{ 2t \sin t \cdot (-\sin t) + 2t \cos t \cdot \cos t + 2t \cdot 2 \right\} dt$$

$$= \int_0^\pi \left(2t \cos 2t + 4t \right) dt = 2\pi^2. \quad \text{(refer to the Check)}$$

Answer $2\pi^2$

Check **Integration by parts**

To integrate $t \cos 2t$, let's apply the formula for integration by parts:

$$\int t \cos 2t \, dt = t \frac{\sin 2t}{2} - \int \frac{\sin 2t}{2} dt = t \frac{\sin 2t}{2} + \frac{\cos 2t}{4}.$$

Then we get

$$\int_0^\pi t \cos 2t \, dt = t \frac{\sin 2t}{2} + \frac{\cos 2t}{4} \bigg|_0^\pi = 0.$$

9.1.2 Path independence of line integrals

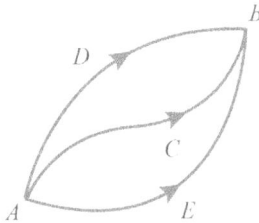

Figure 9.4 Path independence.

Figure 9.4 shows several paths from point A to point B.

In general, the line integral typically yields different values depending on the chosen path. However, if the line integral given by Eq. (9.2) yields the same value regardless of the path taken in space, it indicates that the value of the integral is independent of the specific path chosen. This property is termed *path independence*.

Remark   Path independence

The path independence of Eq. (9.2) holds if and only if:

  i. $F(r) = \text{grad } f$ (where grad f is the gradient of f)
  ii. Integral around closed curves C always gives 0.
  iii. curl $F = 0$

Proof of (i)

Since $F = \text{grad } f = \nabla f = \dfrac{\partial f}{\partial x}\mathbf{i} + \dfrac{\partial f}{\partial y}\mathbf{j} + \dfrac{\partial f}{\partial z}\mathbf{k}$, we get

$$F(r)\cdot dr = \nabla f \cdot dr$$

$$= \left(\frac{\partial f}{\partial x}\mathbf{i} + \frac{\partial f}{\partial y}\mathbf{j} + \frac{\partial f}{\partial z}\mathbf{k}\right)\cdot(dx\,\mathbf{i} + dy\,\mathbf{j} + dz\,\mathbf{k})$$

$$= F_x dx + F_y dy + F_z dz.$$

where $F_x = \dfrac{\partial f}{\partial x}$, $F_y = \dfrac{\partial f}{\partial y}$, and $F_z = \dfrac{\partial f}{\partial z}$.

In Figure 9.4, let C be any path from the initial position $A(x(t_1), y(t_1), z(t_1))$ at $t = t_1$ to the terminal position $B(x(t_2), y(t_2), z(t_2))$ at $t = t_2$.
Then we obtain

$$\int_C F(r)\cdot dr = \int_C (F_x dx + F_y dy + F_z dz)$$

$$= \int_C \left(\frac{\partial f}{\partial x}dx + \frac{\partial f}{\partial y}dy + \frac{\partial f}{\partial z}dz\right)$$

$$= \int_C \left(\frac{\partial f}{\partial x}\frac{dx}{dt} + \frac{\partial f}{\partial y}\frac{dy}{dt} + \frac{\partial f}{\partial z}\frac{dz}{dt}\right)dt$$

$$= \int_C \left(\frac{df}{dt}\right)dt = f(x(t), y(t), z(t))\Big|_{t=t_1}^{t=t_2}$$

$$= f(x(t_2), y(t_2), z(t_2)) - f(x(t_1), y(t_1), z(t_1))$$

$$= f(B) - f(A).$$

Then it implies path independence.

Proof of (ii)

In Figure 9.4, the line integrals along paths C and D yield the same value (a constant k) regardless of the path, i.e.,

$$\int_C \mathbf{F}(\mathbf{r}) \cdot d\mathbf{r} = \int_D \mathbf{F}(\mathbf{r}) \cdot d\mathbf{r} = k.$$

For the closed loop departing from point A to point B along path C, and returning from point B to point A along path D (in the opposite direction), its line integrals evaluate to 0.

$$\oint \mathbf{F}(\mathbf{r}) \cdot d\mathbf{r} = \int_C \mathbf{F}(\mathbf{r}) \cdot d\mathbf{r} + \left(-\int_D \mathbf{F}(\mathbf{r}) \cdot d\mathbf{r} \right) = k - k = 0$$

Of course, conversely, if the calculation of the integral along a closed loop, starting from point A, along path C, passing through point B, and returning via path D (in the reverse direction), yields a value of 0, then the integrals along paths C and D are equal. This implies path independence.

Proof of (iii)

If we substitute the vector function $\mathbf{F} = grad\ f = \nabla f = \dfrac{\partial f}{\partial x}\mathbf{i} + \dfrac{\partial f}{\partial y}\mathbf{j} + \dfrac{\partial f}{\partial z}\mathbf{k}$ derived in (i) into curl $\mathbf{F}$, we get

$$curl\ \mathbf{F} = \nabla \times \mathbf{F} = \nabla \times (\nabla f) = 0$$

or

$$curl\ \mathbf{F} = \nabla \times \mathbf{F} = \left(\frac{\partial}{\partial x}\mathbf{i} + \frac{\partial}{\partial y}\mathbf{j} + \frac{\partial}{\partial z}\mathbf{k} \right) \times \left(F_x\mathbf{i} + F_y\mathbf{j} + F_z\mathbf{k} \right)$$

$$= \begin{vmatrix} \mathbf{i} & \mathbf{j} & \mathbf{k} \\ \dfrac{\partial}{\partial x} & \dfrac{\partial}{\partial y} & \dfrac{\partial}{\partial z} \\ F_x & F_y & F_z \end{vmatrix} = \mathbf{i}\left(\frac{\partial F_z}{\partial y} - \frac{\partial F_y}{\partial z} \right) + \mathbf{j}\left(\frac{\partial F_x}{\partial z} - \frac{\partial F_z}{\partial x} \right) + \mathbf{k}\left(\frac{\partial F_y}{\partial x} - \frac{\partial F_x}{\partial y} \right) = 0.$$

Therefore, if the following conditions are satisfied, it implies path independence.

$$\frac{\partial F_x}{\partial y} = \frac{\partial F_y}{\partial x}, \frac{\partial F_y}{\partial z} = \frac{\partial F_z}{\partial y}, \frac{\partial F_z}{\partial x} = \frac{\partial F_x}{\partial z}$$

Remark   Path independence

If the line integral in the xy-plane satisfies the following condition, it implies path independence.

$$\frac{\partial F_x}{\partial y} = \frac{\partial F_y}{\partial x} \qquad\qquad (9.4a)$$

If the line integral in xyz-space satisfies the following condition, it implies path independence. In this case, it is said to be *exact* in space.

$$\frac{\partial F_x}{\partial y} = \frac{\partial F_y}{\partial x}, \; \frac{\partial F_y}{\partial z} = \frac{\partial F_z}{\partial y}, \; \frac{\partial F_z}{\partial x} = \frac{\partial F_x}{\partial z} \qquad\qquad (9.4b)$$

Example 9.3

Show that the following line integral is independent of the path and also find the value of the integral from point $A(0, 0)$ to point $B(0, \pi/2)$.

$$I = \int_C \left[\{x + \sin(x+y)\} dx + \{y^2 + \sin(x+y)\} dy \right]$$

Solution

From the given equation, we get

$$F_x = x + \sin(x+y) \qquad\qquad ①$$

$$F_y = y^2 + \sin(x+y) \qquad\qquad ②$$

and

$$\frac{\partial F_x}{\partial y} = \frac{\partial}{\partial y}\{x + \sin(x+y)\} = \cos(x+y)$$

$$\frac{\partial F_y}{\partial x} = \frac{\partial}{\partial x}\{y^2 + \sin(x+y)\} = \cos(x+y).$$

Then since we obtain

$$\frac{\partial F_x}{\partial y} = \frac{\partial F_y}{\partial x},$$

it is independent of the path.

If we integrate equations ① and ② separately, we obtain

$$f = \frac{x^2}{2} - \cos(x+y) + h(y),$$

$$f = \frac{y^3}{3} - \cos(x+y) + g(x).$$

By comparing the functions $h(y)$ and $g(x)$ in equations ① and ②, respectively, we can complete the function f as follows:

$$f = f(x, y) = \frac{x^2}{2} - \cos(x+y) + \frac{y^3}{3}.$$

Therefore, we obtain the value of the integral

$$f(B) - f(A) = \left[\frac{x^2}{2} - \cos(x+y) + \frac{y^3}{3} \right]_{(0,\,\pi/2)} - \left[\frac{x^2}{2} - \cos(x+y) + \frac{y^3}{3} \right]_{(0,\,0)}$$

$$= \frac{\pi^3}{24} + 1.$$

$$\textbf{Answer } \frac{\pi^3}{24} + 1$$

Example 9.4

Show that the following line integral is independent of the path and also find the value of the integral from point $A(0, 0, 0)$ to point $B(1, 0, \pi/2)$.

$$I = \int_C \left[\left\{ y^2 z^2 - y\sin(xy+z) \right\} dx + \left\{ 2xyz^2 - x\sin(xy+z) \right\} dy + \left\{ 2xy^2 z - \sin(xy+z) \right\} dz \right]$$

Solution

From the given equation, we get

$$F_x = y^2 z^2 - y\sin(xy+z) \qquad\qquad ①$$

$$F_y = 2xyz^2 - x\sin(xy+z) \qquad\qquad ②$$

$$F_z = 2xy^2 z - \sin(xy+z) \qquad\qquad ③$$

Then since we obtain

$$\frac{\partial F_x}{\partial y} = \frac{\partial F_y}{\partial x} = 2yz^2 - \sin(x+y) - xy\cos(xy+z),$$

$$\frac{\partial F_y}{\partial z} = \frac{\partial F_z}{\partial y} = 4xyz - x\cos(xy+z),$$

$$\frac{\partial F_z}{\partial x} = \frac{\partial F_x}{\partial z} = 2y^2z - y\cos(xy+z),$$

it is independent of the path.

If we integrate equations ①, ②, and ③ separately, we obtain

$$f = xy^2z^2 + \cos(xy+z) + g_1(y,z),$$

$$f = xy^2z^2 + \cos(xy+z) + g_2(x,z),$$

$$f = xy^2z^2 + \cos(xy+z) + g_3(x,y).$$

By comparing the functions $g_1(y,z)$, $g_2(x,z)$, and $g_3(x,y)$, respectively, we can complete the function f as follows:

$$f = f(x,\ y,\ z) = xy^2z^2 + \cos(xy+z).$$

Therefore, we obtain the value of the integral

$$f(B) - f(A) = \left[xy^2z^2 + \cos(xy+z)\right]_{(1,\ 0,\ \pi/2)} - \left[xy^2z^2 + \cos(xy+z)\right]_{(0,0,0)}$$

$$= -1.$$

Answer −1

Problem 9.1

Find the value of the line integral $\int_C \mathbf{F}(\mathbf{r}) \cdot d\mathbf{r}$ along the following path C from the point A to the point B. [1 ~ 2]

1. $\mathbf{F}(\mathbf{r}) = [2xy,\ xy^2]$, C_1 and C_2 are paths in Figure 9.5
   a. path C_1 (straight line),
   b. path C_2 ($y = x^2$ curve)

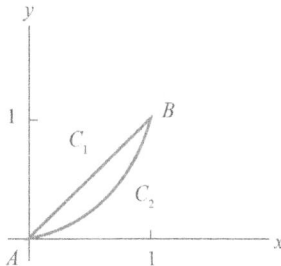

Figure 9.5 Paths.

2. $F(r) = [x, y^2]$, C_1 and C_2 are paths in Figure 9.6
   a. path C_1 (straight line)
   b. path C_2 ($x^2 + y^2 = 4$ circle)

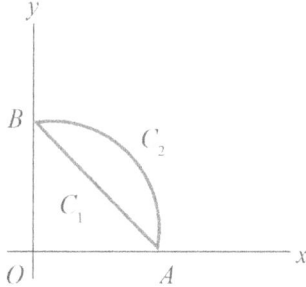

Figure 9.6 Paths.

Find the value of the line integral $\int_C F(r) \cdot dr$. [3 ~ 6]

3. $F(r) = [x^2, y^2, 2z]$, $r(t) = [2\cos t, \ 2\sin t, \ e^t]$ $\left(0 \le t \le \dfrac{\pi}{2}\right)$

4. $F(r) = [x+y, \ y+z, \ z+x]$, $r(t) = [2t, \ 3t, \ t]$ $(0 \le t \le 1)$

5. $F(r) = [x^2, \ y, \ 2z]$, $r(t) = [t, \ t^2, \ t^3]$ $(-1 \le t \le 1)$

6. $F(r) = [x^2, \ xy, \ y^2]$, $r(t) = [\cos t, \ \sin t, \ t]$ $(0 \le t \le \pi)$

Show that the following line integral is independent of the path and also find the value of the integral from point A to point B. [7 ~ 14]

7. $I = \displaystyle\int_C \left\{ \left(2e^{2x}\sin 2y\right)dx + \left(2e^{2x}\cos 2y\right)dy \right\}$, $A(0, \ 0)$, $B(1, \ \pi)$

8. $I = \displaystyle\int_C \left[\left\{(x+1)e^x + e^{2y}\right\}dx + 2xe^{2y}dy \right]$, $A(0, \ 0)$, $B(1, \ 1)$

9. $I = \displaystyle\int_C \left\{ \left(2xye^{x^2y}\right)dx + x^2e^{x^2y}dy \right\}$, $A(-1, \ -1)$, $B(1, \ 1)$

10. $I = \displaystyle\int_C \left[\left\{-2x\sin\left(x^2+y\right)\right\}dx - \sin\left(x^2+y\right)dy \right]$, $A(0, \ 0)$, $B(0, \ \pi/2)$

11. $I = \displaystyle\int_C \left[ydx + (x+z)dy + (y-2z)dz \right]$, $A(0, \ 0, \ 0)$, $B(1, \ 1, \ 0)$

12. $I = \displaystyle\int_C \left(yz\cos xy \ dx + zx\cos xy \ dy + \sin xy \ dz \right)$, $A(0, \ 0, \ 0)$, $B(1, \ \pi/2, \ 1)$

13. $I = \displaystyle\int_C e^{2x}\left(2\sin yz \ dx + z\cos yz \ dy + y\cos yz \ dz \right)$, $A(0, \ 0, \ 0)$, $B(1, \ \pi/2, \ 1)$

14. $I = \displaystyle\int_C \left(yz\cosh zx \ dx + \sinh zx \ dy + xy\cosh zx \ dz \right)$, $A(0, \ 0, \ 0)$, $B(1, \ 1, \ 1)$

9.2 Double integrals

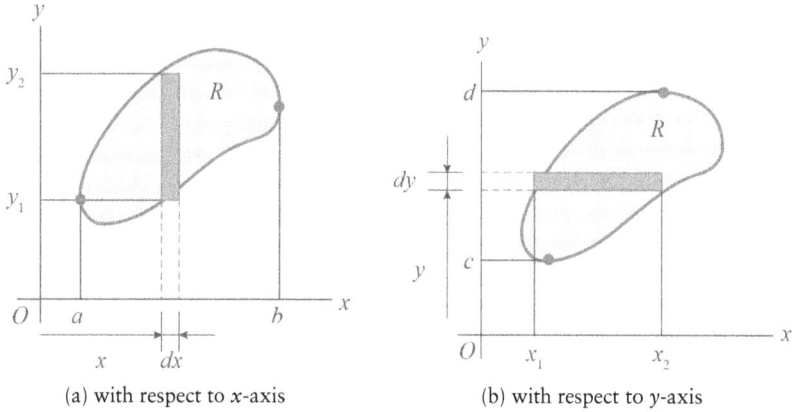

(a) with respect to x-axis   (b) with respect to y-axis

Figure 9.7 Region R.

Double integrals have various physical and geometric applications.

The area A of region R on the xy-plane in Figure 9.7(a) is calculated as

$$A = \int_a^b dA = \int_a^b (y_2 - y_1)dx. \qquad (a \le x \le b) \qquad (9.5a)$$

And the area A of region R on the xy-plane in Figure 9.7(b) is calculated as

$$A = \int_c^d dA = \int_c^d (x_2 - x_1)dy \qquad (c \le y \le d) \qquad (9.5b)$$

When the region R on the xy-plane in Figure 9.8 is taken as the base area and the volume of the solid with height $z = f(x, y)$ is calculated, it becomes as follows:

$$V = \iint_R f(x, y)dA. \qquad (9.6)$$

As seen in Figure 9.8, let's calculate Eq. (9.6) separately for the regions with respect to the x-axis and the y-axis.

i. with respect to the x-axis

When applying Eq. (9.5a), the calculation of the double integral over the region $R: y_1(x) \le y \le y_2(x)$, $a \le x \le b$ yields the following result.

$$\iint_R f(x, y)dA = \int_a^b \left\{ \int_{y_1(x)}^{y_2(x)} f(x, y)dy \right\} dx \qquad (9.7a)$$

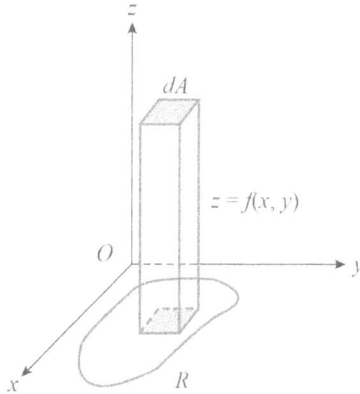

Figure 9.8 Volume $dV = f(x, y)dA$.

ii. with respect to the y-axis

Similarly, when applying Eq. (9.5b), the calculation of the double integral over the region $R: x_1(y) \leq x \leq x_2(y), c \leq y \leq d$ yields the following result.

$$\iint_R f(x, y)dA = \int_c^d \left\{ \int_{x_1(y)}^{x_2(y)} f(x, y)dx \right\} dy \qquad (9.7b)$$

Remark   Double integrals

Using double integrals, we can find the area S and the volume V of the shape.

i. when the height $f(x, y) = 1$, $S = \displaystyle\iint_R dA$ (9.8a)

ii. when the height $z = f(x, y)(> 0)$, $V = \displaystyle\iint_R f(x, y)dA$ (9.8b)

Example 9.5

Evaluate the following double integral over the region R enclosed by $y = 0$, $x = 1$ and $y = x^2$, as shown in Figure 9.9.

$$\iint_R x^2 y^3 \, dA$$

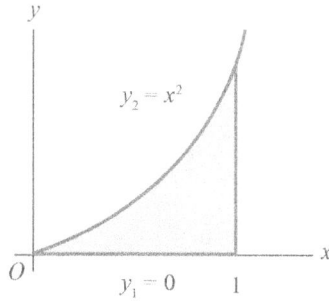

Figure 9.9 Region R.

Solution

When representing the region R with respect to the x-axis, we get

$$y_1 = 0 \quad \text{and} \quad y_2 = x^2.$$

Now we calculate the double integral over the region R.

$$\iint_R x^2 y^3 \, dA = \int_0^1 \left(\int_{y_1}^{y_2} x^2 y^3 \, dy \right) dx$$

$$= \int_0^1 x^2 \left(\int_0^{x^2} y^3 \, dy \right) dx$$

$$= \int_0^1 x^2 \left[\frac{y^4}{4} \right]_0^{x^2} dx$$

$$= \int_0^1 \frac{1}{4} x^{10} \, dx = \frac{1}{44}$$

Answer $\dfrac{1}{44}$

Another Solution

When representing the region R with respect to the y-axis, we get

$$x_1 = \sqrt{y} \quad \text{and} \quad x_2 = 1.$$

Now we calculate the double integral over the region R.

$$\iint_R x^2 y^3 \, dA = \int_0^1 y^3 \left(\int_{\sqrt{y}}^1 x^2 \, dx \right) dy$$

$$= \int_0^1 y^3 \left[\frac{x^3}{3} \right]_{\sqrt{y}}^1 dy$$

$$= \int_0^1 \frac{1}{3}\left(y^3 - y^{9/2}\right) dy = \frac{1}{44}$$

9.2.1 Variation of parameters

We often encounter practical problems that necessitate a change of the variables of integration in double integrals.

In a double integral, as shown in the following equation, it is possible to perform a change of variables to transform the integration variables x and y to u and v.

$$\iint_R f(x, y)\,dx\,dy = \iint_{R^*} f\{x(u, v), y(u, v)\} \, J \, du\,dv \qquad (9.9)$$

where the region R represents the area defined by variables x and y, while the region R^* represents the area defined by variables u and v. Additionally, the Jacobian is calculated as follows:

$$J = \frac{\partial(x, y)}{\partial(u, v)} = \begin{vmatrix} \dfrac{\partial x}{\partial u} & \dfrac{\partial x}{\partial v} \\ \dfrac{\partial y}{\partial u} & \dfrac{\partial y}{\partial v} \end{vmatrix} = \frac{\partial x}{\partial u}\frac{\partial y}{\partial v} - \frac{\partial y}{\partial u}\frac{\partial x}{\partial v}. \qquad (9.10)$$

If we set $x = x(r, \theta) = r\cos\theta$ and $y = y(r, \theta) = r\sin\theta$, then the Jacobian becomes

$$J = \frac{\partial(x, y)}{\partial(r, \theta)} = \frac{\partial x}{\partial r}\frac{\partial y}{\partial \theta} - \frac{\partial y}{\partial r}\frac{\partial x}{\partial \theta} = \cos\theta \cdot r\cos\theta - \sin\theta \cdot (-r\sin\theta) = r. \qquad (9.11)$$

Therefore, when transforming the double integral to polar coordinates, it becomes as follows:

$$\iint_R f(x, y)\,dx\,dy = \iint_{R^*} f\left(r\cos\theta, r\sin\theta\right) r \, dr \, d\theta. \qquad (9.12)$$

Example 9.6

Using polar coordinates, evaluate the following double integral in the region R, as shown in Figure 9.10.

$$\int_0^{\sqrt{3}} \int_{x/\sqrt{3}}^{\sqrt{4-x^2}} \frac{1}{3+x^2+y^2}\, dy\, dx$$

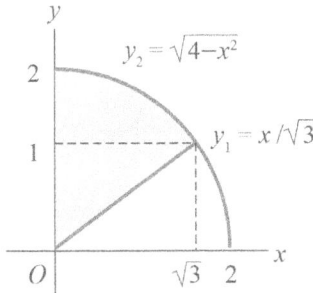

Figure 9.10 Region R.

Solution

When we set $x = x(r, \theta) = r\cos\theta$, $y = y(r, \theta) = r\sin\theta$, we get

$$\int_0^{\sqrt{3}} \int_{x/\sqrt{3}}^{\sqrt{4-x^2}} \frac{1}{3+x^2+y^2}\, dy\, dx = \int_{\pi/6}^{\pi/2} \int_0^2 \frac{1}{3+r^2}\, r\, dr\, d\theta$$

$$= \int_{\pi/6}^{\pi/2} \left[\frac{1}{2}\ln(3+r^2) \right]_0^2 d\theta$$

$$= \int_{\pi/6}^{\pi/2} \frac{1}{2}\ln\frac{7}{3}\, d\theta = \frac{\pi}{6}\ln\frac{7}{3}.$$

Answer $\dfrac{\pi}{6}\ln\dfrac{7}{3}$

Problem 9.2

Evaluate the following double integral in the region R. [1 ~ 8]

1. $\displaystyle\iint_R (y + 4y^3)\, dA$, $y = \sqrt{x}$, $y = 0$, $x = 2$ (refer to Figure 9.11)

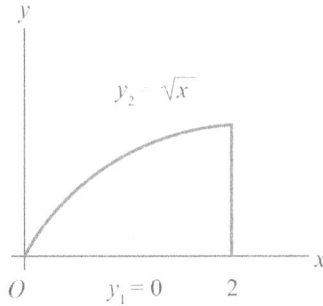

Figure 9.11 Region R.

2. $\iint_R (x^2 + 3y^2)\,dA,$ $y = x,\ y = 0,\ x = 1$ (refer to Figure 9.12)

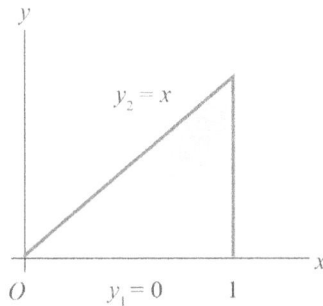

Figure 9.12 Region R.

3. $\iint_R xy\,dA,$ $y = x,\ y = \sqrt{x+2},\ x = 0$ (refer to Figure 9.13)

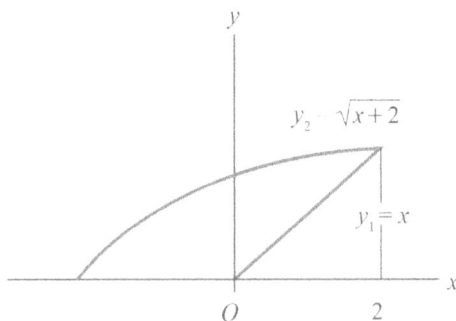

Figure 9.13 Region R.

4. $\iint_R 2(x+2)y\,dA,$ $y = -x,\ y = x^2,\ x = 1$ (refer to Figure 9.14)

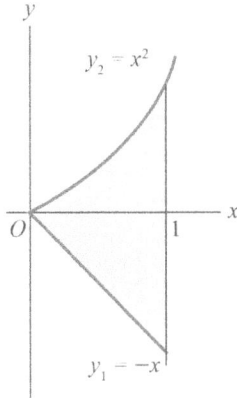

Figure 9.14 Region R.

5. $\iint_R (x+2)dA$, $y = x$, $x + y = 4$, $y = 1$, $y = 0$ (refer to Figure 9.15)

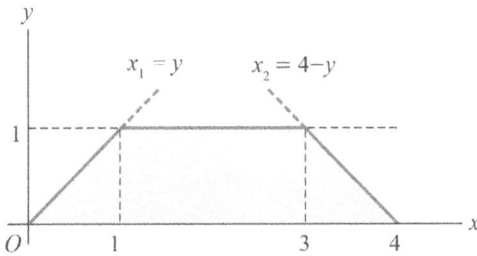

Figure 9.15 Region R.

6. $\iint_R e^{x+2y} dA$, $y = 0, y = 1, y = x, y = -\dfrac{x}{2} + 2$ (refer to Figure 9.16)

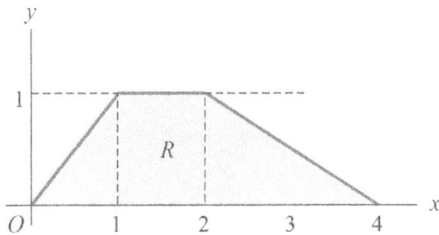

Figure 9.16 Region R.

7. $\iint_R xy\, dA$, $y = x^2$, $y = x + 2$, $y = 0$ (refer to Figure 9.17)

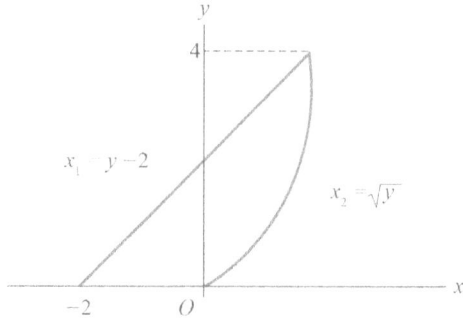

Figure 9.17 Region R.

8. $\iint_R (2x + y) dA,$ $y = \sqrt{x},\ y = x - 2,\ y = 0$ (refer to Figure 9.18)

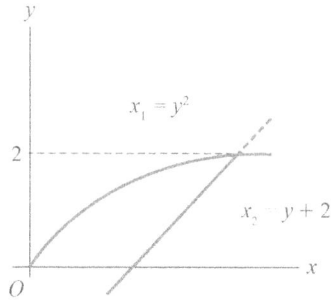

Figure 9.18 Region R.

Using polar coordinates, evaluate the following double integral in the region R. [9 ~ 12]

9. $\int_{-2}^{2} \int_{0}^{\sqrt{4-x^2}} (x^2 + y^2) dy\,dx$ (refer to Figure 9.19)

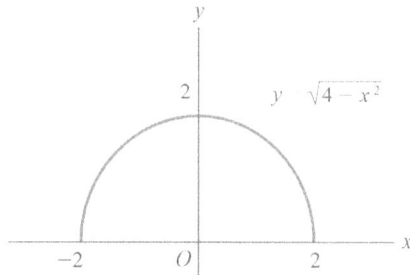

Figure 9.19 Region R.

10. $\int_0^1 \int_0^{\sqrt{1-x^2}} e^{x^2+y^2} \, dy \, dx$ (refer to Figure 9.20)

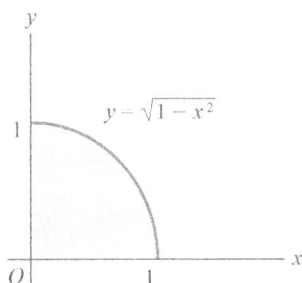

Figure 9.20 Region R.

11. $\int_0^{2\sqrt{2}} \int_0^{\sqrt{8-y^2}} \frac{y}{x^2+y^2} \, dx \, dy$ (refer to Figure 9.21)

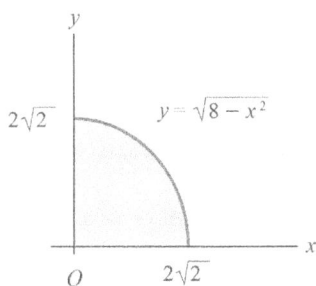

Figure 9.21 Region R.

12. $\int_0^1 \int_{\sqrt{1-x^2}}^{\sqrt{2-x^2}} \frac{y^2}{x^2+y^2} \, dy \, dx + \int_1^{\sqrt{2}} \int_0^{\sqrt{2-x^2}} \frac{y^2}{x^2+y^2} \, dy \, dx$ (refer to Figure 9.22)

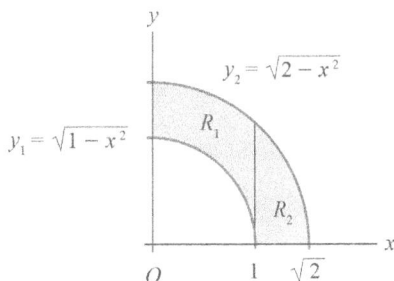

Figure 9.22 Region R.

9.3   Green's theorem in the plane

Using *Green's theorem* in the two-dimensional plane, double integrals over a plane region may be converted into line integrals over the boundary of the region and conversely.

Remark   Green's theorem in the plane

Let R be a closed bounded region in the xy-plane whose boundary C consists of finitely many smooth curves. And let $P(x, y)$ and $Q(x, y)$ be functions that are continuous and have continuous derivatives $\dfrac{\partial P}{\partial y}$ and $\dfrac{\partial Q}{\partial x}$ everywhere in some domain containing the region R. Then

$$\iint_R \left(\frac{\partial Q}{\partial x} - \frac{\partial P}{\partial y} \right) dA = \oint_C (P\,dx + Q\,dy). \qquad (9.13)$$

Here, the region R is located to the left of the direction in which the boundary C progresses. Therefore, if the region R is closed, the boundary C always forms a counterclockwise direction.

Let's say that in Figure 9.23, the region R exists as follows:

$$R: a \le x \le b, \quad y_1(x) \le y \le y_2(x)$$

$$R: x_1(y) \le x \le x_2(y), \quad c \le y \le d.$$

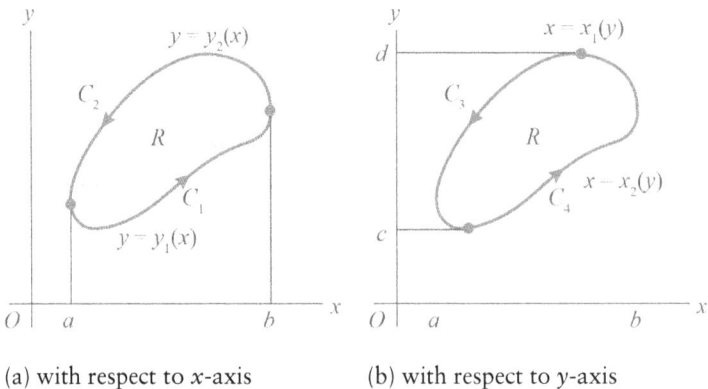

(a) with respect to x-axis   (b) with respect to y-axis

Figure 9.23 Region R whose boundary C.

In Figure 9.23(a), the boundary curve C_1 represents the curve moving along $y = y_1(x)$ over the interval $a \le x \le b$, while the boundary curve C_2 represents the curve moving along $y = y_2(x)$ over the interval $a \le x \le b$. Thus, by combining boundary curves C_1 and C_2, we form the closed curve C, indicating that the region R lies to the left of the direction in which each boundary curve C_1, C_2 progresses.

Using Eq. (9.7a), we obtain for the second term on the Eq. (9.13)

$$-\iint_R \frac{\partial P}{\partial y}\,dA = -\int_a^b \int_{y_1(x)}^{y_2(x)} \frac{\partial P}{\partial y}\,dy\,dx$$

$$= -\int_a^b \left[P(x, y_2(x)) - P(x, y_1(x)) \right] dx$$

$$= \int_a^b P(x, y_1(x))\,dx + \int_b^a P(x, y_2(x))\,dx$$

$$= \oint_C P(x, y)\,dx. \tag{9.14a}$$

This proves Eq. (9.13) in Green's theorem if $Q = 0$.

Similarly, in Figure 9.23(b), the boundary curve C_3 represents the curve moving along $x = x_1(y)$ over the interval $c \le y \le d$, while the boundary curve C_4 represents the curve moving along $x = x_2(y)$ over the interval $c \le y \le d$. Thus, by combining boundary curves C_3 and C_4, we form the closed curve C, indicating that the region R lies to the left of the direction in which each boundary curve C_3, C_4 progresses.

Using Eq. (9.7b), we obtain for the first term on the Eq. (9.13)

$$\iint_R \frac{\partial Q}{\partial x}\,dA = \int_c^d \int_{x_1(y)}^{x_2(y)} \frac{\partial Q}{\partial x}\,dx\,dy$$

$$= \int_c^d \left[Q(x_2(y), y) - Q(x_1(y), y) \right] dy$$

$$= \int_c^d Q(x_2(y), y)\,dy + \int_d^c Q(x_1(y), y)\,dy$$

$$= \oint_C Q(x, y)\,dy. \tag{9.14b}$$

Combining Eq. (9.14a) and Eq. (9.14b) yields Green's theorem (9.13). However, this does not hold for complex-shaped regions.

Remark   Green's theorem in the plane

Let R be a closed bounded region in the xy-plane whose boundary C consists of finitely many smooth curves. And let $P(x, y)$ and $Q(x, y)$ be functions that are continuous and have continuous derivatives $\dfrac{\partial P}{\partial y}$ and $\dfrac{\partial Q}{\partial x}$ everywhere in some domain containing the region R.

When we denote $\mathbf{F} = P(x, y)\mathbf{i} + Q(x, y)\mathbf{j}$ and $\mathbf{r} = x\mathbf{i} + y\mathbf{j}$, then Green's theorem can be expressed by the following equation:

$$\iint_R (\operatorname{curl}\mathbf{F}) \cdot \mathbf{k} \, dx\,dy = \oint_C \mathbf{F} \cdot d\mathbf{r} \qquad (9.15)$$

where $\operatorname{curl}\mathbf{F} = \nabla \times \mathbf{F} = \left(\dfrac{\partial}{\partial x}\mathbf{i} + \dfrac{\partial}{\partial y}\mathbf{j}\right) \times (P\mathbf{i} + Q\mathbf{j}) = \left(\dfrac{\partial Q}{\partial x} - \dfrac{\partial P}{\partial y}\right)\mathbf{k}$ and the region R lies to the left of the direction in the boundary C.

In other words, the line integral of the tangential component of $\mathbf{F}$ is equal to the double integral of the normal component of curl $\mathbf{F}$.

Example 9.7

Given that C encloses the region R as a closed boundary as shown in Figure 9.24, calculate the following:

$$\oint_C \left\{\left(x^3 - y^3\right)dx + \left(4y - x\right)dy\right\}.$$

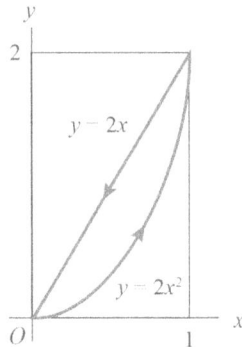

Figure 9.24 Region R.

Solution

Since $P = P(x, y) = x^3 - y^3$, $Q = Q(x, y) = 4y - x$, we get

$$\frac{\partial P}{\partial y} = -3y^2 \text{ and } \frac{\partial Q}{\partial x} = -1.$$

Therefore,

$$\oint_C (P\,dx + Q\,dy) = \iint_R \left(\frac{\partial Q}{\partial x} - \frac{\partial P}{\partial y}\right) dA$$

$$= \iint_R \left(-1 + 3y^2\right) dA$$

$$= \int_0^1 \int_{2x^2}^{2x} \left(-1 + 3y^2\right) dy\, dx$$

$$= \int_0^1 \left[-y + y^3\right]_{2x^2}^{2x} dx$$

$$= \int_0^1 \left\{\left(-2x + 8x^3\right) - \left(-2x^2 + 8x^6\right)\right\} dx = \frac{11}{21}.$$

Answer $\dfrac{11}{21}$

Example 9.8

When the boundary C forms the circle $(x-1)^2 + (y-2)^2 = 4$, calculate the following:

$$\oint_C \left\{\left(x^2 + 2y\right)dx + \left(4x - y^2\right)dy\right\}.$$

Solution

Since $P = P(x, y) = x^2 + 2y$ and $Q = Q(x, y) = 4x - y^2$, we get

$$\frac{\partial P}{\partial y} = 2 \text{ and } \frac{\partial Q}{\partial x} = 4.$$

Therefore, we obtain

$$\oint_C (P\,dx + Q\,dy) = \iint_R \left(\frac{\partial Q}{\partial x} - \frac{\partial P}{\partial y}\right) dA$$

$$= \iint_R (4 - 2)\,dA$$

$$= 2\iint_R dA$$

$$= 8\pi. \quad \text{(because the circle's area is } 4\pi)$$

Answer 8π

Example 9.9

When the region R is a rectangle with vertices at $(0, 0)$, $(2, 0)$, $(2, 1)$, and $(0, 1)$, evaluate $\oint_C \mathbf{F} \cdot d\mathbf{r}$ counterclockwise along the boundary C of region R.

$$\mathbf{F} = \left(e^{2x} - y\right)\mathbf{i} + \left(2x - e^y\right)\mathbf{j}$$

Solution

Since $P = e^x - y$ and $Q = 2x - e^y$, we get

$$\frac{\partial P}{\partial y} = -1 \quad \text{and} \quad \frac{\partial Q}{\partial x} = 2.$$

Therefore, we obtain

$$\oint_C (P\,dx + Q\,dy) = \iint_R \left(\frac{\partial Q}{\partial x} - \frac{\partial P}{\partial y}\right) dA$$

$$= \iint_R \{2 - (-1)\} dA$$

$$= 3\iint_R dA = 6. \qquad \left(\text{because the rectangular area is 2}\right)$$

Answer 6

Problem 9.3

When the boundary curve C encloses the region R counterclockwise, calculate the following. [1 ~ 6]

1. $\oint_C \{(x - y)dx + x^2 y\,dy\}$

   C: $y = x$, $y = 0$, $x = 1$ (refer to Figure 9.25)

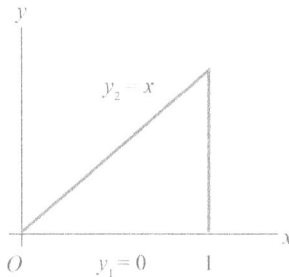

Figure 9.25 Region R.

2. $\oint_C \left(x^2 dx + 2xy\, dy \right)$

C: $y = x$, $y = \sqrt{x+2}$, $x = 0$ (refer to Figure 9.26)

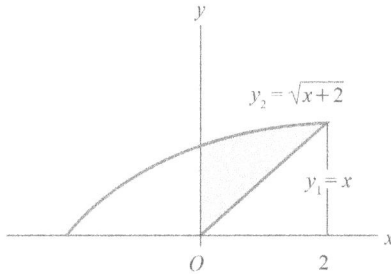

Figure 9.26 Region R.

3. $\oint_C \left(y^2 dx + x^2 dy \right)$,

C: $y = x^2$, $y = x+2$, $y = 0$ (refer to Figure 9.27)

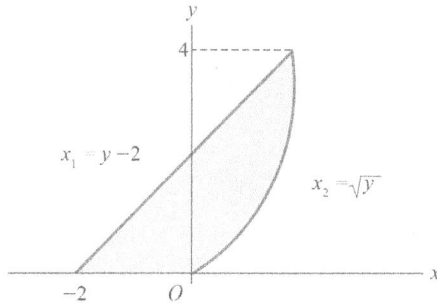

Figure 9.27 Region R.

4. $\oint_C \left\{ xy^2 dx + \left(x^2 + y \right) dy \right\}$

C: $y = \sqrt{x}$, $y = x-2$, $y = 0$ (refer to Figure 9.28)

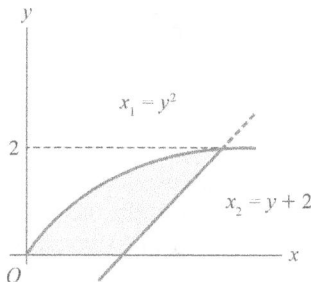

Figure 9.28 Region R.

5. $\oint_C \left\{ \left(2x^2 - y \right) dx + \left(3x + y^2 \right) dy \right\}$

C: circle $(x+1)^2 + (y+1)^2 = 1$

6. $\oint_C \left(-y\, dx + x\, dy \right)$

C: ellipse $\dfrac{(x-2)^2}{4} + \dfrac{(y+1)^2}{9} = 1$

When the boundary curve C encloses the region R counterclockwise, calculate the $\oint_C \mathbf{F} \cdot d\mathbf{r}$, using Green's theorem in the plane. [7 ~ 10]

7. $\mathbf{F} = xe^y \mathbf{i} + ye^x \mathbf{j}$,
   R: rectangle with vertices at $(0, 0)$, $(1, 0)$, $(1, 2)$, and $(0, 2)$

8. $\mathbf{F} = \left(2x^2 + 3y \right) \mathbf{i} + \left(4x - 2y^2 \right) \mathbf{j}$,
   R: rectangle with vertices at $(1, 1)$, $(3, 1)$, $(3, 3)$, and $(1, 3)$

9. $\mathbf{F} = \left(x^2 + y^2 \right) \mathbf{i} + \left(x^2 - y^2 \right) \mathbf{j}$,
   $R: 1 \le x^2 + y^2 \le 4, \ x \ge 0, \ y \ge x$

10. $\mathbf{F} = \left(2x - y \right) \mathbf{i} + \left(3x + y \right) \mathbf{j}$,
    $R: 0 \le x^2 + y^2 \le 3, \ y \ge -x, \ y \ge x$

9.4   Surface integrals

Line integrals involve integrating over curves in space, whereas *surface integrals* involve integrating over surfaces in space. In this section, we will familiarize ourselves with the parametric representation of surfaces and learn about surface integrals using this representation. Additionally, we will explore how to use surface integrals to calculate the area of surfaces in three-dimensional space.

9.4.1   Surface in surface integrals

A surface S in space can be represented by a two-variable function as follows.

$$z = f(x, y) \quad \text{or} \quad g(x, y, z) = 0$$

For surface integrals, as with curves in the plane, it is more practical to use a parametric representation for the surface S. Since a surface is two-dimensional, it can be represented using two parameters, u and v, as follows:

$$\mathbf{r}(u, v) = \left[x(u, v),\, y(u, v),\, z(u, v) \right] = x(u, v)\mathbf{i} + y(u, v)\mathbf{j} + z(u, v)\mathbf{k}. \quad (9.16)$$

Example 9.10

Consider a cylinder with a base radius a and height h, centered around the z-axis, as shown in Figure 9.29. Represent the vector **r** describing the side of this cylinder using angular parameter u and height parameter v.

$$x^2 + y^2 = a^2, \qquad 0 \le z \le h$$

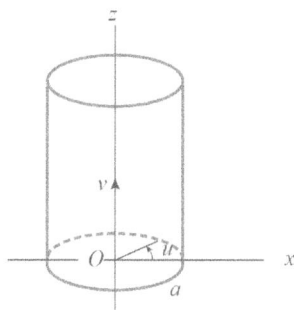

Figure 9.29 Cylinder.

Solution

First, let's express the equation of the circle $x^2 + y^2 = a^2$ using the parameter u. Then

$$x = a\cos u, \quad y = a\sin u. \qquad (0 \le u \le 2\pi)$$

Furthermore, let's substitute the height h with the parameter v.

$$z = v. \qquad (0 \le v \le h)$$

Therefore, the surface vector **r** of the cylinder can be expressed as follows:

$$\mathbf{r}(u, v) = [a\cos u, \ a\sin u, \ v] = a\cos u\,\mathbf{i} + a\sin u\,\mathbf{j} + v\,\mathbf{k},$$

where the parameters u and v vary within a rectangle R: $0 \le u \le 2\pi,\ 0 \le v \le h$ in the uv-plane.

Answer $\mathbf{r}(u, v) = [a\cos u, \ a\sin u, \ v]$ $(0 \le u \le 2\pi,\ 0 \le v \le h)$

Example 9.11

Express the equation of a sphere with radius a as $x^2 + y^2 + z^2 = a^2$ using two angular parameters u and v, as shown in Figure 9.30.

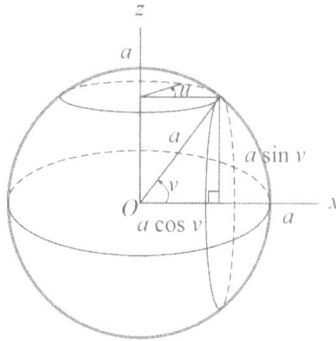

Figure 9.30 Sphere.

Solution

When expressing the equation of the sphere parametrically, we get

$$\mathbf{r}(u, v) = a\cos v(\cos u\,\mathbf{i} + \sin u\,\mathbf{j}) + a\sin v\,\mathbf{k}$$

$$= a\cos v\cos u\,\mathbf{i} + a\cos v\sin u\,\mathbf{j} + a\sin v\,\mathbf{k}$$

where the parameters u and v vary within a rectangle R: $0 \le u \le 2\pi$, $-\pi/2 \le v \le \pi/2$ in the uv-plane.

$$\textbf{Answer } \mathbf{r}(u, v) = a\cos v\cos u\,\mathbf{i} + a\cos v\sin u\,\mathbf{j} + a\sin v\,\mathbf{k}$$
$$(0 \le u \le 2\pi,\ -\pi/2 \le v \le \pi/2)$$

9.4.2 Surface integrals

If we define a surface S as in Eq. (9.16) to define surface integrals, the surface S has a unit normal vector at every point (excluding vertices of polyhedra or apexes of cones, etc.) given by the following. If the equation of the surface S is given by $g(x, y, z) = 0$, then the unit normal vector $\mathbf{n}$ is as follows:

$$n = \frac{\nabla g}{|\nabla g|} \qquad (9.17)$$

where ∇g is represented as grad g, meaning $\nabla g = \dfrac{\partial g}{\partial x}\mathbf{i} + \dfrac{\partial g}{\partial y}\mathbf{j} + \dfrac{\partial g}{\partial z}\mathbf{k}$.

Furthermore, if the surface S is represented by $\mathbf{r}(u, v) = x(u, v)\mathbf{i} + y(u, v)\mathbf{j} + z(u, v)\mathbf{k}$, then the unit normal vector $\mathbf{n}$ of the surface S at an arbitrary point P on the surface is given as:

$$\mathbf{n} = \frac{\mathbf{N}}{|\mathbf{N}|} \qquad (9.18)$$

where the normal vector is $\mathbf{N} = \mathbf{r}_u \times \mathbf{r}_v$, and $\mathbf{r}_u$, $\mathbf{r}_v$ denotes the partial derivatives with respect to each subscript u and v. And $|\mathbf{N}|$ is equal to the area of the parallelogram formed by the two vectors $\mathbf{r}_u$ and $\mathbf{r}_v$.

In this case, the surface integral of a vector function $\mathbf{F}$ over a surface with a unit normal vector $\mathbf{n}$ can be defined as:

$$\iint_S \mathbf{F} \cdot \mathbf{n}\, dA = \iint_R \mathbf{F}(\mathbf{r}(u, v)) \cdot \mathbf{N}(u, v)\, du\, dv \tag{9.19}$$

where $\mathbf{n}\, dA = \mathbf{n}|\mathbf{N}|\, du\, dv = \mathbf{N}\, du\, dv$. This integral is commonly referred to as the flux of $\mathbf{F}$ across the surface S.

On the other hand, another form of surface integral that does not consider direction is as:

$$\iint_S G(\mathbf{r})\, dA = \iint_R G(\mathbf{r}(u, v))|\mathbf{N}(u, v)|\, du\, dv \tag{9.20}$$

As a prominent application of this, if $G(\mathbf{r})$ represents the mass density (mass per unit area) of the surface S, then Eq. (9.20) becomes the total mass of the surface S.

If $G(\mathbf{r}) = 1$, it reduces to

$$A(S) = \iint_S dA = \iint_R |\mathbf{r}_u \times \mathbf{r}_v|\, du\, dv, \tag{9.21}$$

which is the area of the surface S.

Example 9.12

Find the flux of the vector field $\mathbf{F}(x, y, z) = [\, 0, x, 0\,]$ crossing the first octant $(x \geq 0,\ y \geq 0,\ z \geq 0)$ of a sphere $x^2 + y^2 + z^2 = 1$ with radius 1.

Solution

When expressing the equation of the sphere parametrically, we get

$$\mathbf{r}(u, v) = [\cos u \cos v, \sin u \cos v, \sin v]. \quad \left(0 \leq u \leq \frac{\pi}{2},\ 0 \leq v \leq \frac{\pi}{2} \right)$$

We get the partial derivatives with respect to each subscript u and v.

$$\mathbf{r}_u = [-\sin u \cos v,\ \cos u \cos v,\ 0],$$

$$\mathbf{r}_v = [-\cos u \sin v,\ -\sin u \sin v,\ \cos v].$$

Then the normal vector is

$$\mathbf{N} = \mathbf{r}_u \times \mathbf{r}_v = \left[\cos u \cos^2 v, \ \sin u \cos^2 v, \ \sin v \cos v\right].$$

And from $\mathbf{F}(x, y, z) = [0, x, 0]$, we get

$$\mathbf{F} = \mathbf{F}(\mathbf{r}(u, v)) = [0, \ \cos u \cos v, \ 0].$$

Then

$$\mathbf{F} \cdot \mathbf{N} = \left[\cos u \cos^2 v, \ \sin u \cos^2 v, \ \sin v \cos v\right] \cdot [0, \ \cos u \cos v, \ 0]$$

$$= \sin u \cos u \cos^3 v.$$

Therefore, we obtain the area of the surface S.

$$\iint_S \mathbf{F} \cdot \mathbf{n} \, dA = \iint_R \mathbf{F} \cdot \mathbf{N} \, du \, dv$$

$$= \int_0^{\pi/2} \int_0^{\pi/2} \left(\sin u \cos u \cos^3 v\right) du \, dv$$

$$= \int_0^{\pi/2} \sin u \cos u \, du \int_0^{\pi/2} \cos^3 v \, dv$$

$$= \int_0^{\pi/2} \frac{1}{2}\sin 2u \, du \int_0^{\pi/2} \cos^3 v \, dv \quad (\text{refer to the Check})$$

$$= \frac{1}{2} \cdot \frac{2}{3} = \frac{1}{3}.$$

$$\text{Answer } \iint_S \mathbf{F} \cdot \mathbf{n} \, dA = \frac{1}{3}$$

Check

$$\int_0^{\pi/2} \cos^n v \, dv = \frac{n-1}{n} \cdot \frac{n-3}{n-2} \cdots \cdot \frac{2}{3} \ (n \text{ odd number})$$

$$\int_0^{\pi/2} \sin^n v \, dv = \frac{n-1}{n} \cdot \frac{n-3}{n-2} \cdots \cdot \frac{2}{3} \ (n \text{ odd number})$$

Example 9.13

Find the lateral surface area of a cylinder with a radius of r and a height of h, as shown in Figure 9.31.

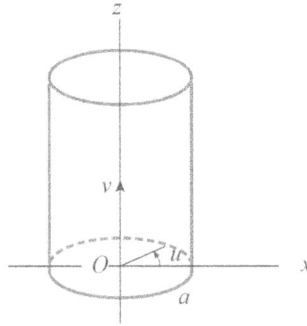

Figure 9.31 Cylinder.

Solution

When expressing the equation of the cylinder parametrically, we get

$$\mathbf{r}(u,v) = [a\cos u, \ a\sin u, \ v], \quad (0 \le u \le 2\pi, \ 0 \le v \le h)$$

We get the partial derivatives with respect to each subscript u and v.

$$\mathbf{r}_u = [-a\sin u, \ a\cos u, \ 0],$$

$$\mathbf{r}_v = [0, \ 0, \ 1].$$

Then

$$\mathbf{N} = \mathbf{r}_u \times \mathbf{r}_v = [a\cos u, \ a\sin u, \ 0].$$

Therefore, we obtain the area of the vector F as:

$$A(S) = \iint_S dA = \iint_R |\mathbf{r}_u \times \mathbf{r}_v| \, du \, dv$$

$$= \int_0^h \int_0^{2\pi} \sqrt{a^2\cos^2 u + a^2\sin^2 u} \ du \, dv$$

$$= \int_0^h \int_0^{2\pi} a \ du \, dv = 2\pi a h.$$

Answer $A(S) = 2\pi a h$

Example 9.14

Find the surface area of a sphere with a radius of a, as shown in Figure 9.32.

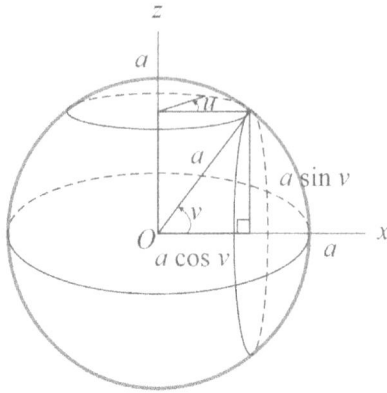

Figure 9.32 Sphere.

Solution

The surface S of a sphere can be parametrized as

$$\mathbf{r}(u,v) = [a\cos u\cos v,\ a\sin u\cos v,\ a\sin v].\quad \left(0 \le u \le 2\pi,\ -\frac{\pi}{2} \le v \le \frac{\pi}{2}\right)$$

The partial derivatives with respect to u and v are as follows:

$$\mathbf{r}_u = [-a\sin u\cos v,\ a\cos u\cos v,\ 0],$$

$$\mathbf{r}_v = [-a\cos u\sin v,\ -a\sin u\sin v,\ a\cos v].$$

Then we get

$$\mathbf{N} = \mathbf{r}_u \times \mathbf{r}_v = a^2\left(\cos u\cos^2 v\,\mathbf{i} + \sin u\cos^2 v\,\mathbf{j} + \sin v\cos v\,\mathbf{k}\right)$$

and

$$|\mathbf{N}| = a^2|\cos v|.$$

Therefore, we obtain the surface area of a sphere as

$$A(S) = a^2\int_{-\pi/2}^{\pi/2}\int_0^{2\pi}|\cos v|\,du\,dv = 2\pi a^2\int_{-\pi/2}^{\pi/2}\cos v\,dv = 4\pi a^2.$$

Answer $A(S) = 4\pi a^2$

Problem 9.4

For the vector field **F** and the surface S, evaluate the surface integral $\iint_S \mathbf{F} \cdot \mathbf{n}\, dA$. [1 ~ 6]

1. $\mathbf{F} = \left[-x^2, y^2, 0\right]$, $S: \mathbf{r} = \left[u, v, 3u - 2v\right]$ $(0 \le u \le 1, -3 \le v \le 3)$

2. $\mathbf{F} = \left[x, y, z\right]$, $S: \mathbf{r} = \left[u\cos v, u\sin v, u^2\right]$ $(0 \le u \le 2, -\pi \le v \le \pi)$

3. $\mathbf{F} = \left[\cosh y, 0, \sinh x\right]$, $S: z = x + y^2$ $(0 \le y \le x, 0 \le x \le 1)$

4. $\mathbf{F} = \left[\tan xy, x, y\right]$, $S: y^2 + z^2 = 1$ $(0 \le x \le 2, y \ge 0, z \ge 0)$

5. $\mathbf{F} = \left[1, 1, 1\right]$, $S: x^2 + y^2 + 4z^2 = 4$ $(z \ge 0)$

6. $\mathbf{F} = \left[x, xy, z\right]$, $S: x^2 + y^2 = 1$ $(0 \le z \le 2)$

For a given surface S, evaluate the surface integral $\iint_S G(\mathbf{r})\, dA$. [7 ~ 10]

7. $G = x$, $S: x + y + z = 1$ $(x > 0, y > 0, z > 0)$

8. $G = x + y + z$, $S: z = x + 2y$ $\left(0 \le x \le \sqrt{6}, 0 \le y \le x\right)$

9. $G = ax + by + cz$, $S: x^2 + y^2 + z^2 = 1, y = 0, z = 0$

10. $G = (1 + 9xz)^{3/2}$, $S: \mathbf{r} = \left[u, v, u^3\right]$ $(0 \le u \le 1, 0 \le v \le 2)$

9.5 Triple integral and Gauss's divergence theorem

Triple integral is an extension of double integral to one additional dimension, meaning the integration of a given function $f(x, y, z)$ over a region V in 3-dimensional space. Therefore, triple integral can be expressed as follows:

$$\iiint_T f(x, y, z)\, dV = \iiint_T f(x, y, z)\, dx\, dy\, dz. \qquad (9.22)$$

As seen in Eq. (9.22), triple integral involves the successive repetition of integration three times; hence, its operation is analogous to double integral, which involves successive repetition of integration twice.

The divergence theorem, proposed and proven by Gauss, deals with the conversion of surface integrals over a surface S into triple integrals over a volume V and vice versa.

As defined in Chapter 8, the divergence of an arbitrary vector function $\mathbf{F} = \left[F_1, F_2, F_3\right] = F_1\mathbf{i} + F_2\mathbf{j} + F_3\mathbf{k}$ is as follows:

$$\operatorname{div} \mathbf{F} = \frac{\partial F_1}{\partial x} + \frac{\partial F_2}{\partial y} + \frac{\partial F_3}{\partial z}.$$

Remark   Gauss's divergence theorem

Gauss's divergence theorem is given by the following equation:

$$\iiint_T \text{div}\, \mathbf{F}\, dV = \iint_S \mathbf{F} \cdot \mathbf{n}\, dA. \tag{9.23}$$

When calculating surface integrals in Eq. (9.23), the direction of the vector perpendicular to the surface is chosen to point outward from the volume. Eq. (9.23) can be expressed in terms of vector components as follows:

$$\iiint_T \left(\frac{\partial F_1}{\partial x} + \frac{\partial F_2}{\partial y} + \frac{\partial F_3}{\partial z} \right) dx\, dy\, dz = \iint_S \left(F_1\, dydz + F_2\, dzdx + F_3\, dxdy \right).$$

$$\tag{9.24}$$

Example 9.15

Evaluate the surface integral $\iint_S \mathbf{F} \cdot \mathbf{n}\, dA$ by the Gauss's divergence theorem for the given vector field $\mathbf{F}(x,\, y,\, z) = x^3\mathbf{i} + y^3\mathbf{j} + z^3\mathbf{k}$ and surface S of the sphere with radius R.

Solution

Since $\text{div}\, \mathbf{F} = 3x^2 + 3y^2 + 3z^2 = 3\left(x^2 + y^2 + z^2\right)$, we obtain

$$\iint_S \mathbf{F} \cdot \mathbf{n}\, dA = \iiint_T 3\left(x^2 + y^2 + z^2\right) dV$$

$$= \int_{-\frac{\pi}{2}}^{\frac{\pi}{2}} \int_0^{2\pi} \int_0^R \left(3r^2\right)\left(dr\right)\left(r\cos\varphi\, d\theta\right)\left(r d\varphi\right)$$

$$= \int_{-\frac{\pi}{2}}^{\frac{\pi}{2}} \left(\frac{3}{5}R^5\right)\left(\int_0^{2\pi} d\theta\right)\cos\varphi d\varphi$$

$$= \frac{3}{5}R^5 \cdot 2\pi \cdot \int_{-\frac{\pi}{2}}^{\frac{\pi}{2}} \cos\varphi\, d\varphi = \frac{12}{5}\pi R^5.$$

$$\textbf{Answer}\ \frac{12}{5}\pi R^5$$

Example 9.16

Evaluate the surface integral $\iint_S \mathbf{F} \cdot \mathbf{n} \, dA$ by the Gauss's divergence theorem for the given vector field $\mathbf{F}(x, y, z) = 2x^2\mathbf{i} + y^2\mathbf{j} + z^2\mathbf{k}$ over the surface S, which is the boundary of the region $0 \le x \le 1, 0 \le y \le 1, 0 \le z \le 1$.

Solution

Since $\operatorname{div} \mathbf{F} = 4x + 2y + 2z$, we obtain

$$\iint_S \mathbf{F} \cdot \mathbf{n} \, dA = \iiint_T (4x + 2y + 2z) \, dx \, dy \, dz$$

$$= 2 \int_0^1 \int_0^1 \int_0^1 (2x + y + z) \, dx \, dy \, dz$$

$$= 2 \int_0^1 \int_0^1 (1 + y + z) \, dy \, dz$$

$$= 2 \int_0^1 \left(\frac{3}{2} + z \right) dz = 4.$$

Answer 4

Problem 9.5

Evaluate the surface integral $\iint_S \mathbf{F} \cdot \mathbf{n} \, dA$ by the Gauss's divergence theorem, for the given vector field $\mathbf{F}$ over the surface S. [1 ~ 10]

1. $\mathbf{F} = \left[x^2, \, 0, \, z^2 \right]$
   $S: |x| \le 1, |y| \le 1, |z| \le 1$ (surface of the box)
2. $\mathbf{F} = \left[x^3 - y^3, \, y^3 - z^3, \, z^3 - x^3 \right]$
   $S: x^2 + y^2 + z^2 \le 1, \, z \ge 0$ (surface of the half-sphere)
3. $\mathbf{F} = \left[\sin y, \, \cos x, \, \cos z \right]$
   $S: x^2 + y^2 \le 4, |z| \le 1$ (surface of the circular cylinder)
4. $\mathbf{F} = \left[2x^2, \, \frac{1}{2} y^2, \, \sin \pi z \right]$
   S: the surface of the tetrahedron with vertices $(0,0,0), (1,0,0), (0,1,0), (0,0,1)$
5. $\mathbf{F} = \left[x^2, \, y^2, \, z^2 \right]$
   $S: x^2 + y^2 \le z^2, 0 \le z \le 1$ (surface of the cone)

6. $F = [xy, yz, zx]$

   S: $x^2 + y^2 \leq 4z^2$, $0 \leq z \leq 2$ (surface of the cone)

7. $F = [ax, by, cz]$

   S: $x^2 + y^2 + z^2 = 9$ (surface of the sphere)

8. $F = [x + y^2, y + z^2, z + x^2]$

   S: $\dfrac{x^2}{a^2} + \dfrac{y^2}{b^2} + \dfrac{z^2}{c^2} = 1$ (surface of the ellipsoid)

9. $F = [y + z, 20y, 2z^3]$

   S: $0 \leq x \leq 1$, $0 \leq y \leq 2$, $0 \leq z \leq y$ (surface)

10. $F = [x + y, y + z, z + x]$,

    S: surface of sphere with the origin O and radius 3

9.6   Stokes's theorem in the space

In the previous section, we extended Green's theorem from the plane to three dimensions, resulting in *Stokes's theorem*.

Green's theorem allows the transformation of line integrals along a boundary of a region into double integrals, and vice versa. Similarly, Stokes's theorem enables the transformation of surface integrals into triple integrals, and vice versa.

Remark   Stokes's theorem

Let S be a smooth oriented surface in space and let the boundary of S be a smooth closed curve C. Let $F(x, y, z) = P(x, y, z)i + Q(x, y, z)j + R(x, y, z)k$ be a continuous vector function that has continuous first partial derivatives in a domain in space containing S. Then

$$\oint_C F \cdot dr = \iint_S (\text{curl}\, F) \cdot n\, dA \qquad (9.25)$$

where $\oint_C F \cdot dr = \oint_C F \cdot r'(t)\, dt$,

$$\text{curl}\, F = \begin{vmatrix} i & j & k \\ \dfrac{\partial}{\partial x} & \dfrac{\partial}{\partial y} & \dfrac{\partial}{\partial z} \\ P & Q & R \end{vmatrix} = \left(\dfrac{\partial R}{\partial y} - \dfrac{\partial Q}{\partial z} \right)i + \left(\dfrac{\partial P}{\partial z} - \dfrac{\partial R}{\partial x} \right)j + \left(\dfrac{\partial Q}{\partial x} - \dfrac{\partial P}{\partial y} \right)k,$$

and n is a unit normal vector of S. (refer Eq. (9.17))

For example, if a smooth simple closed curve C bounds a smooth oriented surface S, expressed as $z = g(x, y)$, then setting $f = z - g(x, y) = 0$ yields $\nabla f = -\frac{\partial g}{\partial x}\mathbf{i} - \frac{\partial g}{\partial y}\mathbf{j} + \mathbf{k}$. Consequently, the unit normal vector $\mathbf{n}$ on the surface S is calculated by the following expression:

$$\mathbf{n} = \frac{\nabla f}{|\nabla f|} = \frac{-\dfrac{\partial g}{\partial x}\mathbf{i} - \dfrac{\partial g}{\partial y}\mathbf{j} + \mathbf{k}}{\sqrt{\left(\dfrac{\partial g}{\partial x}\right)^2 + \left(\dfrac{\partial g}{\partial y}\right)^2 + 1}}. \tag{9.26}$$

Furthermore, Stokes's theorem is transformed into the following expression:

$$\iint_S (\operatorname{curl}\mathbf{F}) \cdot \mathbf{n}\, dA = \iint_R (\operatorname{curl}\mathbf{F}) \cdot \mathbf{N}\, dx\, dy \tag{9.27}$$

where the normal vector over the region R is calculated by $\mathbf{N} = \nabla f$.

Example 9.17

Evaluate $\oint_C \mathbf{F} \cdot d\mathbf{r}$ using Stokes's theorem for the vector field $\mathbf{F}$ over the smooth oriented surface S in space.

$$\mathbf{F}(x, y, z) = y^3\mathbf{i} + z^3\mathbf{j} + x^3\mathbf{k},$$

$$S:\ z = g(x, y) = 4 - \left(x^2 + y^2\right), \quad z \geq 0.$$

Solution

The boundary C of the surface S becomes $z = 0$, forming a circle defined by $x^2 + y^2 = 4$. Then

$$\mathbf{r}(v) = (2\cos v, 2\sin v, 0). \quad (0 \leq v \leq 2\pi)$$

Its derivative is

$$\mathbf{r}'(v) = (-2\sin v, 2\cos v, 0).$$

And we get

$$F(x, y, z) = \left(y^3, z^3, x^3\right) = \left(8\sin^3 v, 0, 8\cos^3 v\right).$$

Therefore, we obtain

$$\oint_C \mathbf{F} \cdot d\mathbf{r} = \oint_C \mathbf{F} \cdot \mathbf{r}'(v) \, dv$$

$$= \int_0^{2\pi} \left(8\sin^3 v, \, 0, \, 8\cos^3 v\right) \cdot \left(-2\sin v, \, 2\cos v, \, 0\right) dv$$

$$= -16 \int_0^{2\pi} \sin^4 v \, dv$$

$$= -16 \cdot 4 \int_0^{\pi/2} \sin^4 v \, dv \quad (\text{refer to the Check})$$

$$= -16 \cdot 4 \cdot \left(\frac{3}{4} \cdot \frac{1}{2} \cdot \frac{\pi}{2}\right) = -12\pi.$$

Answer -12π

Check

$$\int_0^{\pi/2} \cos^n v \, dv = \frac{n-1}{n} \cdot \frac{n-3}{n-2} \cdot \dots \cdot \frac{1}{2} \cdot \frac{\pi}{2} \quad (n \text{ even number})$$

$$\int_0^{\pi/2} \sin^n v \, dv = \frac{n-1}{n} \cdot \frac{n-3}{n-2} \cdot \dots \cdot \frac{1}{2} \cdot \frac{\pi}{2} \quad (n \text{ even number})$$

Example 9.18

Evaluate $\iint_S (\text{curl}\,\mathbf{F}) \cdot \mathbf{n} \, dS$ using Stokes's theorem for the vector field $\mathbf{F}$ over the smooth oriented surface S in space.

$$\mathbf{F}(x, y, z) = y^3 \mathbf{i} + z^3 \mathbf{j} + x^3 \mathbf{k}$$

$$S: z = g(x, y) = 4 - \left(x^2 + y^2\right), \quad z \geq 0$$

Solution

In the smooth surface S, we get

$$f = z + x^2 + y^2 - 4 = 0.$$

And the normal vector is

$$\mathbf{N} = \nabla f = 2x\mathbf{i} + 2y\mathbf{j} + \mathbf{k}.$$

From the vector field $F(x, y, z) = y^3\mathbf{i} + z^3\mathbf{j} + x^3\mathbf{k}$, we get

$$P = y^3, \quad Q = z^3, \quad R = x^3.$$

Then

$$\operatorname{curl} F = \begin{vmatrix} \mathbf{i} & \mathbf{j} & \mathbf{k} \\ \dfrac{\partial}{\partial x} & \dfrac{\partial}{\partial y} & \dfrac{\partial}{\partial z} \\ y^3 & z^3 & x^3 \end{vmatrix} = -3z^2\mathbf{i} - 3x^2\mathbf{j} - 3y^2\mathbf{k}.$$

Since the boundary C of the surface S becomes $z = 0$, we get

$$\operatorname{curl} F = -3x^2\mathbf{j} - 3y^2\mathbf{k}$$

and

$$\operatorname{curl} F \cdot N = (0, \ -3x^2, \ -3y^2) \cdot (2x, \ 2y, \ 1) = -6x^2y - 3y^2.$$

Therefore, we obtain

$$\iint_S (\operatorname{curl} F) \cdot n \, dS = \iint_R (\operatorname{curl} F) \cdot N \, dx \, dy$$

$$= \iint_R \left(-6x^2y - 3y^2 \right) dx \, dy.$$

Let's substitute x and y with r and θ, as the region R lies inside of the circle $x^2 + y^2 = 4$.

$$\iint_S (\operatorname{curl} F) \cdot n \, dS$$

$$= \int_0^{2\pi} \int_0^2 \left(-6 \cdot r^3 \cos^2\theta \sin\theta - 3 \cdot r^2 \sin^2\theta \right) r \, dr \, d\theta$$

$$= \int_0^{2\pi} \cos^2\theta \sin\theta \, d\theta \int_0^2 \left(-6r^4 \right) dr + \int_0^{2\pi} \sin^2\theta \, d\theta \int_0^2 \left(-3r^3 \right) dr$$

$$= \int_0^{2\pi} \left(-\frac{1}{3}\cos^3\theta \right) d\theta \int_0^2 \left(-6r^4 \right) dr + \int_0^{2\pi} \frac{1}{2}(1 - \cos 2\theta) \, d\theta \int_0^2 \left(-3r^3 \right) dr$$

$$= 0 + \pi \cdot (-12) = -12\pi.$$

$$\text{Answer } -12\pi$$

Remark Stokes's theorem

For a vector field $\mathbf{F}(x, y, z) = P(x, y, z)\mathbf{i} + Q(x, y, z)\mathbf{j} + R(x, y, z)\mathbf{k}$, if the projection C_{xy} of the boundary C onto the xy-plane is parameterized by $x = x(t)$, $y = y(t)$ for $a \le t \le b$, then the equation of the boundary C is expressed as $x = x(t)$, $y = y(t)$, $z = g(x(t), y(t))$ for $a \le t \le b$. In this case, Stokes's theorem is expressed as follows:

$$\oint_C \mathbf{F} \cdot d\mathbf{r} = \oint_C (P\,dx + Q\,dy + R\,dz)$$

$$= \iint_R \left[-\left(\frac{\partial R}{\partial y} - \frac{\partial Q}{\partial z} \right)\frac{\partial g}{\partial x} - \left(\frac{\partial P}{\partial z} - \frac{\partial R}{\partial x} \right)\frac{\partial g}{\partial y} + \left(\frac{\partial Q}{\partial x} - \frac{\partial P}{\partial y} \right) \right] dA. \quad (9.28)$$

By the chain rule, we get

$$dz = \left(\frac{\partial g}{\partial x}\frac{dx}{dt} + \frac{\partial g}{\partial y}\frac{dy}{dt} \right) dt.$$

Then Eq. (9.28) becomes

$$\oint_C \mathbf{F} \cdot d\mathbf{r} = \oint_C (P\,dx + Q\,dy + R\,dz)$$

$$= \int_a^b \left[P\frac{dx}{dt} + Q\frac{dy}{dt} + R\left(\frac{\partial g}{\partial x}\frac{dx}{dt} + \frac{\partial g}{\partial y}\frac{dy}{dt} \right) \right] dt$$

$$= \oint_C \left\{ \left(P + R\frac{\partial g}{\partial x} \right) dx + \left(Q + R\frac{\partial g}{\partial y} \right) dy \right\}$$

$$= \iint_R \left\{ \frac{\partial}{\partial x}\left(Q + R\frac{\partial g}{\partial y} \right) - \frac{\partial}{\partial y}\left(P + R\frac{\partial g}{\partial x} \right) \right\} dA. \quad \text{(by Green's theorem)}$$

$$(9.29)$$

The first term inside the double integral can be rearranged as follows:

$$\frac{\partial}{\partial x}\left(Q + R\frac{\partial g}{\partial y} \right) = \frac{\partial}{\partial x}\left[Q(x, y, g(x, y)) + R(x, y, g(x, y))\frac{\partial g}{\partial y} \right]$$

$$= \frac{\partial Q}{\partial x} + \frac{\partial Q}{\partial z}\frac{\partial g}{\partial x} + \left(\frac{\partial R}{\partial x} + \frac{\partial R}{\partial z}\frac{\partial g}{\partial x} \right)\frac{\partial g}{\partial y} + R\frac{\partial^2 g}{\partial x \partial y}$$

$$= \frac{\partial Q}{\partial x} + \frac{\partial Q}{\partial z}\frac{\partial g}{\partial x} + \frac{\partial R}{\partial x}\frac{\partial g}{\partial y} + \frac{\partial R}{\partial z}\frac{\partial g}{\partial x}\frac{\partial g}{\partial y} + R\frac{\partial^2 g}{\partial x \partial y}. \quad (9.30)$$

Similarly, the second term inside the double integral can be rearranged as follows:

$$\frac{\partial}{\partial y}\left(P + R\frac{\partial g}{\partial x}\right) = \frac{\partial P}{\partial y} + \frac{\partial P}{\partial z}\frac{\partial g}{\partial y} + \frac{\partial R}{\partial y}\frac{\partial g}{\partial x} + \frac{\partial R}{\partial z}\frac{\partial g}{\partial y}\frac{\partial g}{\partial x} + R\frac{\partial^2 g}{\partial y \partial x}. \qquad (9.31)$$

When substituting Eq. (9.30) and Eq. (9.31) into Eq. (9.29), we obtain

$$\oint_C \mathbf{F}\cdot d\mathbf{r} = \oint_C (P\,dx + Q\,dy + R\,dz)$$

$$= \iint_R \left[-\left(\frac{\partial R}{\partial y} - \frac{\partial Q}{\partial z}\right)\frac{\partial g}{\partial x} - \left(\frac{\partial P}{\partial z} - \frac{\partial R}{\partial x}\right)\frac{\partial g}{\partial y} + \left(\frac{\partial Q}{\partial x} - \frac{\partial P}{\partial y}\right)\right]dA. \qquad (9.28)$$

Therefore, Stokes's theorem is proven.

Problem 9.6

Evaluate $\oint_C \mathbf{F}\cdot d\mathbf{r}$ or $\iint_S (\text{curl}\,\mathbf{F})\cdot \mathbf{n}\,dS$ using Stokes's theorem for the vector field $\mathbf{F}$ over the smooth oriented surface S in space or its boundary C. [1 ~ 10]

1. $\mathbf{F}(x, y, z) = y\mathbf{i} + 2x\mathbf{j} + 3z\mathbf{k}$,
    C: the tetrahedron with vertices $(0, 0, 0)$, $(2, 0, 0)$, $(0, 3, 3)$, $(2, 3, 3)$
2. $\mathbf{F}(x, y, z) = e^{-z}\mathbf{i} + 2x\mathbf{j} + x\mathbf{k}$,
    C: the tetrahedron with vertices $(0, 0, 2)$, $(4, 0, 2)$, $(4, 3, 2)$, $(0, 3, 2)$
3. $\mathbf{F}(x, y, z) = -2y\mathbf{i} + 3x\mathbf{j} + z\mathbf{k}$,
    C: circle of $x^2 + y^2 = 4$, $z = 1$
4. $\mathbf{F}(x, y, z) = z^2\mathbf{i} + 2x\mathbf{j}$,
    S: $x^2 + y^2 \le 9$, $z = 0$
5. $\mathbf{F}(x, y, z) = z^2\mathbf{i} + x^2\mathbf{j} + y^2\mathbf{k}$,
    S: $z = x^2$, $(0 \le x \le 1, \quad 0 \le y \le 3)$
6. $\mathbf{F}(x, y, z) = x^2\mathbf{i} + xy\mathbf{j} + z^2\mathbf{k}$,
    S: $z = xy$, $(0 \le x \le 1, \quad 0 \le y \le 2)$
7. $\mathbf{F}(x, y, z) = z^2\mathbf{i} + x\mathbf{j} + y\mathbf{k}$,
    S: $z = x^2 + y^2$, $(0 \le z \le 4)$
8. $\mathbf{F}(x, y, z) = -y\mathbf{i} + 3x\mathbf{j} + 2z\mathbf{k}$,
    S: $z = \sqrt{x^2 + y^2}$, $(0 \le z \le 3, y \ge 0)$
9. $\mathbf{F}(x, y, z) = yz\mathbf{i} + x^2\mathbf{j} + 2z^2\mathbf{k}$,
    S: $r = (u, u^2, v)$, $0 \le u \le 3$, $-2 \le v \le 2$
10. $\mathbf{F}(x, y, z) = xz\mathbf{i} + yz\mathbf{j} + z^2\mathbf{k}$,
    S: $r = (u, v, uv)$, $0 \le u \le 2, 0 \le v \le 1$

Answer

Problem 9.1

1. a. 11/12,
   b. 11/14
2. a. 2/3,
   b. 2/3
3. $e^{\pi} - 1$
4. 25/2
5. 2/3
6. $\pi/2$
7. $e^2 - 1$
8. $e + e^2$
9. $e - 1/e$
10. -1
11. 1
12. 1
13. e^2
14. $\left(e - e^{-1}\right)/2$

Problem 9.2

1. 11/3
2. 1/2
3. 4/3
4. $-7/20$
5. 12
6. $e^4 - e^3/3 + 1/3$
7. 16/3
8. 224/15
9. 4π
10. $\pi(e - 1)/4$
11. $2\sqrt{2}$
12. $\pi/8$

Problem 9.3

1. 3/4
2. 10/3
3. $-184/15$
4. 4/15
5. 4π
6. 12π
7. $2e - e^2/2 - 3/2$
8. 4
9. $-14\left(1 + \sqrt{2}\right)/3$
10. 3π

Problem 9.4

1. 42
2. -8π
3. $1 - \sinh 1$
4. -3
5. 4π
6. 2π
7. $\sqrt{3}/6$
8. 42
9. $(b + c)\pi/2$
10. 136/5

Problem 9.5

1. 2
2. $6\pi/5$
3. 0
4. $5/24 + 1/\pi$
5. $\pi/2$
6. 16π
7. $36\pi(a + b + c)$
8. $\pi abc/2$
9. 48
10. 72π

Problem 9.6

1. 6
2. 24
3. 20π
4. 18π
5. -6
6. 2
7. 4π
8. 18π
9. 36
10. -2

10 Fourier analysis

Fourier series is a method of representing a periodic function as an infinite sum of sine and cosine functions. On the other hand, Fourier integral is a method of representing a non-periodic function as an infinite sum of sine and cosine functions.

In the case of a periodic function, we can assume that it repeats over a certain interval in time or space. Therefore, we can express it as a combination of trigonometric functions, sine and cosine, which have periodic properties.

However, for non-periodic functions, this direct approach doesn't work. Instead, we extend the non-periodic function to an infinite period and then apply the concept of Fourier series. This extension to infinity allows us to represent the non-periodic function as an infinite sum of sine and cosine functions, which is the essence of the Fourier integral.

10.1 Fourier series

10.1.1 Function with period 2π

Fourier series represents an arbitrary function $f(x)$ with period 2π as an infinite series of sine and cosine functions, as shown in Eq. (10.1). In the context of Fourier series, a_0, a_k, and b_k are referred to as the Fourier coefficients. These coefficients represent the amplitudes of the constant term, the cosine terms, and the sine terms, respectively, in the Fourier series representation of a function.

Remark   Fourier series of $f(x)$ with period 2π

$$f(x) = a_0 + \sum_{k=1}^{\infty}(a_k \cos kx + b_k \sin kx) \tag{10.1}$$

where $\displaystyle a_0 = \frac{1}{2\pi}\int_{-\pi}^{\pi} f(x)dx,$ $\tag{10.2a}$

$\displaystyle a_k = \frac{1}{\pi}\int_{-\pi}^{\pi} f(x)\cos kx\,dx, \qquad (k=1,\ 2,\ 3,\ \cdots) \tag{10.2b}$

$\displaystyle b_k = \frac{1}{\pi}\int_{-\pi}^{\pi} f(x)\sin kx\,dx, \qquad (k=1,\ 2,\ 3,\ \cdots) \tag{10.2c}$

DOI: 10.1201/9781003608912-10

i. Proof of Eq. (10.2a)
When we integrate Eq. (10.1) over the interval $[-\pi, \pi]$, we get:

LHS: $\displaystyle\int_{-\pi}^{\pi} f(x)\,dx$

RHS: $\displaystyle a_0 \int_{-\pi}^{\pi} dx + \sum_{k=1}^{\infty} a_k \int_{-\pi}^{\pi} \cos kx\,dx + \sum_{k=1}^{\infty} b_k \int_{-\pi}^{\pi} \sin kx\,dx = 2\pi a_0$

(refer to the Check)

Therefore, we obtain

$$a_0 = \frac{1}{2\pi}\int_{-\pi}^{\pi} f(x)\,dx. \tag{10.2a}$$

Check

$$\int_{-\pi}^{\pi} \cos kx\,dx = 0,$$

$$\int_{-\pi}^{\pi} \sin kx\,dx = 0.$$

ii. Proof of Eq. (10.2b)
When we multiply Eq. (10.1) by $\cos(lx)$ (where l is a positive integer) and integrate over the interval $[-\pi, \pi]$, we get:

LHS: $\displaystyle\int_{-\pi}^{\pi} f(x)\cos lx\,dx$

RHS: $\displaystyle a_0 \int_{-\pi}^{\pi} \cos lx\,dx + \sum_{k=1}^{\infty} a_k \int_{-\pi}^{\pi} \cos kx \cos lx\,dx + \sum_{k=1}^{\infty} b_k \int_{-\pi}^{\pi} \sin kx \cos lx\,dx$

$$= a_k \pi \qquad\qquad\text{(refer to the Check)}$$

Therefore, we obtain

$$a_k = \frac{1}{\pi}\int_{-\pi}^{\pi} f(x)\cos kx\,dx. \tag{10.2b}$$

Check

$$\int_{-\pi}^{\pi} \cos lx\, dx = 0,$$

$$\int_{-\pi}^{\pi} \cos kx \cos lx\, dx = \frac{1}{2}\int_{-\pi}^{\pi} \cos(k+l)x\, dx + \frac{1}{2}\int_{-\pi}^{\pi} \cos(k-l)x\, dx$$

$$= 0 + \frac{1}{2}(2\pi) = \pi, \quad (\text{only when } k = l)$$

$$\int_{-\pi}^{\pi} \sin kx \cos lx\, dx = \frac{1}{2}\int_{-\pi}^{\pi} \sin(k+l)x\, dx + \frac{1}{2}\int_{-\pi}^{\pi} \sin(k-l)x\, dx$$

$$= 0.$$

iii. Proof of Eq. (10.2c)

When we multiply Eq. (10.1) by sinlx (where l is a positive integer) and integrate over the interval $[-\pi, \pi]$, we get:

LHS: $\int_{-\pi}^{\pi} f(x)\sin lx\, dx$

RHS: $a_0\int_{-\pi}^{\pi} \sin lx\, dx + \sum_{k=1}^{\infty} a_k \int_{-\pi}^{\pi} \cos kx \sin lx\, dx + \sum_{k=1}^{\infty} b_k \int_{-\pi}^{\pi} \sin kx \sin lx\, dx$

$= b_k \pi$ (refer to the Check)

Therefore, we obtain

$$b_k = \frac{1}{\pi}\int_{-\pi}^{\pi} f(x)\sin kx\, dx. \qquad (10.2c)$$

Check

$$\int_{-\pi}^{\pi} \sin lx\, dx = 0,$$

$$\int_{-\pi}^{\pi} \cos kx \sin lx\, dx = \frac{1}{2}\int_{-\pi}^{\pi} \sin(k+l)x\, dx - \frac{1}{2}\int_{-\pi}^{\pi} \sin(k-l)x\, dx$$

$$= 0,$$

$$\int_{-\pi}^{\pi} \sin kx \sin lx\, dx = -\frac{1}{2}\int_{-\pi}^{\pi} \cos(k+l)x\, dx + \frac{1}{2}\int_{-\pi}^{\pi} \cos(k-l)x\, dx$$

$$= 0 + \frac{1}{2}(2\pi) = \pi. \quad (\text{only when } k = l)$$

Example 10.1

Find the Fourier series of the function $f(x)$ shown in Figure 10.1.

$$f(x) = \begin{cases} 0 & (-\pi < x < 0) \\ 1 & (0 < x < \pi) \end{cases} \quad \text{and} \quad f(x+2\pi) = f(x).$$

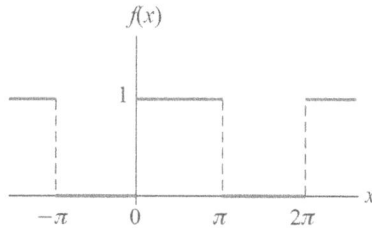

Figure 10.1 Periodic function $f(x)$.

Solution

In the function $f(x) = a_0 + \sum_{k=1}^{\infty}(a_k \cos kx + b_k \sin kx)$, we get the Fourier coefficients

$$a_0 = \frac{1}{2\pi}\int_{-\pi}^{\pi} f(x)dx = \frac{1}{2\pi}\int_{0}^{\pi} 1\,dx = \frac{1}{2},$$

$$a_k = \frac{1}{\pi}\int_{-\pi}^{\pi} f(x)\cos kx\,dx = \frac{1}{\pi}\int_{0}^{\pi} \cos kx\,dx = \frac{1}{\pi}\left[\frac{\sin kx}{k}\right]_0^{\pi} = 0,$$

$$b_k = \frac{1}{\pi}\int_{-\pi}^{\pi} f(x)\sin kx\,dx = \frac{1}{\pi}\int_{0}^{\pi} \sin kx\,dx = \frac{1}{\pi}\left[-\frac{\cos kx}{k}\right]_0^{\pi} = \frac{1}{\pi}\frac{1-(-1)^k}{k}.$$

Therefore, we obtain

$$f(x) = \frac{1}{2} + \frac{1}{\pi}\sum_{k=1}^{\infty}\left(\frac{1-(-1)^k}{k}\sin kx\right).$$

$$\textbf{Answer } f(x) = \frac{1}{2} + \frac{2}{\pi}\left(\sin x + \frac{\sin 3x}{3} + \frac{\sin 5x}{5} + \cdots\right)$$

The figures in Figure 10.2 show the partial sums of the corresponding Fourier series.

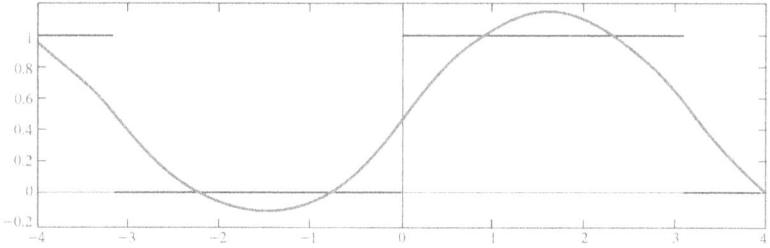

(a) $S_1 = \dfrac{1}{2} + \dfrac{2}{\pi} \sin x$

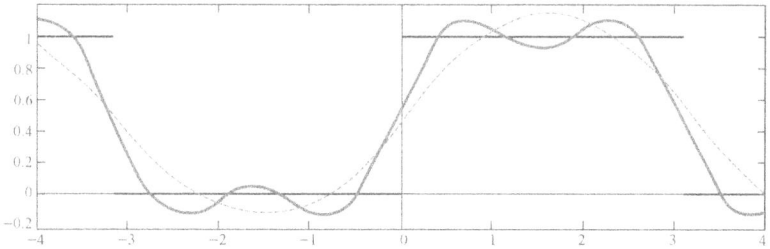

(b) $S_3 = \dfrac{1}{2} + \dfrac{2}{\pi}\left(\sin x + \dfrac{\sin 3x}{3} \right)$

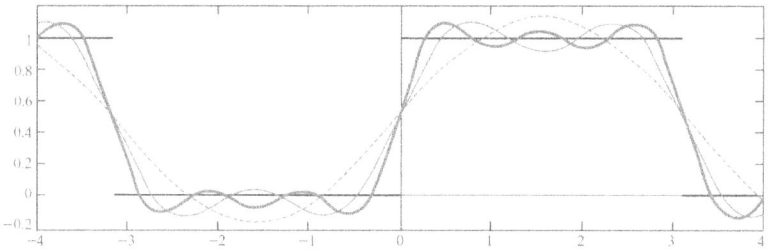

(c) $S_5 = \dfrac{1}{2} + \dfrac{2}{\pi}\left(\sin x + \dfrac{\sin 3x}{3} + \dfrac{\sin 5x}{5} \right)$

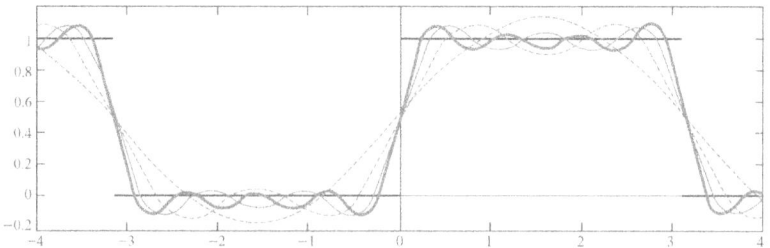

(d) $S_7 = \dfrac{1}{2} + \dfrac{2}{\pi}\left(\sin x + \dfrac{\sin 3x}{3} + \dfrac{\sin 5x}{5} + \dfrac{\sin 7x}{7} \right)$

Figure 10.2 Partial sums of the corresponding Fourier series in Example 10.1.

10.1.2 *Function with period 2L*

For a function $f(x)$ with period $2L$, the integration interval in Eq. (10.1) and Eq. (10.2) changes from $[-\pi, \pi]$ to $[-L, L]$, and kx becomes $\dfrac{k\pi}{L}x$.

Remark **Fourier series of $f(x)$ with period 2L**

$$f(x) = a_0 + \sum_{k=1}^{\infty}\left(a_k \cos\frac{k\pi x}{L} + b_k \sin\frac{k\pi x}{L}\right) \qquad (10.3)$$

where $a_0 = \dfrac{1}{2L}\displaystyle\int_{-L}^{L} f(x)\,dx,$ \hfill (10.4a)

$$a_k = \frac{1}{L}\int_{-L}^{L} f(x)\cos\frac{k\pi x}{L}\,dx, \qquad (k = 1,\, 2,\, 3,\, \cdots) \qquad (10.4b)$$

$$b_k = \frac{1}{L}\int_{-L}^{L} f(x)\sin\frac{k\pi x}{L}\,dx, \qquad (k = 1,\, 2,\, 3,\, \cdots) \qquad (10.4c)$$

Example 10.2

Find the Fourier series of the function $f(x)$ shown in Figure 10.3.

$$f(x) = \begin{cases} x & (0 < x < 1) \\ 0 & (1 < x < 2) \end{cases} \quad \text{and} \quad f(x+2) = f(x).$$

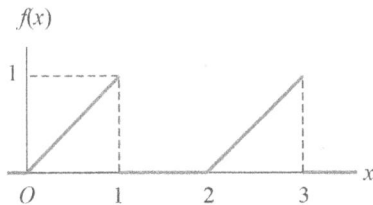

Figure 10.3 Periodic function $f(x)$.

Solution

Since $L = 1$ in the function $f(x) = a_0 + \sum_{k=1}^{\infty}\left(a_k \cos\frac{k\pi x}{L} + b_k \sin\frac{k\pi x}{L}\right)$, we get the Fourier coefficients

$$a_0 = \frac{1}{2L}\int_{-L}^{L} f(x)\,dx = \frac{1}{2}\int_0^1 x\,dx = \frac{1}{4},$$

$$a_k = \frac{1}{L}\int_{-L}^{L} f(x)\cos\frac{k\pi x}{L}\,dx = \int_0^1 x\cos k\pi x\,dx$$

$$= \left[x\frac{\sin k\pi x}{k\pi}\right]_0^1 - \int_0^1 \frac{\sin k\pi x}{k\pi}\,dx$$

$$= 0 + \left[\frac{\cos k\pi x}{(k\pi)^2}\right]_0^1 = \frac{(-1)^k - 1}{(k\pi)^2},$$

(When k is odd, then $a_k = -\dfrac{2}{(k\pi)^2}$, and when k is even, then $a_k = 0$.)

$$b_k = \frac{1}{L}\int_{-L}^{L} f(x)\sin\frac{k\pi}{L} x\,dx = \int_0^1 x\sin k\pi x\,dx$$

$$= \left[x\left(-\frac{\cos k\pi x}{k\pi}\right)\right]_0^1 - \int_0^1 \left(-\frac{\cos k\pi x}{k\pi}\right)dx$$

$$= \frac{-(-1)^k}{k\pi} + \left[\frac{\sin k\pi x}{(k\pi)^2}\right]_0^1 = \frac{-(-1)^k}{k\pi}.$$

Therefore, we obtain

$$f(x) = \frac{1}{4} + \sum_{k=1}^{\infty}\left(\frac{(-1)^k - 1}{(k\pi)^2}\cos\frac{k\pi x}{L} + \frac{-(-1)^k}{k\pi}\sin\frac{k\pi x}{L}\right).$$

Answer $f(x) = \dfrac{1}{4} - \dfrac{2}{\pi^2}\left(\cos\pi x + \dfrac{\cos 3\pi x}{3^2} + \cdots\right) + \dfrac{1}{\pi}\left(\sin\pi x - \dfrac{\sin 2\pi x}{2} + \dfrac{\sin 3\pi x}{3} - + \cdots\right)$

The figures in Figure 10.4 show the partial sums of the corresponding Fourier series.

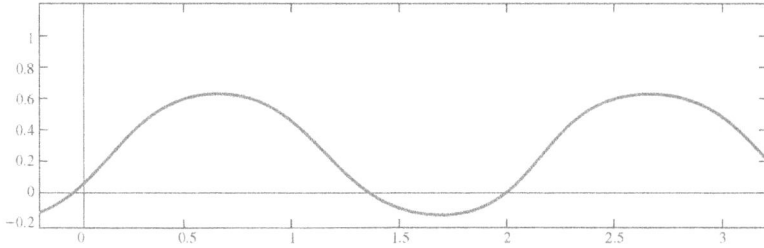

$$\text{(a)} \ \ S_1 = \frac{1}{4} - \frac{2}{\pi^2}(\cos \pi x) + \frac{1}{\pi}(\sin \pi x)$$

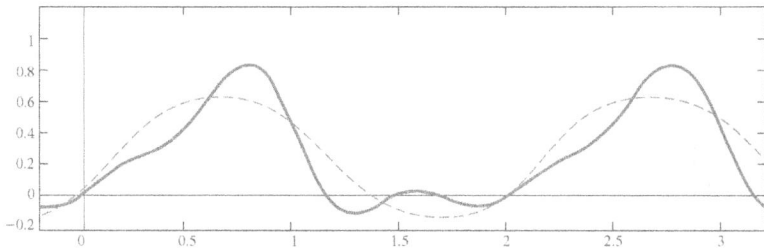

$$\text{(b)} \ \ S_3 = S_1 - \frac{2}{\pi^2}\left(\frac{\cos 3\pi x}{3^2}\right) + \frac{1}{\pi}\left(-\frac{\sin 2\pi x}{2} + \frac{\sin 3\pi x}{3}\right)$$

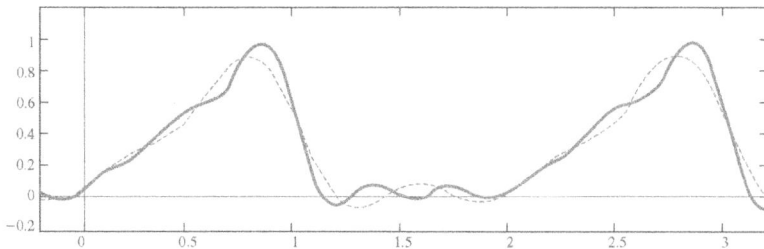

$$\text{(c)} \ \ S_5 = S_3 - \frac{2}{\pi^2}\left(\frac{\cos 5\pi x}{5^2}\right) + \frac{1}{\pi}\left(-\frac{\sin 4\pi x}{4} + \frac{\sin 5\pi x}{5}\right)$$

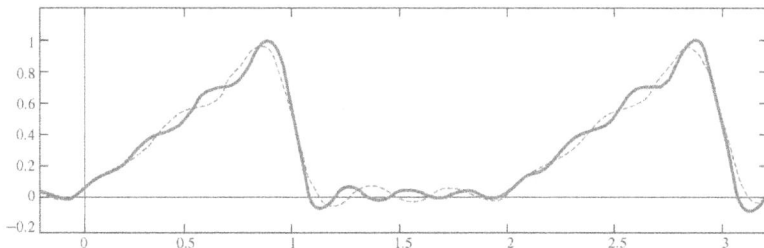

$$\text{(d)} \ \ S_7 = S_5 - \frac{2}{\pi^2}\left(\frac{\cos 7\pi x}{7^2}\right) + \frac{1}{\pi}\left(-\frac{\sin 6\pi x}{6} + \frac{\sin 7\pi x}{7}\right)$$

Figure 10.4 Partial sums of the corresponding Fourier series in Example 10.2.

10.1.3 *Even function and odd function*

An *even function* is a function that is symmetric about the y-axis, as shown in Figure 10.5(a). Typical examples of even functions include $\cos x$, $|x|$, x^2, x^4, and so on, and they satisfy the following equation:

$$f(-x) = f(x). \tag{10.5}$$

And an *odd function* is a function that is symmetric about the origin, as shown in Figure 10.5(b). Typical examples of odd functions include $\sin x$, $\tan x$, x, x^3, and so on, and they satisfy the following equation:

$$f(-x) = -f(x). \tag{10.6}$$

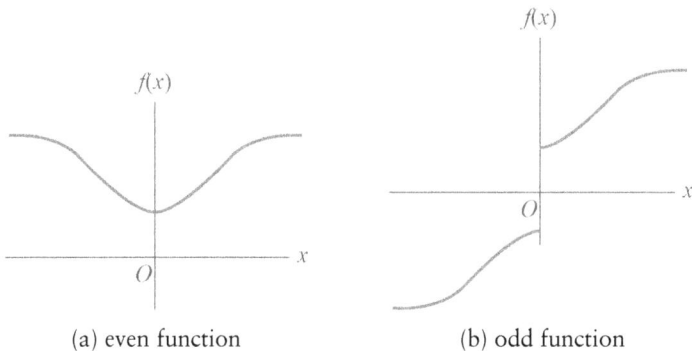

(a) even function   (b) odd function

Figure 10.5 Even function and odd function.

Moreover, the product of an even function and an odd function results in an odd function, while the product of two even functions or two odd functions results in an even function.

i. $f(x)$ is an even function
   In Eq. (10.4a), we get

$$a_0 = \frac{1}{2L}\int_{-L}^{L} f(x)\,dx = \frac{1}{L}\int_{0}^{L} f(x)\,dx,$$

and since $f(x)\cos\dfrac{k\pi x}{L}$, which is the product of two even functions, is even in Eq. (10.4b), we get

$$a_k = \frac{1}{L}\int_{-L}^{L} f(x)\cos\frac{k\pi x}{L}\,dx = \frac{2}{L}\int_{0}^{L} f(x)\cos\frac{k\pi x}{L}\,dx.$$

On the other hand, since $f(x)\sin\dfrac{k\pi x}{L}$, which is the product of an even function and an odd function, is an odd function in Eq. (10.4c), we get

$$b_k = \frac{1}{L}\int_{-L}^{L} f(x)\sin\frac{k\pi x}{L}\,dx = 0.$$

ii. $f(x)$ is an odd function
In Eq. (10.4a), we get

$$a_0 = \frac{1}{2L}\int_{-L}^{L} f(x)\,dx = 0,$$

and since $f(x)\cos\dfrac{k\pi x}{L}$ is odd in Eq. (10.4b), we get

$$a_k = \frac{1}{L}\int_{-L}^{L} f(x)\cos\frac{k\pi x}{L}\,dx = 0.$$

On the other hand, since $f(x)\sin\dfrac{k\pi x}{L}$ is an odd function in Eq. (10.4c), we get

$$b_k = \frac{1}{L}\int_{-L}^{L} f(x)\sin\frac{k\pi x}{L}\,dx = \frac{2}{L}\int_{0}^{L} f(x)\sin\frac{k\pi x}{L}\,dx.$$

Therefore, summarizing these, we have the following:

Remark **Fourier series of an even function with period 2π**

$$f(x) = a_0 + \sum_{k=1}^{\infty} a_k \cos kx \tag{10.7}$$

where $a_0 = \dfrac{1}{\pi}\displaystyle\int_{0}^{\pi} f(x)\,dx,$ $\tag{10.8a}$

$$a_k = \frac{2}{\pi}\int_{0}^{\pi} f(x)\cos kx\,dx, \qquad (k = 1,\ 2,\ 3,\ \cdots) \tag{10.8b}$$

Remark   Fourier series of an odd function with period 2π

$$f(x) = \sum_{k=1}^{\infty} b_k \sin kx \tag{10.9}$$

where $b_k = \dfrac{2}{\pi} \displaystyle\int_0^{\pi} f(x) \sin kx\, dx, \qquad (k = 1,\ 2,\ 3,\ \cdots) \tag{10.10}$

Remark   Fourier series of an even function with period $2L$

$$f(x) = a_0 + \sum_{k=1}^{\infty} a_k \cos \frac{k\pi x}{L} \tag{10.11}$$

where $a_0 = \dfrac{1}{L} \displaystyle\int_0^{L} f(x)\, dx, \tag{10.12a}$

$$a_k = \frac{2}{L} \int_0^{L} f(x) \cos \frac{k\pi x}{L}\, dx, \qquad (k = 1,\ 2,\ 3,\ \cdots) \tag{10.12b}$$

Remark   Fourier series of an odd function with period $2L$

$$f(x) = \sum_{k=1}^{\infty} b_k \sin \frac{k\pi x}{L} \tag{10.13}$$

where $b_k = \dfrac{2}{L} \displaystyle\int_0^{L} f(x) \sin \frac{k\pi x}{L}\, dx \qquad (k = 1,\ 2,\ 3,\ \cdots) \tag{10.14}$

Example 10.3   Find the Fourier series of the even function $f(x)$ shown in Figure 10.6.

$$f(x) = \begin{cases} -1 & (-1 < x < 1) \\ 1 & (1 < x < 3) \end{cases} \quad \text{and} \quad f(x+4) = f(x).$$

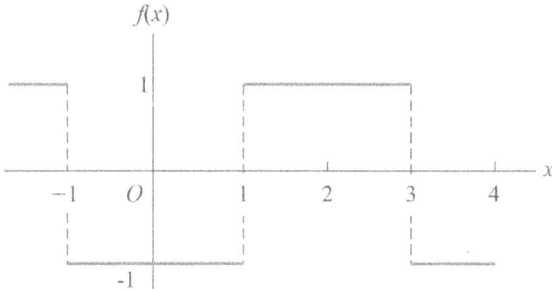

Figure 10.6 Periodic function $f(x)$.

Solution

Since $L = 2$ in the function $f(x) = a_0 + \sum_{k=1}^{\infty} a_k \cos \dfrac{k\pi x}{L}$, we get the Fourier coefficients

$$a_0 = \frac{1}{L} \int_0^L f(x)\,dx$$

$$= \frac{1}{2} \left(\int_0^1 (-1)\,dx + \int_1^2 1\,dx \right) = 0$$

$$a_k = \frac{2}{L} \int_0^L f(x) \cos \frac{k\pi x}{L}\,dx$$

$$= \int_0^1 (-1)\cdot \cos \frac{k\pi x}{2}\,dx + \int_1^2 (1)\cdot \cos \frac{k\pi x}{2}\,dx$$

$$= \left[-\frac{2}{k\pi} \sin \frac{k\pi x}{2} \right]_0^1 + \left[\frac{2}{k\pi} \sin \frac{k\pi x}{2} \right]_1^2$$

$$= -\frac{4}{k\pi} \sin \frac{k\pi}{2}.$$

Therefore, we obtain

$$f(x) = -\frac{4}{\pi} \sum_{k=1}^{\infty} \frac{1}{k} \sin \frac{k\pi}{2} \cos \frac{k\pi x}{2}.$$

Answer $f(x) = -\dfrac{4}{\pi} \left(\cos \dfrac{\pi x}{2} - \dfrac{1}{3} \cos \dfrac{3\pi x}{2} + \dfrac{1}{5} \cos \dfrac{5\pi x}{2} + \cdots \right)$

Figure 10.7 shows the partial sums of the corresponding Fourier series in Example 10.3.

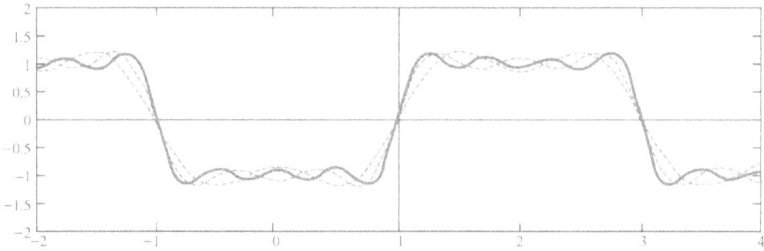

Figure 10.7 Partial sums of the corresponding Fourier series in Example 10.3.

Example 10.4

Find the Fourier series of the odd function $f(x)$ shown in Figure 10.8.

$$f(x) = \begin{cases} -1 & (-1 < x < 0) \\ 1 & (0 < x < 1) \end{cases} \quad \text{and} \quad f(x+2) = f(x).$$

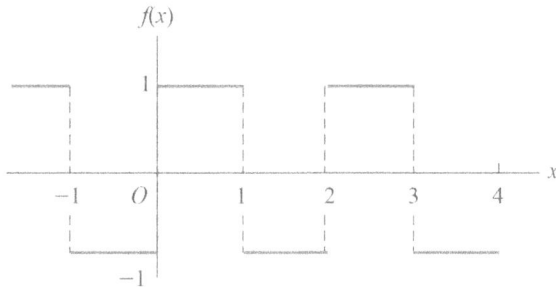

Figure 10.8 Periodic function $f(x)$.

Solution

Since $L = 1$ in the function $f(x) = \sum\limits_{k=1}^{\infty} b_k \sin\dfrac{k\pi}{L}x$, we get the Fourier coefficients

$$b_k = \frac{2}{L} \int_0^L f(x) \sin\frac{k\pi x}{L}\,dx$$

$$= 2\int_0^1 \sin k\pi x\,dx$$

$$= 2\left[-\frac{\cos k\pi x}{k\pi}\right]_0^1 = \frac{2}{\pi}\frac{1-\cos k\pi}{k} = \frac{2}{\pi}\frac{1-(-1)^k}{k}.$$

Therefore, we obtain

$$f(x) = \frac{2}{\pi} \sum_{k=1}^{\infty} \frac{1-(-1)^k}{k} \sin k\pi x.$$

Answer $f(x) = \frac{4}{\pi} \left(\sin \pi x + \frac{\sin 3\pi x}{3} + \frac{\sin 5\pi x}{5} + \cdots \right)$

Figure 10.9 shows the partial sums of the corresponding Fourier series in Example 10.4.

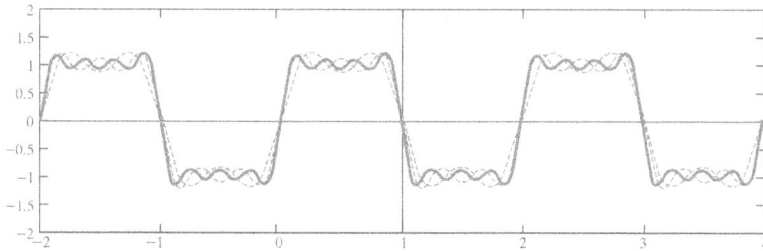

Figure 10.9 Partial sums of the corresponding Fourier series in Example 10.4.

10.1.4 Half-range expansions

Half-range expansions are also Fourier series. When $f(x)$ is a function of period L, we can develop the expanded function into a Fourier series. In other words, a function $f(x)$ with period L, due to boundary conditions, can be represented as an even periodic expansion or an odd periodic expansion, as shown in Figure 10.10. The period after expansion becomes $2L$.

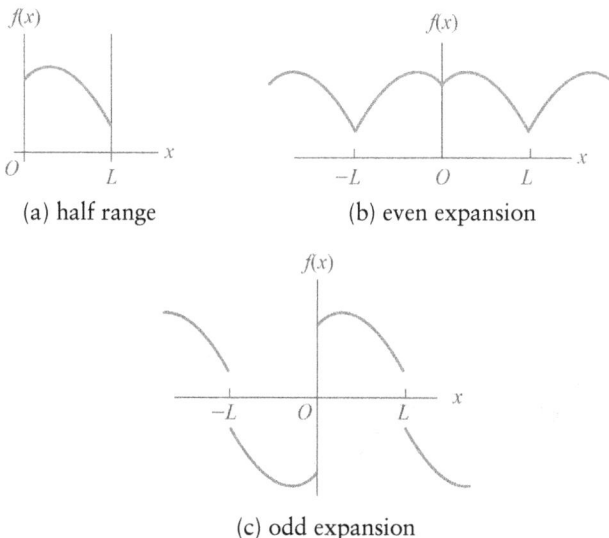

(a) half range

(b) even expansion

(c) odd expansion

Figure 10.10 Even and odd expansions of period $2L$.

Example 10.5

Find the two half-range expansions of the given function shown in Figure 10.11.

$$f(x) = \begin{cases} x & (0 < x < 1) \\ 2-x & (1 < x < 2) \end{cases}$$

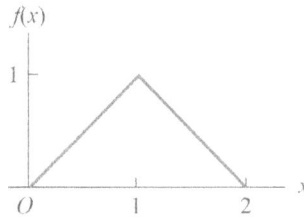

Figure 10.11 Function $f(x)$.

Solution

i. Even periodic expansion

Since $L = 2$ in the function $f(x) = a_0 + \sum_{k=1}^{\infty} a_k \cos \frac{k\pi x}{L}$, we get the Fourier coefficients

$$a_0 = \frac{1}{L} \int_0^L f(x) dx$$

$$= \frac{1}{2} \left(\int_0^1 x \, dx + \int_1^2 (2-x) dx \right) = \frac{1}{2},$$

$$a_k = \frac{2}{L} \int_0^L f(x) \cos \frac{k\pi x}{L} dx$$

$$= \int_0^1 x \cos \frac{k\pi x}{2} dx + \int_1^2 (-x+2) \cos \frac{k\pi x}{2} dx \qquad \text{(refer to the Check)}$$

$$= \left[\frac{x \sin(k\pi x/2)}{k\pi/2} + \frac{\cos(k\pi x/2)}{(k\pi/2)^2} \right]_0^1 + \left[\frac{(-x+2)\sin(k\pi x/2)}{k\pi/2} - \frac{\cos(k\pi x/2)}{(k\pi/2)^2} \right]_1^2$$

$$= \frac{\sin(k\pi/2)}{k\pi/2} + \frac{\cos(k\pi/2)}{(k\pi/2)^2} - \frac{1}{(k\pi/2)^2} - \frac{\cos(k\pi)}{(k\pi/2)^2} - \frac{\sin(k\pi/2)}{k\pi/2} + \frac{\cos(k\pi/2)}{(k\pi/2)^2}$$

$$= \frac{4}{(k\pi)^2} \left(2\cos \frac{k\pi}{2} - 1 - \cos k\pi \right).$$

Therefore, we obtain

$$f(x) = \frac{1}{2} + \frac{4}{\pi^2} \sum_{k=1}^{\infty} \frac{1}{k^2} \left(2\cos\frac{k\pi}{2} - 1 - \cos k\pi \right) \cos\frac{k\pi x}{2}.$$

Answer $f(x) = \frac{1}{2} - \frac{4}{\pi^2} \left(\cos \pi x + \frac{1}{9}\cos 3\pi x + \frac{1}{25}\cos 5\pi x + \cdots \right)$

Figure 10.12 shows the even periodic expansion in Example 10.5.

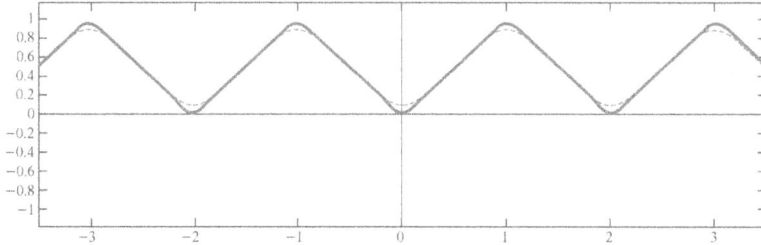

Figure 10.12 Even periodic expansion in Example 10.5.

Check

$$\int x\cos(k\pi x/2)\,dx = x \cdot \frac{\sin(k\pi x/2)}{k\pi/2} - 1 \cdot \left(\frac{-\cos(k\pi x/2)}{(k\pi/2)^2} \right)$$

$$\int (-x+2)\cos(k\pi x/2)\,dx = (-x+2) \cdot \frac{\sin(k\pi x/2)}{k\pi/2} - (-1) \cdot \frac{-\cos(k\pi x/2)}{(k\pi/2)^2}$$

Another Solution

Let's consider $L = 1$ in the given expression.

In $f(x) = a_0 + \sum_{k=1}^{\infty} a_k \cos\frac{k\pi x}{L}$, we get

$$a_0 = \frac{1}{L}\int_0^L f(x)\,dx = \int_0^1 x\,dx = \frac{1}{2},$$

$$a_k = \frac{2}{L}\int_0^L f(x)\cos\frac{k\pi x}{L}\,dx = 2\int_0^1 x\cos k\pi x\,dx \qquad \text{(refer to the Check)}$$

$$= 2\left[\frac{x\sin k\pi x}{k\pi} + \frac{\cos(k\pi x)}{(k\pi)^2} \right]_0^1 = 2\,\frac{\cos k\pi - 1}{(k\pi)^2}.$$

Therefore, we obtain

$$f(x) = \frac{1}{2} + \frac{2}{\pi^2} \sum_{k=1}^{\infty} \frac{\cos k\pi - 1}{k^2} \cos k\pi x.$$

Answer $f(x) = \frac{1}{2} - \frac{4}{\pi^2}\left(\cos \pi x + \frac{1}{9}\cos 3\pi x + \frac{1}{25}\cos 5\pi x + \cdots \right)$

Check

$$\int x \cos k\pi x\, dx = x \cdot \frac{\sin k\pi x}{k\pi} - 1 \cdot \left(\frac{-\cos k\pi x}{(k\pi)^2} \right) = \frac{x\sin k\pi x}{k\pi} + \frac{\cos(k\pi x)}{(k\pi)^2}$$

ii. Odd periodic expansion

Since $L = 2$ in the function $f(x) = \sum_{k=1}^{\infty} b_k \sin \frac{k\pi}{L} x$, we get the Fourier coefficients

$$b_k = \frac{2}{L} \int_0^L f(x)\sin\frac{k\pi x}{L}\, dx$$

$$= \int_0^1 x \sin\frac{k\pi x}{2}\, dx + \int_1^2 (-x+2)\sin\frac{k\pi x}{2}\, dx \qquad \text{(refer to the Check)}$$

$$= \left[\frac{-x\cos(k\pi x/2)}{k\pi/2} + \frac{\sin(k\pi x/2)}{(k\pi/2)^2} \right]_0^1 + \left[\frac{-(-x+2)\cos(k\pi x/2)}{k\pi/2} - \frac{\sin(k\pi x/2)}{(k\pi/2)^2} \right]_1^2$$

$$= -\frac{\cos(k\pi/2)}{k\pi/2} + \frac{\sin(k\pi/2)}{(k\pi/2)^2} - \frac{\sin(k\pi)}{(k\pi/2)^2} + \frac{\cos(k\pi/2)}{k\pi/2} + \frac{\sin(k\pi/2)}{(k\pi/2)^2}$$

$$= \frac{8}{(k\pi)^2}\sin\frac{k\pi}{2}$$

Therefore, we obtain

$$f(x) = \frac{8}{\pi^2} \sum_{k=1}^{\infty} \frac{1}{k^2}\sin\frac{k\pi}{2}\sin\frac{k\pi x}{2}.$$

Answer $f(x) = \frac{8}{\pi^2}\left(\sin\frac{\pi x}{2} - \frac{1}{3^2}\sin\frac{3\pi x}{2} + \frac{1}{5^2}\sin\frac{5\pi x}{2} + -\cdots \right)$

Figure 10.13 shows the odd periodic expansion in Example 10.5.

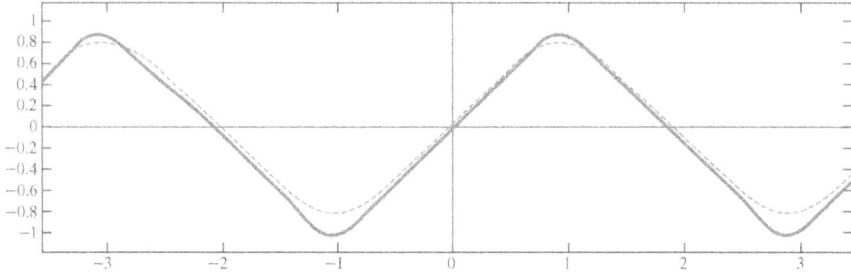

Figure 10.13 Odd periodic expansion in Example 10.5.

Check

$$\int x \sin \frac{k\pi x}{2} \, dx = x \cdot \frac{-\cos(k\pi x/2)}{k\pi/2} - 1 \cdot \frac{-\sin(k\pi x/2)}{(k\pi/2)^2}$$

$$\int (-x+2)\sin(k\pi x/2) \, dx = (-x+2) \cdot \frac{-\cos(k\pi x/2)}{k\pi/2} - (-1) \cdot \frac{-\sin(k\pi x/2)}{(k\pi/2)^2}$$

10.1.5 Orthogonality

The individual trigonometric functions comprising Fourier series equation (10.1) $f(x) = a_0 + \sum_{k=1}^{\infty}(a_k \cos kx + b_k \sin kx)$, satisfy the following equation for arbitrary integers m and n over the interval $-\pi \le x \le \pi$.

$$\int_{-\pi}^{\pi} \cos mx \cos nx \, dx = 0 \qquad (m \ne n) \tag{10.15a}$$

$$\int_{-\pi}^{\pi} \sin mx \sin nx \, dx = 0 \qquad (m \ne n) \tag{10.15b}$$

$$\int_{-\pi}^{\pi} \sin mx \cos nx \, dx = 0 \qquad (m \ne n \text{ or } m = n) \tag{10.15c}$$

Therefore, Fourier series equation (10.1) has *orthogonality*. As a result, Fourier series is referred to as an orthogonal series. Additionally, each trigonometric function is orthogonal to each other. This concept of orthogonality is similar to the notion of vectors being orthogonal when their inner product is zero.

i. Proof of Eq. (10.15a) $(m \neq n)$

$$\int_{-\pi}^{\pi} \cos mx \cos nx\, dx = \frac{1}{2}\int_{-\pi}^{\pi}\cos(m+n)x\, dx + \frac{1}{2}\int_{-\pi}^{\pi}\cos(m-n)x\, dx = 0$$

ii. Proof of Eq. (10.15b) $(m \neq n)$

$$\int_{-\pi}^{\pi} \sin mx \sin nx\, dx = -\frac{1}{2}\int_{-\pi}^{\pi}\cos(m+n)x\, dx + \frac{1}{2}\int_{-\pi}^{\pi}\cos(m-n)x\, dx = 0$$

iii. Proof of Eq. (10.15c) $(m \neq n \text{ or } m = n)$

$$\int_{-\pi}^{\pi} \sin mx \cos nx\, dx = \frac{1}{2}\int_{-\pi}^{\pi}\sin(m+n)x\, dx + \frac{1}{2}\int_{-\pi}^{\pi}\sin(m-n)x\, dx = 0$$

If, for a given interval $a \leq x \leq b$, distinct integers m and n, the functions $y_m(x)$ and $y_n(x)$ satisfy the following equation involving a *weight function* $r(x)\,(>0)$, then these functions $y_m(x)$ and $y_n(x)$ are called *generalized orthogonal functions*:

$$(y_m, y_n) = \int_a^b r(x)y_m(x)y_n(x)\,dx = 0 \qquad (m \neq n). \qquad (10.16)$$

Furthermore, the *norm* of $y_n(x)$ denoted by $\| y_n \|$ is defined as follows.

$$\|y_n\| = \sqrt{(y_n, y_n)} = \sqrt{\int_a^b r(x)y_n^2(x)\,dx} \qquad (10.17)$$

And the functions y_m, y_n are called *orthonormal* on the interval $a \leq x \leq b$ if they are orthogonal on this interval and all have norm 1.

$$(y_m, y_n) = \int_a^b r(x)y_m(x)y_n(x)\,dx = 0 \qquad (m \neq n) \qquad (10.18a)$$

$$(y_m, y_n) = \int_a^b r(x)y_m(x)y_n(x)\,dx = 1 \qquad (m = n) \qquad (10.18b)$$

Example 10.6

a. Show the functions $\cos 2x$ and $\cos 3x$ are orthogonal on the interval $0 \leq x \leq 2\pi$.
b. Find the norm $\|\cos 2x\|$.

Solution

a. $(\cos 2x, \cos 3x) = \int_0^{2\pi} \cos 2x \cos 3x \, dx$

$$= \frac{1}{2}\int_0^{2\pi} \cos(3+2)x \, dx + \frac{1}{2}\int_0^{2\pi} \cos(3-2)x \, dx = 0$$

Answer orthogonal

b. $\|\cos 2x\| = \sqrt{(\cos 2x, \cos 2x)} = \sqrt{\int_0^{2\pi} \cos^2 2x \, dx}$

$$= \sqrt{\int_0^{2\pi} \frac{1+\cos 4x}{2} \, dx} = \sqrt{\frac{1}{2}\left[x + \frac{\sin 4x}{4} \right]_0^{2\pi}} = \sqrt{\pi}$$

Answer $\|\cos 2x\| = \sqrt{\pi}$

10.1.6 (*optional*) Orthogonality of eigenfunctions of Sturm-Liouville problems

To gain a better understanding of generalized orthogonal functions, let's introduce a *Sturm-Liouville problem* expressed as Eq. (10.19). Consider a second-order ODE of the following form on some interval $a \le x \le b$.

$$\left[p(x)y(x)'\right]' + \{q(x) + sr(x)\}y(x) = 0 \tag{10.19}$$

where s represents the eigenvalue (parameter), and this equation is called the Sturm-Liouville equation. Let's assume that it satisfies the following boundary conditions at $x = a$ and $x = b$.

$$c_1 y'(a) + c_2 y(a) = 0 \tag{10.20a}$$

$$d_1 y'(b) + d_2 y(b) = 0 \tag{10.20b}$$

where at least one of the real constants c_1, c_2, d_1, d_2 must not be zero.

For example, from an ODE $y'' + sy = 0$, $y(0) = 0$, $y(\pi) = 0$, let's find eigenvalues and its corresponding eigenvectors.

We can recognize that this ODE is a Sturm-Liouville equation with $p(x) = 1$, $q(x) = 0$, $r(x) = 1$ as given in Eq. (10.19).

Furthermore, $a = 0$ and $b = \pi$, and it has boundary condition with $c_1 = 0$, $c_2 = 1$, $d_1 = 0$, and $d_2 = 1$ as given in Eq. (10.20).

Let $y = Ce^{\lambda x}$ represent the solution to the ODE, and let's obtain the characteristic equation as:

$$\lambda^2 + s = 0. \tag{10.21}$$

Then we obtain the solution of the ODE as:

(Case I) $s < 0$

Since the characteristic equation has two distinct real roots $\lambda = \pm\sqrt{-s}$, the general solution to the equation is given by:

$$y = C_1 e^{\sqrt{-s}\,x} + C_2 e^{-\sqrt{-s}\,x}. \tag{10.22}$$

Incorporating the boundary conditions $y(0) = 0$, $y(\pi) = 0$, we get:

$$C_1 + C_2 = 0, \qquad C_1 e^{\sqrt{-s}\,\pi} + C_2 e^{-\sqrt{-s}\,\pi} = 0$$

or

$$C_1 = 0, \quad C_2 = 0. \tag{10.23}$$

Thus, at $s < 0$, where $y \equiv 0$, there are no eigenfunctions.

(Case II) $s = 0$

Since the characteristic equation has a double root $\lambda = 0$, the general solution to the equation is given by:

$$y = C_1 + C_2 x. \tag{10.24}$$

Incorporating the boundary condition $y(0) = 0$, $y(\pi) = 0$, we get:

$$C_1 = 0, \quad C_2 = 0. \tag{10.25}$$

Thus, at $s = 0$, where $y \equiv 0$, there are no eigenfunctions.

(Case III) $s > 0 \left(s = v^2\right)$

Since the characteristic equation has complex conjugate roots $\lambda = \pm vi$, the general solution to the equation is given by:

$$y(x) = A\cos vx + B\sin vx. \tag{10.26}$$

Incorporating the boundary conditions $y(0) = 0$, $y(\pi) = 0$, we get:

$$A = 0, \qquad B\sin v\pi = 0. \tag{10.27}$$

Satisfying $\sin v\pi = 0$ as in Eq. (10.27), we find

$$v = 1, \ 2, \ 3, \ \cdots.$$

Therefore, the eigenvalues $\left(s=v^2\right)$ are $s=1^2, 2^2, 3^2, \cdots$. By setting $B=1$, the eigenfunctions are

$$y(x)=\sin vx \quad (v=1, 2, 3, \cdots).$$

We can observe from Eq. (10.19) that the Sturm-Liouville equation with respect to $p(x)$, $q(x)$, and $r(x)$ has numerous eigenvalues and their corresponding eigenfunctions. Now, let's verify the orthogonality of these eigenfunctions.

Remark   Orthogonality of eigenfunctions of Sturm-Liouville problems

If for distinct eigenvalues λ_m and λ_n in the interval $a \le x \le b$, the functions y_m and y_n are each solution to the Sturm-Liouville equation, then with respect to the weight function $r(x)$ (>0), the functions y_m and y_n are orthogonal to each other. In other words:

$$\left(y_m, y_n\right)=\int_a^b r(x)y_m(x)y_n(x)dx=0 \qquad (m \ne n) \qquad (10.16)$$

If the functions y_m and y_n satisfy the Sturm-Liouville equation (10.19), then the following expression holds true:

$$\left[py_m'\right]'+\left(q+\lambda_m r\right)y_m=0 \qquad (10.28a)$$

$$\left[py_n'\right]'+\left(q+\lambda_n r\right)y_n=0 \qquad (10.28b)$$

Multiplying y_n to Eq. (10.28a) and y_m to Eq. (10.28b), then subtracting the two equations, we get:

$$\left(\lambda_m-\lambda_n\right)ry_my_n = y_m\left[py_n'\right]'-y_n\left[py_m'\right]'$$

$$=\left\{y_m'\left(py_n'\right)+y_m\left[py_n'\right]'\right\}-\left\{y_n'\left(py_m'\right)+y_n\left[y_m'\right]'\right\}$$

$$=\left[y_m\left(py_n'\right)-y_n\left(py_m'\right)\right]'$$

$$=\left[p\left(y_my_n'-y_m'y_n\right)\right]'$$

Integrating this over the interval $a \leq x \leq b$, we get:

$$\left(\lambda_m - \lambda_n\right) \int_a^b r y_m y_n \, dx = \left[p\left(y_m y_n' - y_m' y_n\right)\right]_a^b$$

$$= p(b)\left\{y_m(b)y_n'(b) - y_m'(b)y_n(b)\right\} - p(a)\left\{y_m(a)y_n'(a) - y_m'(a)y_n(a)\right\}$$

$$(10.29)$$

Since $\lambda_m \neq \lambda_n$, if the right-hand side of Eq. (10.29) becomes zero, orthogonality will be demonstrated. Therefore, let's divide the case based on the conditions of $p(a)$ and $p(b)$.

i. In case of $p(a) = 0$ and $p(b) = 0$, the right-hand side of Eq. (10.29) becomes zero.

ii. In case of $p(a) = 0$ and $p(b) \neq 0$, we need to show that the front term $y_m(b)y_n'(b) - y_m'(b)y_n(b)$ on the right-hand side of Eq. (10.29) becomes zero.

From Eq. (10.20b), we get the following equations:

$$d_1 y_n'(b) + d_2 y_n(b) = 0 \qquad (10.30a)$$

$$d_1 y_m'(b) + d_2 y_m(b) = 0 \qquad (10.30b)$$

First, assuming $d_1 \neq 0$, subtracting the equation obtained by multiplying $y_m(b)$ to Eq. (10.30a) from the equation obtained by multiplying $y_n(b)$ to Eq. (10.30b), we get:

$$d_1\left\{y_m(b)y_n'(b) - y_m'(b)y_n(b)\right\} = 0$$

or

$$y_m(b)y_n'(b) - y_m'(b)y_n(b) = 0. \qquad (10.31)$$

Now, assuming $d_2 \neq 0$, subtracting the equation obtained by multiplying $y_m'(b)$ to Eq. (10.30a) from the equation obtained by multiplying $y_n'(b)$ to Eq. (10.30b), we get:

$$d_2\left\{y_m(b)y_n'(b) - y_m'(b)y_n(b)\right\} = 0$$

or

$$y_m(b)y_n'(b) - y_m'(b)y_n(b) = 0. \qquad (10.32)$$

iii. In case of $p(a) \neq 0$ and $p(b) = 0$, we need to show that the last term $y_m(a)y_n'(a) - y_m'(a)y_n(a)$ on the right-hand side of Eq. (10.29) becomes zero.

This is proven similarly to the case in (ii).

iv. In case of $p(a) \neq 0$ and $p(b) \neq 0$, we need to demonstrate that both terms on the right-hand side of Eq. (10.29) become zero. This is also proven similarly to the cases (ii) and (iii).

Example 10.7

Show the eigenfunctions of the following equation on the interval $-1 \leq x \leq 1$.

$$\left(1 - x^2\right)y'' - 2xy + n(n+1)y = 0$$

Solution

The given Legendre equation can be expressed as follows:

$$[(1 - x^2)y']' + sy = 0 \qquad \text{where } s = n(n+1).$$

This is a Sturm-Liouville equation with $p(x) = 1 - x^2$, $q(x) = 0$, $r(x) = 1$. And we can see that the boundary conditions are $y = 0$ at $x = 1$ and $x = -1$.

Therefore, for $n = 0, 1, 2, 3, \cdots$, i.e., $s = 0, 1\cdot2, 2\cdot3, 3\cdot4, \cdots$, the solutions $y_n(x)$ of the Legendre equation are the eigenfunctions of this problem. In other words, all eigenfunctions of the Legendre equation are orthogonal to each other.

$$\int_{-1}^{1} y_m(x)y_n(x)dx = 0.$$

Problem 10.1

Find the Fourier series of the given function $f(x)$ with period 2π. [1 ~ 2]

1. $f(x) = \begin{cases} 0 & (-\pi < x < 0) \\ \pi - x & (0 < x < \pi) \end{cases}$ (see Figure 10.14)

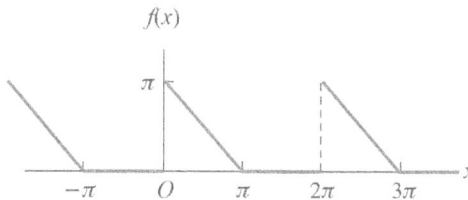

Figure 10.14 Periodic function $f(x)$.

2. $f(x) = \begin{cases} 0 & (-\pi < x < 0) \\ \sin x & (0 < x < \pi) \end{cases}$ (see Figure 10.15)

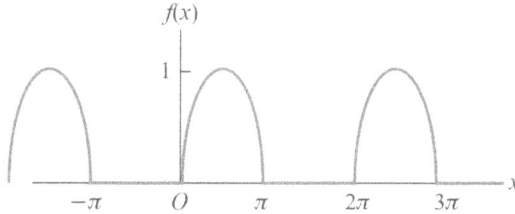

Figure 10.15 Periodic function $f(x)$.

Find the Fourier series of the given function $f(x)$. [3 ~ 4]

3. $f(x) = \begin{cases} 1 & (0 < x < 1) \\ 0 & (1 < x < 2) \end{cases}$ and $f(x+2) = f(x)$ (see Figure 10.16)

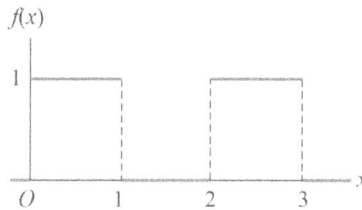

Figure 10.16 Periodic function $f(x)$.

4. $f(x) = \begin{cases} \sin \pi x & (0 < x < 1) \\ 0 & (1 < x < 2) \end{cases}$ and $f(x+2) = f(x)$ (see Figure 10.17)

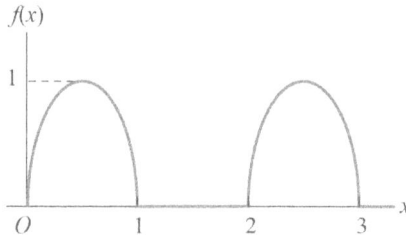

Figure 10.17 Periodic function $f(x)$.

Find the Fourier series of the given function $f(x)$. [5 ~ 10]

5. $f(x) = \begin{cases} 2x & (0 < x < 2) \\ 4 - 2x & (2 < x < 4) \end{cases}$ and $f(x+4) = f(x)$ (see Figure 10.18)

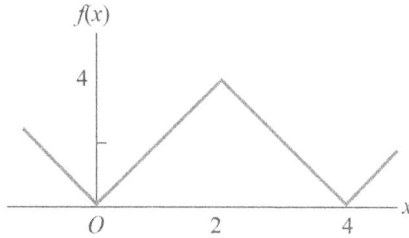

Figure 10.18 Periodic function $f(x)$.

6. $f(x) = \dfrac{x^2}{\pi}$ $(-\pi < x < \pi)$ and $f(x+2\pi) = f(x)$ (see Figure 10.19)

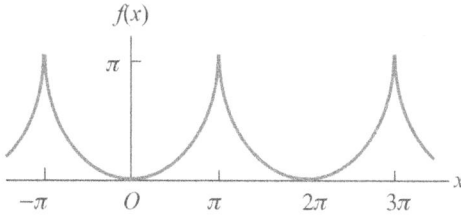

Figure 10.19 Periodic function $f(x)$.

7. $f(x) = x$ $(-\pi < x < \pi)$ and $f(x+2\pi) = f(x)$ (see Figure 10.20)

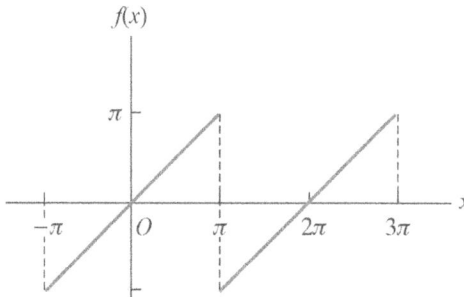

Figure 10.20 Periodic function $f(x)$.

8. $f(x) = \begin{cases} -x^2 & (-2 < x < 0) \\ x^2 & (0 < x < 2) \end{cases}$ and $f(x+4) = f(x)$ (see Figure 10.21)

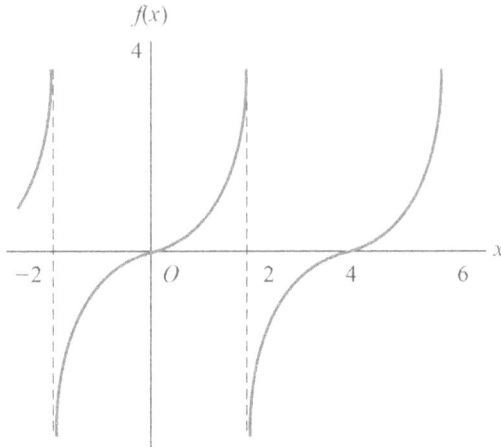

Figure 10.21 Periodic function $f(x)$.

9. $f(x) = x$ $(0 < x < 3)$ and $f(x+3) = f(x)$ (see Figure 10.22)

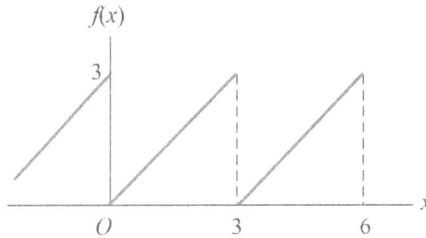

Figure 10.22 Periodic function $f(x)$.

10. $f(x) = x + 1$ $(-1 < x < 1)$ and $f(x+2) = f(x)$ (see Figure 10.23)

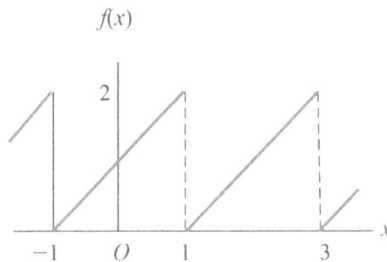

Figure 10.23 Periodic function $f(x)$.

Find the two half-range expansions of the given function $f(x)$. [11 ~ 14]

11. $f(x) = x \qquad (0 < x < 1)$

12. $f(x) = 2 - 2x \qquad (0 < x < 1)$

13. $f(x) = \begin{cases} 0 & (0 < x < 1) \\ 1 & (1 < x < 2) \end{cases}$

14. $f(x) = \begin{cases} 1 & (0 < x < 1) \\ -1 & (1 < x < 2) \end{cases}$

Determine the orthogonality on the interval $0 \le x \le 2\pi$. [15 ~ 18]

15. $\sin 3x, \cos 2x$
16. $\sin 2x, \cos 2x$
17. $\cos x, \cos 3x$
18. $\sin 2x, \sin x$

Find the norm on the interval $0 \le x \le 2\pi$. [19 ~ 20]

19. $\sin 2x$
20. $\cos x$

10.2 Fourier integral

Fourier series is a powerful tool in solving various problems involving periodic functions. However, in many practical problems, a significant number of problems cannot be represented as periodic functions.

While Fourier series expresses periodic functions as infinite sums of sine or cosine functions, the method to represent nonperiodic functions is through *Fourier integrals*. The Fourier integral aids in better understanding and analyzing the characteristics of nonperiodic functions.

10.2.1 Definition of Fourier integral

Let's consider a periodic function with a period of $2L$.

$$f_L(x) = a_0 + \sum_{k=1}^{\infty} (a_k \cos w_k x + b_k \sin w_k x) \tag{10.33}$$

where $w_k = \dfrac{k\pi}{L}, \quad (k = 1, 2, 3, \cdots)$ \hfill (10.34a)

$$a_0 = \frac{1}{2L} \int_{-L}^{L} f_L(v) dv, \tag{10.34b}$$

$$a_k = \frac{1}{L}\int_{-L}^{L} f_L(v)\cos w_k v\, dv, \qquad (10.34c)$$

$$b_k = \frac{1}{L}\int_{-L}^{L} f_L(v)\sin w_k v\, dv. \qquad (10.34d)$$

If we let $L \to \infty$, we would be representing the nonperiodic function $f(x)$ without a defined period.

$$f(x) = \lim_{L \to \infty} f_L(x) \qquad (10.35)$$

Firstly, summarizing Eq. (10.34b), we have the following:

$$\lim_{L \to \infty} a_0 = \lim_{L \to \infty} \frac{1}{2L}\int_{-L}^{L} f_L(v)\, dv = 0. \qquad (10.36)$$

From equation (10.34a), we have $\Delta w = w_{k+1} - w_k = \dfrac{(k+1)\pi}{L} - \dfrac{k\pi}{L} = \dfrac{\pi}{L}$, so applying $\dfrac{1}{L} = \dfrac{\Delta w}{\pi}$, and substituting Eq. (10.35) and Eq. (10.36) into Eq. (10.33), and then simplifying, we get the following:

$$f(x) = \lim_{L \to \infty} \sum_{k=1}^{\infty}\left\{A(w_k)\cos w_k x + B(w_k)\sin w_k x\right\}\Delta w \qquad (10.37)$$

where $A(w_k) = \dfrac{1}{\pi}\lim_{L \to \infty}\int_{-L}^{L} f_L(x)\cos w_k x\, dx$ and $B(w_k) = \dfrac{1}{\pi}\lim_{L \to \infty}\int_{-L}^{L} f_L(x)\sin w_k x\, dx$.

Therefore, we can summarize as follows:

Remark   Fourier integral of nonperiodic function $f(x)$

$$f(x) = \int_{0}^{\infty}\left\{A(w)\cos wx + B(w)\sin wx\right\}dw \qquad (10.38)$$

where $A(w) = \dfrac{1}{\pi}\int_{-\infty}^{\infty} f(x)\cos wx\, dx$ \qquad (10.39a)

$$B(w) = \dfrac{1}{\pi}\int_{-\infty}^{\infty} f(x)\sin wx\, dx \qquad (10.39b)$$

Example 10.8

Find Fourier integral representation of the function shown in Figure 10.24.

$$f(x) = \begin{cases} 1 & (|x| < 1) \\ 0 & (|x| > 1) \end{cases}$$

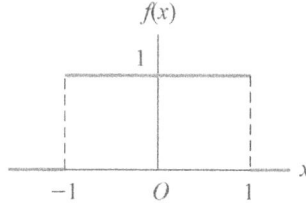

Figure 10.24 Nonperiodic function f(x).

Solution

In $f(x) = \int_0^\infty \{A(w)\cos wx + B(w)\sin wx\} dw$, we get

$$A(w) = \frac{1}{\pi}\int_{-\infty}^{\infty} f(x)\cos wx \, dx = \frac{1}{\pi}\int_{-1}^{1} 1 \cdot \cos wx \, dx = \frac{2\sin w}{\pi w},$$

$$B(w) = \frac{1}{\pi}\int_{-\infty}^{\infty} f(x)\sin wx \, dx = \frac{1}{\pi}\int_{-1}^{1} 1 \cdot \sin wx \, dx = 0.$$

Therefore, we obtain

$$f(x) = \frac{2}{\pi}\int_0^\infty \left(\frac{\sin w}{w}\cos wx\right) dw \qquad \qquad ①$$

$$\textbf{Answer } f(x) = \frac{2}{\pi}\int_0^\infty \left(\frac{\sin w}{w}\cos wx\right) dw$$

Remark   Sinc function

Rephrasing the solution to this problem, we have

$$\frac{2}{\pi}\int_0^\infty \left(\frac{\sin w}{w}\cos wx\right) dw = \begin{cases} 1 & (0 \le x < 1) \\ 1/2 & (x = 1) \\ 0 & (x > 1) \end{cases} \qquad ②$$

At $x = 1$, the average of the left-hand limit of $f(x)$ which is 1, and the right-hand limit of $f(x)$ which is 0 is calculated. And we get

$$\int_0^\infty \left(\frac{\sin w}{w} \cos wx \right) dw = \begin{cases} \pi/2 & (0 \le x < 1) \\ \pi/4 & (x = 1) \\ 0 & (x > 1) \end{cases} \qquad \text{③}$$

Specially, at $x = 0$ in Eq. ③, then we obtain

$$\int_0^\infty \frac{\sin w}{w} dw = \frac{\pi}{2}, \qquad \text{④}$$

where we refer to $g(x) = \dfrac{\sin x}{x}$ as the Sinc function, satisfying $g(0) = 1$.
Figure 10.25 shows the graph of $g(x) = \dfrac{\sin x}{x}$.

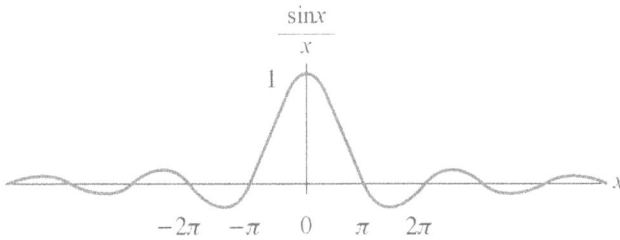

Figure 10.25 Sinc function.

10.2.2 *Fourier cosine integral and Fourier sine integral*

Similar to Example 10.8, if the function $f(x)$ is an even function, meaning $f(-x) = f(x)$ then $B(w) = 0$ in Eq. (10.39b). Furthermore, if the function $f(x)$ is an odd function, meaning $f(-x) = -f(x)$, then $A(w) = 0$ in Eq. (10.39a).

Therefore, summarizing when a function is even and odd, we have the following:

Remark Fourier cosine integral and Fourier sine integral

i. Fourier cosine integral of nonperiodic even function $f(x)$

$$f(x) = \int_0^\infty A(w) \cos wx \, dw \qquad (10.40)$$

where $A(w) = \dfrac{2}{\pi} \int_0^\infty f(x) \cos wx \, dx.$ $\qquad (10.41)$

In the computation of $A(w)$, the $f(x)\cos wx$ (the product of an even function $f(x)$ and an even function $\cos wx$ is even, so the integral over the interval $(-\infty, \infty)$ is twice the integral over the interval $(0, \infty)$.

ii. **Fourier sine integral of nonperiodic odd function $f(x)$**

$$f(x) = \int_0^\infty B(w)\sin wx\, dw \qquad (10.42)$$

$$\text{where } B(w) = \frac{2}{\pi}\int_0^\infty f(x)\sin wx\, dx. \qquad (10.43)$$

In the computation of $B(w)$, the $f(x)\sin wx$ (the product of an odd function $f(x)$ and an odd function $\sin wx$ is even, so the integral over the interval $(-\infty, \infty)$ is twice the integral over the interval $(0, \infty)$.

Example 10.9

Find Fourier integral representation of the nonperiodic even function shown in Figure 10.26.

$$f(x) = \begin{cases} 1-x & (0 \le x < 1) \\ 0 & (x > 1) \end{cases}$$

$$f(-x) = f(x)$$

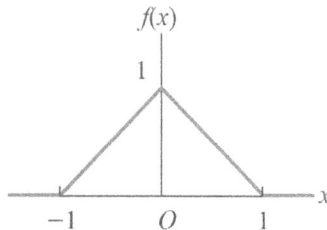

Figure 10.26 Nonperiodic function $f(x)$.

Solution

In Eq. (10.40) and Eq. (10.41)

$$f(x) = \int_0^\infty A(w)\cos wx\, dw$$

$$A(w) = \frac{2}{\pi}\int_0^\infty f(x)\cos wx\, dx$$

$$= \frac{2}{\pi}\int_0^1 (1-x)\cdot \cos wx\, dx = \frac{2(1-\cos w)}{\pi w^2}$$

Therefore, we obtain the following answer.

$$\textbf{Answer } f(x) = \frac{2}{\pi}\int_0^\infty \left(\frac{1-\cos w}{w^2}\right)\cos wx\, dw$$

Example 10.10

Find Fourier integral representation of the nonperiodic odd function shown in Figure 10.27.

$$f(x) = \begin{cases} 1 & (0 \le x < 1) \\ 0 & (x > 1) \end{cases}$$

$$f(-x) = -f(x)$$

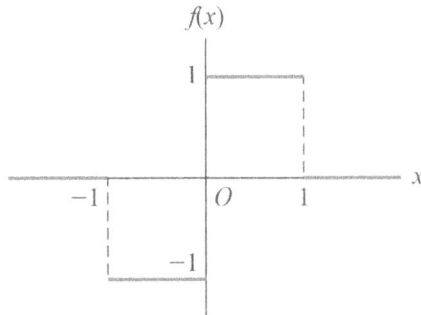

Figure 10.27 Nonperiodic function $f(x)$.

Solution

From Eq. (10.42) and Eq. (10.43), we get

$$f(x) = \int_0^\infty B(w)\sin wx\, dw,$$

$$B(w) = \frac{2}{\pi}\int_0^\infty f(x)\sin wx\, dx$$

$$= \frac{2}{\pi}\int_0^1 1\cdot\sin wx\, dx = \frac{2(1-\cos w)}{\pi w}.$$

Therefore, we obtain the following answer.

$$\textbf{Answer } f(x) = \frac{2}{\pi}\int_0^\infty \left(\frac{1-\cos w}{w}\sin wx\right) dw$$

Example 10.11

Find Fourier integral representation of the nonperiodic even function shown in Figure 10.28.

$$f(x) = e^{-kx} \qquad (x > 0,\ k > 0) \text{ and } f(-x) = f(x)$$

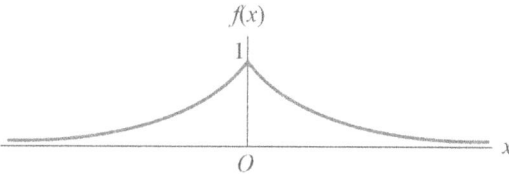

Figure 10.28 Nonperiodic function $f(x)$.

Solution

From Eq. (10.40) and Eq. (10.41), we get

$$A(w) = \frac{2}{\pi}\int_0^\infty f(x)\cos wx\, dx$$

$$= \frac{2}{\pi}\int_0^\infty e^{-kx}\cos wx\, dx = \frac{2k}{\pi(k^2+w^2)}. \qquad \text{(refer to the Check)}$$

Therefore, we obtain

$$f(x) = \frac{2}{\pi} \int_0^\infty \left(\frac{k}{k^2 + w^2} \cos wx \right) dw \qquad \text{①}$$

Answer $f(x) = \dfrac{2}{\pi} \displaystyle\int_0^\infty \left(\dfrac{k}{k^2 + w^2} \cos wx \right) dw$

Check $\displaystyle\int_0^\infty \frac{\cos wx}{k^2 + w^2} dw = \frac{\pi}{2k} e^{-kx} \qquad (x > 0,\ k > 0)$

From Eq. ①, we get

$$\frac{2}{\pi} \int_0^\infty \left(\frac{k}{k^2 + w^2} \cos wx \right) dw = e^{-kx} \qquad \text{②}$$

Therefore, we obtain

$$\int_0^\infty \frac{\cos wx}{k^2 + w^2} dw = \frac{\pi}{2k} e^{-kx}. \qquad (x > 0,\ k > 0) \qquad \text{③}$$

Check $\displaystyle\int_0^\infty e^{-kx} \cos wx \, dx = \frac{k}{k^2 + w^2}$

$$I = 1 \int e^{-kx} \cos wx \, dx$$

$$= \frac{e^{-kx}}{-k} \cdot \cos wx - \int \frac{e^{-kx}}{-k} \cdot (-w \sin wx) \, dx$$

$$= \frac{e^{-kx}}{-k} \cdot \cos wx - \frac{e^{-kx}}{k^2} \cdot (-w \sin wx) + \int \frac{e^{-kx}}{k^2} \cdot (-w^2 \cos wx) \, dx$$

$$= \frac{e^{-kx}}{-k} \cdot \cos wx - \frac{e^{-kx}}{k^2} \cdot (-w \sin wx) - \frac{w^2}{k^2} I.$$

Then we get

$$(k^2 + w^2) I = e^{-kx} (-k \cos wx + w \sin wx).$$

Therefore, we obtain

$$\left(k^2 + w^2\right)\int_0^\infty e^{-kx}\cos wx\, dx = e^{-kx}\left(-k\cos wx + w\sin wx\right)\Big|_0^\infty = k,$$

or

$$\int_0^\infty e^{-kx}\cos wx\, dx = \frac{k}{k^2 + w^2}$$

Example 10.12

Find Fourier integral representation of the nonperiodic odd function shown in Figure 10.29.

$$f(x) = e^{-kx} \qquad (x > 0,\ k > 0) \text{ and } f(-x) = -f(x)$$

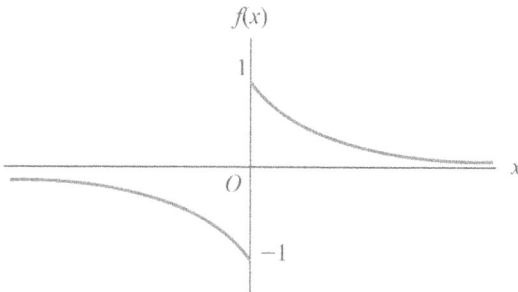

Figure 10.29 Nonperiodic function $f(x)$.

Solution

From Eq. (10.40) and Eq. (10.41), we get

$$f(x) = \int_0^\infty B(w)\sin wx\, dw$$

$$B(w) = \frac{2}{\pi}\int_0^\infty f(x)\sin wx\, dx$$

$$= \frac{2}{\pi}\int_0^\infty e^{-kx}\sin wx\, dx = \frac{2w}{\pi\left(k^2 + w^2\right)}. \qquad \text{(refer to the Check)}$$

Therefore, we obtain

$$f(x) = \frac{2}{\pi} \int_0^\infty \left(\frac{w}{k^2 + w^2} \sin wx \right) dw \qquad \text{①}$$

Answer $f(x) = \frac{2}{\pi} \int_0^\infty \left(\frac{w}{k^2 + w^2} \sin wx \right) dw$

Check $\int_0^\infty \frac{w \sin wx}{k^2 + w^2} dw = \frac{\pi}{2} e^{-kx} \qquad (x > 0,\ k > 0)$

From Eq. ①, we get

$$\frac{2}{\pi} \int_0^\infty \left(\frac{w}{k^2 + w^2} \sin wx \right) dw = e^{-kx} \qquad \text{②}$$

Therefore, we obtain

$$\int_0^\infty \frac{w \sin wx}{k^2 + w^2} dw = \frac{\pi}{2} e^{-kx} \qquad (x > 0,\ k > 0) \qquad \text{③}$$

Check $\int_0^\infty e^{-kx} \sin wx\, dx = \frac{w}{k^2 + w^2}$

$$I = \int e^{-kx} \sin wx\, dx$$

$$= \frac{e^{-kx}}{-k} \cdot \sin wx - \int \frac{e^{-kx}}{-k} \cdot (w \cos wx)\, dx$$

$$= \frac{e^{-kx}}{-k} \cdot \sin wx - \frac{e^{-kx}}{k^2} \cdot (w \cos wx) + \int \frac{e^{-kx}}{k^2} \cdot (-w^2 \sin wx)\, dx$$

$$= \frac{e^{-kx}}{-k} \cdot \sin wx - \frac{e^{-kx}}{k^2} \cdot (w \cos wx) - \frac{w^2}{k^2} I.$$

Then we get

$$(k^2 + w^2) I = -e^{-kx} (k \sin wx + w \cos wx),$$

or

$$\left(k^2+w^2\right)\int_0^\infty e^{-kx}\sin wx\,dx = -e^{-kx}\left(k\sin wx+w\cos wx\right)\Big|_0^\infty = w.$$

Therefore, we obtain

$$\int_0^\infty e^{-kx}\sin wx\,dx = \frac{w}{k^2+w^2}$$

Problem 10.2

Using the Fourier integral representation, show that [1 ~ 4]

1. $\displaystyle\int_0^\infty\left(\frac{\sin w}{w}\cos wx\right)dw = \begin{cases}\dfrac{\pi}{2} & (0\le x<1)\\[2mm]\dfrac{\pi}{4} & (x=1)\\[2mm]0 & (x>1)\end{cases}$ [hint: $f(x)=1$]

2. $\displaystyle\int_0^\infty\frac{\cos(\pi w/2)}{1-w^2}\cos wx\,dw = \begin{cases}\dfrac{\pi}{2}\cos x & (0\le x\le\pi/2)\\[2mm]0 & (x>\pi/2)\end{cases}$ [hint: $f(x)=\cos x$]

3. $\displaystyle\int_0^\infty\left(\frac{-w\cos w+\sin w}{w^2}\sin wx\right)dw = \begin{cases}\pi x/2 & (0\le x<1)\\ \pi/4 & (x=1)\\ 0 & (x>1)\end{cases}$ [hint: $f(x)=x$]

4. $\displaystyle\int_0^\infty\left(\frac{\sin 2\pi w}{w^2-1}\sin wx\right)dw = \begin{cases}\dfrac{\pi}{2}\sin x & (0\le x\le 2\pi)\\[2mm]0 & (x>2\pi)\end{cases}$ [hint: $f(x)=\sin x$]

Find Fourier cosine integral representation of the following function. [5 ~ 8]

5. $f(x)=\begin{cases}x & (0\le x<1)\\0 & (x>1)\end{cases}$

6. $f(x)=\begin{cases}e^{-x} & (0\le x<1)\\0 & (x>1)\end{cases}$

7. $f(x) = \begin{cases} \cos x & (0 < x < \pi/2) \\ 0 & (x > \pi/2) \end{cases}$

8. $f(x) = \begin{cases} \sin x & (0 < x < \pi/2) \\ 0 & (x > \pi/2) \end{cases}$

Find Fourier sine integral representation of the following function. [9 ~ 12]

9. $f(x) = \begin{cases} x & (0 \le x < 1) \\ 0 & (x > 1) \end{cases}$

10. $f(x) = \begin{cases} e^{-x} & (0 \le x < 1) \\ 0 & (x > 1) \end{cases}$

11. $f(x) = \begin{cases} \cos x & (0 < x < \pi/2) \\ 0 & (x > \pi/2) \end{cases}$

12. $f(x) = \begin{cases} \sin x & (0 < x < \pi/2) \\ 0 & (x > \pi/2) \end{cases}$

Answer

Problem 10.1

1. $f(x) = \dfrac{\pi}{4} + \dfrac{2}{\pi}\left(\cos x + \dfrac{\cos 3x}{3^2} + \cdots\right) + \left(\sin x + \dfrac{\sin 2x}{2} + \dfrac{\sin 3x}{3} + \cdots\right)$

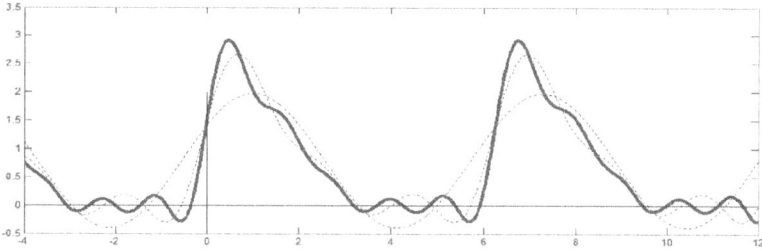

2. $f(x) = \dfrac{1}{\pi} - \dfrac{2}{\pi}\left(\dfrac{\cos 2x}{3\cdot 1} + \dfrac{\cos 4x}{5\cdot 3} + \cdots\right) + \dfrac{1}{2}\sin x$

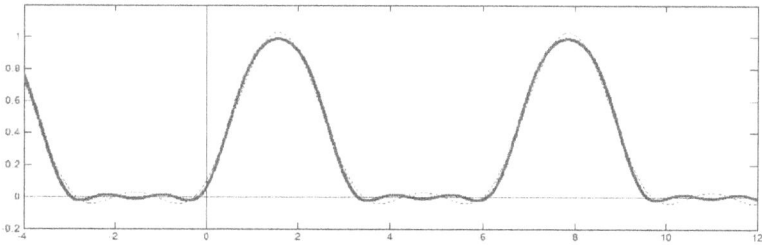

3. $f(x) = \dfrac{1}{2} + \dfrac{2}{\pi}\left(\sin \pi x + \dfrac{\sin 3\pi x}{3} + \dfrac{\sin 5\pi x}{5} + \cdots\right)$

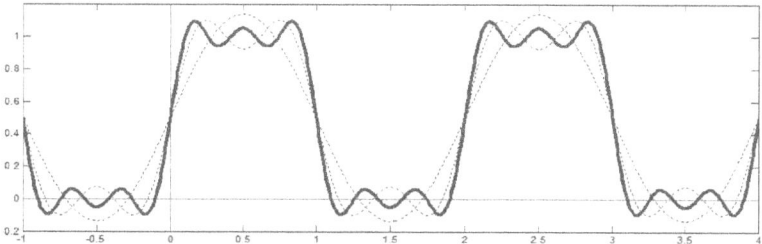

4. $f(x) = \dfrac{1}{\pi} + \dfrac{1}{2}\sin \pi x - \dfrac{2}{\pi}\left(\dfrac{1}{1\cdot 3}\cos 2\pi x + \dfrac{1}{3\cdot 5}\cos 4\pi x + \dfrac{1}{5\cdot 7}\cos 6\pi x + \cdots\right)$

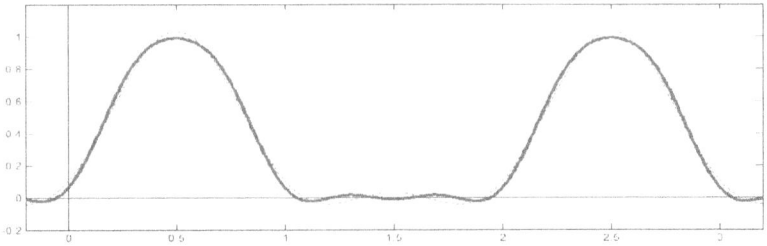

5. $f(x) = 2 - \dfrac{16}{\pi^2}\left(\cos \dfrac{\pi x}{2} + \dfrac{1}{9}\cos \dfrac{3\pi x}{2} + \dfrac{1}{25}\cos \dfrac{5\pi x}{2} + \cdots\right)$

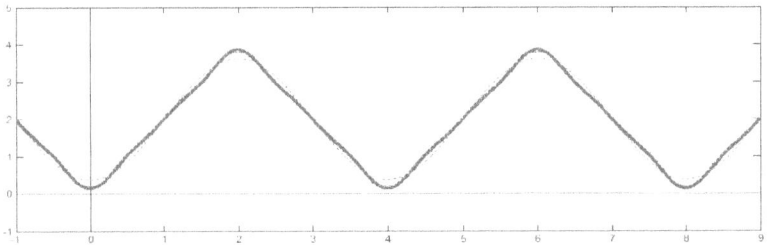

6. $f(x) = \dfrac{\pi}{3} - \dfrac{4}{\pi}\left(\cos x - \dfrac{1}{4}\cos 2x + \dfrac{1}{9}\cos 3x + - \cdots\right)$

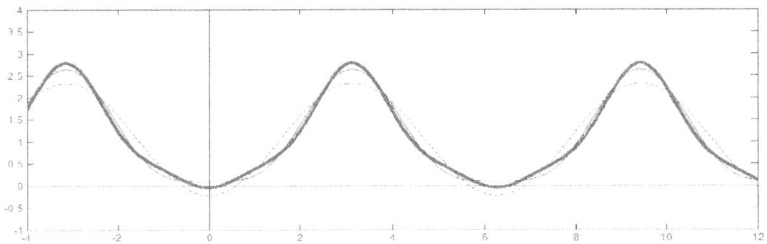

7. $f(x) = 2\left(\sin x - \dfrac{1}{2}\sin 2x + \dfrac{1}{3}\sin 3x + - \cdots\right)$

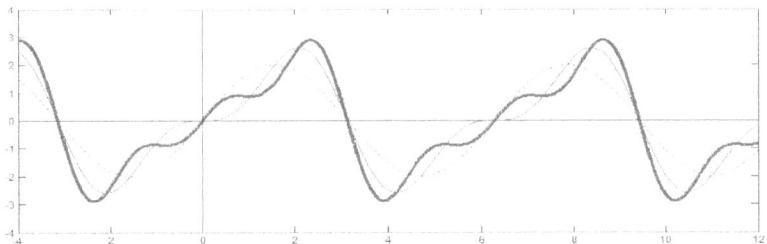

8. $f(x) = \left(\dfrac{8}{\pi} - \dfrac{32}{\pi^3}\right)\sin(\pi x/2) - \dfrac{4}{\pi}\sin\pi x + \left(\dfrac{8}{3\pi} - \dfrac{32}{(3\pi)^3}\right)\sin(3\pi x/2) + \cdots$

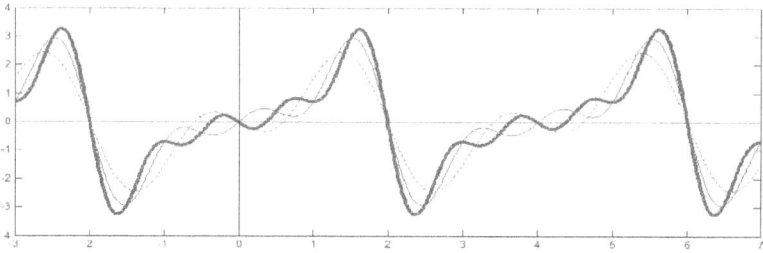

9. $f(x) = \dfrac{3}{2} - \dfrac{3}{\pi}\left\{\sin(\pi x/1.5) + \dfrac{1}{2}\sin(2\pi x/1.5) + \dfrac{1}{3}\sin(3\pi x/1.5) + \cdots\right\}$

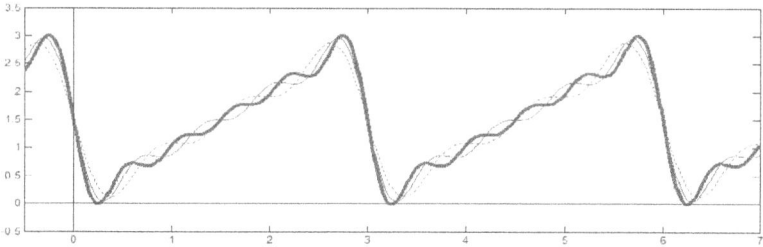

10. $f(x) = 1 + \dfrac{2}{\pi}\left(\sin\pi x - \dfrac{1}{2}\sin 2\pi x + \dfrac{1}{3}\sin 3\pi x + - \cdots\right)$

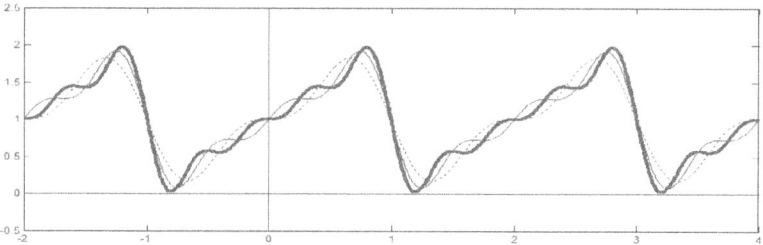

11. i. Even periodic expansion

$$f(x) = \dfrac{1}{2} - \dfrac{4}{\pi^2}\left(\cos\pi x + \dfrac{1}{9}\cos 3\pi x + \dfrac{1}{25}\cos 5\pi x + \cdots\right)$$

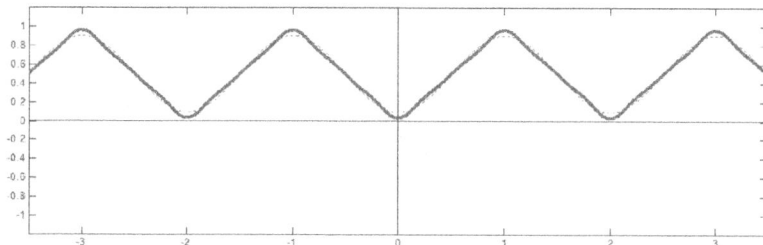

ii. Odd periodic expansion

$$f(x) = \frac{2}{\pi}\left(\sin\pi x - \frac{1}{2}\sin 2\pi x + \frac{1}{3}\sin 3\pi x + -\cdots\right)$$

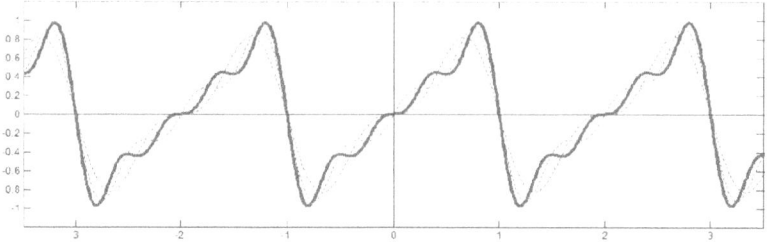

12. i. Even periodic expansion

$$f(x) = 1 + \frac{8}{\pi^2}\left(\cos\pi x + \frac{1}{9}\cos 3\pi x + \frac{1}{25}\cos 5\pi x + \cdots\right)$$

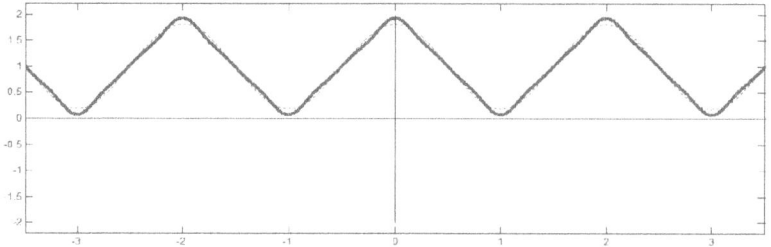

ii. Odd periodic expansion

$$f(x) = \frac{4}{\pi}\left(\sin\pi x + \frac{1}{2}\sin 2\pi x + \frac{1}{3}\sin 3\pi x + \cdots\right)$$

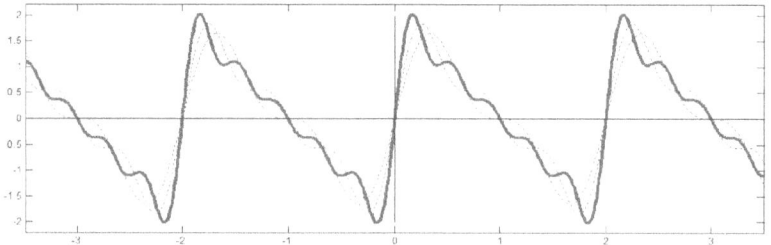

13. i. Even periodic expansion

$$f(x)=\frac{1}{2}-\frac{2}{\pi}\left\{\cos(\pi x/2)-\frac{1}{3}\cos(3\pi x/2)+\frac{1}{5}\cos(5\pi x/2)-+\cdots\right\}$$

ii. Odd periodic expansion

$$f(x)=\frac{2}{\pi}\left\{\sin(\pi x/2)-\sin\pi x+\frac{1}{3}\sin(3\pi x/2)+\frac{1}{5}\sin(5\pi x/2)-\frac{1}{3}\sin(3\pi x)+\cdots\right\}$$

14. i. Even periodic expansion

$$f(x)=\frac{4}{\pi}\left\{\cos(\pi x/2)-\frac{1}{3}\cos(3\pi x/2)+\frac{1}{5}\cos(5\pi x/2)-+\cdots\right\}$$

ii. Odd periodic expansion

$$f(x) = \frac{4}{\pi}\left(\sin \pi x + \frac{1}{3}\sin 3\pi x + \frac{1}{5}\sin 5\pi x + \cdots\right)$$

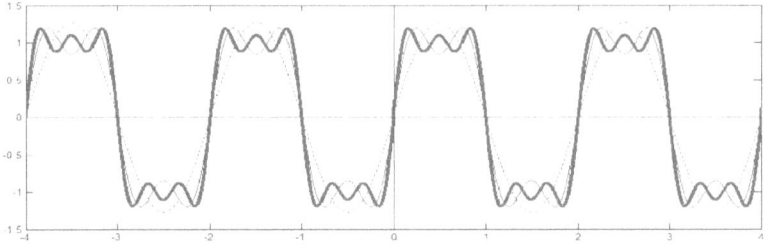

15. Orthogonal functions
16. Orthogonal functions
17. Orthogonal functions
18. Orthogonal functions
19. $\sqrt{\pi}$
20. $\sqrt{\pi}$

Problem 10.2

1. $\displaystyle\int_0^\infty \left(\frac{\sin w}{w}\cos wx\right)dw = \begin{cases} \dfrac{\pi}{2} & (0 \le x < 1) \\[2ex] \dfrac{\pi}{4} & (x = 1) \\[2ex] 0 & (x > 1) \end{cases}$

2. $\displaystyle\int_0^\infty \frac{\cos(\pi w/2)}{1 - w^2}\cos wx\, dw = \begin{cases} \dfrac{\pi}{2}\cos x & (0 \le x \le \pi/2) \\[2ex] 0 & (x > \pi/2) \end{cases}$

3. $\displaystyle\int_0^\infty \left(\frac{-w\cos w + \sin w}{w^2}\sin wx\right)dw = \begin{cases} \pi x/2 & (0 \le x < 1) \\ \pi/4 & (x = 1) \\ 0 & (x > 1) \end{cases}$

4. $\displaystyle\int_0^\infty \left(\frac{\sin 2\pi w}{w^2 - 1}\sin wx\right)dw = \begin{cases} \dfrac{\pi}{2}\sin x & (0 \le x \le 2\pi) \\[2ex] 0 & (x > 2\pi) \end{cases}$

5. $f(x) = \dfrac{2}{\pi}\displaystyle\int_0^\infty \left(\frac{w\sin w + \cos w - 1}{w^2}\cos wx\right)dw$

6. $f(x) = \dfrac{2}{\pi} \displaystyle\int_0^\infty \left(\dfrac{w \sin w - \cos w}{e(w^2 + 1)} + \dfrac{1}{w^2 + 1} \right) \cos wx\, dw$

7. $f(x) = \dfrac{2}{\pi} \displaystyle\int_0^\infty \left(\dfrac{\cos(\pi w/2)}{1 - w^2} \cos wx \right) dw$

8. $f(x) = \dfrac{2}{\pi} \displaystyle\int_0^\infty \left(\dfrac{w \sin(\pi w/2)}{w^2 - 1} \cos wx \right) dw$

9. $f(x) = \dfrac{2}{\pi} \displaystyle\int_0^\infty \left(\dfrac{-w \cos w + \sin w}{w^2} \sin wx \right) dw$

10. $f(x) = \dfrac{2}{\pi} \displaystyle\int_0^\infty \left(-\dfrac{w \cos w + \sin w}{e(w^2 + 1)} + \dfrac{w}{w^2 + 1} \right) \sin wx\, dw$

11. $f(x) = \dfrac{2}{\pi} \displaystyle\int_0^\infty \left(\dfrac{\sin(\pi w/2) - w}{1 - w^2} \right) \sin wx\, dw$

12. $f(x) = \dfrac{2}{\pi} \displaystyle\int_0^\infty \left(\dfrac{w \cos(\pi w/2)}{1 - w^2} \right) \sin wx\, dw$

11 Partial differential equations

A *partial differential equation* (PDE) is an equation that involves two or more independent variables and their partial derivatives.

The one-dimensional heat conduction equation describes how heat is transferred based on temperature differences. Similarly, the two-dimensional Laplace's equation illustrates how potential distributes in static situations. These equations play significant roles in various fields such as electromagnetics, thermodynamics, and fluid mechanics. The two-dimensional wave equation illustrates how waves propagate in a flat plane. These equations play significant roles in various fields such as vibration and acoustics.

PDEs cover a broader range of applications compared to ODEs, encompassing areas like vibration, acoustics, elasticity, heat transfer, fluid mechanics, electromagnetics, and quantum mechanics.

11.1 One-dimensional wave equation

11.1.1 Transverse vibration of string

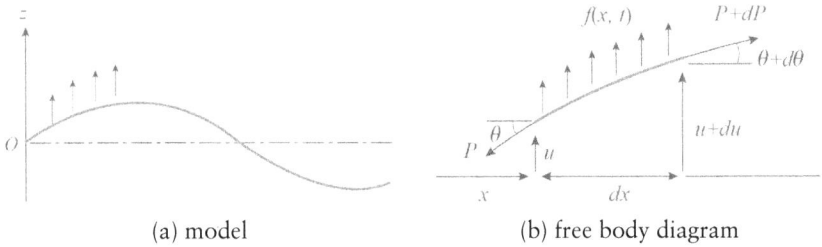

(a) model   (b) free body diagram

Figure 11.1 Transverse vibration of string.

Let's examine a uniform elastic string of length l subjected to a transverse force $f(x, t)$ per unit length, as shown in Figure 11.1. For one element, as shown in Figure 11.1(b), if we denote the transverse displacement of the string as $u(x, t)$, applying Newton's second law yields the following equation of motion:

$$(P + dP)\sin(\theta + d\theta) - P\sin\theta + fdx = \rho dx \frac{\partial^2 u}{\partial t^2} \tag{11.1}$$

where P represents the tension of the string [N], ρ denotes the line density of the string [kg/m], θ is the angle between the string and the x-axis, and $\frac{\partial^2 u}{\partial t^2}$ represents the vertical acceleration. For the length of the element dx, since

$$\sin\theta \cong \theta, \quad \sin(\theta + d\theta) \cong \theta + d\theta,$$

DOI: 10.1201/9781003608912-11

then substituting these into Eq. (11.1) and neglecting infinitesimal terms of $dPd\theta$, we get:

$$Pd\theta + \theta dP + fdx = \rho dx \frac{\partial^2 u}{\partial t^2}. \tag{11.2}$$

Assuming that the tension P along the string is constant and neglecting the transverse force $f(x, t)$ per unit length, applying Eq. (11.2) with

$$d\theta = \frac{\partial \theta}{\partial x} dx,$$

we derive the following equation:

$$P \frac{\partial \theta}{\partial x} = \rho \frac{\partial^2 u}{\partial t^2}. \tag{11.3}$$

Here, applying the geometric condition

$$\theta \cong \tan\theta = \frac{\partial u}{\partial x},$$

we obtain the following equation:

$$\frac{\partial^2 u}{\partial t^2} = c^2 \frac{\partial^2 u}{\partial x^2} \tag{11.4}$$

where $c^2 = P/\rho$ and c represents the *propagation velocity of the wave* [m/s].

Eq. (11.4) is referred to as the one-dimensional wave equation because it is expressed solely with one spatial variable. For reference, introducing the two-dimensional wave equation and the three-dimensional wave equation, we have:

Two-dimensional wave equation:

$$\frac{\partial^2 u}{\partial t^2} = c^2 \left(\frac{\partial^2 u}{\partial x^2} + \frac{\partial^2 u}{\partial y^2} \right). \tag{11.5}$$

Three-dimensional wave equation:

$$\frac{\partial^2 u}{\partial t^2} = c^2 \left(\frac{\partial^2 u}{\partial x^2} + \frac{\partial^2 u}{\partial y^2} + \frac{\partial^2 u}{\partial z^2} \right). \tag{11.6}$$

Here, u represents the displacement or wave function, t denotes time, x, y, and z are spatial variables, and c is the propagation velocity of the wave.

Using the *Laplace operator* ∇^2, it can be expressed as follows:

$$\frac{\partial^2 u}{\partial t^2} = c^2 \nabla^2 u. \tag{11.7}$$

In this chapter, it is sufficient to only know the *wave equation* (11.4), and the derivation process of Eqs. (11.1) to (11.3) can be omitted. A detailed study of the derivation process can be covered in a course on mechanical vibration.

11.1.2   *Method of separation of variables*

Remark   One-dimensional wave equation

The wave equation for a uniform string of length l is as follows:

$$\frac{\partial^2 u}{\partial t^2} = c^2 \frac{\partial^2 u}{\partial x^2} \tag{11.4}$$

where $c^2 = P/\rho$, P represents the tension of the string [N], ρ denotes the line density of the string [kg/m], and c represents the propagation velocity of the wave [m/s].

The solution to the wave equation utilizes *the method of separation of variables*. Let the solution $u(x, t)$ be expressed as the product of a function of displacement $X(x)$ and a function of time $T(t)$ as follows:

$$u(x, t) = X(x)T(t). \tag{11.8}$$

Differentiating this expression with respect to each variable, then we get

$$\frac{\partial^2 u}{\partial t^2} = X \frac{d^2 T}{dt^2}, \tag{11.9a}$$

$$\frac{\partial^2 u}{\partial x^2} = \frac{d^2 X}{dx^2} T. \tag{11.9b}$$

By inserting Eq. (11.9) into Eq. (11.4), we have

$$\frac{1}{T}\frac{d^2 T}{dt^2} = \frac{c^2}{X}\frac{d^2 X}{dx^2}. \tag{11.10}$$

Since the left-hand side of Eq. (11.10) is a function of t only, and the right-hand side is a function of x only, their common value must be a constant (negative, $-\omega^2$). Therefore, we can separate it into two independent equations as follows:

$$\frac{d^2 X}{dx^2} + \frac{\omega^2}{c^2} X = 0, \tag{11.11a}$$

$$\frac{d^2 T}{dt^2} + \omega^2 T = 0. \tag{11.11b}$$

The solutions to these two equations are as follows:

$$X(x) = A\cos\frac{\omega x}{c} + B\sin\frac{\omega x}{c}, \qquad (11.12a)$$

$$T(t) = C\cos\omega t + D\sin\omega t. \qquad (11.12b)$$

11.1.3 Solution of the transverse vibration of string

If the string is fixed at both ends (at $x = 0$ and $x = L$, the boundary conditions at all times $t \geq 0$ are $u(0, t) = 0$ and $u(l, t) = 0$. Applying this to $X(x)$, we get:

$$X(0) = 0, \qquad (11.13a)$$

$$X(l) = 0. \qquad (11.13b)$$

By inserting these into Eq. (11.12a) respectively, we have

$$A = 0, \qquad (11.14a)$$

$$B\sin\frac{\omega l}{c} = 0. \qquad (11.14b)$$

Eq. (11.14b) is called the *frequency equation*. The n-th natural frequency ω_n is expressed as follows:

$$\omega_n = \frac{n\pi c}{l} \quad (n = 1, 2, 3, \cdots). \qquad (11.15)$$

Therefore, the solution u_n of the wave equation for the natural frequency ω_n is as follows:

$$u_n(x, t) = X_n(x)T_n(t) = \sin\frac{n\pi x}{l}\{C_n\cos\omega_n t + D_n\sin\omega_n t\} \qquad (11.16)$$

where C_n, D_n are arbitrary, the solution $u_n(x, t)$ is called the n-th eigenfunction of the vibrating string or the n-th *mode shape* of the vibrating string, and $\omega_n = \frac{n\pi c}{l}$ is the n-th eigenvalue or the n-th *natural frequency*. Figure 11.2 shows the n-th mode shape of the vibrating string.

As observed in Figures (b) and (c), points where the string exhibits no vertical displacement are termed nodes (or nodal points). In the second mode shape, there is one node, and in the third mode shape, there are two nodes. Therefore, the solution of the wave equation (11.4) satisfying the two

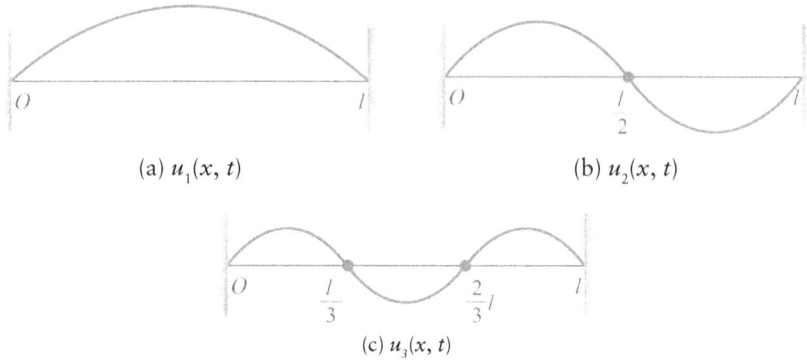

(a) $u_1(x, t)$ (b) $u_2(x, t)$

(c) $u_3(x, t)$

Figure 11.2 Mode shapes of the vibrating string.

boundary condition equations (11.13) is expressed as the following Fourier series, representing the superposition of all displacements $u_n(x, t)$:

$$u(x, t) = \sum_{n=1}^{\infty} u_n(x, t) = \sum_{n=1}^{\infty} \sin \frac{n\pi x}{l} \left\{ C_n \cos \frac{n\pi c}{l} t + D_n \sin \frac{n\pi c}{l} t \right\} \quad (11.17)$$

where the constants C_n and D_n are determined by the initial conditions, and Eq. (11.17) represents the expression for all vibrations of the string.

If the initial conditions (initial displacement and initial velocity conditions) are

$$u(x, 0) = u_0(x), \quad (11.18a)$$

$$\frac{\partial u}{\partial t}(x, 0) = \dot{u}_0(x), \quad (11.18b)$$

then from Eq. (11.17) we get

$$\sum_{n=1}^{\infty} C_n \sin \frac{n\pi x}{l} = u_0(x), \quad (11.19a)$$

$$\sum_{n=1}^{\infty} D_n \frac{n\pi c}{l} \sin \frac{n\pi x}{l} = \dot{u}_0(x). \quad (11.19b)$$

This involves a Fourier sine series expansion over the interval $0 \le x \le l$, hence the coefficients C_n and D_n can be computed using Fourier sine integrals.

$$C_n = \frac{2}{l} \int_0^l u_0(x) \sin \frac{n\pi x}{l} dx, \quad (11.20a)$$

$$D_n = \frac{2}{n\pi c} \int_0^l \dot{u}_0(x) \sin \frac{n\pi x}{l} dx. \quad (11.20b)$$

To simplify the problem, let's consider the case of initial velocity $\dot{u}_0(x)=0$. Applying this to Eq. (11.17), we get:

$$u(x, t) = \sum_{n=1}^{\infty} C_n \sin\frac{n\pi x}{l} \cos\frac{n\pi ct}{l}. \tag{11.21}$$

And since $\sin\frac{n\pi x}{l}\cos\frac{n\pi ct}{l} = \frac{1}{2}\left[\sin\left\{\frac{n\pi}{l}(x+ct)\right\}+\sin\left\{\frac{n\pi}{l}(x-ct)\right\}\right]$, we get

$$u(x, t) = \frac{1}{2}\sum_{n=1}^{\infty} C_n \sin\left\{\frac{n\pi}{l}(x+ct)\right\}+\frac{1}{2}\sum_{n=1}^{\infty} C_n \sin\left\{\frac{n\pi}{l}(x-ct)\right\}. \tag{11.22}$$

Therefore, we obtain

$$u(x, t) = \frac{1}{2}\{u_0(x+ct)+u_0(x-ct)\}, \tag{11.23}$$

where $u_0(x+ct)$ and $u_0(x-ct)$ are obtained by substituting $x+ct$ and $x-ct$ respectively into the initial displacement condition equation (11.22), where $u_0(x+ct)$ represents the function $u_0(x)$ left-shifted by ct, and $u_0(x-ct)$ represents the function $u_0(x)$ right-shifted by ct.

Example 11.1

When a uniform string of length l with both ends fixed is initially positioned as shown in Figure 10.3, with the midpoint of the string raised by h and then released, find the equation of motion $u(x, t)$.

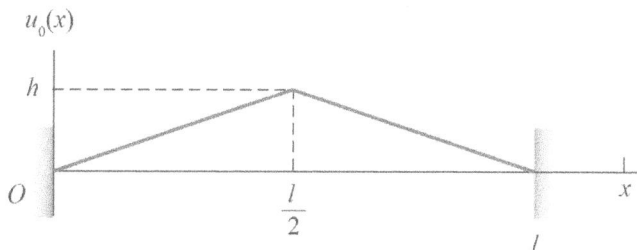

Figure 11.3 String model.

Solution

Displacement equation (11.17) of string is

$$u(x,\, t) = \sum_{n=1}^{\infty} u_n(x,\, t) = \sum_{n=1}^{\infty} \sin\frac{n\pi x}{l}\left\{ C_n \cos\frac{n\pi ct}{l} + D_n \sin\frac{n\pi ct}{l} \right\}. \qquad \text{①}$$

By differentiating it with respect to *t*, we obtain

$$\dot{u}(x,\, t) = \sum_{n=1}^{\infty} \sin\frac{n\pi x}{l}\left\{ -C_n \frac{nc\pi}{l}\sin\frac{n\pi ct}{l} + D_n \frac{nc\pi}{l}\cos\frac{n\pi ct}{l} \right\}.$$

From the initial velocity $\dot{u}(x,\, 0) = 0$, we get

$$D_n = 0.$$

And from the initial displacement $u_n(x,\, 0) = u_0(x)$, we get

$$u(x,\, 0) = \sum_{n=1}^{\infty} u_n(x) = \sum_{n=1}^{\infty} C_n\sin\frac{n\pi x}{l}.$$

Then we get

$$C_n = \frac{2}{l}\int_0^l u_0(x)\sin\frac{n\pi x}{l}\,dx, \qquad \text{②}$$

where initial displacement $u_0(x)$ is

$$u_0(x) = \begin{cases} \dfrac{2hx}{l} & \left(0 \le x < \dfrac{l}{2} \right) \\[3mm] \dfrac{2h(l-x)}{l} & \left(\dfrac{l}{2} \le x \le l \right) \end{cases} \qquad \text{③}$$

Inserting Eq. ③ into Eq. ② yields

$$C_n = \frac{4h}{l^2}\left\{ \int_0^{l/2} x\sin\frac{n\pi x}{l}\,dx + \int_{l/2}^l (l-x)\sin\frac{n\pi x}{l}\,dx \right\},$$

where since $\displaystyle\int_0^{l/2} x\sin\frac{n\pi x}{l}\,dx = -\frac{l^2}{2n\pi}\cos\frac{n\pi}{2} + \left(\frac{l}{n\pi}\right)^2 \sin\frac{n\pi}{2}$ and

$$\int_{l/2}^l (l-x)\sin\frac{n\pi x}{l}\,dx = \frac{l^2}{2n\pi}\cos\frac{n\pi}{2} + \left(\frac{l}{n\pi}\right)^2 \sin\frac{n\pi}{2},$$

we get

$$C_n = \frac{8h}{n^2\pi^2}\sin\frac{n\pi}{2}.$$

Therefore, we obtain

$$u(x, t) = \frac{8h}{\pi^2}\sum_{n=1}^{\infty}\sin\frac{n\pi x}{l}\cos\frac{n\pi ct}{l}\left(\frac{1}{n^2}\sin\frac{n\pi}{2}\right).$$

Answer

$$u(x, t) = \frac{8h}{\pi^2}\left(\sin\frac{\pi x}{l}\cos\frac{\pi ct}{l} - \frac{1}{9}\sin\frac{3\pi x}{l}\cos\frac{3\pi ct}{l} + \frac{1}{25}\sin\frac{5\pi x}{l}\cos\frac{5\pi ct}{l}\mp\cdots\right)$$

Figure 11.4 shows the vibration of string in Example 11.1.

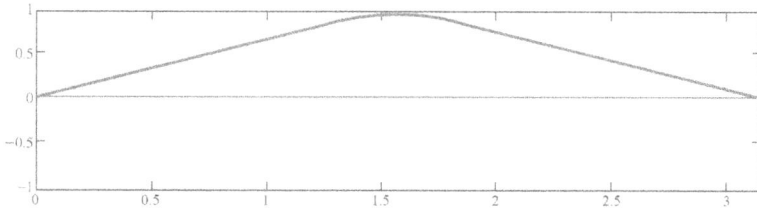

(a) $t = 0$

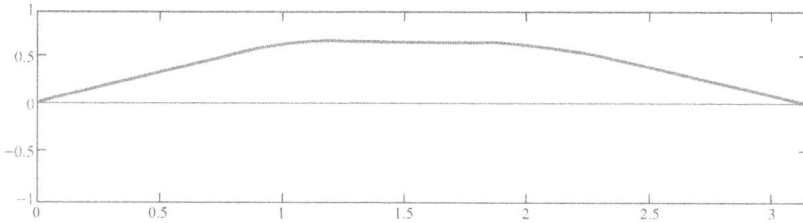

(b) $t = \dfrac{l}{6c}$

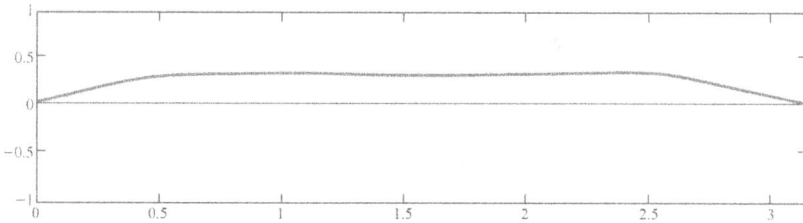

(c) $t = \dfrac{l}{3c}$

Figure 11.4 Vibration of string ($l = \pi$, $h = 1$, $c = 1$) in Example 11.1. *(Continued)*

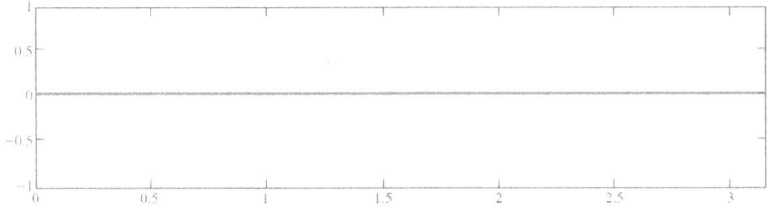

(d) $t = \dfrac{l}{2c}$

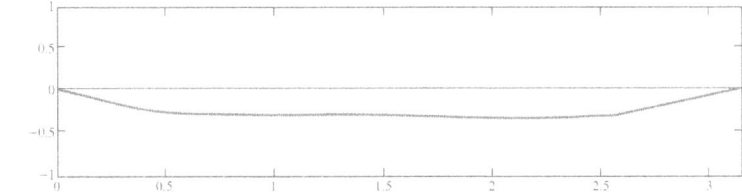

(e) $t = \dfrac{2l}{3c}$

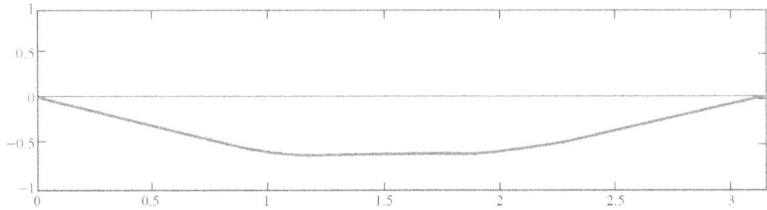

(f) $t = \dfrac{5l}{6c}$

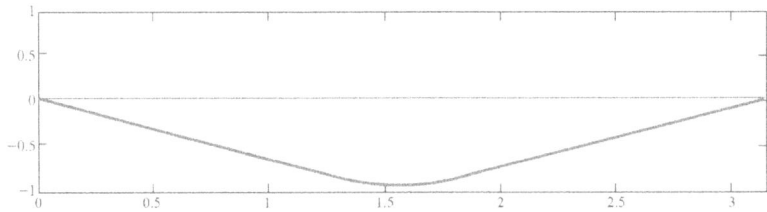

(g) $t = \dfrac{l}{c}$

Figure 11.4 (Continued)

Remark   Superposition

Eq. (11.23) represents a shape formed by superimposing the function $u_0(x)$ left-shifted by ct, denoted as $u_0(x + ct)$, and the function $u_0(x)$ right-shifted by ct, denoted as $u_0(x - ct)$.

Problem 11.1

For the vibration of a uniform string with both ends fixed, with an initial velocity of 0 and an initial displacement as shown in the following diagram, where $l = 1$, $h = 1$, and $c = 1$, find the solution $u(x, t)$. [1 ~ 4]

1. (see Figure 11.5)

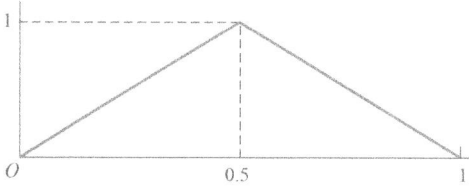

Figure 11.5 String model.

2. (see Figure 11.6)

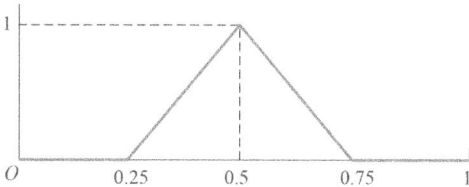

Figure 11.6 String model.

3. (see Figure 11.7)

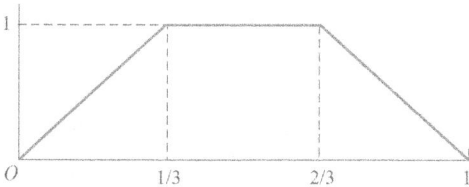

Figure 11.7 String model.

4. (see Figure 11.8)

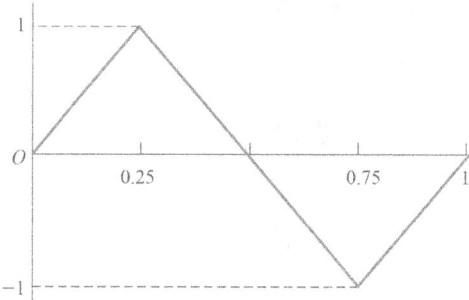

Figure 11.8 String model.

11.2 Two-dimensional wave equation

11.2.1 *Wave equation of a rectangular membrane*

Figure 11.9 Rectangular membrane.

> **Remark   Two-dimensional wave equation in Cartesian coordinates**
>
> The two-dimensional wave equation for a uniform rectangular membrane shown in Figure 11.9, can be expressed as follows:
>
> $$\frac{\partial^2 u}{\partial t^2} = c^2\left(\frac{\partial^2 u}{\partial x^2} + \frac{\partial^2 u}{\partial y^2}\right) \qquad (11.4)$$
>
> where $c^2 = P/\rho$, P represents the tension of the membrane [N], and ρ is the area density [kg/m²].
>
> The solution $u = u(x, y, t)$ of the wave equation is a function of time t at the position (x, y) on the membrane.

In a similar manner to the one-dimensional wave equation, the wave equation for a rectangular membrane with width a and height b also employs the *method of separation of variables*.

Let's express the solution $u(x, y, t)$ as the product of functions $X(x)$ of displacement x, $Y(y)$ of displacement y, and $T(t)$ of time t.

$$u(x, y, t) = X(x)Y(y)T(t). \qquad (11.24)$$

Taking partial derivatives with respect to each variable, we get:

$$\frac{\partial^2 u}{\partial t^2} = XY\frac{d^2 T}{dt^2}, \qquad (11.25a)$$

$$\frac{\partial^2 u}{\partial x^2} = YT\frac{d^2 X}{dx^2}, \qquad (11.25b)$$

$$\frac{\partial^2 u}{\partial y^2} = XT\frac{d^2 Y}{dy^2}. \qquad (11.25c)$$

Substituting these into Eq. (11.24), we obtain the following:

$$\frac{1}{T}\frac{d^2T}{dt^2} = c^2\left(\frac{1}{X}\frac{d^2X}{dx^2} + \frac{1}{Y}\frac{d^2Y}{dy^2}\right). \tag{11.26}$$

Since the left-hand side of Eq. (11.26) is a function of t only, and the right-hand side is a function of x and y, their common value must be a constant (negative, $-\omega^2$). Therefore, we can separate it into three independent equations as:

$$\frac{d^2X}{dx^2} + p^2X = 0 \tag{11.27a}$$

$$\frac{d^2Y}{dy^2} + q^2Y = 0 \tag{11.27b}$$

$$\frac{d^2T}{dt^2} + \omega^2T = 0 \tag{11.27c}$$

where $p^2 + q^2 = \dfrac{\omega^2}{c^2}$. Therefore, the solutions to the three equations are as follows:

$$X(x) = A\cos px + B\sin px \tag{11.28a}$$

$$Y(y) = C\cos qy + D\sin qy \tag{11.28b}$$

$$T(t) = E\cos \omega t + E*\sin \omega t \tag{11.28c}$$

If the four sides ($x = 0$, $x = a$, $y = 0$, and $y = b$) of the rectangular membrane are fixed, the boundary conditions become $u(0, y, t) = 0$, $u(a, y, t) = 0$, $u(x, 0, t) = 0$, and $u(x, b, t) = 0$ for all time $t \geq 0$. Applying this to $X(x)$ and $Y(y)$, we get:

$$X(0) = 0 \tag{11.29a}$$

$$X(a) = 0 \tag{11.29b}$$

$$Y(0) = 0 \tag{11.29c}$$

$$Y(b) = 0 \tag{11.29d}$$

Substituting these into Eq. (11.28a) and Eq. (11.28b), we obtain the following.

$$A = 0 \tag{11.30a}$$

$$B\sin pa = 0 \tag{11.30b}$$

$$C = 0 \tag{11.30c}$$

$$D\sin qb = 0 \tag{11.30d}$$

Eq. (11.30b) and Eq. (11.30d) involve multiple p and q values, so they can be expressed as follows:

$$p_m = \frac{m\pi}{a} \qquad (m = 1, 2, 3, \cdots), \qquad (11.31a)$$

$$q_n = \frac{n\pi}{b} \qquad (n = 1, 2, 3, \cdots). \qquad (11.31b)$$

Therefore, we obtain the solution $X_m(x)Y_n(y)$ to the wave equation of the membrane:

$$X_m(x)Y_n(y) = \sin\frac{m\pi x}{a}\sin\frac{n\pi y}{b}. \qquad (11.32)$$

When applying Eq. (11.31a) and Eq. (11.31b) to the condition $p^2 + q^2 = \frac{\omega^2}{c^2}$ in Eq. (11.27), we get:

$$\omega_{mn} = c\sqrt{p_m^2 + q_n^2} = c\pi\sqrt{\frac{m^2}{a^2} + \frac{n^2}{b^2}}. \quad (m = 1, 2, \cdots, n = 1, 2, \cdots) \quad (11.33)$$

Therefore, the solutions to the wave equation for the membrane can be summarized as follows:

$$u_{mn}(x,\,y,\,t) = \sin\frac{m\pi x}{a}\sin\frac{n\pi y}{b}\left(E_{mn}\cos\omega_{mn}t + E_{mn}^{*}\sin\omega_{mn}t\right) \qquad (11.34)$$

where $u_{mn}(x,\,y,\,t)$ is referred to as the eigenfunction for the membrane, and ω_{mn} is its eigenvalue.

As seen in Figure 11.10, the stationary lines where the membrane's vertical motion is absent are referred to as nodal lines. In Figures (b) and (c), one nodal line is observed, while in Figures (d) to (f), two nodal lines are evident. Therefore, the solutions to the wave equation (11.24) satisfying the boundary condition (11.29) are expressed as a superposition of all displacements $u_{mn}(x,\,y,\,t)$.

$$u(x,\,y,\,t) = \sum_{m=1}^{\infty}\sum_{n=1}^{\infty} u_{mn}(x,\,y,\,t)$$

$$= \sum_{m=1}^{\infty}\sum_{n=1}^{\infty}\sin\frac{m\pi x}{a}\sin\frac{n\pi y}{b}\left(E_{mn}\cos\omega_{mn}t + E_{mn}^{*}\sin\omega_{mn}t\right) \quad (11.35)$$

where $\omega_{mn} = c\pi\sqrt{\frac{m^2}{a^2} + \frac{n^2}{b^2}}$, constants E_{mn} and E_{mn}^{*} are determined by the initial conditions (initial displacement and initial velocity), and Eq. (11.35) represents all vibrations of the membrane. Substituting $t = 0$ into it, the

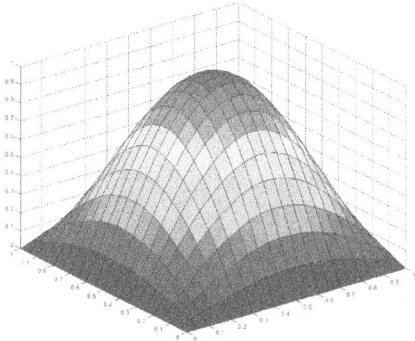

(a) $u_{11}(x,y,0)$

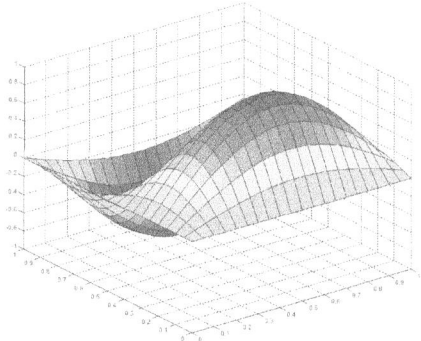

(b) $u_{21}(x,y,0)$

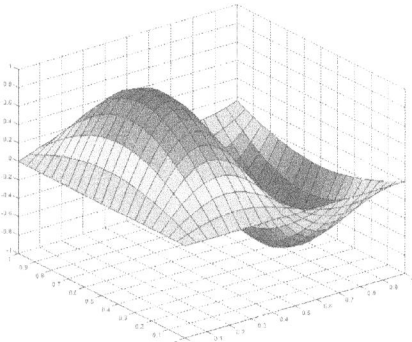

(c) $u_{12}(x,y,0)$

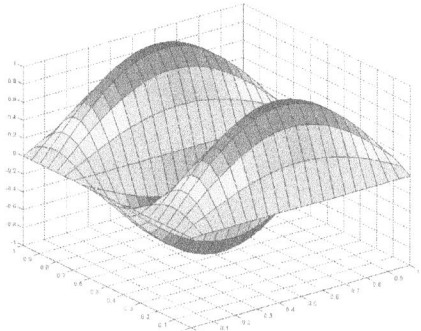

(d) $u_{31}(x,y,0)$

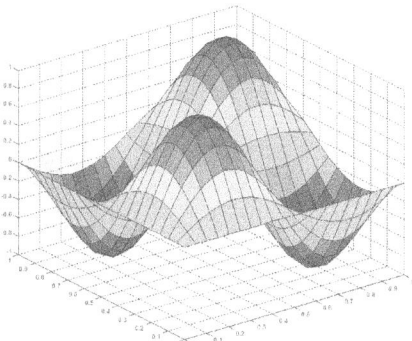

(e) $u_{22}(x,y,0)$

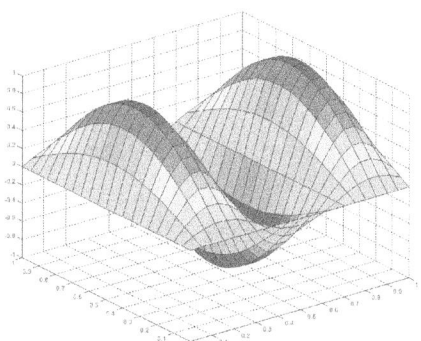

(f) $u_{13}(x,y,0)$

Figure 11.10 The eigenfunctions for the membrane (in case of $a = 1$ and $b = 1$).

initial displacement function $f(x, y)$ is expressed as the following double Fourier series.

$$u(x, y, 0) = \sum_{m=1}^{\infty} \sum_{n=1}^{\infty} E_{mn} \sin \frac{m\pi x}{a} \sin \frac{n\pi y}{b} = f(x, y) \tag{11.36}$$

Here, if we let

$$K_m(y) = \sum_{n=1}^{\infty} E_{mn} \sin \frac{n\pi y}{b}, \tag{11.37}$$

we can obtain the coefficient E_{mn} using Fourier sine integral over the interval $0 \le y \le b$.

$$E_{mn} = \frac{2}{b} \int_0^b K_m(y) \sin \frac{n\pi y}{b} dy \tag{11.38}$$

Furthermore, substituting Eq. (11.37) into Eq. (11.36) yields

$$f(x, y) = \sum_{m=1}^{\infty} K_m(y) \sin \frac{m\pi x}{a}. \tag{11.39}$$

This represents a Fourier sine series expansion of $f(x, y)$ over the interval $0 \le x \le a$ for a fixed y, enabling us to compute the coefficient $K_m(y)$ using Fourier sine integral.

$$K_m(y) = \frac{2}{a} \int_0^a f(x, y) \sin \frac{m\pi x}{a} dx \tag{11.40}$$

Substituting Eq. (11.40) back into Eq. (11.38), we obtain

$$E_{mn} = \frac{4}{ab} \int_0^b \int_0^a f(x, y) \sin \frac{m\pi x}{a} \sin \frac{n\pi y}{b} dx dy. \tag{11.41}$$

$$(m = 1, 2, \cdots, n = 1, 2, \cdots)$$

Once familiar with these calculation steps, one might directly obtain Eq. (11.41) from Eq. (11.36) without intermediate steps.

To find the constant E_{mn}^*, differentiating Eq. (11.35) with respect to time t and then substituting $t = 0$, the initial velocity function $g(x, y)$ is expressed as the following double Fourier series.

$$\frac{\partial u}{\partial t}\bigg|_{t=0} = \sum_{m=1}^{\infty} \sum_{n=1}^{\infty} E_{mn}^* \omega_{mn} \sin \frac{m\pi x}{a} \sin \frac{n\pi y}{b} = g(x, y) \tag{11.42}$$

Therefore, the constant E_{mn}^* is determined as follows:

$$E_{mn}^* = \frac{4}{ab\,\omega_{mn}} \int_0^b \int_0^a g(x,y) \sin\frac{m\pi x}{a} \sin\frac{n\pi y}{b} \, dx\, dy. \qquad (11.43)$$

$$(m = 1, 2,\cdots, n = 1, 2, \cdots)$$

Example 11.2

Find the vibration equation $u(x, y, t)$ for a uniform membrane with fixed edges in the region $0 \le x \le 3$ and $0 \le y \le 2$, given the initial displacement described by the following equation, and with an initial velocity of zero. (Note: The coefficient for the wave equation is $c = 1$.)

$$f(x, y) = (3x - x^2)(2y - y^2)$$

Solution

The solution equation (11.35) of the two-dimensional wave equation is

$$u(x, y, t) = \sum_{m=1}^{\infty} \sum_{n=1}^{\infty} \sin\frac{m\pi x}{a} \sin\frac{n\pi y}{b}\left(E_{mn} \cos\omega_{mn}t + E_{mn}^* \sin\omega_{mn}t\right),$$

where $\omega_{mn} = c\pi\sqrt{\dfrac{m^2}{a^2} + \dfrac{n^2}{b^2}}$, $a = 3$, $b = 2$, and $c = 1$.

Since the initial velocity is 0, we get

$$E_{mn}^* = 0.$$

In the initial displacement equation $f(x, y) = \displaystyle\sum_{m=1}^{\infty}\sum_{n=1}^{\infty} E_{mn} \sin\frac{m\pi x}{a} \sin\frac{n\pi y}{b}$, we get

$$E_{mn} = \frac{4}{ab} \int_0^b \int_0^a f(x, y) \sin\frac{m\pi x}{a} \sin\frac{n\pi y}{b} \, dx\, dy \quad (m = 1, 2,\cdots, n = 1, 2, \cdots)$$

$$= \frac{2}{3} \int_0^2 \int_0^3 (3x - x^2)(2y - y^2) \sin\frac{m\pi x}{3} \sin\frac{n\pi y}{2} \, dx\, dy$$

$$= \frac{2}{3} \int_0^3 (3x - x^2)\sin\frac{m\pi x}{3}\, dx \int_0^2 (2y - y^2)\sin\frac{n\pi y}{2}\, dy \quad (\text{refer to the \textbf{Check}})$$

$$= \frac{2}{3} \frac{54}{m^3\pi^3}(1 - \cos m\pi)\frac{16}{n^3\pi^3}(1 - \cos n\pi)$$

$$= \frac{576}{m^3 n^3 \pi^6}(1 - \cos m\pi)(1 - \cos n\pi).$$

Therefore, we obtain

$$u(x,y,t) = \sum_{m=1}^{\infty}\sum_{n=1}^{\infty} E_{mn} \sin\frac{m\pi x}{3} \sin\frac{n\pi y}{2} \cos\omega_{mn}t$$

$$= \frac{576}{\pi^6}\sum_{m=1}^{\infty}\sum_{n=1}^{\infty}\frac{1}{m^3 n^3}(1-\cos m\pi)(1-\cos n\pi)\sin\frac{m\pi x}{3}\sin\frac{n\pi y}{2}\cos\omega_{mn}t,$$

where $\omega_{mn} = \pi\sqrt{\dfrac{m^2}{9}+\dfrac{n^2}{4}}$.

Answer

$$u(x,y,t) = \frac{576}{\pi^6}\sum_{m=1}^{\infty}\sum_{n=1}^{\infty}\frac{1}{m^3 n^3}(1-\cos m\pi)(1-\cos n\pi)\sin\frac{m\pi x}{3}\sin\frac{n\pi y}{2}\cos\omega_{mn}t$$

Figure 11.11 shows the vibration of membrane in Example 11.2.

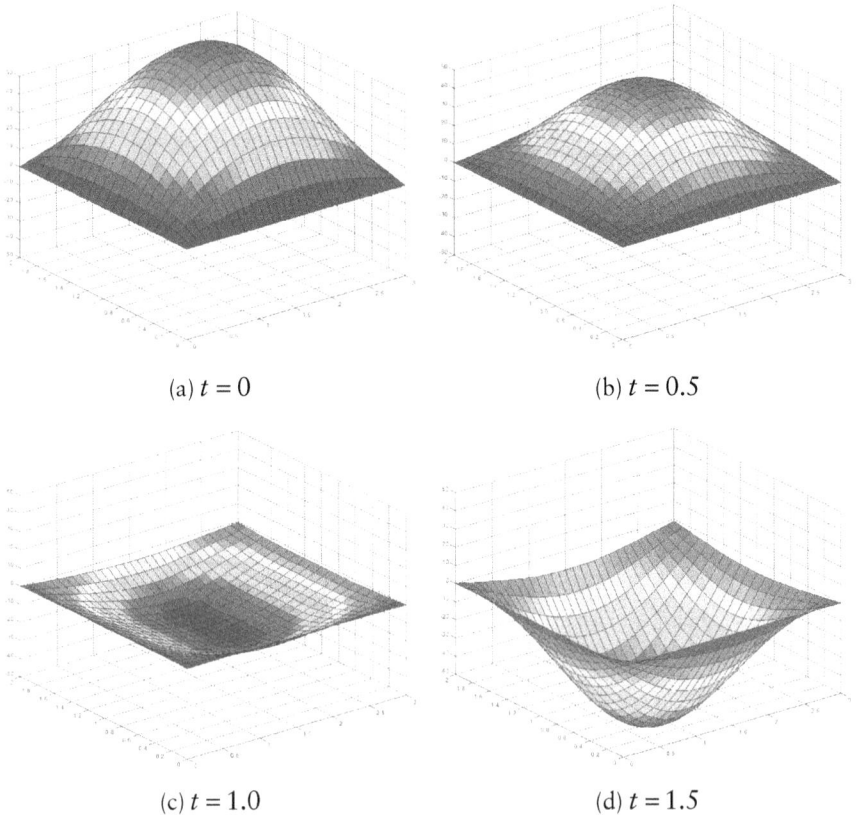

(a) $t = 0$ (b) $t = 0.5$

(c) $t = 1.0$ (d) $t = 1.5$

Figure 11.11 Vibration of membrane in Example 11.2.

Check

$$\int_0^3 \left(3x - x^2\right)\sin\frac{m\pi x}{3}\,dx$$

$$= \left[\left(3x - x^2\right)\left(-\frac{3}{m\pi}\cos\frac{m\pi x}{3}\right) - \left(3 - 2x\right)\left(-\frac{9}{m^2\pi^2}\sin\frac{m\pi x}{3}\right)\right.$$

$$\left.+ (-2)\cdot\left(\frac{27}{m^3\pi^3}\cos\frac{m\pi x}{3}\right)\right]_0^3$$

$$= \frac{54}{m^3\pi^3}\left(1 - \cos m\pi\right)$$

$$\int_0^2 \left(2y - y^2\right)\sin\frac{n\pi y}{2}\,dy$$

$$= \left[\left(2y - y^2\right)\left(-\frac{2}{n\pi}\cos\frac{n\pi y}{2}\right) - \left(2 - 2y\right)\left(-\frac{4}{n^2\pi^2}\sin\frac{n\pi y}{2}\right)\right.$$

$$\left.+ (-2)\cdot\left(\frac{8}{n^3\pi^3}\cos\frac{n\pi y}{2}\right)\right]_0^2$$

$$= \frac{16}{n^3\pi^3}\left(1 - \cos n\pi\right)$$

11.2.2 Wave equation of a circular membrane

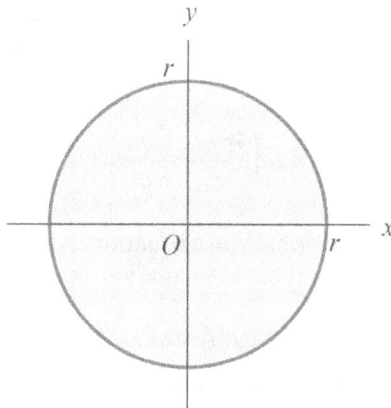

Figure 11.12 Circular membrane.

In Section 11.1, it was recognized that the two-dimensional wave equation is expressed as:

$$\frac{\partial^2 u}{\partial t^2} = c^2 \nabla^2 u. \tag{11.7}$$

Furthermore, to represent a circular membrane shown in Figure 11.12, the Laplace operator can be expressed in polar coordinates as follows [refer Appendix A5].

$$\nabla^2 u = \frac{\partial^2 u}{\partial r^2} + \frac{1}{r}\frac{\partial u}{\partial r} + \frac{1}{r^2}\frac{\partial^2 u}{\partial \theta^2} \tag{A5.3}$$

Remark   Two-dimensional wave equation in polar coordinates

Two-dimensional wave equation for a uniform circular membrane can be expressed as follows:

$$\frac{\partial^2 u}{\partial t^2} = c^2 \left(\frac{\partial^2 u}{\partial r^2} + \frac{1}{r}\frac{\partial u}{\partial r} + \frac{1}{r^2}\frac{\partial^2 u}{\partial \theta^2} \right) \tag{11.44}$$

where $c^2 = P/\rho$, P represents the tension of the membrane [N], and ρ is the area density [kg/m²].

The solution $u = u(r, \theta, t)$ of the wave equation is a function of time t at the position (r, θ) on the membrane.

If the circular membrane is radially symmetric, meaning that it does not vary with angle θ, then $u_{\theta\theta} = 0$, and Eq. (11.44) simplifies to the following concise form.

$$\frac{\partial^2 u}{\partial t^2} = c^2 \left(\frac{\partial^2 u}{\partial r^2} + \frac{1}{r}\frac{\partial u}{\partial r} \right) \tag{11.45}$$

Similar to the one-dimensional wave equation, we also employ the method of separation of variables in the wave equation of a circular membrane with radius r.

Let the solution $u = u(r, t)$ be expressed as the product of a function $R(r)$ of radius r and a function $T(t)$ of time t:

$$u(r, t) = R(r)T(t). \tag{11.46}$$

Differentiating this expression with respect to each variable yields

$$\frac{\partial^2 u}{\partial t^2} = R\frac{d^2 T}{dt^2}, \tag{11.47a}$$

$$\frac{\partial u}{\partial r} = T\frac{dR}{dr}, \tag{11.47b}$$

$$\frac{\partial^2 u}{\partial r^2} = T\frac{d^2 R}{dr^2}. \tag{11.47c}$$

Substituting these into Eq. (11.45), we obtain

$$\frac{1}{c^2 T}\frac{d^2 T}{dt^2} = \frac{1}{R}\left(\frac{d^2 R}{dr^2} + \frac{1}{r}\frac{dR}{dr}\right). \tag{11.48}$$

Since the left-hand side of Eq. (11.48) is a function of t only, and the right-hand side is a function of r only, their common value must be a constant (negative, $-\omega^2$). Therefore, we can separate it into two independent equations as:

$$\frac{d^2 R}{dr^2} + \frac{1}{r}\frac{dR}{dr} + \omega^2 R = 0, \tag{11.49a}$$

$$\frac{d^2 T}{dt^2} + c^2\omega^2 T = 0. \tag{11.49b}$$

Multiplying Eq. (11.49a) by r^2, we get

$$r^2\frac{d^2 R}{dr^2} + r\frac{dR}{dr} + \omega^2 r^2 R = 0. \tag{11.50}$$

Here, if we let $s = \omega r$, we get

$$\frac{dR}{dr} = \frac{dR}{ds}\frac{ds}{dr} = \frac{dR}{ds}\omega, \tag{11.51a}$$

$$\frac{d^2 R}{dr^2} = \frac{d}{dr}\left(\frac{dR}{ds}\omega\right) = \frac{d}{ds}\left(\frac{dR}{ds}\omega\right)\frac{ds}{dr} = \frac{d^2 R}{ds^2}\omega^2. \tag{11.51b}$$

Substituting these into Eq. (11.50) yields the following *Bessel equation* (refer to Section 5.3):

$$s^2\frac{d^2 R}{ds^2} + s\frac{dR}{ds} + s^2 R = 0. \tag{11.52}$$

Remark   Standard form of Bessel's equation (refer to Section 5.3)

$$x^2 y'' + xy' + \left(x^2 - v^2\right)y = 0 \qquad (v \geq 0) \qquad (5.23)$$

Remark   General solution of Bessel's equation (for $v = 0$)

The general solution of Bessel's equation $xy'' + y' + xy = 0$ is

$$y(x) = c_1 J_0(x) + c_2 Y_0(x). \qquad (x \neq 0) \qquad (5.52)$$

where $J_0(x) = \displaystyle\sum_{k=0}^{\infty} \frac{(-1)^k}{2^{2k}(k!)^2} x^{2k} = 1 - \frac{x^2}{2^2(1!)^2} + \frac{x^4}{2^4(2!)^2} - \frac{x^6}{2^6(3!)^2} + - \cdots$

$$(5.39)$$

$$Y_0(x) = \frac{2}{\pi} J_0(x)\left(\ln\frac{x}{2} + \gamma\right) + \frac{2}{\pi}\sum_{k=1}^{\infty} \frac{(-1)^{k-1}}{2^{2k}(k!)^2}\left(1 + \frac{1}{2} + \frac{1}{3} + \cdots + \frac{1}{k}\right)x^{2k}$$

$$(5.51)$$

Eq. (11.52) represents the Bessel equation for $v = 0$, and its general solution consists of a linear combination of $J_0(s)$ and $Y_0(s)$. However, as s approaches 0, $Y_0(s)$ approaches $-\infty$, making it unsuitable since it needs to have a finite value at the center of the circular membrane.

Therefore, the solution $R(r)$ of Eq. (11.50) is as follows:

$$R(r) = J_0(s) = J_0(\omega r)$$

$$= \sum_{k=0}^{\infty} \frac{(-1)^k}{2^{2k}(k!)^2} (\omega r)^{2k} = 1 - \frac{(\omega r)^2}{2^2(1!)^2} + \frac{(\omega r)^4}{2^4(2!)^2} - \frac{(\omega r)^6}{2^6(3!)^2} + - \cdots. \qquad (11.53)$$

Figure 11.13 shows the Bessel function $J_0(s)$ for $v = 0$. The nodal points, where $J_0(s) = 0$, occur at the following values of s:

$$s_1 = 2.405,$$

$$s_2 = 5.520,$$

$$s_3 = 8.654,$$

$$s_4 = 11.792,$$

$$\cdots$$

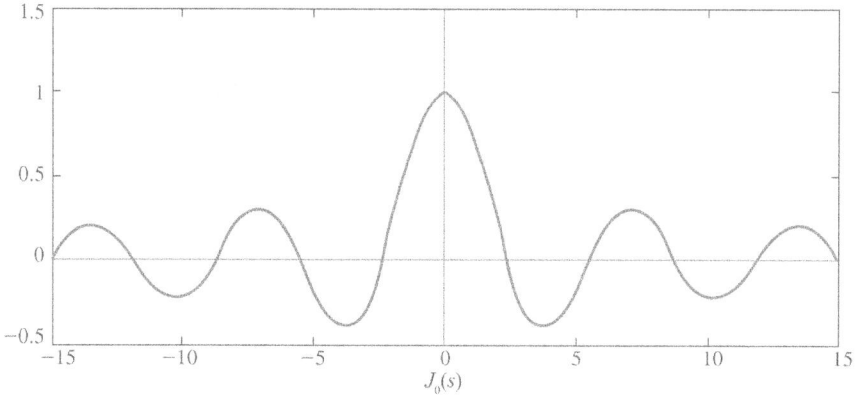

Figure 11.13 Bessel function $J_0(s)$ for $v = 0$.

Since the nodal points should occur at the boundary $r = r_0$ of the circular membrane with radius r_0, they satisfy the following equation:

$$\omega_n = \frac{s_n}{r_0}. \tag{11.54}$$

Therefore, the following equation represents the solution of the wave equation (11.49a) satisfying $J_0(s) = 0$ at the boundary $r = r_0$ of the circular membrane.

$$R_n(r) = J_0(\omega_n r) = J_0\left(\frac{s_n}{r_0} r\right) \qquad (n = 1, 2, 3, \cdots) \tag{11.55}$$

Figure 11.14 shows the eigenfunctions of a circular membrane. The lines where the membrane's vertical motion is absent, known as nodal lines or nodal circles, are shown in Figures (b), (c), and (d) with 1, 2, and 3 circles respectively.

Now, let's find the solution of the time function (11.49b).

$$T_n(t) = C_n \cos c\omega_n t + C_n^* \sin c\omega_n t$$

$$= C_n \cos \frac{cs_n}{r_0} t + C_n^* \sin \frac{cs_n}{r_0} t \tag{11.56}$$

Therefore, the n-th solution of the wave equation for a circular membrane is summarized as:

$$u_n(r, t) = R_n(r) T_n(t) = J_0\left(\frac{s_n}{r_0} r\right)\left(C_n \cos \frac{cs_n}{r_0} t + C_n^* \sin \frac{cs_n}{r_0} t\right), \tag{11.57}$$

$$(n = 1, 2, \cdots)$$

(a) the first eigenfunction

(b) the second eigenfunction

(c) the third eigenfunction

(d) the fourth eigenfunction

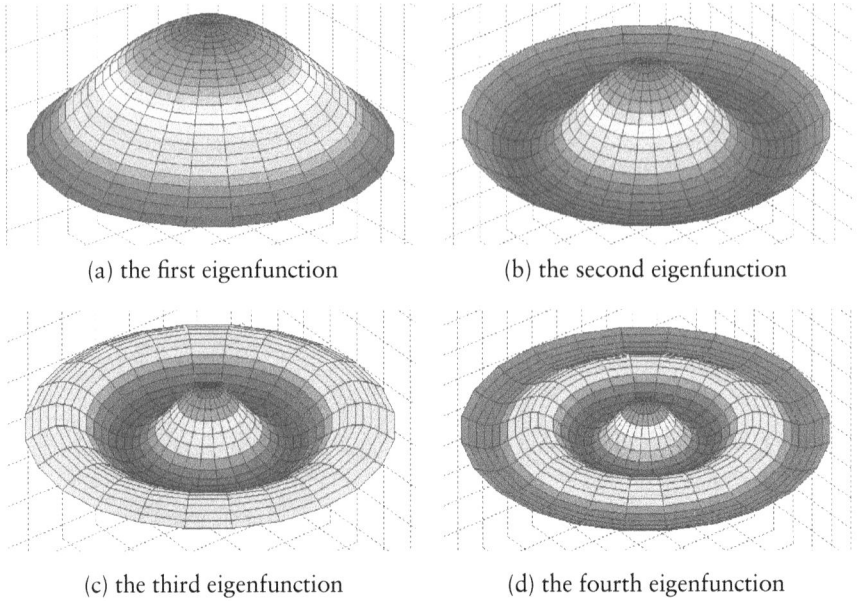

Figure 11.14 The eigenfunctions of a circular membrane.

where the equation $u_n(r, t)$ is referred to as the eigenfunction for the vibration of a circular membrane, and $\dfrac{cs_n}{r_0}$ is referred to as the eigenvalue.

Therefore, the solutions to the wave equation (11.45) for a circular membrane are expressed as a superposition of all displacements $u_n(r, t)$ as follows:

$$u(r, t) = \sum_{n=1}^{\infty} u_n(r, t)$$

$$= \sum_{n=1}^{\infty} J_0\left(\frac{s_n}{r_0} r\right)\left(C_n \cos \frac{cs_n}{r_0} t + C_n^* \sin \frac{cs_n}{r_0} t\right), \qquad (11.58)$$

where constants C_n and C_n^* are determined by initial conditions (initial displacement and initial velocity).

Substituting $t = 0$ into this equation yields

$$u(r, 0) = \sum_{n=1}^{\infty} C_n J_0\left(\frac{s_n}{r_0} r\right) = f(r). \qquad (11.59)$$

The initial displacement function $f(r)$ is expressed as the following Fourier-Bessel series. (The derivation process is complex and will be omitted.)

$$C_n = \frac{2}{r_0^2 J_1^2(s_n)} \int_0^{r_0} rf(r) J_0\left(\frac{s_n}{r_0}r\right) dr \qquad (11.60)$$

Problem 11.2

Find the vibration equation $u(x, y, t)$ for a uniform membrane with fixed edges in the region $0 \le x \le 1$ and $0 \le y \le 1$, given the initial displacement described by the following equation, and with an initial velocity of zero. (Note: The coefficient for the wave equation is $c = 1$.) [1 ~ 4]

1. $f(x, y) = 1$
2. $f(x, y) = x$
3. $f(x, y) = \sin \pi x$
4. $f(x, y) = xy$

11.3   Wave equation of sound pressure

11.3.1   (*optional) Sound pressure equation

The equations representing the process of wave propagation are the continuity equation and the momentum equation.

$$\text{continuity equation: } \frac{\partial \rho}{\partial t} + \nabla \cdot (\rho v) = 0 \qquad (11.61)$$

$$\text{momentum equation: } \rho\left\{\frac{\partial v}{\partial t} + (v \cdot \nabla)v\right\} = -\nabla p \qquad (11.62)$$

where ρ represents the density of air, v represents the velocity of air particles, and p represents the sound pressure. Additionally, ∇ denotes $\frac{\partial}{\partial x}i + \frac{\partial}{\partial y}j + \frac{\partial}{\partial z}k$.

When representing ρ, v, and p as quantities and infinitesimal changes in equilibrium state, they are as follows:

$$\rho = \rho_0 + \rho_1$$
$$v = v_0 + v_1$$
$$p = p_0 + p_1.$$

Therefore, Eq. (11.61) becomes:

$$\frac{\partial(\rho_0 + \rho_1)}{\partial t} + \nabla \cdot \{(\rho_0 + \rho_1)(v_0 + v_1)\} = 0 \qquad (11.63a)$$

or

$$\frac{\partial \rho_0}{\partial t} + \frac{\partial \rho_1}{\partial t} + \nabla \cdot \left(\rho_0 \mathbf{v}_0 + \rho_1 \mathbf{v}_0 + \rho_0 \mathbf{v}_1 + \rho_1 \mathbf{v}_1 \right) = 0 \qquad (11.63b)$$

where $\frac{\partial \rho_0}{\partial t} + \nabla \cdot \left(\rho_0 \mathbf{v}_0 \right) = 0$ and $\rho_1 \mathbf{v}_1 \cong 0$ which is double infinitesimal. And in the equilibrium state where the medium is not moving ($\mathbf{v}_0 = 0$), Eq. (11.63) simplifies to the following reduced form of the continuity equation.

$$\frac{\partial \rho_1}{\partial t} + \rho_0 \nabla \cdot \mathbf{v}_1 = 0 \qquad (11.64)$$

Applying the same method to Eq. (11.62), we obtain the following simplified momentum equation.

$$\rho_0 \frac{\partial \mathbf{v}_1}{\partial t} = -\nabla p_1 \qquad (11.65)$$

Differentiating Eq. (11.64) with respect to time and then substituting Eq. (11.65) to eliminate $\mathbf{v}_1$, we obtain the following relationship between infinitesimal density ρ_1 and infinitesimal sound pressure p_1.

$$\frac{\partial^2 \rho_1}{\partial t^2} - \nabla^2 p_1 = 0 \qquad (11.66)$$

Meanwhile, in thermodynamics, sound pressure p is expressed as a function of density ρ and entropy s, that is, $p = p(\rho, s)$. As sound propagation is known to occur through isentropic processes, expanding sound pressure p into a Taylor series, we have:

$$p = p(\rho, s) = p_0(\rho_0, s_0) + \left(\frac{\partial p}{\partial \rho} \right)_{\rho_0, s_0} (\rho - \rho_0) + \cdots \qquad (11.67)$$

Therefore, we get

$$p_1 = p - p_0 \cong \left(\frac{\partial p}{\partial \rho} \right)_{\rho_0, s_0} \rho_1 \qquad (11.68)$$

Since $\left(\frac{\partial p}{\partial \rho} \right)_{\rho_0, s_0}$ always has a positive value $(= c^2)$ thermodynamically, Eq. (11.68) can be expressed as follows:

$$p_1 = c^2 \rho_1. \qquad (11.69)$$

Then substituting Eq. (11.69) into Eq. (11.66), we obtain the following *sound pressure equation*:

$$\frac{\partial^2 p_1}{\partial t^2} = c^2 \nabla^2 p_1. \tag{11.70}$$

Expressing the infinitesimal sound pressure p_1 from Eq. (11.70) back in terms of the change in sound pressure p, we derive the following wave equation for sound pressure.

Remark   Sound pressure equation

The sound pressure equation $p(x, t)$ can be expressed using the Laplace operator ∇^2 as follows:

$$\frac{\partial^2 p}{\partial t^2} = c^2 \nabla^2 p \tag{11.71}$$

where c is the sound speed [m/s]. On the other hand, the one-dimensional wave equation for sound pressure $p(x, t)$ is as follows:

$$\frac{\partial^2 p}{\partial t^2} = c^2 \frac{\partial^2 p}{\partial x^2}. \tag{11.72}$$

11.3.2   Solution of the wave equation for sound pressure (plane wave)

If sound pressure propagates in one dimension as described in Eq. (11.72), we refer to this wave as a *plane wave*.

Let's find the general solution to the plane wave equation using the *method of separation of variables*. We can express the solution $p(x, t)$ as the product of a function $X(x)$ of displacement x and a function $T(t)$ of time t as follows:

$$p(x, t) = X(x)T(t). \tag{11.73}$$

Differentiating this expression with respect to each variable yields

$$\frac{\partial^2 p}{\partial t^2} = X \frac{d^2 T}{dt^2}, \tag{11.74a}$$

$$\frac{\partial^2 p}{\partial x^2} = \frac{d^2 X}{dx^2} T. \tag{11.74b}$$

Substituting these into Eq. (11.73), we obtain:

$$\frac{1}{T}\frac{d^2T}{dt^2} = \frac{c^2}{X}\frac{d^2X}{dx^2}. \tag{11.75}$$

Since the left-hand side of Eq. (11.75) is a function of t only, and the right-hand side is a function of x only, their common value must be a constant (negative, $-\omega^2$). Therefore, we can separate it into two independent equations as:

$$\frac{d^2X}{dx^2} + \frac{\omega^2}{c^2}X = 0, \tag{11.76a}$$

$$\frac{d^2T}{dt^2} + \omega^2 T = 0. \tag{11.76b}$$

The solutions to these two equations are as follows:

$$X(x) = Ae^{-i\frac{\omega x}{c}} + Be^{i\frac{\omega x}{c}}, \tag{11.77a}$$

$$T(t) = Ce^{-i\omega t} + De^{i\omega t}. \tag{11.77b}$$

Here, assuming the time function $T(t)$ to be a simple harmonic function, i.e., $T(t) = e^{i\omega t}$, the solution to the sound pressure equation is organized as follows:

$$p(x,\ t) = X(x)T(t)$$

$$= \left(Ae^{-i\frac{\omega x}{c}} + Be^{i\frac{\omega x}{c}} \right)e^{i\omega t}$$

$$= Ae^{-i\frac{\omega}{c}(x-ct)} + Be^{i\frac{\omega}{c}(x+ct)}. \tag{11.78}$$

Therefore, the general solution to the sound pressure (plane wave) equation can be expressed as follows:

$$p(x,t) = f(x-ct) + g(x+ct), \tag{11.79}$$

where $f(x-ct) = Ae^{-i\frac{\omega}{c}(x-ct)}$ and $g(x+ct) = Be^{i\frac{\omega}{c}(x+ct)}$.

And $f(x-ct)$ and $g(x+ct)$ represent the sound pressure propagating in the $+x$ direction and the $-x$ direction, respectively.

11.3.3 Solution of sound pressure equation (spherical wave)

In the wave equation given by Eq. (11.71), for a point sound source, the propagation of sound waves forms spherical waves, thus requiring representation in spherical coordinates. In spherical coordinates, the Laplace operator ∇^2 is represented as follows:

$$\nabla^2 = \frac{1}{r^2}\frac{\partial}{\partial r}\left(r^2\frac{\partial}{\partial r}\right) + \frac{1}{r^2\sin\theta}\frac{\partial}{\partial\theta}\left(\sin\theta\frac{\partial}{\partial\theta}\right) + \frac{1}{r^2\sin^2\theta}\frac{\partial^2}{\partial\varphi^2}. \tag{11.80}$$

If we consider only the simple spherical wave that depends solely on r, the following equation is obtained.

$$\nabla^2 = \frac{\partial^2}{\partial r^2} + \frac{2}{r}\frac{\partial}{\partial r} \tag{11.81}$$

Applying this to the wave equation (11.71) yields

$$\frac{\partial^2 p}{\partial t^2} = c^2\left(\frac{\partial^2 p}{\partial r^2} + \frac{2}{r}\frac{\partial p}{\partial r}\right), \tag{11.82}$$

or

$$\frac{\partial^2(rp)}{\partial t^2} = c^2\frac{\partial^2(rp)}{\partial r^2}. \tag{11.83}$$

Therefore, the general solution to the sound pressure (spherical wave) equation is as follows:

$$p(r, t) = \frac{1}{r}f(r - ct) + \frac{1}{r}g(r + ct), \tag{11.84}$$

where, $f(r - ct) = Ae^{-i\frac{\omega}{c}(r-ct)}$ and $g(r + ct) = Be^{i\frac{\omega}{c}(r+ct)}$.

The first term on the right-hand side of Eq. (11.84) represents the diverging wave, propagating outward from the origin, while the second term represents the converging wave, propagating toward the origin. Typically, in an infinite medium with a sound source, there are no waves converging toward the origin, so this term is referred to as the radiation sound.

Problem 11.3

1. In spherical coordinates, the Laplace operator ∇^2 is represented as follows:

$$\nabla^2 = \frac{1}{r^2}\frac{\partial}{\partial r}\left(r^2\frac{\partial}{\partial r}\right) + \frac{1}{r^2\sin\theta}\frac{\partial}{\partial\theta}\left(\sin\theta\frac{\partial}{\partial\theta}\right) + \frac{1}{r^2\sin^2\theta}\frac{\partial^2}{\partial\varphi^2}.$$

For a simple spherical wave with spherical symmetry, derive the following expression.

$$\nabla^2 = \frac{\partial^2}{\partial r^2} + \frac{2}{r}\frac{\partial}{\partial r}$$

2. When applying the result from Problem 11.1 to the wave equation of sound pressure $\frac{\partial^2 p}{\partial t^2} = c^2 \nabla^2 p$, we obtain

$$\frac{\partial^2 p}{\partial t^2} = c^2 \left(\frac{\partial^2 p}{\partial r^2} + \frac{2}{r}\frac{\partial p}{\partial r} \right).$$

Show that this can be expressed as the following equation:

$$\frac{\partial^2 (rp)}{\partial t^2} = c^2 \frac{\partial^2 (rp)}{\partial r^2}.$$

11.4   One-dimensional heat conduction equation

11.4.1   (*optional) Heat conduction equation

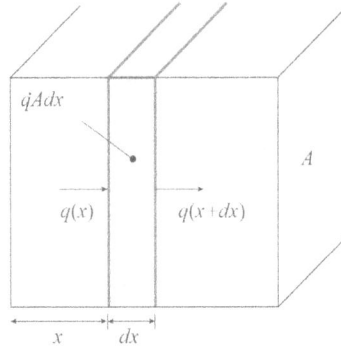

Figure 11.15 Heat conduction model.

Heat conduction refers to the phenomenon where energy is transferred from a region of high temperature to a region of low temperature. The rate of change of energy q [J/s] due to heat conduction is expressed as a function of temperature $u = u(x, t)$ as follows:

$$q = -kA\frac{\partial u}{\partial x} \tag{11.85}$$

where $\frac{\partial u}{\partial x}$ represents the temperature gradient in the direction of heat flow, k represents the heat conductivity of the material [J/(m s °C)], and A represents the cross-sectional area [m^2].

In a one-dimensional heat conduction model with the thickness dx shown in Figure 11.15, the sum of the energy $q(x)$ conducted into the left face and the heat $\dot{q}Adx$ generated within the element equals the sum of the internal energy change $\rho sA\dfrac{\partial u}{\partial t}dx$ and the energy $q(x+dx)$ conducted from the right face. This can be summarized as follows:

$$q(x)+\dot{q}Adx = \rho sA\frac{\partial u}{\partial t}dx + q(x+dx), \qquad (11.86a)$$

or

$$-kA\frac{\partial u}{\partial x}+\dot{q}Adx = \rho sA\frac{\partial u}{\partial t}dx - A\left\{k\frac{\partial u}{\partial x}+\frac{\partial}{\partial x}\left(k\frac{\partial u}{\partial x}\right)dx\right\}. \qquad (11.86b)$$

Then we obtain

$$\frac{\partial}{\partial x}\left(k\frac{\partial u}{\partial x}\right)+\dot{q} = \rho s\frac{\partial u}{\partial t}, \qquad (11.87)$$

where ρ represents density [kg/m^3], and s represents specific heat of the material [J/(kg °C)]. In Eq. (11.86), neglecting internal heat generation and assuming uniform thermal conductivity k of the material, the following heat conduction equation is derived:

$$\frac{\partial u}{\partial t} = c^2\frac{\partial^2 u}{\partial x^2} \qquad (11.88)$$

where $c^2 = \dfrac{k}{\rho s}$ means the thermal diffusivity. If we extend this to two-dimensional equation and three-dimensional equation, it becomes as follows:

$$\frac{\partial u}{\partial t} = c^2\left(\frac{\partial^2 u}{\partial x^2}+\frac{\partial^2 u}{\partial y^2}\right) \qquad (11.89)$$

$$\frac{\partial u}{\partial t} = c^2\left(\frac{\partial^2 u}{\partial x^2}+\frac{\partial^2 u}{\partial y^2}+\frac{\partial^2 u}{\partial z^2}\right) \qquad (11.90)$$

Using the Laplace operator ∇^2 in these equations simplifies them as follows:

$$\frac{\partial u}{\partial t} = c^2\nabla^2 u. \qquad (11.91)$$

In this chapter, only the heat conduction equation (11.91) is referenced, and it's acceptable to omit the derivation process of Eq. (11.85) to Eq. (11.87).

Remark   One-dimensional heat conduction equation

The one-dimensional heat conduction equation for a uniform rod of length l is as follows:

$$\frac{\partial u}{\partial t} = c^2 \frac{\partial^2 u}{\partial x^2} \tag{11.88}$$

where $c^2 = \dfrac{k}{\rho s}$ means the thermal diffusivity, k represents the heat conductivity of the material [J/(m s °C)], ρ represents density [kg/m³], and s represents specific heat of the material [J/(kg °C)].

11.4.2   Method of separation of variables

When solving the heat conduction equation, we can also utilize the *method of separation of variables*. We can express the solution $u(x, t)$ as the product of a function $X(x)$ of displacement x and a function $T(t)$ of time t as follows:

$$u(x, t) = X(x)T(t). \tag{11.92}$$

Differentiating this expression with respect to each variable, we have:

$$\frac{\partial u}{\partial t} = X \frac{dT}{dt}, \tag{11.93a}$$

$$\frac{\partial^2 u}{\partial x^2} = \frac{d^2 X}{dx^2} T. \tag{11.93b}$$

Substituting these into Eq. (11.92), we obtain:

$$\frac{1}{c^2 T} \frac{dT}{dt} = \frac{1}{X} \frac{d^2 X}{dx^2}. \tag{11.94}$$

Since the left-hand side of Eq. (11.94) is a function of t only, and the right-hand side is a function of x only, their common value must be a constant (negative, $-\omega^2$). Therefore, we can separate it into two independent equations as:

$$\frac{d^2 X}{dx^2} + \omega^2 X = 0, \tag{11.95a}$$

$$\frac{dT}{dt} + c^2 \omega^2 T = 0. \tag{11.95b}$$

The solutions to these two equations are as follows:

$$X(x) = A\cos\omega x + B\sin\omega x, \qquad (11.96a)$$

$$T(t) = Ce^{-c^2\omega^2 t}. \qquad (11.96b)$$

11.4.3 Solution of the heat conduction equation

If the temperature at both ends $(x = 0, x = l)$ of the uniform rod of length l is maintained at 0, the boundary conditions for all times $t \geq 0$ become $u(0, t) = 0$ and $u(l, t) = 0$. Applying this to $X(x)$ yields

$$X(0) = 0, \qquad (11.97a)$$

$$X(l) = 0. \qquad (11.97b)$$

Substituting these into Eq. (11.96a) yields

$$A = 0, \qquad (11.98a)$$

$$B\sin\omega l = 0. \qquad (11.98b)$$

Since there are multiple ω in Eq. (11.98b), the n-th natural frequency ω_n is expressed as follows.

$$\omega_n = \frac{n\pi}{l} \qquad (n = 1, 2, 3, \cdots). \qquad (11.99)$$

Therefore, the n-th solution u_n corresponding to the n-th natural frequency ω_n for the heat conduction equation is as follows:

$$u_n(x, t) = X_n(x)T_n(t) = C_n e^{-\lambda_n^2 t}\sin\frac{n\pi x}{l} \qquad (11.100)$$

where C_n is constant, $\lambda_n = \frac{n\pi c}{l}$, and the n-th solution $u_n(x, t)$ is called as the n-th eigenfunction.

The solution to the one-dimensional heat conduction equation (11.88) is represented by the following Fourier series expression, which is a superposition of all $u_n(x, t)$.

$$u(x, t) = \sum_{n=1}^{\infty} u_n(x, t) = \sum_{n=1}^{\infty} C_n e^{-\lambda_n^2 t}\sin\frac{n\pi x}{l} \qquad (11.101)$$

Here, the constant C_n is determined by the initial conditions, and Eq. (11.101) represents the temperature distribution of the rod. If the initial temperature condition is given as

$$u(x,\ 0) = u_0(x),\tag{11.102}$$

from Eq. (11.102), we get

$$\sum_{n=1}^{\infty} C_n \sin\frac{n\pi x}{l} = u_0(x).\tag{11.103}$$

This is a Fourier sine series expansion over the interval $0 \le x \le l$, so we can calculate the coefficient C_n.

$$C_n = \frac{2}{l}\int_0^l u_0(x)\sin\frac{n\pi x}{l}dx\tag{11.104}$$

Example 11.3

When the initial temperature distribution of a uniform rod of length l with insulated ends is as shown in Figure 11.16, find the solution $u(x,\ t)$ of the heat conduction equation.

Here, where $c^2 = \dfrac{k}{\rho s}$ means the thermal diffusivity, k represents the heat conductivity of the material [J/(m s °C)], ρ represents density [kg/m³], and s represents specific heat of the material [J/(kg °C)].

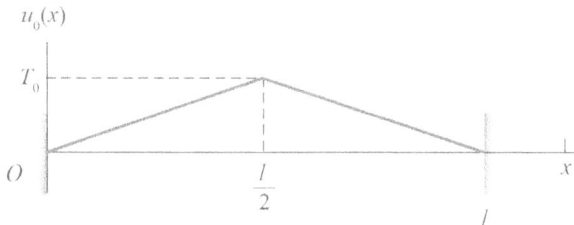

Figure 11.16 Initial temperature distribution.

Solution

From the heat conduction equation (11.28), we get

$$u(x,\ t) = \sum_{n=1}^{\infty} u_n(x,\ t) = \sum_{n=1}^{\infty} C_n e^{-\lambda_n^2 t}\sin\frac{n\pi x}{l},\tag{a}$$

where $\lambda_n = \dfrac{n\pi c}{l}$.

Applying the initial condition $u_n(x, 0) = u_0(x)$ yields

$$u(x, 0) = \sum_{n=1}^{\infty} u_n(x) = \sum_{n=1}^{\infty} C_n \sin \frac{n\pi x}{l}.$$

Then we get

$$C_n = \frac{2}{l} \int_0^l u_0(x) \sin \frac{n\pi x}{l} dx, \tag{b}$$

where the initial temperature $u_0(x)$ is

$$u_0(x) = \begin{cases} \dfrac{2T_0}{l} x & \left(0 \le x < \dfrac{l}{2}\right) \\[3mm] \dfrac{2T_0}{l}(l-x) & \left(\dfrac{l}{2} \le x \le l\right) \end{cases}. \tag{c}$$

Substituting Eq. (c) into Eq. (b) yields

$$C_n = \frac{4T_0}{l^2} \left\{ \int_0^{\frac{l}{2}} x \sin \frac{n\pi x}{l} dx + \int_{\frac{l}{2}}^{l} (l-x) \sin \frac{n\pi x}{l} dx \right\}.$$

Here, since $\displaystyle\int_0^{l/2} x \sin \frac{n\pi x}{l} dx = -\frac{l^2}{2n\pi} \cos \frac{n\pi}{2} + \left(\frac{l}{n\pi}\right)^2 \sin \frac{n\pi}{2}$,

and $\displaystyle\int_{l/2}^{l} (l-x) \sin \frac{n\pi x}{l} dx = \frac{l^2}{2n\pi} \cos \frac{n\pi}{2} + \left(\frac{l}{n\pi}\right)^2 \sin \frac{n\pi}{2}$,

we get

$$C_n = \frac{8T_0}{n^2\pi^2} \sin \frac{n\pi}{2}.$$

Therefore, we obtain

$$u(x, t) = \frac{8T_0}{\pi^2} \sum_{n=1}^{\infty} \left(\frac{1}{n^2} \sin \frac{n\pi}{2}\right) e^{-\lambda_n^2 t} \sin \frac{n\pi x}{l} \quad \left(\lambda_n = \frac{n\pi c}{l}\right)$$

$$= \frac{8T_0}{\pi^2} \left(e^{-\pi^2 c^2 t/l^2} \sin \frac{\pi x}{l} - \frac{1}{9} e^{-9\pi^2 c^2 t/l^2} \sin \frac{3\pi x}{l} + \frac{1}{25} e^{-25\pi^2 c^2 t/l^2} \sin \frac{5\pi x}{l} - + \cdots \right).$$

Answer

$$u(x, t) = \frac{8T_0}{\pi^2} \left(e^{-\pi^2 c^2 t/l^2} \sin \frac{\pi x}{l} - \frac{1}{9} e^{-9\pi^2 c^2 t/l^2} \sin \frac{3\pi x}{l} + \frac{1}{25} e^{-25\pi^2 c^2 t/l^2} \sin \frac{5\pi x}{l} - + \cdots \right)$$

Figure 11.17 shows the temperature distribution in Example 11.3.

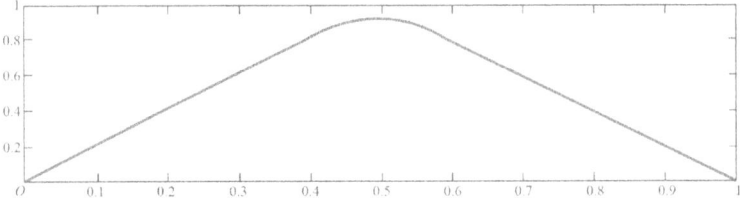

(a) $t = 0$

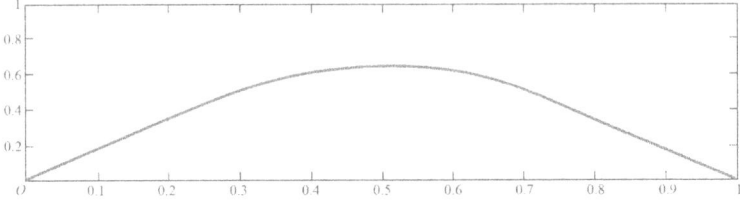

(b) $t = \dfrac{l^2}{40c^2\pi^2}$

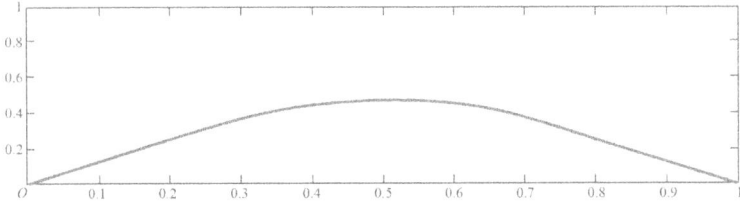

(c) $t = \dfrac{l^2}{20c^2\pi^2}$

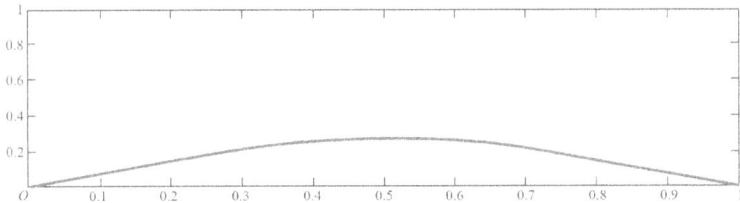

(d) $t = \dfrac{l^2}{10c^2\pi^2}$

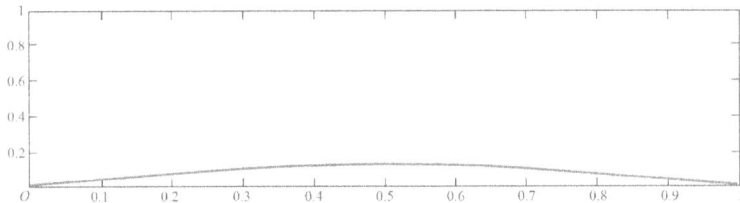

(e) $t = \dfrac{l^2}{5c^2\pi^2}$

Figure 11.17 Temperature distribution ($l = 1$, $T_0 = 1$, $c = 1$) in Example 11.3.

Problem 11.4

When the initial temperature distribution of a uniform rod of length l with insulated ends is as the equation, find the solution $u(x, t)$ of the heat conduction equation. (Here, the thermal diffusivity $c^2 = \dfrac{K}{\rho} \sigma = 1$.) [1 ~ 6]

1. $u(x, 0) = \sin \pi x$
2. $u(x, 0) = 1$
3. $u(x, 0) = x$
4. $u(x, 0) = 4x(1 - x)$
5. $u(x, 0) = 1 - \sin \pi x$
6. $u(x, 0) = |2x - 1|$

11.5 Two-dimensional heat conduction equation

11.5.1 (*optional) Heat conduction equation

The heat conduction equation, which extends the one-dimensional heat conduction equation learned in Section 11.4 to two dimensions, is as follows:

$$\frac{\partial u}{\partial t} = c^2 \left(\frac{\partial^2 u}{\partial x^2} + \frac{\partial^2 u}{\partial y^2} \right). \tag{11.89}$$

If we consider steady-state heat conduction, where $\dfrac{\partial u}{\partial t} = 0$, it results in the following Laplace equation.

Remark   Steady-state two-dimensional heat conduction equation

$$\nabla^2 u = \frac{\partial^2 u}{\partial x^2} + \frac{\partial^2 u}{\partial y^2} = 0 \tag{11.105}$$

11.5.2 Method of separation of variables

In the rectangular region R with boundary temperature distributions u_1, u_2, u_3, u_4 as shown in Figure 11.18, let's express the solution $u = u(x, y)$ using the method of separation of variables as a product of functions $X(x)$ of displacement x and $Y(y)$ of displacement y as follows:

$$u(x, y) = X(x)Y(y). \tag{11.106}$$

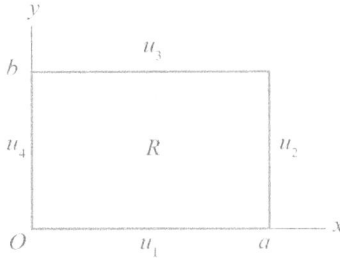

Figure 11.18 Heat conduction region R.

After differentiating this expression with respect to each variable and substituting them into Eq. (11.106), we get:

$$\frac{1}{X}\frac{d^2X}{dx^2} = -\frac{1}{Y}\frac{d^2Y}{dy^2} = -k^2. \qquad (11.107)$$

The solutions to these two equations are as follows:

$$X(x) = A\cos kx + B\sin kx, \qquad (11.108a)$$

$$Y(y) = C\cosh ky + D\sinh ky. \qquad (11.108b)$$

11.5.3 Solution of the two-dimensional heat conduction equation

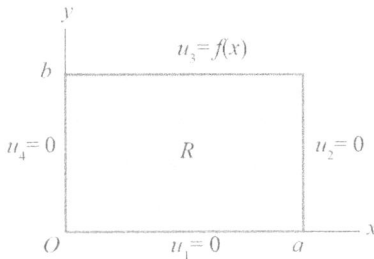

Figure 11.19 Steady-state temperature distribution.

In Figure 11.19, since the steady-state temperature distribution is zero at the left and right boundaries, namely, $u_4 = u(0, y) = 0$ and $u_2 = u(a, y) = 0$, applying $X(x)$ yields

$$X(0) = 0, \qquad (11.109a)$$

$$X(a) = 0. \qquad (11.109b)$$

Substituting these into Eq. (11.108a) yields

$$A = 0, \tag{11.110a}$$

$$B \sin ka = 0. \tag{11.110b}$$

Since there are multiple k in Eq. (11.108b), the n-th k is expressed as follows:

$$k_n = \frac{n\pi}{a} \quad (n = 1, 2, 3, \cdots). \tag{11.111}$$

On the other hand, in Figure 11.19, since the temperature distribution is zero at the bottom boundary, namely, $u_1 = u(x, 0) = 0$, applying this to $Y(y)$ results in

$$Y(0) = 0. \tag{11.112}$$

Substituting this into Eq. (11.108b) yields

$$C = 0. \tag{11.113}$$

Therefore, from Eq. (11.106), the temperature distribution solution $u = u(x, y)$ is summarized as follows:

$$u_n(x, y) = X_n(x)Y_n(y) = A_n^* \sin\frac{n\pi x}{a}\sinh\frac{n\pi y}{a}. \tag{11.114}$$

The solution to the two-dimensional heat conduction equation (11.114) is represented by the following Fourier series expression, which is a superposition of all $u_n(x, y)$.

$$u(x, y) = \sum_{n=1}^{\infty} u_n(x, y) = \sum_{n=1}^{\infty} A_n^* \sin\frac{n\pi x}{a}\sinh\frac{n\pi y}{a} \tag{11.115}$$

Applying the top boundary condition $u_3 = u(x, b) = f(x)$ to Eq. (11.115) yields the following expression:

$$u(x, b) = f(x) = \sum_{n=1}^{\infty} \left(A_n^* \sinh\frac{n\pi b}{a} \right)\sin\frac{n\pi x}{a}. \tag{11.116}$$

Since this is a Fourier sine series expansion over the interval $0 \le x \le a$, we can apply the Fourier sine series learned in Chapter 10 to Eq. (11.116) to calculate the coefficient A_n^*.

$$A_n^* \sinh\frac{n\pi b}{a} = \frac{2}{a}\int_0^a f(x)\sin\frac{n\pi x}{a}dx \tag{11.117}$$

Therefore, the solution $u(x, y)$ of the two-dimensional heat conduction equation is as follows:

$$u(x, y) = \sum_{n=1}^{\infty} u_n(x, y) = \sum_{n=1}^{\infty} A_n^* \sin\frac{n\pi x}{a} \sinh\frac{n\pi y}{a}, \qquad (11.118)$$

where $A_n^* = \dfrac{2}{a\sinh\dfrac{n\pi b}{a}} \displaystyle\int_0^a f(x)\sin\frac{n\pi x}{a}\,dx.$

Example 11.4

Given the steady-state temperature distribution on the plate as shown in Figure 11.20, find the solution $u(x, y)$ of the two-dimensional heat conduction equation. Here $f(x) = T_0 \sin\dfrac{\pi x}{a}$, $a = \pi$ and $b = 1$.

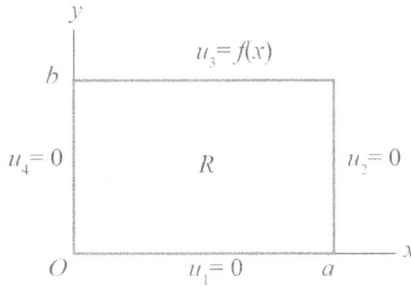

Figure 11.20 Steady-state temperature distribution.

Solution

The solution of the two-dimensional heat conduction equation is

$$u(x, y) = \sum_{n=1}^{\infty} u_n(x, y) = \sum_{n=1}^{\infty} A_n^* \sin nx \sinh ny \qquad (a)$$

where $A_n^* = \dfrac{2}{\pi \sinh n} \displaystyle\int_0^\pi f(x)\sin nx\,dx$

or $A_n^* = \dfrac{2}{\pi \sinh n} \displaystyle\int_0^\pi T_0 \sin x \sin nx\,dx.$

Then at $n \neq 1$, we get

$$A_n^* = 0.$$

And at $n = 1$, we get

$$A_1^* = \frac{2}{\pi \sinh 1} \int_0^\pi T_0 \sin x \sin x \, dx = \frac{T_0}{\sinh 1}.$$

Therefore, we obtain

$$u(x, y) = \frac{T_0}{\sinh 1} \sin x \sinh y.$$

Answer $u(x, y) = \dfrac{T_0}{\sinh 1} \sin x \sinh y$

Problem 11.5

Given the steady-state temperature distribution on the plate as shown in Figure 11.21, find the solution $u(x, y)$ of the two-dimensional heat conduction equation. $[1 \sim 2]$

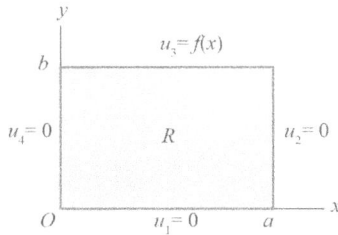

Figure 11.21 Steady-state temperature distribution.

1. $f(x) = T_0, a = \pi, b = 1$

2. $f(x) = \begin{cases} \dfrac{2T_0}{a} x, & 0 < x < \dfrac{a}{2} \\ \dfrac{2T_0}{a}(a - x), & \dfrac{a}{2} < x < a \end{cases}, a = \pi, b = 1$

Given the steady-state temperature distribution on the plate as shown Figure 11.22, find the solution $u(x, y)$ of the two-dimensional heat conduction equation.

3. $u_2 = T_0 \sin y, a = 1, b = \pi$

4. $u_2 = T_0, a = 1, b = \pi$

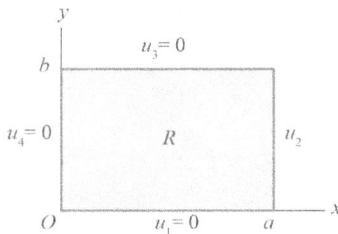

Figure 11.22 Steady-state temperature distribution.

Answer

Problem 11.1

1. $u(x, t) = \dfrac{8}{\pi^2}\left(\sin \pi x \cos \pi t - \dfrac{1}{9}\sin 3\pi x \cos 3\pi t + \dfrac{1}{25}\sin 5\pi x \cos 5\pi t - +\cdots\right)$

2. $u(x, t) = \displaystyle\sum_{n=1}^{\infty}\sin n\pi x \cos n\pi t\left(\dfrac{16}{n^2\pi^2}\sin 0.5n\pi - \dfrac{8}{n^2\pi^2}\sin 0.75n\pi - \dfrac{8}{n^2\pi^2}\sin 0.25n\pi\right)$

3. $u(x, t) = \dfrac{6}{\pi^2}\displaystyle\sum_{n=1}^{\infty}\sin n\pi x \cos n\pi t\left[\dfrac{1}{n^2}\left\{\sin\left(\dfrac{n\pi}{3}\right) + \sin(2n\pi/3) - \sin n\pi\right\}\right]$

4. $u(x, t) = \dfrac{8}{\pi^2}\displaystyle\sum_{n=1}^{\infty}\sin n\pi x \cos n\pi t\left[\dfrac{1}{n^2}(2\sin 0.25n\pi - 2\sin 0.75n\pi + \sin n\pi)\right]$

Problem 11.2

1. $u(x, y, t) = \dfrac{4}{\pi^2}\displaystyle\sum_{m=1}^{\infty}\sum_{n=1}^{\infty}\left(\dfrac{1-\cos m\pi}{m}\right)\left(\dfrac{1-\cos n\pi}{n}\right)\sin m\pi x \sin n\pi y \cos \omega_{mn}t$

2. $u(x, y, t) = -\dfrac{4}{\pi^2}\displaystyle\sum_{m=1}^{\infty}\sum_{n=1}^{\infty}\dfrac{\cos m\pi}{m}\left(\dfrac{1-\cos n\pi}{n}\right)\sin m\pi x \sin n\pi y \cos \omega_{mn}t$

3. $u(x, y, t) = \dfrac{2}{\pi}\displaystyle\sum_{n=1}^{\infty}\left(\dfrac{1-\cos n\pi}{n}\right)\sin \pi x \sin n\pi y \cos \omega_n t$

4. $u(x, y, t) = \dfrac{4}{\pi^2}\displaystyle\sum_{m=1}^{\infty}\sum_{n=1}^{\infty}\dfrac{\cos m\pi}{m}\dfrac{\cos n\pi}{n}\sin m\pi x \sin n\pi y \cos \omega_{mn}t$

Problem 11.3

1. $\nabla^2 = \dfrac{\partial^2}{\partial r^2} + \dfrac{2}{r}\dfrac{\partial}{\partial r}$

2. $\dfrac{\partial^2(rp)}{\partial t^2} = c^2\dfrac{\partial^2(rp)}{\partial r^2}$

Problem 11.4

1. $u(x, t) = e^{-\pi^2 t}\sin \pi x$

2. $u(x, t) = \dfrac{4}{\pi}\left(e^{-\pi^2 t}\sin \pi x + \dfrac{1}{3}e^{-9\pi^2 t}\sin 3\pi x + \dfrac{1}{5}e^{-25\pi^2 t}\sin 5\pi x + \cdots\right)$

3. $u(x, t) = -2\displaystyle\sum_{n=1}^{\infty}\dfrac{\cos n\pi}{n\pi}e^{-n^2\pi^2 t}\sin n\pi x$

4. $u(x, t) = \dfrac{16}{\pi^3}\displaystyle\sum_{n=1}^{\infty}\left(\dfrac{1-\cos n\pi}{n^3}\right)e^{-n^2\pi^2 t}\sin n\pi x$

5. $u(x, t) = \left(\dfrac{4}{\pi} - 1\right)e^{-\pi^2 t}\sin\pi x + \dfrac{2}{\pi}\displaystyle\sum_{n=2}^{\infty}\left(\dfrac{1-\cos n\pi}{n}\right)e^{-n^2\pi^2 t}\sin n\pi x$

6. $u(x, t) = \dfrac{2}{\pi}\displaystyle\sum_{n=2}^{\infty}\left(\dfrac{1-\cos n\pi}{n} - \dfrac{4\sin 0.5n\pi}{n^2\pi}\right)e^{-n^2\pi^2 t}\sin n\pi x$

Problem 11.5

1. $u(x, y) = \dfrac{2T_0}{\pi}\displaystyle\sum_{n=1}^{\infty}\left[\dfrac{1-(-1)^n}{n\sinh n}\right]\sin nx\sinh ny$

2. $u(x, y) = \dfrac{4T_0}{\pi^2}\displaystyle\sum_{n=1}^{\infty}\dfrac{1}{\sinh n}\left(-\dfrac{\pi}{n}\cos\dfrac{n\pi}{2} + \dfrac{2}{n^2}\sin\dfrac{n\pi}{2}\right)\sin nx\sinh ny$

3. $u(x, y) = \dfrac{T_0}{\sinh 1}\sinh x\sin y$

4. $u(x, y) = \dfrac{2T_0}{\pi}\displaystyle\sum_{n=1}^{\infty}\left[\dfrac{1-(-1)^n}{n\sin hn}\right]\sinh nx\sin ny$

12 Complex numbers and functions

Complex numbers are one of the fundamental concepts in mathematics. A complex number consists of a real part and an imaginary part, and the complex plane is a tool where the real part is represented on the x-axis and the imaginary part on the y-axis. Complex numbers are depicted as points on the coordinate plane in the complex plane. In this chapter, we delve into understanding complex numbers through their correlation with Cartesian coordinates and the complex plane. We explore various operations and differentiation involving complex numbers.

12.1 Complex number

12.1.1 *Rectangular form of complex number*

In Cartesian coordinates, the position of a point P is represented by the values a on the x-axis and b on the y-axis. Thus, it is denoted as $P(a, b)$ (where a and b are real numbers).

Similarly, in complex coordinates, a *complex number* z is represented by a real coordinate value a and an imaginary coordinate value b. That is, it is expressed as $z = a + bi$.

In other words, as seen in Figure 12.1, a point $P(a, b)$ in Cartesian coordinates corresponds to a complex number $z = a + bi$ in *complex coordinates*.

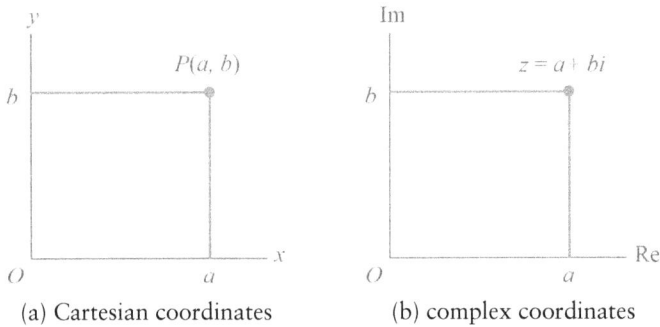

(a) Cartesian coordinates   (b) complex coordinates

Figure 12.1 Cartesian coordinates and complex coordinates.

The addition and subtraction operations of two complex numbers, $z_1 = a_1 + b_1 i$ and $z_2 = a_2 + b_2 i$, are as follows.

$$z_1 + z_2 = (a_1 + b_1 i) + (a_2 + b_2 i) = (a_1 + a_2) + (b_1 + b_2)i \qquad (12.1a)$$

$$z_1 - z_2 = (a_1 + b_1 i) - (a_2 + b_2 i) = (a_1 - a_2) + (b_1 - b_2)i \qquad (12.1b)$$

$$cz_1 = c(a_1 + b_1 i) = ca_1 + cb_1 i \qquad (c \text{ arbitrary}) \qquad (12.1c)$$

DOI: 10.1201/9781003608912-12

Furthermore, using $i = \sqrt{-1}$, or $i^2 = -1$, the multiplication and division operations of two complex numbers, $z_1 = a_1 + b_1 i$ and $z_2 = a_2 + b_2 i$, are as follows.

$$z_1 z_2 = (a_1 + b_1 i)(a_2 + b_2 i) = (a_1 a_2 - b_1 b_2) + (a_1 b_2 + a_2 b_1) i \qquad (12.1\text{d})$$

$$\frac{z_1}{z_2} = \frac{a_1 + b_1 i}{a_2 + b_2 i} = \frac{(a_1 + b_1 i)(a_2 - b_2 i)}{(a_2 + b_2 i)(a_2 - b_2 i)} = \frac{a_1 a_2 + b_1 b_2}{a_2^2 + b_2^2} + \frac{-a_1 b_2 + a_2 b_1}{a_2^2 + b_2^2} i \qquad (12.1\text{e})$$

Also, when denoting the *complex conjugate* of a complex number as $\bar{z} = x - yi$, we get

$$r^2 = |z|^2 = z\bar{z}. \qquad (12.2)$$

And x is called the real part, and y is the imaginary part of $z = x + yi$ as

$$x = \text{Re}(z) = \frac{1}{2}(z + \bar{z}), \qquad (12.3\text{a})$$

$$y = \text{Im}(z) = \frac{1}{2i}(z - \bar{z}). \qquad (12.3\text{b})$$

The properties of *complex conjugates* are as follows.

$$\overline{z_1 + z_2} = \overline{z_1} + \overline{z_2} \qquad (12.4\text{a})$$

$$\overline{z_1 - z_2} = \overline{z_1} - \overline{z_2} \qquad (12.4\text{b})$$

$$\overline{z_1 z_2} = \overline{z_1}\, \overline{z_2} \qquad (12.4\text{c})$$

$$\overline{\left(\frac{z_1}{z_2}\right)} = \frac{\overline{z_1}}{\overline{z_2}} \qquad (12.4\text{d})$$

And it should also be noted that $z^2 \neq |z|^2$.

Check $\quad z^2 \neq |z|^2$

When the complex $z = x + iy$, we get

$$z^2 = zz = (x + iy)(x + iy) = (x^2 - y^2) + i(2xy)$$

$$|z|^2 = z\bar{z} = (x + iy)(x - iy) = x^2 + y^2$$

Answer $z^2 \neq |z|^2$

Example 12.1

For $z_1 = 2 + 3i$ and $z_2 = 1 - 2i$, show the details.

a. $2z_1 + z_2$ b. $z_1 - 2z_2$ c. $z_1 z_2$

d. $\dfrac{z_1}{z_2}$ e. $\overline{z_1}$ f. $\dfrac{1}{z_2}$

Answer a. $5 + 4i$, b. $7i$, c. $8 - i$,
d. $(-4 + 7i)/5$, e. $2 - 3i$, f. $(1 - 2i)/5$

12.1.2   *Polar form of complex numbers*

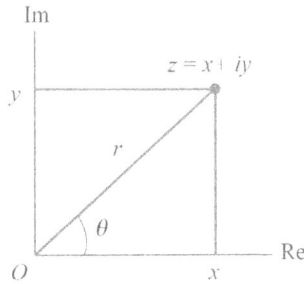

Figure 12.2 Polar form of a complex number.

As seen in Figure 12.2, a complex number $z = x + iy$ can be expressed in terms of distance r and angle θ, and this is referred to as the *polar form* of the complex number.

Remark   Polar form of a complex number

$$z = r(\cos\theta + i\sin\theta) \qquad (12.5)$$

where $x = r\cos\theta$ and $y = r\sin\theta$.

If we denote r as the absolute value of complex number z and θ as the argument of complex number z, then expressing Eq. (12.5) in the exponential form of Euler's formula yields the following.

$$z = re^{i\theta}, \qquad (12.6)$$

where

$$r = |z|, \tag{12.7a}$$

$$\theta = \arg(z) = \arctan(y/x). \tag{12.7b}$$

$$\arg(z) = \text{Arg}(z) + 2n\pi, \qquad (n = 0,\ 1,\ 2,\ \cdots)$$

Here, $\text{Arg}(z)$ is the principal value of $\arg(z)$ and $-\pi < \text{Arg}(z) \le \pi$.

When simplifying the multiplication and division of two complex numbers, $z_1 = r_1(\cos\theta_1 + i\sin\theta_1)$ and $z_2 = r_2(\cos\theta_2 + i\sin\theta_2)$, expressed in complex form, we get

$$z_1 z_2 = r_1 r_2 \{\cos(\theta_1 + \theta_2) + i\sin(\theta_1 + \theta_2)\}, \tag{12.8a}$$

$$\frac{z_1}{z_2} = \frac{r_1}{r_2}\{\cos(\theta_1 - \theta_2) + i\sin(\theta_1 - \theta_2)\}. \tag{12.8b}$$

Furthermore, when expressed in the exponential form of *Euler's formula*, the simplification of multiplication and division of two complex numbers, $z_1 = r_1 e^{i\theta_1}$ and $z_2 = r_2 e^{i\theta_2}$, yields

$$z_1 z_2 = r_1 r_2\, e^{i(\theta_1 + \theta_2)}, \tag{12.9a}$$

$$\frac{z_1}{z_2} = \frac{r_1}{r_2}\, e^{i(\theta_1 - \theta_2)}. \tag{12.9b}$$

When expressing only the absolute value and argument in Eq. (12.8) and Eq. (12.9), we have the following.

$$|z_1 z_2| = |z_1||z_2| \tag{12.10a}$$

$$\left|\frac{z_1}{z_2}\right| = \frac{|z_1|}{|z_2|} \tag{12.10b}$$

$$\arg(z_1 z_2) = \arg(z_1) + \arg(z_2) \tag{12.10c}$$

$$\arg\left(\frac{z_1}{z_2}\right) = \arg(z_1) - \arg(z_2) \tag{12.10d}$$

Check $\arg(z_1 z_2) = \arg(z_1) + \arg(z_2)$

$$(\cos\theta_1 + i\sin\theta_1)(\cos\theta_2 + i\sin\theta_2)$$
$$= (\cos\theta_1 \cos\theta_2 - \sin\theta_1 \sin\theta_2) + i(\cos\theta_1 \sin\theta_2 + \sin\theta_1 \cos\theta_2)$$
$$= \cos(\theta_1 + \theta_2) + i\sin(\theta_1 + \theta_2),$$

then we get

$$\arg(z_1 z_2) = \arg(z_1) + \arg(z_2). \tag{12.10c}$$

However, using Euler's formula provides a proof in itself.

$$e^{i\theta_1} e^{i\theta_2} = e^{i(\theta_1 + \theta_2)}.$$

Therefore, we get

$$\arg(z_1 z_2) = \arg(z_1) + \arg(z_2). \tag{12.10c}$$

Check $\arg\left(\dfrac{z_1}{z_2}\right) = \arg(z_1) - \arg(z_2)$

$$\frac{\cos\theta_1 + i\sin\theta_1}{\cos\theta_2 + i\sin\theta_2} = \frac{(\cos\theta_1 + i\sin\theta_1)(\cos\theta_2 - i\sin\theta_2)}{(\cos\theta_2 + i\sin\theta_2)(\cos\theta_2 - i\sin\theta_2)}$$

$$= \frac{(\cos\theta_1 \cos\theta_2 + \sin\theta_1 \sin\theta_2) + i(\sin\theta_1 \cos\theta_2 - \cos\theta_1 \sin\theta_2)}{\cos^2\theta_2 + \sin^2\theta_2}$$

$$= \cos(\theta_1 - \theta_2) + i\sin(\theta_1 - \theta_2),$$

then we get

$$\arg\left(\frac{z_1}{z_2}\right) = \arg(z_1) - \arg(z_2). \tag{12.10d}$$

However, using Euler's formula provides a proof in itself.

$$\frac{e^{i\theta_1}}{e^{i\theta_2}} = e^{i(\theta_1 - \theta_2)}.$$

Therefore, we get

$$\arg\left(\frac{z_1}{z_2}\right) = \arg(z_1) - \arg(z_2) \tag{12.10d}$$

Example 12.2

For $z_1 = 1 + \sqrt{3}i$, $z_2 = 1 + i$, show the details.

a. $|z_1|$ b. $\arg(z_1)$ c. $|z_2|$ d. $\arg(z_2)$

e. $|z_1 z_2|$ f. $\left|\dfrac{z_1}{z_2}\right|$ g. $\text{Arg}(z_1 z_2)$ h. $\text{Arg}\left(\dfrac{z_1}{z_2}\right)$

Answer a. 2, b. $\dfrac{\pi}{3} + 2n\pi$, c. $\sqrt{2}$, d. $\dfrac{\pi}{4} + 2n\pi$,

e. $2\sqrt{2}$, f. $\sqrt{2}$, g. $\dfrac{7\pi}{12}$, h. $\dfrac{\pi}{12}$

12.1.3 De Moivre's theorem

De Moivre's theorem is a formula that facilitates the calculation of powers of complex numbers. It allows us to express trigonometric functions and exponential functions in complex forms. Applying Eq. (12.10a) and Eq. (12.10c) when $z = z_1 = z_2 = r(\cos\theta + i\sin\theta)$ in the multiplication equation (12.8a), we can obtain the following equation for $n = 0, 1, 2, \cdots$.

$$z^n = r^n(\cos n\theta + i\sin n\theta). \tag{12.11}$$

Here, when $r = |z| = 1$, we obtain the following De Moivre's equation.

$$(\cos\theta + i\sin\theta)^n = \cos n\theta + i\sin n\theta \tag{12.12}$$

Check   De Moivre's theorem using mathematical induction

Let's prove it using *mathematical induction*, which is a method taught in high school mathematics

$$(\cos\theta + i\sin\theta)^n = \cos n\theta + i\sin n\theta \tag{12.12}$$

i. when $n = 1$, LHS = RHS.
ii. We establish the inductive hypothesis: Assume that Eq. (12.12) holds for $n = k$,

$$(\cos\theta + i\sin\theta)^k = \cos k\theta + i\sin k\theta.$$

Now, we prove that the equations also hold for $n = k + 1$.

Multiplying $\cos\theta + i\sin\theta$ in Eq. (12.12) yields

LHS: $= \cos(k+1)\theta + i\sin(k+1)\theta$

RHS: $= (\cos k\theta + i\sin k\theta)(\cos\theta + i\sin\theta)$

$\quad = (\cos k\theta \cos\theta - \sin k\theta \sin\theta) + i(\sin k\theta \cos\theta + \cos k\theta \sin\theta)$

$\quad = \cos(k+1)\theta + i\sin(k+1)\theta$

Then LHS = RHS.

From (i) and (ii), we prove that the equations also hold for all natural numbers $n = 1, 2, \dots$.

iii. For $n = 0$, Eq. (12.12) holds.

iv. Now, we prove that Eq. (12.12) holds for negative integer n.
For $n = -m$ (m natural number), we get

$$(\cos\theta + i\sin\theta)^n = \frac{1}{(\cos\theta + i\sin\theta)^m} \qquad (m = 1, 2, \cdots)$$

$$= \frac{1}{\cos m\theta + i\sin m\theta}$$

$$= \frac{\cos m\theta - i\sin m\theta}{(\cos m\theta + i\sin m\theta)(\cos m\theta - i\sin m\theta)}$$

$$= \cos(-m)\theta + i\sin(-m)\theta = \cos n\theta + i\sin n\theta.$$

Then, Eq. (12.12) also holds for negative integer n.

Therefore, from (i), (ii), (iii), and (iv), Eq. (12.12) holds for all integers.

Furthermore, we can extend the proof to demonstrate that Eq. ① holds for all real numbers.

Check   De Moivre's theorem using Euler's formula

Using Euler's formula simplifies the proof of De Moivre's theorem.

$$(\cos\theta + i\sin\theta)^n = \cos n\theta + i\sin n\theta \qquad (12.12)$$

LHS: $(\cos\theta + i\sin\theta)^n = (e^{i\theta})^n = e^{i(n\theta)}$

RHS: $\cos n\theta + i\sin n\theta = e^{i(n\theta)}$

Then, LHS = RHS.

Therefore, Eq. (12.12) holds for all real numbers.

De Moivre's theorem can be applied to finding the n-th roots of a given complex number $Z = R(\cos\varphi + i\sin\varphi)$ as well.

$$z^n = Z = R(\cos\varphi + i\sin\varphi) \tag{12.13a}$$

or

$$z = \sqrt[n]{Z} \tag{12.13b}$$

Applying De Moivre's theorem to Eq. (12.13a), we get:

$$r^n(\cos n\theta + i\sin n\theta) = R(\cos\varphi + i\sin\varphi),$$

or

$$r = \sqrt[n]{R} \quad \text{and} \quad n\theta = \varphi + 2k\pi.$$

Therefore, we can obtain n different roots as follows.

$$z_k = \sqrt[n]{R}\left\{\cos\left(\frac{\varphi+2k\pi}{n}\right) + i\sin\left(\frac{\varphi+2k\pi}{n}\right)\right\} \quad (k = 0, 1, 2, \cdots, n-1) \tag{12.14}$$

Figure 12.3 shows the n-th root of $z^3 = 1$, $z^4 = 1$, and $z^5 = 1$, respectively.

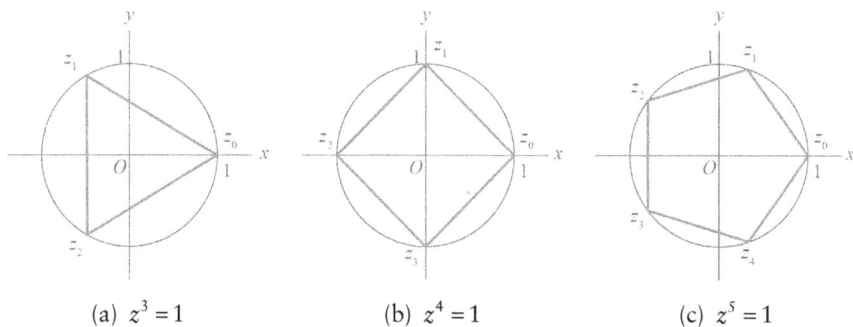

(a) $z^3 = 1$ (b) $z^4 = 1$ (c) $z^5 = 1$

Figure 12.3 n-th root of z.

Example 12.3

Solve the complex z.

$$z^3 = i \quad \left(z = \sqrt[3]{i}\right)$$

Solution

Applying De Moivre's theorem to $z = r(\cos\theta + i\sin\theta)$ yields:

$$\text{LHS: } z^3 = r^3(\cos 3\theta + i\sin 3\theta).$$

If we express the term i on the right-hand side in polar form, we get:

$$|i| = 1,$$

$$\arg(i) = \frac{\pi}{2} + 2k\pi. \qquad (k = 0,\ 1,\ 2)$$

Then from LHS = RHS, we get:

$$r = 1,$$

$$3\theta = \frac{\pi}{2} + 2k\pi.$$

Therefore, we obtain the k-th root as follows.

$$z_k = \cos\left(\frac{\pi}{6} + \frac{2k\pi}{3}\right) + i\sin\left(\frac{\pi}{6} + \frac{2k\pi}{3}\right)$$

$$\textbf{Answer } z_k = \cos\left(\frac{\pi}{6} + \frac{2k\pi}{3}\right) + i\sin\left(\frac{\pi}{6} + \frac{2k\pi}{3}\right) \qquad (k = 0,\ 1,\ 2)$$

Figure 12.4 shows the n-th root of z in Example 12.3.

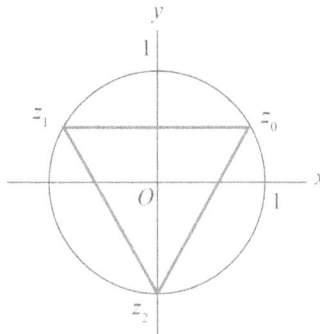

Figure 12.4 n-th root of z in Example 12.3.

Problem 12.1

For $z_1 = 3 + i$ and $z_2 = 1 - 2i$, show details. [1 ~ 6]

1. $z_1 z_2, z_1^2, |z_1|^2$

2. $\overline{z_1 z_2}, z_2^2, |z_2|^2$

3. $\dfrac{z_1}{z_2}, \dfrac{z_2}{z_1}$

4. $\text{Re}\left(2z_1 + \overline{z_2}\right), \text{Im}\left(2z_1 + \overline{z_2}\right)$

5. $\text{Re}\left(z_2^2\right), \left(\text{Re}\left(z_2\right)\right)^2$

6. $\text{Im}\left(z_2^2\right), \left(\text{Im}\left(z_2\right)\right)^2$

Represent in polar form and graph in the complex plane. [7 ~ 10]

7. $3 + \sqrt{3}i$

8. $\sqrt{2} - \sqrt{2}i$

9. $-\sqrt{3} + i$

10. $-3 - 4i$

Show details. [11 ~ 16]

11. If $z_1 = 2\left(\cos\dfrac{\pi}{3} + i\sin\dfrac{\pi}{3}\right)$ and $z_2 = \sqrt{3}\left(\cos\dfrac{\pi}{4} + i\sin\dfrac{\pi}{4}\right)$, solve $|z_1 z_2|$ and $\text{Arg}(z_1 z_2)$.

12. If $z_1 = \sqrt{2}\left(\cos\dfrac{\pi}{4} + i\sin\dfrac{\pi}{4}\right)$ and $z_2 = 2\left(\cos\dfrac{\pi}{6} - i\sin\dfrac{\pi}{6}\right)$, solve $\left|\dfrac{z_1}{z_2}\right|$ and $\text{Arg}\left(\dfrac{z_1}{z_2}\right)$.

13. If $z_1 = -\cos\dfrac{\pi}{6} + i\sin\dfrac{\pi}{6}$ and $z_2 = \sqrt{3}\left(\cos\dfrac{\pi}{4} - i\sin\dfrac{\pi}{4}\right)$, solve $|z_1 z_2|$ and $\text{Arg}(z_1 z_2)$.

14. If $z_1 = -2\left(\cos\dfrac{\pi}{3} + i\sin\dfrac{\pi}{3}\right)$ and $z_2 = \sqrt{3}\left(\cos\dfrac{\pi}{2} - i\sin\dfrac{\pi}{2}\right)$, solve $\left|\dfrac{z_1}{z_2}\right|$ and $\text{Arg}\left(\dfrac{z_1}{z_2}\right)$.

15. If $z_1 = \sqrt{2}\left(\sin\dfrac{\pi}{3} + i\cos\dfrac{\pi}{3}\right)$ and $z_2 = -2\left(\cos\dfrac{\pi}{6} + i\sin\dfrac{\pi}{6}\right)$, solve $|z_1 z_2|$ and $\text{Arg}(z_1 z_2)$.

16. If $z_1 = 3\left(\sin\dfrac{\pi}{3} - i\cos\dfrac{\pi}{3}\right)$ and $z_2 = -\left(\sin\dfrac{\pi}{4} + i\cos\dfrac{\pi}{4}\right)$, solve $\left|\dfrac{z_1}{z_2}\right|$ and $\text{Arg}\left(\dfrac{z_1}{z_2}\right)$.

Solve the complex z. [17 ~ 22]

17. $z^3 = 2$
18. $z^2 = i$
19. $z^4 = 1 + i$
20. $z^3 = 1 + \sqrt{3}i$
21. $z^3 = -2$
22. $z^4 = 2i$

12.2   Analytic functions and Cauchy-Riemann equation

12.2.1   Complex equation

We can represent a circle using *complex equation*. For example, we can write

$$|z| = 1. \tag{12.15}$$

As seen in Figure 12.5, this equation represents the unit circle in the complex plane centered at the origin. When $z = x + iy$, its magnitude becomes

$$|z|^2 = z\bar{z} = (x + iy)(x - iy) = x^2 + y^2 = 1.$$

Then it represents the circle $x^2 + y^2 = 1$ in Cartesian coordinates.

Furthermore, in the complex plane, a circle centered at $z_0 = x_0 + iy_0$ with radius r is represented as follows (refer to Figure 12.6).

$$|z - z_0| = r \tag{12.16a}$$

or

$$(x - x_0)^2 + (y - y_0)^2 = r^2 \tag{12.16b}$$

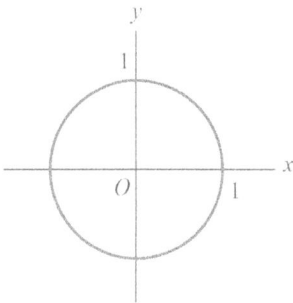

Figure 12.5 Unit circle.

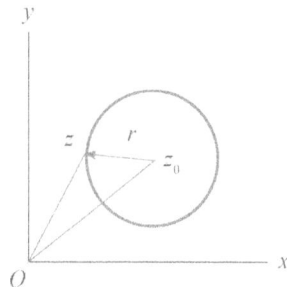

Figure 12.6 Circle with radius r.

We can represent the interior of a circle using inequalities, as shown in Figure 12.7, where a closed circle $\left(|z - z_0| \leq r\right)$ and an open circle $\left(|z - z_0| < r\right)$ denote a closed and open region, respectively. And Figure 12.8 shows a closed annulus and an open annulus.

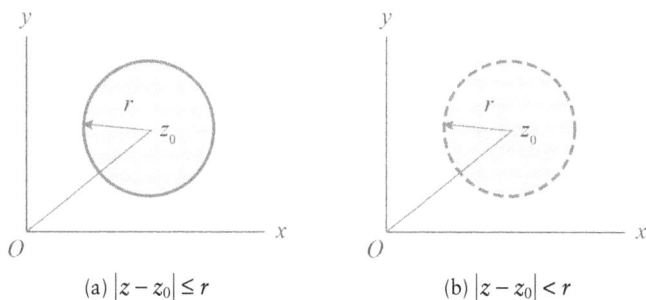

(a) $|z - z_0| \leq r$ (b) $|z - z_0| < r$

Figure 12.7 A closed circle and an open circle.

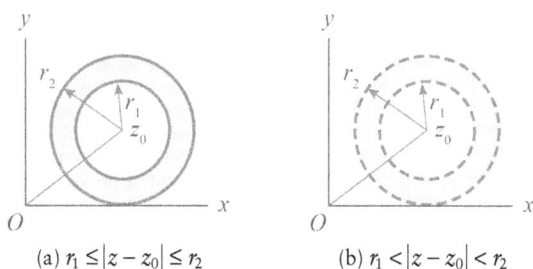

(a) $r_1 \leq |z - z_0| \leq r_2$ (b) $r_1 < |z - z_0| < r_2$

Figure 12.8 A closed annulus and an open annulus.

12.2.2 Complex function

The complex number $z = x + iy$, and a *complex function* w defined as $w = f(z)$, can be expressed as follows. That is,

$$w = f(z) = u(x,\ y) + iv(x,\ y) \tag{12.17}$$

where $u = u(x,\ y)$ and $v = v(x,\ y)$ are the real part and the imaginary part of a complex function w, respectively.

For example, a complex number is $z = x + iy$ and a complex function is $w = z^2 - 3z$, the real part and the imaginary part of a complex function are

$$u = x^2 - y^2 - 3x \quad \text{and} \quad v = 2xy - 3y.$$

That is,

$$w = z^2 - 3z = (x + iy)^2 - 3(x + iy) = \left(x^2 - y^2 - 3x\right) + i\left(2xy - 3y\right).$$

12.2.3   *Derivative of a complex function, analytic functions*

At a point z_0 in the complex plane, the value of the derivative $f'(z_0)$ of the complex function $f(z)$ is defined as follows, and when it exists, the complex function $f(z)$ is said to be differentiable at $z = z_0$.

$$f'(z_0) = \lim_{\Delta z \to 0} \frac{f(z_0 + \Delta z) - f(z_0)}{\Delta z}, \qquad (12.18a)$$

or

$$f'(z_0) = \lim_{\Delta z \to 0} \frac{f(z) - f(z_0)}{z - z_0}. \qquad (12.18b)$$

The derivative $f'(z)$ of a complex function $f(z)$ for all complex numbers z is defined as follows.

Remark   Derivative of a complex function

$$f'(z) = \lim_{\Delta z \to 0} \frac{f(z + \Delta z) - f(z)}{\Delta z} \qquad (12.19)$$

The forms of Eq. (12.18) and Eq. (12.19) resemble the expressions for the value of the derivative $f'(x_0)$ at a real number $x = x_0$, and the derivative $f'(x)$ of a real function $f(x)$, respectively. That is,

Value of the derivative: $f'(x_0) = \lim_{\Delta x \to 0} \dfrac{f(x_0 + \Delta x) - f(x_0)}{\Delta x}$

or

$$f'(x_0) = \lim_{\Delta x \to 0} \frac{f(x) - f(x_0)}{x - x_0}$$

Derivative: $f'(x) = \lim_{\Delta x \to 0} \dfrac{f(x + \Delta x) - f(x)}{\Delta x}$

Therefore, derivative of a complex function can be obtained using the same methods as derivative of a real function.

Remark   Analytic functions

A complex function $f(z)$ is called to be *analytic* in an open domain D if $f(z)$ is defined and differentiable at all points of D. And the complex function $f(z)$ is called to be analytic at a point $z = z_0$ in D if $f(z)$ is analytic in a neighborhood of $z = z_0$.

If a complex function $f(z)$ is said to be analytic, then for $\Delta z = \Delta x + i\Delta y$, regardless of the path of $\Delta z \to 0$, i.e., along the following two paths, it should have the same convergence value (value of the derivative).

i. A path where it converges first along $\Delta y \to 0$ and later along $\Delta x \to 0$.
ii. A path where it converges first along $\Delta x \to 0$ and later along $\Delta y \to 0$.

Figure 12.9 shows two paths of $\Delta z \to 0$.

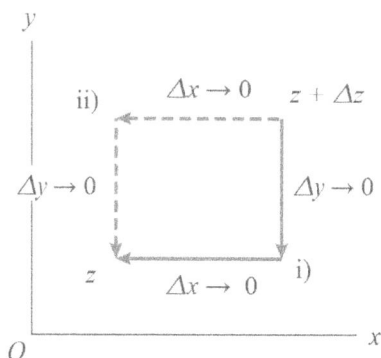

Figure 12.9 Path of $\Delta z \to 0$.

Example 12.4

For the complex function $w = f(z) = z^2 - 3z$, find $f'(z)$.

Solution

$$f'(z) = \lim_{\Delta z \to 0} \frac{f(z + \Delta z) - f(z)}{\Delta z}$$

$$= \lim_{\Delta z \to 0} \frac{\left\{(z + \Delta z)^2 - 3(z + \Delta z)\right\} - \left(z^2 - 3z\right)}{\Delta z}$$

$$= \lim_{\Delta z \to 0} \frac{2z\Delta z + (\Delta z)^2 - 3\Delta z}{\Delta z} = 2z - 3$$

Answer $f'(z) = 2z - 3$

12.2.4 *Cauchy-Riemann equations*

The *Cauchy-Riemann equations* are the criteria for determining whether a complex function is analytic.

Remark   Cauchy-Riemann equations in Cartesian coordinates

Let the complex function $w = f(z) = u(x, y) + iv(x, y)$ be defined and continuous in a domain D and differentiable. Then, the first-order partial derivatives of u and v in a domain D exist and satisfy the Cauchy-Riemann equations.

$$\frac{\partial u}{\partial x} = \frac{\partial v}{\partial y}, \quad \frac{\partial v}{\partial x} = -\frac{\partial u}{\partial y} \qquad (12.20)$$

The derivative of $f(z)$ is as follows:

$$f'(z) = \lim_{\Delta z \to 0} \frac{f(z + \Delta z) - f(z)}{\Delta z}. \qquad (12.19)$$

In Eq. (12.19), let $\Delta z = \Delta x + i\Delta y$.

$$f'(z) = \lim_{\Delta z \to 0} \frac{\{u(x + \Delta x, y + \Delta y) + iv(x + \Delta x, y + \Delta y)\} - \{u(x, y) + iv(x, y)\}}{\Delta x + i\Delta y}. \qquad (12.21)$$

In Figure 12.9, for the complex function $f(z)$ to be analytic, the values along the two paths must be equal.

Let's first examine path (i) in Figure 12.9. If we first converge along $\Delta y \to 0$, we get:

$$f'(z) = \lim_{\Delta x \to 0} \frac{\{u(x + \Delta x, y) + iv(x + \Delta x, y)\} - \{u(x, y) + iv(x, y)\}}{\Delta x}$$

$$= \lim_{\Delta x \to 0} \frac{u(x + \Delta x, y) - u(x, y)}{\Delta x} + i \lim_{\Delta x \to 0} \frac{v(x + \Delta x, y) - v(x, y)}{\Delta x}$$

$$= \frac{\partial u}{\partial x} + i \frac{\partial v}{\partial x}. \qquad (12.22a)$$

Now, let's examine path (ii) in Figure 12.9. If we converge along $\Delta x \to 0$, we get:

$$f'(z) = \lim_{\Delta y \to 0} \frac{\{u(x, y + \Delta y) + iv(x, y + \Delta y)\} - \{u(x, y) + iv(x, y)\}}{\Delta y}$$

$$= \lim_{\Delta y \to 0} \frac{u(x, y + \Delta y) - u(x, y)}{i\Delta y} + i \lim_{\Delta y \to 0} \frac{v(x, y + \Delta y) - v(x, y)}{i\Delta y}$$

$$= \frac{\partial v}{\partial y} - i \frac{\partial u}{\partial y}. \qquad (12.22b)$$

Therefore, in the equality relationship between Eq. (12.22a) and Eq. (12.22b), the Cauchy-Riemann equations hold.

$$\frac{\partial u}{\partial x} = \frac{\partial v}{\partial y}, \quad \frac{\partial v}{\partial x} = -\frac{\partial u}{\partial y} \qquad (12.20)$$

Example 12.5

Determine whether the following complex function is analytic.

$$w(x, y) = e^x (\cos y + i \sin y)$$

Solution

In $w = u + iv$, we get:

$$u = e^x \cos y \quad \text{and} \quad v = e^x \sin y.$$

Applying Cauchy-Riemann equation yields:

$$\frac{\partial u}{\partial x} = e^x \cos y \quad \text{and} \quad \frac{\partial v}{\partial y} = e^x \cos y$$

$$\therefore \quad \frac{\partial u}{\partial x} = \frac{\partial v}{\partial y}$$

$$\frac{\partial v}{\partial x} = e^x \sin y \quad \text{and} \quad -\frac{\partial u}{\partial y} = e^x \sin y$$

$$\therefore \quad \frac{\partial v}{\partial x} = -\frac{\partial u}{\partial y}.$$

Therefore, we conclude that the given function $w(x, y) = e^x (\cos y + i \sin y)$ is analytic, as we know that the Cauchy-Riemann equations hold.

Answer analytic

Remark   Cauchy-Riemann equations in polar coordinates

Let the complex function $w = f(z) = u(r, \theta) + iv(r, \theta)$ be defined and continuous in a domain D and differentiable. Then, the first-order partial derivatives of u and v in a domain D exist and satisfy the Cauchy-Riemann equations.

$$\frac{\partial u}{\partial r} = \frac{1}{r}\frac{\partial v}{\partial \theta}, \quad \frac{\partial v}{\partial r} = -\frac{1}{r}\frac{\partial u}{\partial \theta} \qquad (12.23)$$

The derivative of $f(z)$ is as follows:

$$f'(z) = \lim_{\Delta z \to 0} \frac{f(z + \Delta z) - f(z)}{\Delta z}. \tag{12.19}$$

In Eq. (12.19), let $\Delta z = (z + \Delta z) - z = (r + \Delta r)e^{i(\theta + \Delta\theta)} - re^{i\theta}$.

$$f'(z) = \lim_{\Delta z \to 0} \frac{\{u(r + \Delta r, \theta + \Delta\theta) + iv(r + \Delta r, \theta + \Delta\theta)\} - \{u(r, \theta) + iv(r, \theta)\}}{(r + \Delta r)e^{i(\theta + \Delta\theta)} - re^{i\theta}}$$

$$\tag{12.24}$$

In Figure 12.10, for the complex function $f(z)$ to be analytic, the values along the two paths must be equal.

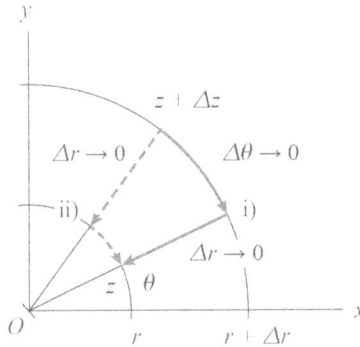

Figure 12.10 Path of $\Delta z \to 0$.

Let's first examine path (i) in Figure 12.10. If we first converge along $\Delta\theta \to 0$, we get:

$$f'(z) = \lim_{\Delta r \to 0} \frac{\{u(r + \Delta r, \theta) + iv(r + \Delta r, \theta)\} - \{u(r, \theta) + iv(r, \theta)\}}{(r + \Delta r)e^{i\theta} - re^{i\theta}}$$

$$= \lim_{\Delta r \to 0} \frac{u(r + \Delta r, \theta) - u(r, \theta)}{\Delta r e^{i\theta}} + i \lim_{\Delta r \to 0} \frac{v(r + \Delta r, \theta) - v(r, \theta)}{\Delta r e^{i\theta}}$$

$$= \frac{\partial u}{\partial r}e^{-i\theta} + i\frac{\partial v}{\partial r}e^{-i\theta}. \tag{12.25a}$$

Now, let's examine path (ii) in Figure 12.10. If we converge along $\Delta r \to 0$, we get:

$$f'(z) = \lim_{\Delta\theta \to 0} \frac{\left\{u(r, \theta + \Delta\theta) + iv(r, \theta + \Delta\theta)\right\} - \left\{u(r, \theta) + iv(r, \theta)\right\}}{r\left\{e^{i(\theta + \Delta\theta)} - e^{i\theta}\right\}}$$

$$= \lim_{\Delta\theta \to 0} \frac{u(r, \theta + \Delta\theta) - u(r, \theta)}{r\left\{e^{i(\theta + \Delta\theta)} - e^{i\theta}\right\}} + i \lim_{\Delta\theta \to 0} \frac{v(r, \theta + \Delta\theta) - v(r, \theta)}{r\left\{e^{i(\theta + \Delta\theta)} - e^{i\theta}\right\}}$$

$$= \lim_{\Delta\theta \to 0} \frac{\left\{u(r, \theta + \Delta\theta) - u(r, \theta)\right\} / \Delta\theta}{r\left\{e^{i(\theta + \Delta\theta)} - e^{i\theta}\right\} / \Delta\theta} + i \lim_{\Delta\theta \to 0} \frac{\left\{v(r, \theta + \Delta\theta) - v(r, \theta)\right\} / \Delta\theta}{r\left\{e^{i(\theta + \Delta\theta)} - e^{i\theta}\right\} / \Delta\theta}$$

$$= \frac{1}{r}\frac{\partial u}{\partial \theta}\frac{1}{ie^{i\theta}} + i\frac{1}{r}\frac{\partial v}{\partial \theta}\frac{1}{ie^{i\theta}}$$

$$= \frac{1}{r}\frac{\partial v}{\partial \theta}e^{-i\theta} - i\frac{1}{r}\frac{\partial u}{\partial \theta}e^{-i\theta}. \tag{12.25b}$$

Therefore, in the equality relationship between Eq. (12.25a) and Eq. (12.25b), the Cauchy-Riemann equations hold.

$$\frac{\partial u}{\partial r} = \frac{1}{r}\frac{\partial v}{\partial \theta}, \quad \frac{\partial v}{\partial r} = -\frac{1}{r}\frac{\partial u}{\partial \theta} \tag{12.23}$$

Example 12.6

Determine whether the following complex function is analytic.

$$w(r, \theta) = \left(r^2 \cos 2\theta - 2r\cos\theta\right) + i\left(r^2 \sin 2\theta - 2r\sin\theta\right)$$

Solution

In $w = u + iv$, we get:

$$u = r^2 \cos 2\theta - 2r\cos\theta \quad \text{and} \quad v = r^2 \sin 2\theta - 2r\sin\theta.$$

Applying Cauchy-Riemann equation yields:

$$\frac{\partial u}{\partial r} = 2r\cos 2\theta - 2\cos\theta, \quad \frac{1}{r}\frac{\partial v}{\partial \theta} = 2r\cos 2\theta - 2\cos\theta$$

$$\therefore \frac{\partial u}{\partial r} = \frac{1}{r}\frac{\partial v}{\partial \theta}.$$

$$\frac{\partial v}{\partial r} = 2r\sin 2\theta - 2\sin\theta, \qquad -\frac{1}{r}\frac{\partial u}{\partial \theta} = 2r\sin 2\theta - 2\sin\theta$$

$$\therefore \frac{\partial v}{\partial r} = -\frac{1}{r}\frac{\partial u}{\partial \theta}.$$

Therefore, as we know that the Cauchy-Riemann equations hold, we conclude that the given function $w(r,\ \theta) = \left(r^2\cos 2\theta - 2r\cos\theta\right) + i\left(r^2\sin 2\theta - 2r\sin\theta\right)$ is analytic.

Answer analytic

12.2.5 Laplace's equation and harmonic function

As learned in Chapter 11, there is an important equation called the *Laplace's equation*, which is extensively employed in various fields such as vibration dynamics, heat transfer, electricity, fluid dynamics, and others. A solution to the Laplace's equation, where both the real and imaginary parts of the analytic function satisfy the Laplace's equation, is termed a *harmonic function*.

Remark   Laplace's equation

If a complex function $f(z) = u(x,\ y) + iv(x,\ y)$ is analytic in a domain D, then both the real part $u = u(x,\ y)$ and the imaginary part $v = v(x,\ y)$ satisfy the following Laplace's equations in D.

$$\nabla^2 u = \frac{\partial^2 u}{\partial x^2} + \frac{\partial^2 u}{\partial y^2} = 0, \qquad (12.26a)$$

$$\nabla^2 v = \frac{\partial^2 v}{\partial x^2} + \frac{\partial^2 v}{\partial y^2} = 0, \qquad (12.26b)$$

where u and v are called *harmonic conjugate functions*.

Example 12.7

Determine whether $u(x,\ y) = x^2 - y^2$ is an analytic function, and if it is a harmonic function, find the harmonic conjugate function. And find the corresponding analytic function $f(z) = u(x,\ y) + iv(x,\ y)$.

Solution

The given function $u(x, y) = x^2 - y^2$ satisfies the Laplace's equation

$$\nabla^2 u = \frac{\partial^2 u}{\partial x^2} + \frac{\partial^2 u}{\partial y^2} = 0.$$

Let's apply the Cauchy-Riemann equation.

From $\dfrac{\partial u}{\partial x} = \dfrac{\partial v}{\partial y} = 2x$, we get:

$$v = 2xy + g(x). \qquad\qquad ①$$

From $\dfrac{\partial v}{\partial x} = -\dfrac{\partial u}{\partial y} = 2y$, we get:

$$v = 2xy + h(y). \qquad\qquad ②$$

From Eq. ① and Eq. ②, we can find the harmonic conjugate function:

$$v = 2xy + c \qquad (c \text{ any real constant})$$

Therefore, we obtain the analytic function:

$$f(z) = u + iv = \left(x^2 - y^2\right) + i\left(2xy + c\right)$$

Since $z = x + iy$, we obtain

$$f(z) = z^2 + ci.$$

$$\textbf{Answer } v = 2xy + c \text{ (}c\text{ any real constant)}, f(z) = z^2 + ci$$

Remark   Laplace's equation in polar form

If a complex function $f(z) = u(r, \theta) + iv(r, \theta)$ is analytic in a domain D, then both the real part $u = u(r, \theta)$ and the imaginary part $v = v(r, \theta)$ satisfy the following Laplace's equations in D.

$$\nabla^2 u = \frac{\partial^2 u}{\partial r^2} + \frac{1}{r}\frac{\partial u}{\partial r} + \frac{1}{r^2}\frac{\partial^2 u}{\partial \theta^2} = 0, \qquad (12.27\text{a})$$

$$\nabla^2 v = \frac{\partial^2 v}{\partial r^2} + \frac{1}{r}\frac{\partial v}{\partial r} + \frac{1}{r^2}\frac{\partial^2 v}{\partial \theta^2} = 0, \qquad (12.27\text{b})$$

where u and v be called *harmonic conjugate functions*.

Example 12.8

Determine whether $v = r^3 \sin 3\theta$ is a harmonic function, and if it is a harmonic function, find the harmonic conjugate function. And find the corresponding analytic function $f(z) = u(r, \theta) + iv(r, \theta)$.

Solution

From the function $v = r^3 \sin 3\theta$, we get:

$$\frac{\partial^2 v}{\partial r^2} = 6r \sin 3\theta,$$

$$\frac{1}{r}\frac{\partial v}{\partial r} = 3r \sin 3\theta,$$

$$\text{and } \frac{1}{r^2}\frac{\partial^2 v}{\partial \theta^2} = -9r \sin 3\theta.$$

Then we can check if the function $v = r^3 \sin 3\theta$ satisfies the Laplace's equation

$$\nabla^2 v = \frac{\partial^2 v}{\partial r^2} + \frac{1}{r}\frac{\partial v}{\partial r} + \frac{1}{r^2}\frac{\partial^2 v}{\partial \theta^2} = 0.$$

Now, let's apply the Cauchy-Riemann equation.

From $\dfrac{\partial u}{\partial r} = \dfrac{1}{r}\dfrac{\partial v}{\partial \theta}$, we get:

$$\frac{\partial u}{\partial r} = 3r^2 \cos 3\theta,$$

or

$$u = r^3 \cos 3\theta + g(\theta). \qquad \qquad ①$$

From $\dfrac{\partial v}{\partial r} = -\dfrac{1}{r}\dfrac{\partial u}{\partial \theta}$, we get:

$$3r^2 \sin 3\theta = -\frac{1}{r}\frac{\partial u}{\partial \theta},$$

or

$$u = r^3 \cos 3\theta + h(r). \qquad \qquad ②$$

From Eq. ① and Eq. ②, we can find the harmonic conjugate function:

$$u = r^3 \cos 3\theta + c. \qquad (c \text{ any real constant})$$

Therefore, we obtain the analytic function:

$$f(z) = u + iv = (r^3 \cos 3\theta + c) + ir^3 \sin 3\theta = c + r^3 (\cos 3\theta + i \sin 3\theta)$$

Since $z = re^{i\theta}$, we obtain

$$f(z) = z^3 + c.$$

Answer $u = r^3 \cos 3\theta + c$, $f(z) = z^3 + c$ (c any real constant)

Problem 12.2

Determine whether a given function is analytic or not. [1 ~ 8]

1. $f(z) = e^x (\cos y - i \sin y)$
2. $f(z) = e^{-2x} (\sin 2y + i \cos 2y)$
3. $f(z) = -x^2 + y^2 - 2xyi$
4. $f(z) = x^3 - 2y + 3x^2 yi$
5. $f(z) = r^2 \sin 2\theta - ir^2 \cos 2\theta$
6. $f(z) = \dfrac{\cos\theta - i\sin\theta}{r}$
7. $f(z) = z^2 - 2z$
8. $f(z) = z\bar{z}$

Determine whether a given function is analytic or not, and if it is analytic, find its corresponding analytic function $f(z) = u(x, y) + iv(x, y)$. [9 ~ 14]

9. $u(x, y) = 2xy$
10. $u(x, y) = e^{-x} \sin y$
11. $v(x, y) = \cos x \cos hy$
12. $v(x, y) = 3x^2 y - y^3$
13. $u(r, \theta) = r^2 \cos 2\theta$
14. $v(r, \theta) = \dfrac{\cos\theta}{r}$

Determine a so that the given function is analytic and find the conjugate analytic function. [15 ~ 18]

15. $u(x, y) = e^{-2x} \cos ay$
16. $u(x, y) = \cos x \sinh ay$
17. $u(x, y) = ay^3 + 4xy$
18. $u(r, \theta) = \dfrac{\sin a\theta}{r}$

12.3   Several complex functions

In this section, we will learn various forms of functions involving complex numbers, such as exponential functions, trigonometric functions, hyperbolic functions, logarithmic functions, etc. All the properties and equations that held true for real number x in functions like e^x, $\cos x$, $\sin x$, $\cosh x$, $\sinh x$, $\ln x$, and so on, can be appropriately extended by substituting the real variable x with a complex number z.

12.3.1   *Complex exponential functions and complex logarithms*

The *complex exponential function* e^z for a complex number z can be expressed as follows:

$$e^z = e^{x+iy} = e^x e^{iy} = e^x \left(\cos y + i \sin y \right). \tag{12.28}$$

Moreover, the complex exponential function e^z satisfies the following equations, just as in the case of the real exponential function e^x.

$$e^{z_1 + z_2} = e_1{}^z e_2{}^z \tag{12.29}$$

$$\left(e^z \right)' = e^z \tag{12.30}$$

In particular, when $z = yi$, we obtain the following Euler's formula from Eq. (12.28).

$$e^{yi} = \cos y + i \sin y \tag{12.31}$$

Then we get $e^{0 \cdot i} = 1$, $e^{\frac{\pi}{2} i} = i$, $e^{\pi i} = -1$, $e^{\frac{3\pi}{2} i} = -i$, $e^{2\pi i} = 1$, $e^{\left(2\pi + \frac{\pi}{2} \right) i} = i$, $e^{(2\pi + \pi)i} = -1$, $\cdots$.

That is, in the complex exponential function e^z where $z = x + iy$, y represents the angular component, thus exhibiting periodicity as described by the following equation:

$$e^{z + 2\pi i} = e^z. \tag{12.32}$$

On the other hand, the *complex logarithmic function* $\ln z$ for a complex number $z \left(= re^{i\theta} \right)$ is expressed as follows:

$$\ln z = \ln r + i\theta \tag{12.33}$$

where $r = |z| > 0$ and $\theta = \arg(z)$.

The argument $\arg(z)$ of a complex number z can take infinitely many values by adding integer multiples of 2π to the *principal argument* $\text{Arg}(z)$. In other words,

$$\arg(z) = \text{Arg}(z) + 2n\pi. \qquad (n = 0, 1, 2, ...) \qquad (12.34)$$

The value of $\ln(z)$ corresponding to the principle value $\text{Arg}(z)$ is donated by $\text{Ln}(z)$ and is called the principal value of $\ln(z)$. That is,

$$\ln(z) = \text{Ln}(z) + 2n\pi i. \qquad (n = 0, 1, 2, ...) \qquad (12.35)$$

Example 12.9

Show details.

a. For $z = 4 + 3i$, find e^z in the form $u + vi$, where u and v are real numbers.
b. Represent $z = \ln(1 + \sqrt{3}i)$ in the form $u + vi$, where u and v are real numbers.

Solution

a. $e^z = e^{4+3i} = e^4 e^{3i} = e^4(\cos 3 + i\sin 3)$

$$\textbf{Answer } e^4(\cos 3 + i \sin 3)$$

b. From $z = \ln(1 + \sqrt{3}i)$, we get

$$e^z = 1 + \sqrt{3}i.$$

Since LHS is $e^z = e^{x+iy} = e^x e^{iy} = e^x(\cos y + i\sin y)$, and

RHS is $1 + \sqrt{3}i = 2\left(\cos\dfrac{\pi}{3} + i\sin\dfrac{\pi}{3}\right)$, we get:

$$e^x = 2 \quad \text{and} \quad y = \frac{\pi}{3} + 2n\pi. \qquad (n = 0, 1, 2, ...)$$

Therefore, we obtain:

$$x = \ln 2 \quad \text{and} \quad y = \frac{\pi}{3} + 2n\pi.$$

$$\textbf{Answer } z = \ln 2 + \left(\frac{\pi}{3} + 2n\pi\right)i, \ (n = 0, 1, 2, ...)$$

12.3.2   *Complex trigonometric functions and complex hyperbolic functions*

Complex trigonometric functions $\cos z$, $\sin z$, and $\tan z$ for a complex number z are expressed as follows.

$$\cos z = \frac{1}{2}\left(e^{iz} + e^{-iz}\right) \tag{12.36}$$

$$\sin z = \frac{1}{2i}\left(e^{iz} - e^{-iz}\right) \tag{12.37}$$

$$\tan z = \frac{\sin z}{\cos z} \tag{12.38}$$

Moreover, the derivative of the complex trigonometric function $\cos z$, $\sin z$, and $\tan z$ also takes a form similar to that of the real trigonometric function $\cos x$, $\sin x$, and $\tan x$.

$$\left(\cos z\right)' = -\sin z \tag{12.39}$$

$$\left(\sin z\right)' = \cos z \tag{12.40}$$

$$\left(\tan z\right)' = \sec^2 z \tag{12.41}$$

For all complex numbers z, the following Euler's formula holds true.

$$e^{iz} = \cos z + i \sin z \tag{12.42}$$

On the other hand, *complex hyperbolic functions* such as $\cosh z$, $\sinh z$, and $\tanh z$ for a complex number z are expressed as follows.

$$\cosh z = \frac{1}{2}\left(e^z + e^{-z}\right) \tag{12.43}$$

$$\sinh z = \frac{1}{2}\left(e^z - e^{-z}\right) \tag{12.44}$$

$$\tanh z = \frac{\sinh z}{\cosh z} \tag{12.45}$$

Moreover, the derivative of the complex hyperbolic functions, such as $\cosh z$, $\sinh z$, and $\tanh z$ also takes a form similar to that of the real function $\cosh x$, $\sinh x$, and $\tanh x$.

$$\left(\cosh z\right)' = \sinh z \tag{12.46}$$

$$\left(\sinh z\right)' = \cosh z \tag{12.47}$$

$$\left(\tanh z\right)' = \frac{1}{\cosh^2 z} \tag{12.48}$$

Complex trigonometric functions and complex hyperbolic functions are related to each other as follows.

$$\cos iz = \cosh z \qquad\qquad (12.49)$$

$$\sin iz = i \sinh z \qquad\qquad (12.50)$$

$$\cosh iz = \cos z \qquad\qquad (12.51)$$

$$\sinh iz = i \sin z \qquad\qquad (12.52)$$

Example 12.10

Represent in the form $u + vi$.

a. $\sin(4 + 3i)$
b. $\cosh(2 + \pi i)$

Solution

a. $\sin(4 + 3i) = \sin 4 \cdot \cos 3i + \cos 4 \cdot \sin 3i$

$\qquad = \sin 4 \cdot \cosh 3 + \cos 4 \cdot i \sinh 3$

Answer $\sin 4 \cosh 3 + i \cos 4 \sinh 3$

b. $\cosh(2 + \pi i) = \cos(i(2 + \pi i)) = \cos(-\pi + 2i)$

$\qquad = \cos(-\pi) \cdot \cos(2i) - \sin(-\pi) \cdot \sin(2i)$

$\qquad = (-1) \cdot \cosh 2 + 0 \cdot i \sinh 2$

$\qquad = -\cosh 2$

Answer $-\cosh 2$

Problem 12.3

Solve the complex z. [1 ~ 4]

1. $\ln z = 1 + 2i$
2. $\ln z = \pi i$
3. $\ln z = 1 - \dfrac{\pi}{2} i$
4. $\ln z = 2 + \dfrac{\pi}{3} i$

Represent in the form of $u + vi$. [5 ~ 8]

5. $z = \ln(3i)$

6. $z = \ln(e^{2i})$

7. $z = \ln(1+i)$

8. $z = \ln(\sqrt{3} - i)$

Represent in the form of $u + vi$. [9 ~ 14]

9. $\sin(2i)$

10. $\cos(\pi i)$

11. $\sin\left(\dfrac{\pi}{6} + i\right)$

12. $\cos(1 - i)$

13. $\sinh(2 + \pi i)$

14. $\cosh\dfrac{\pi}{2}(1 - i)$

Answer

Problem 12.1

1. $5 - 5i$, $8 + 6i$, 10
2. $5 + 5i$, $-3 - 4i$, 5
3. $\dfrac{1+7i}{5}$, $\dfrac{1-7i}{10}$
4. 7, 4
5. -3, 1
6. -4, 4
7. $2\sqrt{3}\left(\cos\dfrac{\pi}{6} + i\sin\dfrac{\pi}{6}\right)$
8. $2\left\{\cos\left(-\dfrac{\pi}{4}\right) + i\sin\left(-\dfrac{\pi}{4}\right)\right\}$
9. $2\left(\cos\dfrac{5\pi}{6} + i\sin\dfrac{5\pi}{6}\right)$
10. $5\{\cos(-\varphi) + i\sin(-\varphi)\}$, where $\varphi = \arctan(4/3)$
11. $2\sqrt{3}$, $\dfrac{7\pi}{12}$
12. $\dfrac{\sqrt{2}}{2}$, $\dfrac{5\pi}{12}$
13. $\sqrt{3}$, $\dfrac{7\pi}{12}$
14. $\sqrt{2}$, $-\dfrac{\pi}{6}$
15. $2\sqrt{2}$, $-\dfrac{2\pi}{3}$
16. 3, $\dfrac{7\pi}{12}$
17. $z_k = \sqrt[3]{2}\left\{\cos\left(\dfrac{2k\pi}{3}\right) + i\sin\left(\dfrac{2k\pi}{3}\right)\right\}$ $\quad (k = 0, 1, 2)$
18. $z_k = \cos\left(\dfrac{\pi}{4} + k\pi\right) + i\sin\left(\dfrac{\pi}{4} + k\pi\right)$ $\quad (k = 0, 1)$
19. $z_k = \sqrt[8]{2}\left\{\cos\left(\dfrac{\pi}{16} + \dfrac{k\pi}{2}\right) + i\sin\left(\dfrac{\pi}{16} + \dfrac{k\pi}{2}\right)\right\}$ $\quad (k = 0, 1, 2, 3)$
20. $z_k = \sqrt[3]{2}\left\{\cos\left(\dfrac{\pi}{9} + \dfrac{2k\pi}{3}\right) + i\sin\left(\dfrac{\pi}{9} + \dfrac{2k\pi}{3}\right)\right\}$ $\quad (k = 0, 1, 2)$

21. $z_k = \sqrt[3]{2} \left\{ \cos\left(\dfrac{\pi}{3} + \dfrac{2k\pi}{3} \right) + i\sin\left(\dfrac{\pi}{3} + \dfrac{2k\pi}{3} \right) \right\}$ $\quad (k = 0, 1, 2)$

22. $z_k = \sqrt[4]{2} \left\{ \cos\left(\dfrac{\pi}{8} + \dfrac{k\pi}{2} \right) + i\sin\left(\dfrac{\pi}{8} + \dfrac{k\pi}{2} \right) \right\}$ $\quad (k = 0, 1, 2, 3)$

Problem 12.2

1. not analytic.
2. analytic.
3. analytic.
4. not analytic.
5. analytic.
6. analytic.
7. analytic.
8. not analytic.
9. $v = -x^2 + y^2 + c,$ $\quad f(z) = \bar{z}^2 + ci$ $\quad$ (c any real constant)
10. $v = e^{-x}\cos y + c,$ $\quad f(z) = ie^{-z} + ci$ $\quad$ (c any real constant)
11. $u = \sin x \sinh y + h(x),$ $\quad f(z) = (\sin x \sinh y + c) + i(\cos x \cosh y)$
    (c any real constant),
12. $u = x^3 - 3xy^2 + c,$ $\quad f(z) = z^3 + c$ $\quad$ (c any real constant)
13. $v = r^2 \sin 2\theta + c,$ $\quad f(z) = z^2 + ci$ $\quad$ (c any real constant),
14. $u = \dfrac{\sin\theta}{r} + c,$ $\quad f(z) = iz^{-1} + c$ $\quad$ (c any real constant),
15. $a = 2, v = -e^{-2x}\sin 2y + c$ $\quad$ (c any real constant)
16. $a = \pm 1, v = \mp\sin x \cos hy + c$ $\quad$ (c any real constant)
17. $a = 0, v = -2x^2 + 2y^2 + c$ $\quad$ (c any real constant)
18. $a = \pm 1, v = \pm\dfrac{\cos\theta}{r} + c$ $\quad$ (c any real constant)

Problem 12.3

1. $e(\cos 2 + i\sin 2)$
2. -1
3. $-ei$
4. $\dfrac{e^2}{2}\left(1 - \sqrt{3}i\right)$
5. $\ln 3 + \left(\dfrac{\pi}{2} + 2n\pi \right)i$
6. $2i$

7. $\dfrac{1}{2}\ln 2 + \left(\dfrac{\pi}{4} + 2n\pi\right)i$

8. $\ln 2 + \left(-\dfrac{\pi}{6} + 2n\pi\right)i$

9. $i\sinh 2$

10. $\cosh\pi$

11. $\dfrac{1}{2}\left(\cosh 1 + i\sqrt{3}\sinh 1\right)$

12. $\left(\cos 1\ \cosh 1\right) + i\left(\sin 1\ \sinh 1\right)$

13. $-\cos 2\ \sinh\pi + i\sin 2\ \cosh\pi$

14. $i\sinh\dfrac{\pi}{2}$

13 Complex integration

In this chapter, we will learn about complex integration based on complex numbers and complex functions learned in Chapter 12, focusing particularly on the Cauchy integral theorem. The importance of complex integration lies in the fact that integrals involving physics and other applications, which were not easily approached in real calculus, can be easily solved through complex integration.

13.1 Line integral in the complex plane

13.1.1 Complex line integral

When integrating a function $f(z)$ along a given path C in the complex plane, the *complex line integral* is expressed as follows:

$$\int_C f(z)\,dz. \tag{13.1}$$

In particular, when the path C is a *closed path* (a path where the starting point and the ending point are the same), it is denoted as follows:

$$\oint_C f(z)\,dz. \tag{13.2}$$

Here, the complex function $z(t) = x(t) + iy(t)\ (-2\pi \le t \le 2\pi)$ varies with the parameter t, composed of real components $x(t)$ and imaginary components $y(t)$. The direction in which the parameter t increases is called the positive sense of the path C.

For example, $z(t) = \cos t + i\sin t$ represents the unit circle in the complex plane, where $|z| = 1$.

If the complex function $z(t)$ is continuous and differentiable at every point along the path C, then it satisfies the following equation:

$$\dot{z}(t) = \frac{d}{dt}z(t) = \frac{d}{dt}x(t) + i\frac{d}{dt}y(t) = \dot{x} + i\dot{y}. \tag{13.3}$$

The complex line integral satisfies the following distributive property.

$$\int_C \{k_1 f_1(z) + k_2 f_2(z)\}\,dz = k_1 \int_C f_1(z)\,dz + k_2 \int_C f_2(z)\,dz \tag{13.4}$$

DOI: 10.1201/9781003608912-13

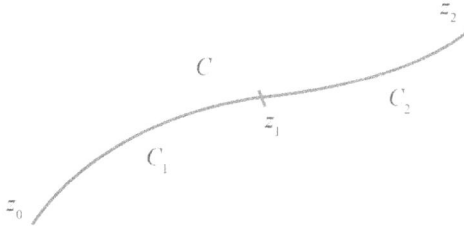

Figure 13.1 The starting point z_0 and the ending point z_2 in a path C.

This means that the integral of the sum of two complex functions along a path C is equal to the sum of the integrals of each function along the same path C.

As seen in Figure 13.1, when the starting point of the path C is z_0 and the ending point is z_2, if the path is reversed, i.e., starting from z_2 and ending at z_0, then the sign of the line integral value changes. A path where the starting and ending points are different is called an *open path*.

$$\int_{z_0}^{z_2} f(z)dz = -\int_{z_2}^{z_0} f(z)dz \tag{13.5}$$

Furthermore, as seen in Figure 13.1, if the path C is divided into paths C_1 (from starting point z_0 to z_1) and C_2 (from z_1 to z_2), then the line integral along path C is equal to the sum of the line integrals along paths C_1 and C_2.

$$\int_{z_0}^{z_2} f(z)dz = \int_{z_0}^{z_1} f(z)dz + \int_{z_1}^{z_2} f(z)dz \tag{13.6a}$$

$$\int_C f(z)dz = \int_{C_1} f(z)dz + \int_{C_2} f(z)dz \tag{13.6b}$$

Remark   Complex line integral

Let $f(z)$ be analytic in a simple connected domain D. If we denote an indefinite integral of the function $f(z)$ in the domain D as $F(z)$, then for any two points z_0 and z_2 connected by any path in the domain D, the following equation holds:

$$\int_C f(z)\,dz = \int_{z_0}^{z_2} f(z)dz = F(z_2) - F(z_0). \tag{13.7}$$

Example 13.1

Find the value of the complex line integral.

a. $\int_{1-2i}^{1+2i} z\, dz$

b. $\int_{0}^{\pi i} \sin z\, dz$

c. $\int_{2-i}^{2+i} e^z\, dz$

Solution

a. $\int_{1-2i}^{1+2i} z\, dz = \left[\dfrac{z^2}{2}\right]_{1-2i}^{1+2i} = \dfrac{1}{2}\left\{(1+2i)^2 - (1-2i)^2\right\} = 4i$

b. $\int_{0}^{\pi i} \sin z\, dz = -\left[\cos z\right]_{0}^{\pi i} = 1 - \cos(\pi i) = 1 - \cosh\pi$

c. $\int_{2-i}^{2+i} e^z\, dz = \left[e^z\right]_{2-i}^{2+i} = e^{2+i} - e^{2-i}$

$= e^2(\cos 1 + i\sin 1) - e^2(\cos 1 - i\sin 1) = 2e^2 i\sin 1$

Answer a. 4i, b. $1 - \cosh\pi$, c. $i2e^2\sin 1$

13.1.2   Representation of parameters for a complex path

To represent a *complex path* in terms of parameters, we typically use a parametric equation. A parametric equation expresses the coordinates of a point on the path as functions of one or more parameters. For a complex path in the complex plane, we can use parametric equations for the real and imaginary parts of the path.

Example 13.2

Represent the following complex path in terms of parameters:

a. The path C is the shortest path from the origin to (3, 2)
b. The path C is the semicircle from (0, 1) to (2, 1) and $y \leq 1$
c. The path C is counterclockwise along the circle with the center at the origin and radius 2.

Solution

a. The shortest path from the origin to (3, 2), as shown in Figure 13.2, can be represented parametrically as follows:

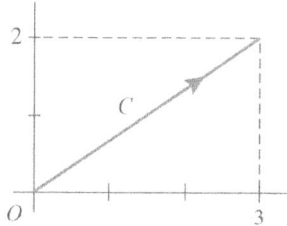

Figure 13.2 Path C.

$$z = 3t + 2ti, \quad (0 \le t \le 1).$$

Answer $z = 3t + 2ti, \left(0 \le t \le 1\right)$

b. The parametric representation of the path C, which is the semicircle from (0, 1) to (2, 1) with $y \le 1$, as shown in Figure 13.3, can be expressed as follows:

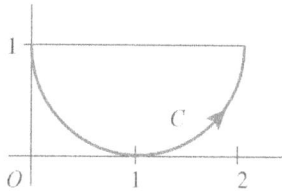

Figure 13.3 Path C.

Since the path *C* represents a circle with a center at (1, 1) and a radius of 1, we can express it as follows:

$$\left|z - (1+i)\right| = 1 \quad \text{and} \quad y \le 1.$$

Then, we can obtain the parametric representation:

$$z = (1 + \cos t) + i(1 + \sin t), \quad (\pi \le t \le 2\pi)$$

Answer $z = (1 + \cos t) + i(1 + \sin t), (\pi \le t \le 2\pi)$

c. The parametric representation of the counterclockwise path on the circle with center at the origin and radius 2, as shown in Figure 13.4, is as follows:

$$z = 2(\cos t + i \sin t), \quad (0 \le t \le 2\pi).$$

Answer $z = 2(\cos t + i \sin t), \quad (0 \le t \le 2\pi)$

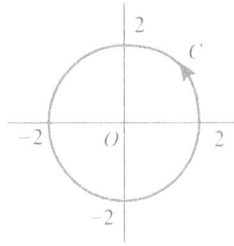

Figure 13.4 Path C.

13.1.3 Representation of parameters for a complex line integral

Remark   Complex line integral

If the continuous complex function z along path C is represented by the parameter t, i.e., $z = z(t) = x(t) + i y(t)$, then the following equation holds:

$$\int_C f(z)\,dz = \int_a^b f\big[z(t)\big]\dot{z}(t)\,dt. \qquad (13.8)$$

Since $z = z(t) = x(t) + i y(t)$, we get

$$dz = dx + i\,dy.$$

where $dx = \dot{x}\,dt$ and $dy = \dot{y}\,dt$.

When applying function $f(z) = u(z) + iv(z)$ to the RHS in Eq. (13.8), we derive the following:

$$\int_C f(z)\,dz = \int_C (u + iv)(dx + i\,dy)$$

$$= \int_a^b (u + iv)(\dot{x} + i\dot{y})\,dt$$

$$= \int_a^b f\big[z(t)\big]\dot{z}(t)\,dt$$

Example 13.3

Calculate the complex line integral along the given path C.

a. $\displaystyle\int_C \mathrm{Re}(z)\,dz$,

   path C: the shortest path from the origin to $2 + 2i$

b. $\int_C \cos z\, dz,$

path C: the semicircle from (2, 0) to (−2, 0) and $y \geq 0$

c. $\oint_C \frac{1}{z}\, dz,$

path C: counterclockwise along the circle with center at the origin and radius 1.

Solution

a. The shortest path from the origin to $2 + 2i$, as shown in Figure 13.5, can be represented parametrically as follows:

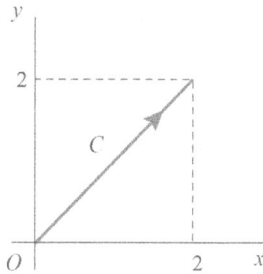

Figure 13.5 Path C.

$$z = t + it, \quad (0 \leq t \leq 2).$$

Then we get

$$dz = (1+i)dt \quad \text{and} \quad \text{Re}(z) = t.$$

Therefore, the answer is

$$\int_C \text{Re}(z)\, dz = \int_0^2 t \cdot (1+i)\, dt$$

$$= (1+i)\frac{t^2}{2}\Big|_0^2 = 2(1+i).$$

Answer $2(1+i)$

b. The parametric representation of the path C, which is the semicircle from (2, 0) to (−2, 0) with $y \geq 0$, as shown in Figure 13.6, can be expressed as follows:

Since it is a circle with a center at $(0, 0)$ and a radius of 2, we can express it as follows:

$$|z| = 2,$$

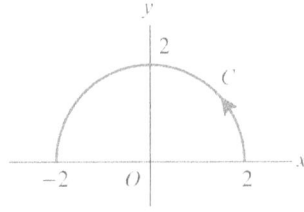

Figure 13.6 Path C.

or

$$z = 2(\cos t + i\sin t) = 2e^{it}. \quad (0 \le t \le \pi)$$

And then

$$dz = 2i\,e^{it}\,dt.$$

Therefore, the answer is

$$\int_C \cos z\,dz = \int_2^{-2} \cos z\,dz$$

$$= \sin z\Big|_2^{-2} = \sin(-2) - \sin(2)$$

$$= -2\sin 2.$$

Answer $-2\sin 2$

c. The parametric representation of the path C, which is counterclockwise along the circle with center at the origin and radius 1, as shown in Figure 13.7, can be expressed as follows:

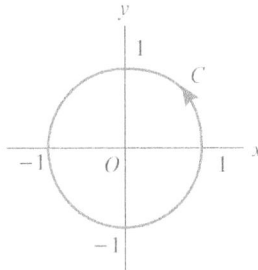

Figure 13.7 Path C.

$$z = \cos t + i\sin t = e^{it} \quad (0 \le t \le 2\pi).$$

Then we get

$$dz = ie^{it}\,dt.$$

Therefore, the answer is

$$\oint_C \frac{1}{z} dz = \int_0^{2\pi} \frac{ie^{it}}{e^{it}} dt$$

$$= i \int_0^{2\pi} dt = 2\pi i.$$

Answer $2\pi i$

Example 13.4

Calculate the complex line integral $\int_C \text{Im}(z) dz$ along the given path shown in Figure 13.8.

a. path C_1: The shortest path from the origin to $2 + 4i$
b. path C_2: The curved path from the origin to $2 + 4i$

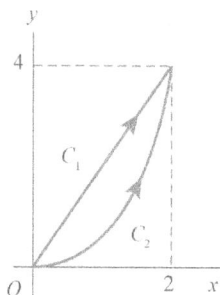

Figure 13.8 Paths C.

Solution

a. The parametric representation of the path C_1, which is the shortest path from the origin to $2 + 4i$, can be expressed as follows:

$$z = t + 2ti, \quad (0 \le t \le 2).$$

Then we get

$$dz = (1 + 2i)dt \quad \text{and} \quad \text{Im}(z) = 2t.$$

Therefore, the answer is

$$\int_C \text{Im}(z) dz = \int_0^2 2t \cdot (1 + 2i) dt$$

$$= (1 + 2i) \cdot t^2 \Big|_0^2 = 4 + 8i.$$

Answer $4 + 8i$

b. The parametric representation of the path C_2, which is the curved path from the origin to $2 + 4i$, can be expressed as follows:

$$z = t + t^2 i, \quad (0 \leq t \leq 2).$$

Then we get

$$dz = (1 + 2ti)dt \quad \text{and} \quad \text{Im}(z) = t^2.$$

Therefore, the answer is

$$\int_C \text{Im}(z)dz = \int_0^2 t^2 \cdot (1 + 2ti)dt$$

$$= \left[\frac{t^3}{3} + \frac{t^4}{2}i \right]_0^2 = \frac{8}{3} + 8i.$$

Answer $\dfrac{8}{3} + 8i$

Remark   Dependence on path

We can observe that the line integral results vary with different paths.

Example 13.5

Calculate the complex line integral along the given path C, which is counterclockwise along the circle with center at z_0 and radius ρ_0, as shown in Figure 13.9.

$$\oint_C (z - z_0)^m \, dz \quad (m \text{ integer})$$

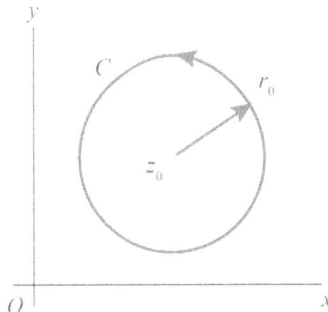

Figure 13.9 Path C.

Solution

The equation of the circle with center at z_0 and radius ρ_0 is

$$|z - z_0| = \rho_0,$$

or

$$z = z_0 + \rho(\cos t + i\sin t) = z_0 + \rho_0 e^{it}. \quad (0 \le t \le 2\pi)$$

Then we get

$$dz = \rho_0 i e^{it} dt.$$

Therefore, the answer is

$$\oint_C (z - z_0)^m \, dz = \int_0^{2\pi} \rho_0^m e^{imt} \cdot \rho_0 i e^{it} \, dt$$

$$= \rho_0^{m+1} i \int_0^{2\pi} e^{i(m+1)t} \, dt$$

$$= \rho_0^{m+1} i \int_0^{2\pi} \left\{\cos(m+1)t + i\sin(m+1)t\right\} dt$$

$$= \begin{cases} i \int_0^{2\pi} dt & (m = -1) \\[2mm] \rho_0^{m+1} i \left[\dfrac{\sin(m+1)t}{m+1} - i\dfrac{\cos(m+1)t}{m+1}\right]_0^{2\pi} & (m \ne -1) \end{cases}$$

$$= \begin{cases} 2\pi i & (m = -1) \\ 0 & (m \ne -1 \text{ integer}). \end{cases}$$

$$\textbf{Answer} \quad \oint_C (z - z_0)^m \, dz = \begin{cases} 2\pi i & (m = -1) \\ 0 & (m \ne -1 \text{ integer}) \end{cases}$$

13.1.4   *Upper bound of a complex line integral*

There might be cases where we cannot precisely compute the complex line integral. In such situations, we can estimate the magnitude of the complex line integral approximately to infer the range of possible values for the complex line integral.

> **Remark   Upper bound of a complex line integral**
>
> If the length of path C is denoted by l, and if it holds that $|f(z)| \leq M$ for all points on the path, then the following equation is satisfied.
>
> $$\left| \int_C f(z)dz \right| \leq M \cdot L \tag{13.9}$$

Applying the general triangle inequality to the integral $\int_C f(z)dz$ partitioned into n segments denoted as $S_n = \sum_{m=1}^{n} f(z_m)\Delta z_m$, we obtain the following inequality.

$$|S_n| = \left| \sum_{m=1}^{n} f(z_m)\Delta z_m \right| \leq \sum_{m=1}^{n} |f(z_m)| \cdot |\Delta z_m| \leq M \cdot \sum_{m=1}^{n} |\Delta z_m| = M \cdot L$$

Example 13.6

When the path C is the shortest path from the origin to $2+2i$, show the details.

a. the complex line integral $\int_C z\,dz$

b. upper bound of the complex line integral

Solution

The parametric representation of the path, which is the shortest path from the origin to $2+2i$, as shown in Figure 13.10, can be expressed as follows:

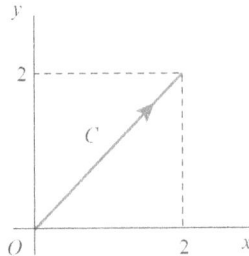

Figure 13.10 Path C.

Then we get

$$dz = (1+i)dt.$$

a. $\displaystyle \int_C z\,dz = \int_0^2 (t+it)\cdot(1+i)\,dt$

$$= (1+i)^2 \int_0^2 t\,dt = 4i$$

Answer $4i$

b. $\left|f(z)\right| = |z| \le |2+2i| = 2\sqrt{2} = M$

The length of the path is

$$L = 2\sqrt{2}.$$

Therefore, we obtain

$$\left|\int_C z\,dz\right| \le M\cdot L = 2\sqrt{2}\cdot 2\sqrt{2} = 8.$$

Answer 8

Problem 13.1

Represent the following complex path in terms of parameters. [1 ~ 10]

1. The straight path from $(0, 1)$ to $(2, 4)$
2. The straight path from $(0, 1)$ to $(3, 1)$
3. The curved path along $y = 1/x$ from $(1, 1)$ to $(3, 1/3)$
4. The curved path along $y = x^2$ from $(1, 1)$ to $(2, 4)$
5. $|z-i| = 1$, counterclockwise
6. $|z-(2+i)| = 2$, clockwise
7. $\dfrac{x^2}{4} + \dfrac{y^2}{9} = 1$, counterclockwise
8. $\dfrac{(x-1)^2}{4} - \dfrac{(y-2)^2}{9} = 1$, counterclockwise
9. unit circle, counterclockwise
10. $-\dfrac{x^2}{4} + \dfrac{y^2}{9} = 1$, clockwise

Calculate the following line integral. [11 ~16]

11. $\displaystyle \int_{-i}^{i} 2z\,dz$

12. $\displaystyle \int_0^{1+i} z^2\,dz$

13. $\int_{-\frac{\pi}{2}i}^{\frac{\pi}{2}i} \cos z \, dz$

14. $\int_{1}^{1+i} \sin 2z \, dz$

15. $\int_{1-i}^{1+i} e^{-z} \, dz$

16. $\int_{-i}^{i} z e^{z^2} \, dz$

Calculate the following line integral. [17 ~ 26]

17. $\int_{C} \operatorname{Im}(z) \, dz$,

   path C: the shortest path from $1+i$ to $2+2i$

18. $\int_{C} \operatorname{Re}(z) \, dz$,

   path C: the curved path along $y = x^2$ from $(0, 0)$ to $(1, 1)$

19. $\int_{C} z \, dz$,

   path C: the straight line from $(0, 1)$ to $(3, 1)$

20. $\int_{C} z \, dz$,

   path C: counterclockwise along the unit circle

21. $\int_{C} z \, dz$,

   path C: counterclockwise along $\dfrac{x^2}{9} + \dfrac{y^2}{4} = 1$,

22. $\int_{C} \cos z \, dz$,

   path C: the semicircle from $(1, 0)$ to $(-1, 0)$, $y \geq 0$

23. $\int_{C} \sin z \, dz$,

   path C: the straight line from $(0, 1)$ to $(0, 2)$

24. $\int_{C} \sec^2 z \, dz$,

   path C: the quarter circle from $(1, 0)$ to $(0, 1)$, $x \geq 0$, $y \geq 0$

25. $\oint_{C} \dfrac{1}{z^2} \, dz$,

   path C: counterclockwise along the circle with center at the origin and radius 1.

26. $\oint_{C} z^2 \, dz$,

   path C: counterclockwise along the circle with center at the origin and radius 1.

13.2   Cauchy's integral theorem I and II

This section is the most important part of this chapter on complex integration.

In the preceding section, we confirmed that the complex line integral of a function $f(z)$ depends on the path taken. We defined a closed path C as one where the starting and ending points coincide, and as seen in Figure 13.11, a closed path that does not intersect or touch itself along the path is called a *simple closed path*.

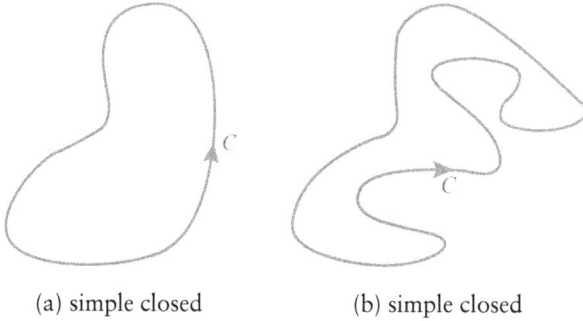

(a) simple closed   (b) simple closed

Figure 13.11 Simple closed paths.

A *simply connected domain* in the complex plane refers to a region where every simple closed path within the region does not leave the region. As seen in Figure 13.12(b), if there is another region within the domain, it cannot be considered a doubly connected domain.

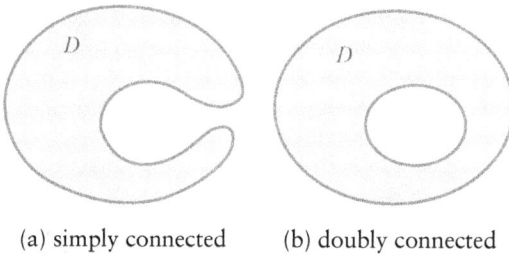

(a) simply connected   (b) doubly connected

Figure 13.12 Simply connected domain and a doubly connected domain.

Remark   Cauchy's integral theorem I

As seen in Figure 13.13, if the complex function $f(z)$ is analytic in a simply connected domain D, it satisfies the following equation for every simple closed path C within the domain D.

$$\oint_C f(z)\,dz = 0 \tag{13.10}$$

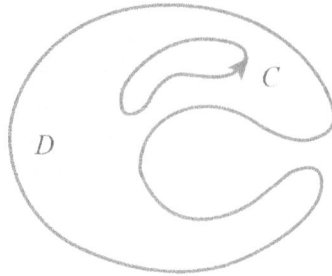

Figure 13.13 Simple closed path C within a simply connected domain D.

Example 13.7

Calculate the following equation when C is a simple closed path within the simply connected domain D.

a. $\oint_C \left(z^2 + z \right) dz$

b. $\oint_C e^z\, dz$

c. $\oint_C \sin z\, dz$

Solution

a. If $f(z) = z^2 + z$ is analytic for all z on the simple closed path C in the simply connected domain D, then $\oint_C \left(z^2 + z \right) dz = 0$.

    Answer 0

b. If $f(z) = e^z$ is analytic for all z on the simple closed path C in the simply connected domain D, then $\oint_C e^z\, dz = 0$.

    Answer 0

c. If $f(z) = \sin z$ is analytic for all z on the simple closed path C in the simply connected domain D, then $\oint_C \sin z\, dz = 0$.

    Answer 0

Example 13.8

If the path is a unit circle, calculate the following equation:

a. $\oint_C \dfrac{1}{\cos z}\,dz$

b. $\oint_C \dfrac{1}{z}\,dz$

c. $\oint_C \dfrac{1}{z^2}\,dz$

d. $\oint_C z\,dz$

e. $\oint_C \dfrac{1}{z^2+9}\,dz$

Solution

a. The function $f(z)=\dfrac{1}{\cos z}$ is not analytic at $=\pm\dfrac{\pi}{2},\pm\dfrac{3\pi}{2},\cdots$.

However, since all these points lie outside the unit circle, we can consider the complex function $f(z)=\dfrac{1}{\cos z}$ to be analytic for all z on the path C.

Therefore, by Cauchy's integral theorem, we obtain

$$\oint_C \dfrac{1}{\cos z}\,dz = 0.$$

Answer 0

b. The function $f(z)=\dfrac{1}{z}$ is not analytic at $z=0$. However, since this point lies inside the unit circle, we cannot use Cauchy's integral theorem.

From

$$z = e^{it}\ \ (0\le t\le 2\pi),$$

we get

$$dz = ie^{it}\,dt.$$

Therefore, we obtain

$$\oint_C \dfrac{1}{z}\,dz = \int_0^{2\pi}\dfrac{1}{e^{it}}ie^{it}\,dt = \int_0^{2\pi} i\,dt = 2\pi i.$$

Answer $2\pi i$

c. The function $f(z) = \dfrac{1}{z^2}$ is not analytic at $z = 0$. However, since this point lies inside the unit circle, we cannot use Cauchy's integral theorem.
   From

$$z = e^{it} \quad (0 \le t \le 2\pi),$$

we get

$$dz = ie^{it}\, dt.$$

Therefore, we obtain

$$\oint_C \frac{1}{z^2}\, dz = \int_0^{2\pi} \frac{1}{e^{2it}} ie^{it}\, dt = \int_0^{2\pi} ie^{-it}\, dt = -e^{-it}\Big|_0^{2\pi} = 0.$$

Answer 0

d. The function $f(z) = z$ is analytic.
   Therefore, by Cauchy's integral theorem, we obtain

$$\oint_C z\, dz = 0.$$

Answer 0

e. The function $f(z) = \dfrac{1}{z^2 + 9}$ is not analytic at $z = \pm 3i$.
   However, since all these points lie outside the unit circle, we can consider the complex function $f(z) = \dfrac{1}{z^2 + 9}$ to be analytic for all z on the path C.
   Therefore, by Cauchy's integral theorem, we obtain

$$\oint_C \frac{1}{z^2 + 9}\, dz = 0.$$

Answer 0

Remark   Independence of path

If the complex function $f(z)$ is analytic in the simply connected domain D, then the integral of $f(z)$ is independent of the path taken within the domain. In other words, as seen in Figure 13.14, the integral values for different paths C_1 and C_2, with starting point z_0 and ending point z_1, are equal.

$$\int_{C_1} f(z)\, dz = \int_{C_2} f(z)\, dz \qquad (13.11)$$

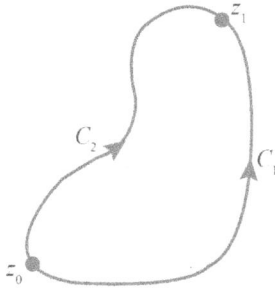

Figure 13.14 Path with starting point z_0 and ending point z_1.

Since $\oint_C f(z)dz = 0$ in Eq. (13.10), if in Figure 13.14, the simple closed path C is composed of the paths C_1 and the reverse path of C_2 (i.e., C_2 *), then the following equation holds:

$$\oint_C f(z)dz = \int_{C_1} f(z)dz + \int_{C_2*} f(z)dz = 0. \tag{13.12}$$

Here, since $\int_{C_2*} f(z)dz = -\int_{C_2} f(z)dz$, substituting this into Eq. (13.12) yields:

$$\int_{C_1} f(z)dz - \int_{C_2} f(z)dz = 0. \tag{13.13}$$

Remark Cauchy's integral theorem II

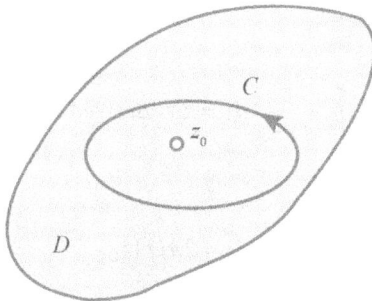

Figure 13.15 Simple closed path C in a domain D that encloses the point z_0.

Let $f(z)$ be analytic in a simply connected domain D. Then for any point z_0 in the domain D and any simple closed path C in D that encloses z_0. (refer to Figure 13.15)

$$\oint_C \frac{f(z)}{z - z_0} dz = 2\pi i f(z_0) \tag{13.14}$$

where the path is counterclockwise.

Dividing both sides of equation (13.14) by $2\pi i$, the equation transforms as follows:

$$f(z_0) = \frac{1}{2\pi i} \oint_C \frac{f(z)}{z - z_0} dz. \tag{13.15}$$

Even if a complex function $f(z)$ is analytic in a simply connected region D, the function $\dfrac{f(z)}{z - z_0}$ becomes nonanalytic at $z = z_0$, exhibiting discontinuity.

By addition and subtraction, $f(z) = \{f(z) - f(z_0)\} + f(z_0)$. Then we get:

$$\oint_C \frac{f(z)}{z - z_0} dz = \oint_C \frac{f(z) - f(z_0)}{z - z_0} dz + \oint_C \frac{f(z_0)}{z - z_0} dz. \tag{13.16}$$

The second term on the RHS of Eq. (13.16) becomes

$$\oint_C \frac{f(z_0)}{z - z_0} dz = f(z_0) \oint_C \frac{1}{z - z_0} dz,$$

and applying the result $\oint_C \dfrac{1}{z - z_0} dz = 2\pi i$ derived in Example 13.5 yields the following:

$$\oint_C \frac{f(z_0)}{z - z_0} dz = 2\pi i f(z_0). \tag{13.17}$$

Therefore, by demonstrating that the first term on the RHS of Eq. (13.16) is zero, Cauchy's integral theorem II is proven.

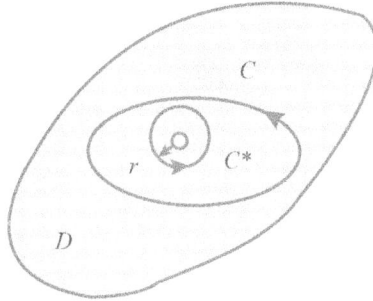

Figure 13.16 The circle C* with center at z_0 and radius ρ.

As seen in Figure 13.16, for every z inside the disk $|z - z_0| \leq \delta$, there exists δ such that $|f(z) - f(z_0)| < \varepsilon$ ($\varepsilon > 0$). If we make the radius ρ of circle C* smaller than δ, then the following inequality holds for every point on circle C*:

$$\left|\frac{f(z) - f(z_0)}{z - z_0}\right| < \frac{\varepsilon}{\rho}. \tag{13.18}$$

Applying the upper bound of the complex line integral in Eq. (13.18) to Eq. (13.9), we have the following equation, since the path length of the circle is $2\pi\rho$.

$$\oint_{C^*}\left|\frac{f(z) - f(z_0)}{z - z_0}\right| dz < \frac{\varepsilon}{\rho} \cdot 2\pi\rho = 2\pi\varepsilon \tag{13.19}$$

Choosing a value infinitesimally small for ε ($\varepsilon > 0$), the value of Eq. (13.19) approaches 0, and thus the first term of Eq. (13.16) becomes 0. Therefore, we get Eq. (13.20).

$$\oint_C \frac{f(z) - f(z_0)}{z - z_0} dz = 0 \tag{13.20}$$

Example 13.9

For any contour C enclosing the point $z_0 = 1$, as shown in Figure 13.17, show the details.

$$\oint_C \frac{z^2}{z - 1} dz$$

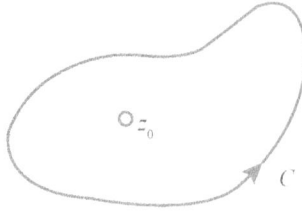

Figure 13.17 The contour C.

Solution

The function $\dfrac{z^2}{z-1}$ is not analytic at the point $z_0 = 1$.

Let $f(z) = z^2$.

By applying Cauchy's integral theorem II, we obtain

$$\oint_C \frac{f(z)}{z-1} dz = 2\pi i f(z)\Big|_{z=1} = 2\pi i.$$

Another Solution

Since $\dfrac{z^2}{z-1} = z+1+\dfrac{1}{z-1}$, we get

$$\oint_C \frac{z^2}{z-1} dz = \oint_C \left(z+1+\frac{1}{z-1}\right) dz = \oint_C (z+1)dz + \oint_C \frac{1}{z-1} dz.$$

Now, the function $z+1$ is analytic.

By applying Cauchy's integral theorem I, we obtain

$$\oint_C (z+1)dz = 0.$$

And as the result derived in Example 13.5, we get

$$\oint_C \frac{1}{z-1} dz = 2\pi i.$$

Therefore, we obtain

$$\oint_C \frac{z^2}{z-1} dz = \oint_C (z+1)dz + \oint_C \frac{1}{z-1} dz = 0 + 2\pi i.$$

Answer $2\pi i$

Example 13.10

For a unit circle with the center $z_0 = -1$ and radius 1 (counterclockwise), as shown in Figure 13.18, show the details.

$$\oint_C \frac{z^2}{z-1}\,dz$$

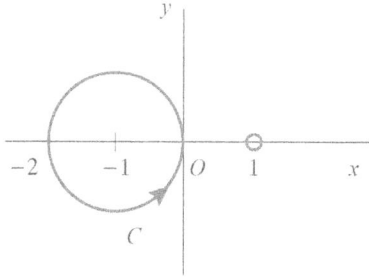

Figure 13.18 The contour C.

Solution

The function $\dfrac{z^2}{z-1}$ is not analytic at the point $z = 1$. But the point $z_0 = 1$ is outside of the path C.

By applying Cauchy's integral theorem I, we obtain

$$\oint_C \frac{z^2}{z-1}\,dz = 0.$$

Answer 0

Problem 13.2

For a unit circle with the center $z_0 = 1$ and radius 2 (counterclockwise), show the details. [1 ~ 6]

1. $\oint_C \dfrac{z}{z^2-4}\,dz$

2. $\oint_C \dfrac{e^z}{z}\,dz$

3. $\oint_C \dfrac{z^2}{z-i}\,dz$

4. $\oint_C \dfrac{z}{z+2i}\,dz$

5. $\oint_C \dfrac{\cos z}{z+2} dz$

6. $\oint_C \dfrac{\sin z}{z - \pi/2} dz$

For the counterclockwise contour C, show the details. [7 ~ 12]

7. $\oint_C \dfrac{z^2}{z+1} dz$ C: $x^2 + \dfrac{(y-1)^2}{4} = 1$

8. $\oint_C \dfrac{z}{z+2i} dz$ C: $\dfrac{(x-1)^2}{4} + (y+1)^2 = 1$

9. $\oint_C \dfrac{\sin z}{z+i} dz$ C: $|z| = 2$

10. $\oint_C \dfrac{z-1}{z+2} dz$ C: $|z+1| = \sqrt{2}$

11. $\oint_C \dfrac{e^{-z}}{z^2 - 9} dz$ C: $|z-1| = 1$

12. $\oint_C \dfrac{z^2}{z^2 + 2} dz$ C: $|z-i| = 1$

13.3 Cauchy's integral theorem III (derivative of an analytic function)

Let's now apply a modified form of Cauchy's integral theorem II, learned in the previous section, to calculate the derivative value at an arbitrary point z_0 within the domain D.

Remark **Cauchy's integral theorem III (derivative of an analytic function)**

If the complex function $f(z)$ is analytic in a domain D, then it has derivatives of all orders in D, which are also analytic functions in D. The values of these derivatives at a point z_0 in D are given by the following formulas:

$$f'(z_0) = \frac{1}{2\pi i} \oint_C \frac{f(z)}{(z - z_0)^2} dz \qquad (13.21)$$

$$f''(z_0) = \frac{2!}{2\pi i} \oint_C \frac{f(z)}{(z - z_0)^3} dz \qquad (13.22)$$

---

and in general,

$$f^{(n)}(z_0) = \frac{n!}{2\pi i}\oint_C \frac{f(z)}{(z-z_0)^{n+1}}dz. \qquad (n=1,2,\cdots) \qquad (13.23)$$

Here, the path C is a counterclockwise and simple closed path in a simply connected region D enclosing the point z_0.

Therefore, transforming the above equations yields the following:

$$\oint_C \frac{f(z)}{(z-z_0)^{n+1}}dz = \frac{2\pi i}{n!}f^{(n)}(z_0). \qquad (n=1,2,\cdots) \qquad (13.24)$$

Using the theorem for derivatives, the value of the derivative at an arbitrary point z_0 within the region D can be expressed as follows:

$$f'(z_0) = \lim_{\Delta z \to 0} \frac{f(z_0+\Delta z)-f(z_0)}{\Delta z}. \qquad (13.25)$$

Applying Cauchy's integral theorem II to $f(z_0+\Delta z)$ and $f(z_0)$ in terms of the numerator of Eq. (13.24), we have:

$$f(z_0+\Delta z) = \frac{1}{2\pi i}\oint_C \frac{f(z)}{z-(z_0+\Delta z)}dz, \qquad (13.26a)$$

$$f(z_0) = \frac{1}{2\pi i}\oint_C \frac{f(z)}{z-z_0}dz. \qquad (13.26b)$$

Then, the fractional expression of Eq. (13.24) is simplified as follows:

$$\frac{f(z_0+\Delta z)-f(z_0)}{\Delta z} = \frac{1}{2\pi i \Delta z}\left\{\oint_C \frac{f(z)}{z-(z_0+\Delta z)}dz - \oint_C \frac{f(z)}{z-z_0}dz\right\}. \qquad (13.27)$$

When we combine the two fractions in the right-hand term of Eq. (13.27) by finding a common denominator, we get:

$$\frac{1}{z-(z_0+\Delta z)} - \frac{1}{z-z_0} = \frac{\Delta z}{(z-z_0-\Delta z)(z-z_0)}.$$

Applying this to Eq. (13.27), we have:

$$\frac{f(z_0+\Delta z)-f(z_0)}{\Delta z} = \frac{1}{2\pi i}\oint_C \frac{f(z)}{(z-z_0-\Delta z)(z-z_0)}dz. \qquad (13.28)$$

Applying $\Delta z \to 0$ to Eq. (13.28) yields Eq. (13.21).

$$f'(z_0) = \lim_{\Delta z \to 0} \frac{f(z_0 + \Delta z) - f(z_0)}{\Delta z} = \frac{1}{2\pi i} \oint_C \frac{f(z)}{(z - z_0)^2} dz \qquad (13.21)$$

Eq. (13.22) and Eq. (13.23) are also derived using a similar approach.

Example 13.11

For any contour C that encloses the point $z_0 = 1$, as shown in Figure 13.19, show the details.

$$\oint_C \frac{z^2}{(z-1)^2} dz$$

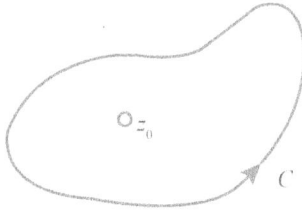

Figure 13.19 The contour C.

Solution

The function $\dfrac{z^2}{(z-1)^2}$ is not analytic at the point $z_0 = 1$.

When we let $f(z) = z^2$, we obtain

$$\oint_C \frac{f(z)}{(z-1)^2} dz = 2\pi i f'(1) = 2\pi i [2z]_{z=1} = 4\pi i.$$

Answer $4\pi i$

Example 13.12

For any contour C that encloses the point $z_0 = \pi/2$, as shown in Figure 13.20, show the details.

$$\oint_C \frac{z \sin z}{(z - \pi/2)^3} dz$$

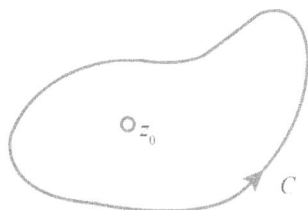

Figure 13.20 The contour C.

Solution

The function $\dfrac{z \sin z}{(z - \pi/2)^3}$ is not analytic at the point $z_0 = \pi/2$.

When we let $f(z) = z \sin z$, we get as follows by Cauchy's integral theorem III:

$$\oint_C \frac{f(z)}{(z - \pi/2)^3} \, dz = \pi i f''\left(\frac{\pi}{2}\right).$$

On the other hand, when we differentiate $f(z) = z \sin z$, we get:

$$f'(z) = \sin z + z \cos z,$$

$$f''(z) = 2 \cos z - z \sin z.$$

And we get

$$f''\left(\frac{\pi}{2}\right) = -\frac{\pi}{2}.$$

Therefore, we obtain

$$\oint_C \frac{f(z)}{(z - \pi/2)^3} \, dz = \pi i \cdot \left(-\frac{\pi}{2}\right) = -\frac{\pi^2}{2} i.$$

$$\text{Answer} \; -\frac{\pi^2}{2} i$$

Example 13.13

For the path C: $4x^2 + y^2 = 1$ (clockwise) shown in Figure 13.21, show the details.

$$\oint_C \frac{\sin 2z}{z^2} \, dz$$

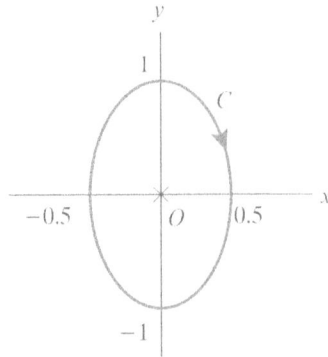

Figure 13.21 The contour C.

Solution

The path C: $\dfrac{x^2}{(1/2)^2} + y^2 = 1$

The function $\dfrac{\sin 2z}{z^2}$ is not analytic at the point $z_0 = 0$. Since the point $z_0 = 0$ lies in the path C, let $f(z) = \sin 2z$ and then $f'(z) = 2\cos 2z$.
And we get

$$f'(0) = 2.$$

Since the direction of the path is clockwise, by applying Cauchy's integral theorem III, which reverses the sign of the integral, the following equation is obtained:

$$\oint_C \frac{f(z)}{z^2}\, dz = -2\pi i f'(0) = -4\pi i.$$

Answer $-4\pi i$

Example 13.14

For the path C: $|z-1| = 1$ (counterlockwise) shown in Figure 13.22, show the details.

$$\oint_C \frac{e^z}{z^3 - z^2 - z + 1}\, dz$$

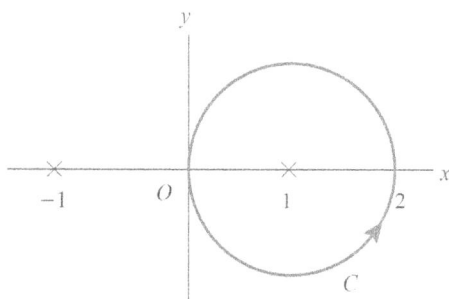

Figure 13.22 The contour C.

Solution

The function $\dfrac{e^z}{z^3 - z^2 - z + 1} = \dfrac{e^z}{(z-1)^2 (z+1)}$ is not analytic at $z_0 = 1$.

Since the point $z_0 = 1$ lies in the path C, let $f(z) = \dfrac{e^z}{z+1}$.

Then, by Cauchy's integral theorem III, we obtain

$$\oint_C \frac{f(z)}{(z-1)^2} dz = 2\pi i f'(1).$$

And the derivative of the function is

$$f'(z) = \frac{ze^z}{(z+1)^2},$$

and

$$f'(1) = \frac{e}{4}.$$

Therefore, we obtain

$$\oint_C \frac{f(z)}{(z-1)^2} dz = \frac{\pi ei}{2}.$$

Answer $\dfrac{\pi ei}{2}$

Problem 13.3

For the unit circle C: $|z| = 1$ (counterclockwise), show the details. [1 ~ 6]

1. $\oint_C \dfrac{1}{z^2(z-2)} dz$

2. $\oint_C \dfrac{e^z}{z^2} dz$

3. $\oint_C \frac{\cos 2z}{z^3} dz$

4. $\oint_C \frac{\cos z}{z^4} dz$

5. $\oint_C \frac{z \cos z}{(z - \pi/4)^2} dz$

6. $\oint_C \frac{e^{2z}}{(z - i/2)^3} dz$

Integrate counterclockwise. Show the details. [7 ~ 14]

7. $\oint_C \frac{z^2}{(z-1)^3} dz$ $\qquad$ C: $x^2 + \frac{(y-1)^2}{4} = 1$

8. $\oint_C \frac{z^2}{(z-i)^3} dz$ $\qquad$ C: $x^2 + \frac{(y-1)^2}{4} = 1$

9. $\oint_C \frac{z}{(z+2i)^2} dz$ $\qquad$ C: $\frac{(x-1)^2}{4} + (y+1)^2 = 1$

10. $\oint_C \frac{e^z}{(z+i)^3} dz$ $\qquad$ C: $\frac{(x-1)^2}{4} + (y+1)^2 = 1$

11. $\oint_C \frac{\sin z}{(z-i)^2} dz$ $\qquad$ C: $|z| = 2$

12. $\oint_C \frac{z^3}{(z+2)^2} dz$ $\qquad$ C: $|z+1| = \sqrt{2}$

13. $\oint_C \frac{e^z}{z^2(z+2)} dz$ $\qquad$ C: $|z| = \sqrt{3}$

14. $\oint_C \frac{e^z \sin z}{(z^2-1)^2} dz$ $\qquad$ C: $|z-i| = 1$

Answer

Problem 13.1

1. $z = t + \left(\dfrac{3}{2}t + 1\right)i$, $(0 \le t \le 2)$

2. $z = i$, $(0 \le t \le 3)$

3. $z = t + \dfrac{1}{t}i$, $(1 \le t \le 3)$

4. $z = t + t^2 i$, $(1 \le t \le 2)$

5. $z = \cos t + (\sin t + 1)i$, $(0 \le t \le 2\pi)$

6. $z = (2\cos t + 2) - (2\sin t + 1)i$, $(0 \le t \le 2\pi)$

7. $z = 2\cos t + (3\sin t)i$, $(0 \le t \le 2\pi)$

8. $z = (2\sec t + 1) + (3\tan t + 2)i$, $(0 \le t \le 2\pi)$

9. $z = \cos t + (\sin t)i$, $(0 \le t \le 2\pi)$

10. $z = (2\tan t) + (-3\sec t)i$, $(0 \le t \le 2\pi)$

11. 0

12. $\dfrac{-2 + 2i}{3}$

13. $2i \sinh \dfrac{\pi}{2}$

14. $\dfrac{1 - \cos 2}{2} \cosh 2 + i\dfrac{\sin 2}{2} \sinh 2$

15. $2ie^{-1} \sin 1$

16. 0

17. $\dfrac{3(1 + i)}{2}$

18. $\dfrac{1}{2} + \dfrac{2}{3}i$

19. $\dfrac{9}{2} + 3i$

20. 0

21. 0

22. $-2 \sin 1$

23. $\cosh 1 - \cosh 2$

24. $-\tan 1 + i \tanh 1$

25. 0

26. 0

Problem 13.2

1. πi

2. $2\pi i$

3. $-2\pi i$
4. 0
5. 0
6. $2\pi i$
7. 0
8. 0
9. $2\pi \sinh 1$
10. $-6\pi i$
11. 0
12. 0

Problem 13.3

1. $-\dfrac{\pi i}{2}$
2. $2\pi i$
3. $-4\pi i$
4. 0
5. $\sqrt{2}\pi i \cdot \left(1 - \dfrac{\pi}{4}\right)$
6. $2\pi(-\sin 1 + i\cos 1)$
7. 0
8. $2\pi i$
9. 0
10. $\pi(\sin 1 + i\cos 1)$
11. $2\pi i \cdot \cosh 1$
12. $24\pi i$
13. $\dfrac{\pi i}{2}$
14. 0

Bibliography

[1] Kreyszig, Erwin. *Advanced Engineering Mathematics*. 10th ed. New York city. New York: John Wiley & Sons, Inc., 2014.

[2] Zill, Dennis G., and Warren S. Wright. *Advanced Engineering Mathematics*. 6th ed. Burlington, Massachusetts: Jones and Bartlett Publishers, Inc., 2018.

[3] Greenberg, Michael D. *Foundations of Applied Mathematics*. Hoboken, New Jersey: Prentice Hall, Inc., 1978.

[4] Yang, Xin-She. *Engineering Mathematics with Examples and Applications*. Cambridge, Massachusetts: Academic Press, 2017.

[5] Palm III, William J. *Matlab for Engineering Applications*. 5th ed. New York city. New York: McGraw Hill, 2018.

[6] Sobot, Robert. *Engineering Mathematics by Example Vol. 1: Algebra and Linear Algebra*. New York city. New York: Springer, 2023.

[7] Stewart, James, Daniel Clegg, and Saleem Watson. *Calculus*. 9th ed. Boston, Massachusetts: Cengage Learning, 2020.

[8] Song, Chul Ki, Kwan Ju Kim, Seok Hyun Kim, Seong Keol Kim, and Dong Ho Nam. *Mechanical Vibration Core*. 3rd ed. Seoul, Korea: Kyobo Press, 2021.

[9] Song, Chul Ki, and Jang Pyo Hong. *Dynamics with Applications*. 5th ed. Seoul, Korea: First Book, 2024.

[10] Holman, J. P. *Heat Transfer*. 10th ed. New York city. New York: McGraw Hill, 2009.

Appendix A
Basic formulas

A1. Trigonometric functions
1. Basic trigonometric formulas

$$\sin(\alpha + \beta) = \sin\alpha\cos\beta + \cos\alpha\sin\beta \tag{A1.1}$$

$$\sin(\alpha - \beta) = \sin\alpha\cos\beta - \cos\alpha\sin\beta \tag{A1.2}$$

$$\cos(\alpha + \beta) = \cos\alpha\cos\beta - \sin\alpha\sin\beta \tag{A1.3}$$

$$\cos(\alpha - \beta) = \cos\alpha\cos\beta + \sin\alpha\sin\beta \tag{A1.4}$$

2. Trigonometric transformation formulas (from sum/difference to product)

$$\sin\alpha + \sin\beta = 2\,\sin\left(\frac{\alpha+\beta}{2}\right)\cos\left(\frac{\alpha-\beta}{2}\right) \tag{A1.5}$$

$$\sin\alpha - \sin\beta = 2\cos\left(\frac{\alpha+\beta}{2}\right)\sin\left(\frac{\alpha-\beta}{2}\right) \tag{A1.6}$$

$$\cos\alpha + \cos\beta = 2\cos\left(\frac{\alpha+\beta}{2}\right)\cos\left(\frac{\alpha-\beta}{2}\right) \tag{A1.7}$$

$$\cos\alpha - \cos\beta = -2\sin\left(\frac{\alpha+\beta}{2}\right)\sin\left(\frac{\alpha-\beta}{2}\right) \tag{A1.8}$$

3. Trigonometric transformation formulas (from product to sum/difference)

$$\sin A\cos B = \frac{1}{2}\{\sin(A+B) + \sin(A-B)\} \tag{A1.9}$$

$$\cos A\sin B = \frac{1}{2}\{\sin(A+B) - \sin(A-B)\} \tag{A1.10}$$

$$\cos A\cos B = \frac{1}{2}\{\cos(A+B) + \cos(A-B)\} \tag{A1.11}$$

$$\sin A\sin B = -\frac{1}{2}\{\cos(A+B) - \cos(A-B)\} \tag{A1.12}$$

A2. **Derivatives** (when $f = f(x)$, $g = g(x)$ and $r = r(x)$)

$$\frac{d}{dx} x^n = nx^{n-1} \tag{A2.1}$$

$$\frac{d}{dx}(fg) = f'g + fg' \tag{A2.2}$$

$$\frac{d}{dx}(fgr) = f'gr + fg'r + fgr' \tag{A2.3}$$

$$\frac{d}{dx}\left(\frac{f}{g}\right) = \frac{f'g - fg}{g^2} \tag{A2.4}$$

$$\frac{d}{dx}(u(t)) = \frac{du}{dt}\frac{dt}{dx} \tag{A2.5}$$

$$\frac{d}{dx}(u(t))^n = v^{n-1}\frac{du}{dx} \tag{A2.6}$$

$$\frac{d}{dx}(e^u) = e^u \frac{du}{dx} \tag{A2.7}$$

$$\frac{d}{dx}(a^u) = a^u \ln a \frac{du}{dx} \tag{A2.8}$$

$$\frac{d}{dx}(\ln u) = \frac{1}{u}\frac{du}{dx} \tag{A2.9}$$

$$\frac{d}{dx}(\sin u) = \cos u \frac{du}{dx} \tag{A2.10}$$

$$\frac{d}{dx}(\cos u) = -\sin u \frac{du}{dx} \tag{A2.11}$$

$$\frac{d}{dx}(\tan u) = \sec^2 u \frac{du}{dx} \tag{A2.12}$$

$$\frac{d}{dx}(\cot u) = -\csc^2 u \frac{du}{dx} \tag{A2.13}$$

$$\frac{d}{dx}(\sec u) = \sec u \tan u \frac{du}{dx} \tag{A2.14}$$

$$\frac{d}{dx}(\csc u) = -\csc u \cot u \frac{du}{dx} \tag{A2.15}$$

$$\frac{d}{dx}\left(\tan^{-1}u\right) = \frac{1}{1+u^2}\frac{du}{dx} \tag{A2.16}$$

$$\frac{d}{dx}\left(\sinh u\right) = \cosh u\frac{du}{dx} \tag{A2.17}$$

$$\frac{d}{dx}\left(\cosh u\right) = \sinh u\frac{du}{dx} \tag{A2.18}$$

A3. Integrals (C arbitrary)

$$\int x^n\,dx = \frac{1}{n+1}x^{n+1} + C \qquad (n \neq -1) \tag{A3.1}$$

$$\int \frac{1}{x}\,dx = \ln|x| + C \tag{A3.2}$$

$$\int e^x\,dx = e^x + C \tag{A3.3}$$

$$\int a^x\,dx = a^x \ln a + C \tag{A3.4}$$

$$\int \sin x\,dx = -\cos x + C \tag{A3.5}$$

$$\int \cos x\,dx = \sin x + C \tag{A3.6}$$

$$\int \tan x\,dx = -\ln|\cos x| + C \tag{A3.7}$$

$$\int \cot x\,dx = \ln|\sin x| + C \tag{A3.8}$$

$$\int \sec^2 x\,dx = \tan x + C \tag{A3.9}$$

$$\int \csc^2 x\,dx = -\cot x + C \tag{A3.10}$$

$$\int \sinh x\,dx = \cosh x + C \tag{A3.11}$$

$$\int \cosh x\,dx = \sinh x + C \tag{A3.12}$$

$$\int f'g\,dx = fg - \int fg'\,dx \qquad \text{(Integration by parts)} \tag{A3.13}$$

A4. Gamma function

Figure A.1 shows the graph of the gamma function, $\Gamma(\alpha)$.

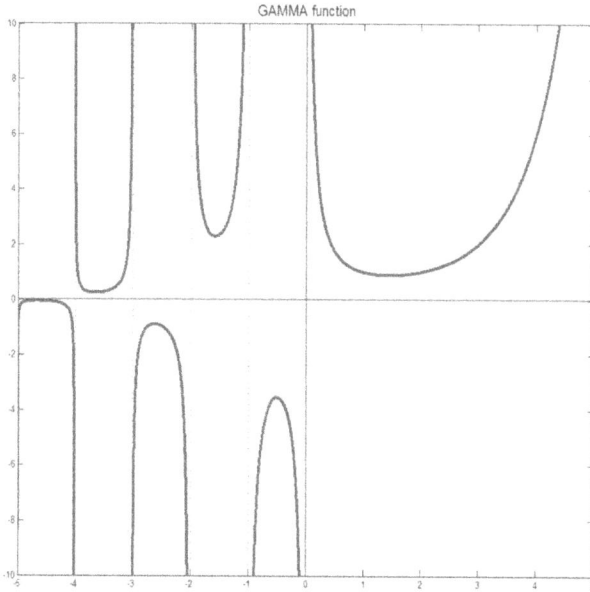

Figure A.1 Gamma function, $\Gamma(\alpha)$.

$$\Gamma(\alpha) = \int_0^\infty e^{-t} t^{\alpha-1}\, dt \qquad\qquad (A4.1)$$

$$\Gamma(\alpha+1) = \alpha\,\Gamma(\alpha) \qquad\qquad (A4.2)$$

Check $\Gamma(\alpha+1) = \alpha\,\Gamma(\alpha)$

$$\Gamma(\alpha+1) = \int_0^\infty e^{-t} t^{\alpha}\, dt$$

$$= \left(-e^{-t}\right) t^{\alpha}\Big|_0^\infty - \int_0^\infty \left(-e^{-t}\right)\left(\alpha t^{\alpha-1}\right) dt$$

$$= \left(-e^{-t}\right) t^{\alpha}\Big|_0^\infty - \int_0^\infty \left(-e^{-t}\right)\left(\alpha t^{\alpha-1}\right) dt$$

$$= \alpha \int_0^\infty e^{-t} t^{\alpha-1}\, dt$$

$$= \alpha\,\Gamma(\alpha)$$

$$\Gamma(1) = 1 \qquad\qquad \text{(A4.3)}$$

Check $\Gamma(1) = 1$

$$\Gamma(1) = \int_0^\infty e^{-t}\,dt$$

$$= -e^{-t}\Big|_0^\infty = 1$$

$$\Gamma(n+1) = n! \qquad\qquad \text{(A4.4)}$$

Check $\Gamma(n+1) = n!$

Since Eq. (A4.2) $\Gamma(n+1) = n\Gamma(n)$, we obtain:

$$\Gamma(n+1) = n(n-1)\Gamma(n-1) = \cdots = n(n-1)(n-2)\cdots 2\cdot 1\Gamma(1) = n!.$$

$$\Gamma(1/2) = \sqrt{\pi} \qquad\qquad \text{(A4.5)}$$

Check $\Gamma(1/2) = \sqrt{\pi}$

In Eq. (A4.1) $\Gamma(1/2) = \int_0^\infty e^{-t}t^{-1/2}\,dt$,

substituting $t = p^2$ yields:

$$\Gamma(1/2) = 2\int_0^\infty e^{-p^2}\,dp.$$

Then since $\int_0^\infty e^{-p^2}\,dp = \int_0^\infty e^{-q^2}\,dq$, we get:

$$\left[\Gamma(1/2)\right]^2 = 4\int_0^\infty e^{-p^2}\,dp \cdot \int_0^\infty e^{-q^2}\,dq$$

$$= 4\int_0^\infty\int_0^\infty e^{-\left(p^2+q^2\right)}\,dp\,dq.$$

Here, substituting $p = r\cos\theta$ and $q = r\sin\theta$ yields:

$$\left[\Gamma(1/2)\right]^2 = 4\int_0^{\pi/2}\int_0^{\infty} e^{-r^2} r\, dr\, d\theta = 4\cdot\frac{\pi}{2}\cdot\left[\frac{e^{-r^2}}{2}\right]_0^{\infty} = \pi.$$

Therefore, we obtain

$$\Gamma(1/2) = \sqrt{\pi}.$$

A5. Laplace operator ∇^2

1. in Cartesian coordinates
   The representation of the Laplace operator ∇^2 in Cartesian coordinates is as follows:

   i. two-dimensional $\qquad \nabla^2 u = \dfrac{\partial^2 u}{\partial x^2} + \dfrac{\partial^2 u}{\partial y^2}$ (A5.1)

   ii. three-dimensional $\qquad \nabla^2 u = \dfrac{\partial^2 u}{\partial x^2} + \dfrac{\partial^2 u}{\partial y^2} + \dfrac{\partial^2 u}{\partial z^2}$ (A5.2)

2. in polar coordinates
   The representation of the Laplace operator ∇^2 in polar coordinates is as follows:

   i. two-dimensional $\quad \nabla^2 u = \dfrac{\partial^2 u}{\partial r^2} + \dfrac{1}{r}\dfrac{\partial u}{\partial r} + \dfrac{1}{r^2}\dfrac{\partial^2 u}{\partial \theta^2}$ (A5.3)

   ii. three-dimensional (cylinder) $\nabla^2 u = \dfrac{\partial^2 u}{\partial r^2} + \dfrac{1}{r}\dfrac{\partial u}{\partial r} + \dfrac{1}{r^2}\dfrac{\partial^2 u}{\partial \theta^2} + \dfrac{\partial^2 u}{\partial z^2}$ (A5.4)

   iii. three-dimensional (sphere)

   $$\nabla^2 u = \frac{\partial^2 u}{\partial r^2} + \frac{2}{r}\frac{\partial u}{\partial r} + \frac{1}{r^2}\frac{\partial^2 u}{\partial \varphi^2} + \frac{\cot\varphi}{r^2}\frac{\partial u}{\partial \varphi} + \frac{1}{r^2\sin^2\varphi}\frac{\partial^2 u}{\partial \theta^2}$$ (A5.5)

Check $\quad \nabla^2 u = \dfrac{\partial^2 u}{\partial r^2} + \dfrac{1}{r}\dfrac{\partial u}{\partial r} + \dfrac{1}{r^2}\dfrac{\partial^2 u}{\partial \theta^2}$

The relationship between Cartesian coordinates and polar coordinates is as follows:

$$x = x(r,\ \theta) = r\cos\theta,$$

$$y = y(r,\ \theta) = r\sin\theta,$$

$$r = r(x, y) = \sqrt{x^2 + y^2},$$

$$\theta = \theta(x, y) = \tan^{-1}\left(\frac{y}{x}\right).$$

If we denote the partial derivative of the function u with respect to the variable x using a subscript, we can express it using the chain rule as follows. For instance, appending the subscript x to the function u implies taking the partial derivative of u with respect to x.

$$u_x = u_r r_x + u_\theta \theta_x \tag{a}$$

$$u_{xx} = (u_r r_x + u_\theta \theta_x)_x$$

$$= (u_r r_x)_x + (u_\theta \theta_x)_x$$

$$= (u_r)_x r_x + u_r (r_x)_x + (u_\theta)_x \theta_x + u_\theta (\theta_x)_x$$

$$= \{u_{rr} r_x + u_{\theta r} \theta_x\} r_x + u_r r_{xx} + \{u_{r\theta} r_x + u_{\theta\theta} \theta_x\} \theta_x + u_\theta \theta_{xx} \tag{b}$$

Here, $r_x = \dfrac{\partial}{\partial x}\sqrt{x^2 + y^2} = \dfrac{x}{\sqrt{x^2 + y^2}} = \dfrac{x}{r}$,

$$\theta_x = \frac{\partial \tan^{-1}(y/x)}{\partial x} = \frac{1}{1 + (y/x)^2}\left(-\frac{y}{x^2}\right) = -\frac{y}{r^2},$$

$$r_{xx} = \frac{\partial}{\partial x}\left(\frac{x}{r}\right) = \frac{r - x r_x}{r^2} = \frac{r^2 - x^2}{r^3} = \frac{y^2}{r^3},$$

$$\theta_{xx} = \frac{\partial}{\partial x}\left(-\frac{y}{r^2}\right) = -y\left(-2r^{-3} r_x\right) = \frac{2xy}{r^4}.$$

Since $u_{\theta r} = u_{r\theta}$, substituting these inti Eq. (A.7) yields

$$u_{xx} = u_{rr}\frac{x^2}{r^2} - 2u_{r\theta}\frac{xy}{r^3} + u_{\theta\theta}\frac{y^2}{r^4} + u_r\frac{y^2}{r^3} + u_\theta\frac{2xy}{r^4}. \tag{c}$$

Similarly, u_{yy} becomes

$$u_{yy} = u_{rr}\frac{y^2}{r^2} + 2u_{r\theta}\frac{xy}{r^3} + u_{\theta\theta}\frac{x^2}{r^4} + u_r\frac{x^2}{r^3} - u_\theta\frac{2xy}{r^4}. \tag{d}$$

Adding equations (c) and (d) leads to the derivation of the Laplace operator.

$$\nabla^2 u = u_{rr} + \frac{1}{r}u_r + \frac{1}{r^2}u_{\theta\theta}. \tag{e}$$

Appendix B
MATLAB®

Open MATLAB®

1. Double-click the MATLAB® icon on the main screen of the computer where the program is installed.
2. Start your work in the Command Window.

B1. Matrix

1. Input

$$A = \begin{bmatrix} 1 & 1 & 1 \\ 1 & 2 & 3 \\ 1 & 3 & 6 \end{bmatrix}.$$

Solution

```
a=[1 1 1; 1 2 3; 1 3 6]          % semicolon(;)
```

---

2. Transpose

```
b=[1+2i 3+4i]
b'                    % b': complex conjugate
b.'                   % b.': transpose
```

---

3. Calculation

```
x=[−1.3 sqrt(3) (1+2+3)*4/5]
x=1:10                        % from 1 to 10
x=1:0.2:10                    % increment 0.2

x=magic(3)                    % magic number
x=ones(2,3)                   % all elements are 1
I=eye(3)                      % identity matrix

c=1:3; d=2:4
x=c.*d                        % multiplication
x=c./d                        % division
x=a.^2                        % squared of each element
x=a^2                         % a^2=a*a
```

4. Basic commands

```
det(a)            % determinant
inv(a)            % inverse matrix
eig(a)            % eigenvalue and eigenvector
```

5. Basic formats

```
format long
format long e
format short
```

6. Special numbers

```
pi        3.1415926535897...
i, j      imaginary
inf       infinity
ans       answer
```

7. Multiplication of polynomials (conv.m)

Expand the product of $a(s) = s^2 + 2s + 3$ and $b(s) = 4s^2 + 5s + 6$.

Solution

---

a=[1 2 3];   % $a(s) = s^2 + 2s + 3$
b=[4 5 6];   % $b(s) = 4s^2 + 5s + 6$
c=conv(a,b)   % $c(s) = (s^2 + 2s + 3)(4s^2 + 5s + 6)$

---

c=4 13 28 27 18

Answer $c(s) = 4s^4 + 13s^3 + 28s^2 + 27s + 18$

8. Division of polynomials (deconv.m)

Find the quotient and remainder of $d(s)$ divided by $a(s)$.

$$d(s) = 4s^4 + 13s^3 + 28s^2 + 27s + 18$$
$$a(s) = s^2 + 2s + 3$$

Solution

---

d=[4 13 28 27 18];
a=[1 2 3];
[q,r]=deconv(d,a)   % $\dfrac{d(s)}{a(s)} = q(s) + \dfrac{r(s)}{a(s)}$, $q(s)$: quotient, $r(s)$: remainder

---

q=4 5 6
r=0 0 0 0 0

Answer $\dfrac{4s^4 + 13s^3 + 28s^2 + 27s + 18}{s^2 + 2s + 3} = 4s^2 + 5s + 6$

B2. Symbolic calculus

1. 'Symbolic Math'

---

1. MATLAB® Toolbox

To perform symbolic processing in MATLAB®, the 'Symbolic Math' Toolbox is required.

Input the following matrix by using the 'Symbolic Math' Toolbox.

$$\begin{bmatrix} a & b \\ c & d \end{bmatrix}$$

Solution

```
>> syms a b c d
>>A=[a b ; c d]
```

ans =
a b
c d

Answer $\begin{bmatrix} a & b \\ c & d \end{bmatrix}$

2. Root of equation

a. (solve.m)

Find the root of $ax + b = 0$.

Solution

```
>> syms a, b, x;
>> eq1='a*x + b=0';
>> solve(eq1)                   % solve.m
```

ans =
−b/a

Answer $x = -\dfrac{b}{a}$

b. (solve.m)

Find the roots of $ax^2 + bx + c = 0$.

Solution

```
>> syms a, b, c, x;
>> eq1='a*x.^2 + b*x +c=0';
>> solve(eq1)                   % solve.m
```

ans =
(−b+sqrt(b^2−4*a*c))/2/a
(−b−sqrt(b^2−4*a*c))/2/a

Answer $x = \dfrac{-b \pm \sqrt{b^2 - 4ac}}{2a}$

3. Differentiation

Differentiate the following function:

a. x^n
b. $\sin^3 x$
c. $x \ln x$

Solution

a. (diff.m)

```
>> syms n, x;
>> diff(x^n)            % diff.m
```

ans =
x^n*n/x

```
>> simplify(ans)        % simplify.m
```

ans =
x^(n−1)*n

Answer nx^{n-1}

b. (diff.m)

```
>> syms x;
>> diff((sin(x))^3)     % diff.m
```

ans =
3*(sin(x))^2*cos(x)

Answer $3\sin^2 x \cos x$

c. (diff.m)

```
>> syms x;
>> diff(x* log(x))      % diff.m
```

ans =
log(x)+1

Answer $\ln x + 1$

4. Partial differentiation

Differentiate the following function:

a. $\dfrac{\partial\left(x^3y^2\right)}{\partial x}$

b. $\dfrac{\partial\left(x^3y^2\right)}{\partial y}$

c. $\dfrac{\partial\left(x\sin xy\right)}{\partial x}$

d. $\dfrac{\partial^2\left(x\sin xy\right)}{\partial x^2}$

Solution

a. (diff.m)

```
>> syms x, y
>> diff(x^3*y^2, x)          % diff.m
```

ans =
3*x^2*y^2

Answer $3x^2y^2$

b. (diff.m)

```
>> syms x, y
>> diff(x^3*y^2, y)          % diff.m
```

ans =
2*x^3*y

Answer $2x^3y$

c. (diff.m)

```
>> syms x, y
>> diff(x*sin(x*y), x)          % diff.m
```

ans =
sin(x*y)+x*y*cos(x*y)

Answer $\sin xy + xy\cos\left(xy\right)$

d. (diff.m)

```
>> syms x, y
>> diff(x*sin(x*y), x, 2)          % diff.m
```

ans =
2*y*cos(x*y)-x*y^2*sin(x*y)

Answer $2y\cos xy - xy^2\sin(xy)$

5. Integration

Integrate the following function:

a. $\int x^n\,dx$

b. $\int \ln(x+1)dx$

c. $\int e^{ax}\sin bx\,dx$

Solution

a. (int.m)

```
>> syms x, n
>> int(x*n, x)          % int.m
```

ans =
x^(n+1)/(n+1)

Answer $\dfrac{x^{n+1}}{n+1}$

b. (int.m)

```
>> syms x, n
>> int(log(x+1), x)          % int.m
```

ans =
(x+1)*log(x+1)-x

Answer $(x+1)\ln(x+1) - x$

c. (int.m)

```
>> syms x, a, b
>> int(exp(a*x)*sin(b*x), x)                    % int.m
```

---

ans =
exp(a*x)*(−b*cos(b*x)+a*sin(b*x))/(a^2+b^2)

$$\textbf{Answer } \frac{e^{ax}}{a^2+b^2}\left(-b\cos bx + a\sin bx\right)$$

6. Definite integration

Integrate the following function:

a. $\displaystyle\int_0^1 x^n\,dx$

b. $\displaystyle\int_1^2 \ln(x+1)\,dx$

c. $\displaystyle\int_t^{t^2} \cos x\,dx$

Solution

a. (int.m)

```
>> syms x, n
>> int(x*n, x, 0, 1)                             % int.m
```

---

ans =
1/(n+1)

$$\textbf{Answer } \frac{1}{n+1}$$

b. (int.m)

```
>> syms x, n
>> int(log(x+1), x, 1, 2)                        % int.m
```

---

ans =
3*log(3)−2*log(2)−1

$$\textbf{Answer } 3\ln 3 - 2\ln 2 - 1$$

c. (int.m)

```
>> syms x, t
>> int(cos(x), x, t, t^2)                    % int.m
```

---

ans =
sin(t^2)–sin(t)

Answer $\sin(t^2) - \sin t$

B3. Laplace transform and inverse Laplace transform

MATLAB® provides functions for computing both the Laplace transform and the inverse Laplace transform.

B3-1. *Laplace transform*

1. Laplace transform (laplace.m)

Find the Laplace transform of $x(t) = 1 + 3t - t^2$.

Solution

Laplace transform of $x(t) = 1 + 3t - t^2$ is

$$X(s) = \mathcal{L}(1 + 3t - t^2) = \frac{1}{s} + \frac{3}{s^2} - \frac{2}{s^3}.$$

The MATLAB® command for this is as follows:

```
>> syms t
>> laplace(1+3*t−t^2)                        % laplace.m
```

---

ans =
1/s + 3/s^2–2/s^3

Answer $F(s) = \dfrac{1}{s} + \dfrac{3}{s^2} - \dfrac{2}{s^3}$

2. Laplace transform (laplace.m)

Find the Laplace transform of $x(t) = t^n$.

Solution

```
>> syms t n
>> laplace(t^n)                          % laplace.m
```

ans =
n!/s^(n+1)

$$\textbf{Answer } F(s) = \frac{n!}{s^{n+1}}$$

3. Laplace transform (laplace.m)

Find the Laplace transform of $x(t) = e^{at}$.

Solution

```
>> syms t a
>> laplace(e^(a*t))                      % laplace.m
```

ans =
1/(s–a)

$$\textbf{Answer } F(s) = \frac{1}{s-a}$$

4. Laplace transform (laplace.m)

Find the Laplace transform of $x(t) = \cos\omega t$.

Solution

```
>> syms t omega
>> laplace(cos(omega*t))                 % laplace.m
```

ans =
s/(s^2+omega^2)

$$\textbf{Answer } F(s) = \frac{s}{s^2 + \omega^2}$$

5. Laplace transform (laplace.m)

Find the Laplace transform of $x(t) = \sin\omega t$.

Solution

```
>> syms t omega
>> laplace(sin(omega*t))                % laplace.m
```

ans =
omega/(s^2+omega^2)

Answer $F(s) = \dfrac{\omega}{s^2 + \omega^2}$

6. Laplace transform (laplace.m)

Find the Laplace transform of $x(t) = te^{at}$.

Solution

```
>> syms t a
>> laplace(t.*e^(a*t))                  % laplace.m
```

ans =
1/(s–a)^2

Answer $F(s) = \dfrac{1}{(s-a)^2}$

7. Laplace transform (laplace.m)

Find the Laplace transform of $x(t) = t^n e^{at}$.

Solution

```
>> syms t a n
>> laplace(t^n.*e^(a*t))                % laplace.m
```

ans =
n!/(s–a)^(n+1)

Answer $F(s) = \dfrac{n!}{(s-a)^{n+1}}$

8. Laplace transform (laplace.m)

Find the Laplace transform of $x(t) = e^{at} \cos \omega t$.

Solution

```
>> syms t, a, omega
>> laplace(e^(a*t).*cos(omega*t))          % laplace.m
```

ans =
(s−a)/((s−a)^2+omega^2)

$$\text{Answer } F(s) = \frac{s-a}{(s-a)^2 + \omega^2}$$

9. Laplace transform (laplace.m)

Find the Laplace transform of $x(t) = e^{at} \sin \omega t$.

Solution

```
>> syms t a omega
>> laplace(e^(a*t).*sin(omega*t))          % laplace.m
```

ans =
omega/((s−a)^2+omega^2)

$$\text{Answer } F(s) = \frac{\omega}{(s-a)^2 + \omega^2}$$

10. Laplace transform (laplace.m)

Find the Laplace transform of $x(t) = t \cosh \omega t$.

Solution

```
>> syms t omega
>> laplace(t.*cosh(omega*t))               % laplace.m
```

ans =
(s^2 +omega^2)/((s^2−omega^2)^2

$$\text{Answer } F(s) = \frac{s^2 + \omega^2}{(s^2 - \omega^2)^2}$$

11. Laplace transform (laplace.m)

Find the Laplace transform of $x(t) = \delta(t)$.

Solution

```
>> syms t
>> laplace(delta(t))                    % laplace.m
```

ans =
1

Answer $F(s) = 1$

B3-2. *Inverse Laplace transform*

1. Inverse Laplace transform (ilaplace.m)

Find the inverse Laplace transform of $\mathcal{L}\{x(t)\} = \dfrac{1}{s+a} + \dfrac{2b}{s^2+b^2}$.

Solution

The inverse Laplace transform of $\mathcal{L}\{x(t)\} = \dfrac{1}{s+a} + \dfrac{2b}{s^2+b^2}$ is

$$x(t) = e^{-at} + 2\sin bt.$$

The MATLAB® command for this is as follows:

```
>> syms s a b
>> ilaplace(1/(s+a)+(2*b)/(s^2+b^2))          % ilaplace.m
```

ans =
exp(-a*t) + 2*sin(b*t)

Answer $f(t) = e^{-at} + 2\sin bt$

2. Inverse Laplace transform (ilaplace.m)

Find the inverse Laplace transform of $F(s) = \dfrac{s+2}{s^2 - 2s + 10}$.

Solution

```
>> syms s
>> ilaplace((s+2)/(s^2−2*s+10))          % ilaplace.m
```

ans =
exp(t).*(cos(3*t)+sin(3*t))

Answer $f(t) = e^t (\cos 3t + \sin 3t)$

3. Inverse Laplace transform (ilaplace.m)

Find the inverse Laplace transform of $F(s) = \dfrac{2s}{(s+3)^2}$.

Solution

```
>> syms s
>> ilaplace(2*s/(s+3)^2)                  % ilaplace.m
```

ans =
(2−6*t).*exp(−3*t)

Answer $f(t) = (2 - 6t)e^{-3t}$

4. Inverse Laplace transform (ilaplace.m)

Find the inverse Laplace transform of $F(s) = \dfrac{5s+2}{s^2+2s}$.

Solution

```
>> syms s
>> ilaplace((5*s+2)/(s^2+2*s))            % ilaplace.m
```

ans =
1+4*exp(−2*t)

Answer $f(t) = 1 + 4e^{-2t}$

5. Inverse Laplace transform (ilaplace.m)

Find the inverse Laplace transform of $F(s) = \dfrac{2s-2}{(s-2)^3}$.

Solution

```
>> syms s
>> ilaplace((2*s-2)/(s-2)^3)          % ilaplace.m
```

ans =
exp(2*t).*(2*t+t.^2)

Answer $f(t) = e^{2t}\left(2t + t^2\right)$

6. Inverse Laplace transform (ilaplace.m)

Find the inverse Laplace transform of $F(s) = \dfrac{s^2 - 10s + 20}{s(s^2 - 6s + 10)}$.

Solution

```
>> syms s
>> ilaplace((s^2-10*s+20)/s/(s^2-6*s+10))          % ilaplace.m
```

ans =
2-exp(3*t).*(cos(t)+sin(t))

Answer $f(t) = 2 - e^{3t}\left(\cos t + \sin t\right)$

7. Inverse Laplace transform (ilaplace.m)

Find the inverse Laplace transform of $F(s) = \dfrac{1}{s(s^2 + \omega^2)}$.

Solution

```
>> syms s omega
>> ilaplace(1/s/(s^2+omega^2)          % ilaplace.m
```

ans =
(1−cos(omega*t))/omega^2

Answer $f(t) = \dfrac{1 - \cos \omega t}{\omega^2}$

8. Inverse Laplace transform (ilaplace.m)

Find the inverse Laplace transform of $F(s) = \dfrac{1}{s\left(s^2 - 4\right)}$.

Solution

```
>> syms s
>> ilaplace(1/s/(s^2−4)                %. m
```

ans =
(cosh(2*t)−1)/4

Answer $f(t) = \dfrac{1}{4}\left(\cosh 2t - 1\right)$

9. Inverse Laplace transform (ilaplace.m)

Find the inverse Laplace transform of $F(s) = \ln\left(\dfrac{s+2}{s-1}\right)$.

Solution

```
>> syms s
>> ilaplace(ln((s+2)/(s−1))             % ilaplace.m
```

ans =
(exp(t)−exp(−2*t))./t

Answer $f(t) = \dfrac{e^t - e^{-2t}}{t}$

10. Inverse Laplace transform (ilaplace.m)

Find the inverse Laplace transform of $F(s) = \ln\left(s + \dfrac{\omega^2}{s}\right)$.

Solution

```
>> syms s omega
>> ilaplace(ln(s+omega^2/s))                    % ilaplace.m
```

```
ans =
(1−2*cos(omega*t))./t
```

$$\text{Answer } f(t) = \frac{1 - 2\cos\omega t}{t}$$

11. Inverse Laplace transform (ilaplace.m)

Find the inverse Laplace transform of $F(s) = \ln\left(1 + \dfrac{4}{s} + \dfrac{5}{s^2}\right)$.

Solution

```
>> syms s
>> ilaplace(ln(1+4/s+5/s^2))                    % ilaplace.m
```

```
ans =
2*(1−exp(−2*t).*cos(t))./t
```

$$\text{Answer } f(t) = \frac{2\left(1 - e^{-2t}\cos t\right)}{t}$$

12. Inverse Laplace transform (ilaplace.m)

Find the inverse Laplace transform of $F(s) = \ln\left\{\dfrac{s+3}{(s^2 + 2s + 5)^2}\right\}$.

Solution

```
>> syms s
>> ilaplace(ln((s+3)/(s^2 +2*S+5)^2))          % ilaplace.m
```

ans =
(4*exp(–t).*cos(2*t)–exp(–3*t))./t

$$\text{Answer } f(t) = \frac{4e^{-t}\cos 2t - e^{-3t}}{t}$$

B4.   Linear algebra (Eigenvalue problem)

1. Find the determinant of the following matrix.

$$a = \begin{bmatrix} 1 & 2 & -3 \\ -1 & 1 & 2 \\ 1 & 1 & 0 \end{bmatrix}$$

Solution

```
a=[1 2 –3 ; –1 1 2 ; 1 1 0]
det(a)
```

Answer 8

2. Find the inverse matrix of the following matrix.

$$a = \begin{bmatrix} 1 & 2 & -3, \\ -1 & 1 & 2 \\ 1 & 1 & 0 \end{bmatrix}$$

Solution

```
a=[1 2 –3 ; –1 1 2 ; 1 1 0]
inv(a)
```

Answer −0.25   −0.375   0.875
        0.25   0.375   0.125
       −0.25   0.125   0.375

3. Find the characteristic equation of the following matrix.   (poly.m)

$$a = \begin{bmatrix} 1 & 2 & 3 \\ 4 & 5 & 6 \\ 7 & 8 & 0 \end{bmatrix}$$

Solution

```
a=[1 2 3; 4 5 6; 7 8 9];
p=poly(a)                                    % poly.m
```

p=1 −6 −72 −27 $\therefore \; \lambda^3 - 6\lambda^2 - 72\lambda - 27 = 0$

Check

$$\begin{vmatrix} 1-\lambda & 2 & 3 \\ 4 & 5-\lambda & 6 \\ 7 & 8 & -\lambda \end{vmatrix} = -\lambda^3 + 6\lambda^2 + 72\lambda + 27 = 0$$

4. Find the roots of the following equation.   (roots.m)

$$\lambda^3 - 6\lambda^2 - 72\lambda - 27 = 0$$

Solution

```
p=[ 1 −6 −72 −27];
r=roots(p)                                   % roots.m
```

r=12.1229,   −6.7345,   −0.3884

5. Find the eigenvalues and its corresponding eigenvectors for the basic form I of linear algebra.

$$a = \begin{bmatrix} 1 & 2 \\ 3 & 4 \end{bmatrix}$$

Solution

```
a=[1 2; 3 4];
[v, d]=eig(a)                    % v: eigenvector, d: eigenvalue
```

v=
    -0.8246   -0.4160
     0.5658   -0.9094

d=
    -0.3723   0
          0   6.3723

---

6. Find the eigenvalues and its corresponding eigenvectors for the basic
   form II of linear algebra.

$$a = \begin{bmatrix} 1 & 2 \\ 3 & 4 \end{bmatrix}, \quad b = \begin{bmatrix} 4 & 0 \\ 0 & 2 \end{bmatrix}$$

Solution

```
a=[1 2; 3 4]; b=[4 0; 0 2];
[v, d]=eig(a, b)                 % v: eigenvector, d: eigenvalue
```

v=
     1.0000   0.2374
    -0.7122   1.0000

d=
    -0.1061   0
          0   2.3561

---

7. Find the eigenvalues and its corresponding eigenvectors for the basic
   form I of linear algebra.

$$a = \begin{bmatrix} 1 & 2 & 3 \\ 1 & 5 & 6 \\ 1 & 4 & 9 \end{bmatrix}$$

Solution

```
a=[1 2 3; 1 5 6; 1 4 9];
[v, d]=eig(a)                    % eig.m
```

v=

−0.2927	−0.9838	−0.2492
−0.6104	0.1763	−0.8332
−0.7360	0.0329	0.4937

d=

12.7148	0	0
0	0.5411	0
0	0	1.7441

---

8. Find the eigenvalues and its corresponding eigenvectors for the basic form II of linear algebra.

$$a = \begin{bmatrix} 4 & -2 & 0 \\ -2 & 4 & -2 \\ 0 & -2 & 4 \end{bmatrix}, \quad b = \begin{bmatrix} 1 & 0 & 0 \\ 0 & 2 & 0 \\ 0 & 0 & 1 \end{bmatrix}$$

Solution

```
a=[4 −2 0; −2 4 −2; 0 −2 4]; b=[1 0 0; 0 2 0; 0 0 1];
[v, d]=eig(a, b)                                    % eig.m
```

v=

0.3717	−0.7071	−0.6015
0.6015	0.0000	0.3717
0.3717	0.7071	−0.6015

d=

0.7639	0	0
0	4.0000	0
0	0	5.2361

B5. Drawing a graph

1. Linear fitting by Least Mean Square method:
Suppose we obtained data from a spring tension experiment as follows. Draw a linear graph based on this data.

Force (N)	0	50	100	150	200	250	300	350
Displacement (mm)	0	1.7	2.6	3.7	5.7	7.3	9.8	11.9

Solution

```
% M5_1
close all; clear all
y=[ 0:50:350];
x=[0 1.7 2.6 3.7 5.7 7.3 9.8 11.9];
p=polyfit(x,y,1)                          % linear fitting
xi=linspace(0, 12, 100);
z=polyval(p,xi);

plot(x, y, '-o', 'linewidth', 2), hold on
plot(xi, z, ':')
xlabel('Deformation (mm)', 'fontsize', 13)
ylabel('Force (N)', 'fontsize', 13)
```

Figure B.1 shows an example of linear fitting.

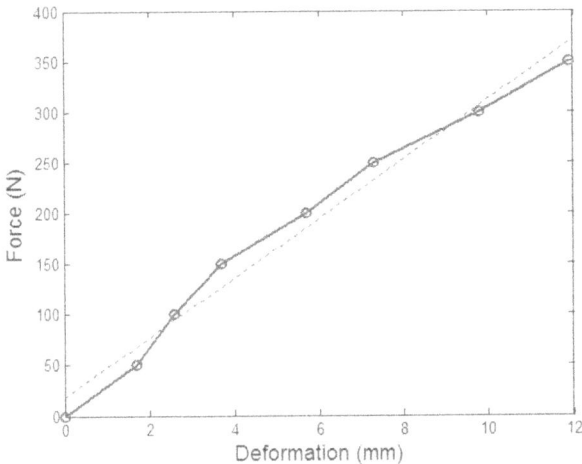

Figure B.1 Linear fitting.

p=29.4257   17.9404

Therefore, the function of the linear model is

$$f = 29.4x + 17.9 \, [\text{N}].$$

For example, instead of p=polyfit(x, y, 1), if we use p=polyfit(x, y, 3), it performs a 3rd-degree polynomial fitting.

2. Drawing the graph
   Write an m-file to plot the graph of $y = t \sin(2t-0.3)$ (where the unit of time t is in seconds and the unit of displacement y is in meters) using MATLAB® as obtained below.

Solution

```
% M5_2
close all; clear all;
t=0:0.01:6;
y=t.* sin(2*t−0.3);                    % '.*'
plot(t, y, 'linewidth', 2 )
xlabel('Time (s)','Fontsize',13)
ylabel('Displacement (m)','Fontsize',13)
title('Graph of y=t sin(2t−0.3)','Fontsize',13)
grid
```

Figure B.2 shows the graph of $y = t \sin(2t-0.3)$.

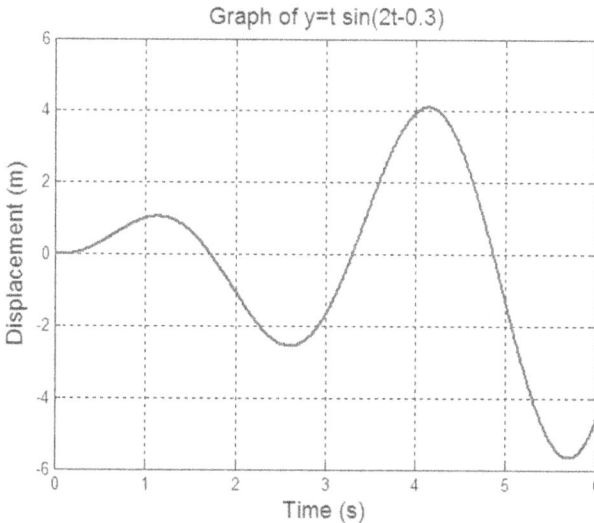

Figure B.2 $y = t \sin(2t-0.3)$.

B6. Plotting the solution of an ODE using the Runge-Kutta method

---

1. Plot the solution of a following ODE:

$$\dot{x} + 3x = 0, \quad x(0) = 2.$$

Solution

Letting $y = \dot{x}$ yields

$$y = -3x.$$

```
% song6_1.m
function y=song6_1(t,x)
y=zeros(1,1);
y=-3*x;
```

```
% main file M6_1.m
close all; clear all;
t=0:0.01:2;
x00=2;                          % initial condition
[t, x]=ode23('song6_1', t, x00);   % ode23.m (ordinary differential
                                     equation)

plot(t, x, 'linewidth', 2)
xlabel('Time (s)', 'fontsize', 13);
ylabel('Displacement (m)', 'fontsize', 13); grid
```

Figure B.3 shows the solution of an ODE, $\dot{x} + 3x = 0$, $x(0) = 2$.

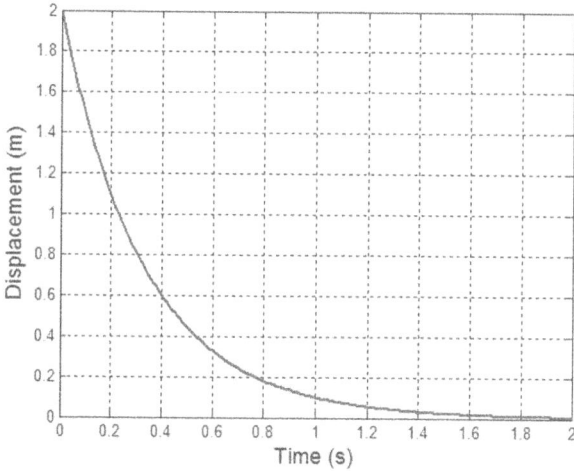

Figure B.3 Solution in this example.

2. Plot the solution of a following ODE

$$\ddot{x} + 0.2\dot{x} + 50x = 20,\ x(0) = 0.1,\ \dot{x}(0) = 0.$$

Solution

Letting $\mathbf{x} = \left\{ \begin{array}{c} x \\ \dot{x} \end{array} \right\}$ and $\mathbf{y} = \left\{ \begin{array}{c} \dot{x} \\ \ddot{x} \end{array} \right\}$ yields

$$y(1) = x(2)$$

$$y(2) = -0.2x(2) - 50x(1) + 20.$$

```
% song6_2.m
function y=song6_2(t,x)
y=zeros(2,1);
y(1)=x(2);
y(2)=-0.2*x(2)-50*x(1)+20;
```

```
% main file M6_2.m
close all; clear all;
t=0:0.01:20;
x00=[0.1; 0];                          % initial condition
[t, x]=ode23('song6_2', t, x00);       % ode23.m (ordinary differential
                                         equation)
plot(t, x(:,1), 'linewidth', 2)
xlabel('Time (s)', 'fontsize', 13);
ylabel('Displacement (m)', 'fontsize', 13); grid
```

Figure B.4 shows the solution of an ODE, $\ddot{x}+0.2\dot{x}+50x=20$, $x(0)=0.1$, $\dot{x}(0)=0$.

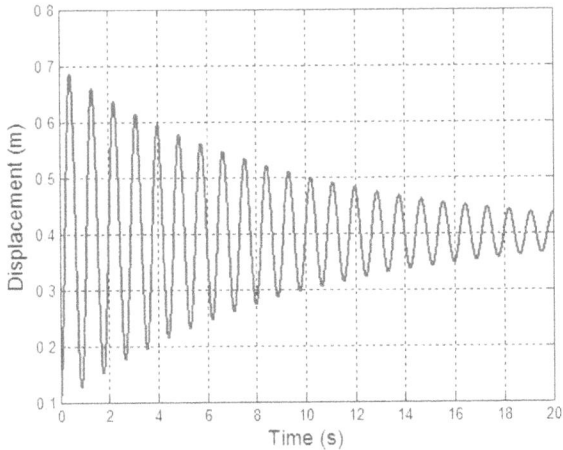

Figure B.4 Solution in this example.

Index

For Product Safety Concerns and Information please contact our EU
representative GPSR@taylorandfrancis.com
Taylor & Francis Verlag GmbH, Kaufingerstraße 24, 80331 München, Germany

www.ingramcontent.com/pod-product-compliance
Lightning Source LLC
Chambersburg PA
CBHW060416220326
41598CB00021BA/2192